Manual of Cardiovascular Proteomics

Giulio Agnetti • Merry L. Lindsey
D. Brian Foster
Editors

Manual of Cardiovascular Proteomics

Springer

Editors
Giulio Agnetti
Division of Cardiology
Johns Hopkins University School of Medicine
Baltimore
Maryland
USA

Department of Biomedical and
Neuromotor Sciences (DIBINEM)
University of Bologna
Bologna
Italy

Merry L. Lindsey
Dept. of Physiology & Biophysics
University Mississippi Medical Center
Jackson
Mississippi
USA

D. Brian Foster
Division of Cardiology
Johns Hopkins University School of Medicine
Baltimore
Maryland
USA

ISBN 978-3-319-31826-4 ISBN 978-3-319-31828-8 (eBook)
DOI 10.1007/978-3-319-31828-8

Library of Congress Control Number: 2016947834

Printed on acid-free paper

This Springer imprint is published by Springer Nature
The registered company is Springer International Publishing AG Switzerland

To Giulio Stoppioni, Alicia Taylor, and all others who have died of cardiovascular disease. To our families who have supported our pursuit of a career in cardiovascular proteomics, and to our colleagues who have helped advance the field.

Preface

The proteome, originally defined as the protein complement of the genome, can be studied by a complex array of technologies commonly encapsulated under the term *proteomics*. Of these technologies, mass spectrometry has single handedly revolutionized our ability to explore biological systems and provides a major advancement for research efforts. The realization that protein post-translational modifications are heterogeneous, ubiquitous, and dynamic offers an expanded view to examine pathophysiological processes. This concept has compelling implications for the understanding of diseases, particularly those that are sporadic in nature and for which a genetic determinant is not readily identified. The low correlation between gene expression and phenotype in health and disease can be explained by changes at the protein post-transcriptional and post-translational levels. This is particularly relevant in the settings of cardiovascular disease. For instance, increasing evidence supports the idea that cardiovascular diseases, such as heart failure, are proteinopathies. Therefore, there is a clear need to evaluate proteins and their modifications, to better understand cardiovascular disease at the molecular level.

This book is a response to that clarion call. We highlight the remarkable advances that have contributed to the development of proteomics over the last two decades. Enormous leaps have accompanied the implementation of new instruments and analytical approaches. As a result, hurdles along the path from data acquisition to biological insight have been dramatically lowered. The aims of this manual are to make these new methods and technologies understandable to scientists who are new to the proteomics arena and to accelerate knowledge dissemination. This book covers the full measure of topics that define modern cardiovascular proteomics, including but not limited to experimental design, sample preparation and separation, mass spectrometry technologies, protein identification and quantitation, as well as statistical, pathway, and network analyses of proteomics results.

While many of the concepts, methods, and technologies presented here can be widely applied across fields of biomedical research, including cancer and brain research, this book has been tailored for the cardiovascular researcher; the many facets of cardiovascular disease are reflected in these chapters. All proteomes are not created equal, and cardiovascular proteomics presents a host of unique challenges, not the least of which is the wide dynamic range of protein abundances present in the cardiac muscle and in plasma. This book addresses many of the technical subtleties to be considered when embarking on a cardiovascular proteomic study, which serves as a guide for scientists entering the field.

This is the dawn of a new century where technology serves biology in an unprecedented way, and the tools currently available in the toolbox provide a rapid means to impact cardiovascular research. The contributing authors are among the most established scientists in the field of cardiovascular proteomics and have experienced the proteomic revolution first hand. We are extremely grateful to all of the authors for dedicating their precious time to share their deep knowledge and help us compile this manual.

Finally, we are immensely grateful for the strong historical foundation on which these chapters are built, laid by scientists pursuing their passion before the term proteomics was even coined. Among them, we would like acknowledge Dr. Jennifer E. Van Eyk, Dr. Peipei Ping, and Dr. Michael J. Dunn for their rich scientific contributions, their enthusiastic mentoring of new generations, and their tireless dedication to the advancement of proteomics as an essential field of cardiovascular discovery. Today, the state of the field is stronger for their efforts, and its vitality is directly reflected in this book. We hope that you find it useful for your research.

Baltimore, MD, USA Giulio Agnetti
Baltimore, MD, USA D. Brian Foster
Jackson, MS, USA Merry L. Lindsey

Contents

Contributors

Giulio Agnetti, PhD, FAHA Laboratory of Cardiovascular Biochemistry, Division of Cardiology, Department of Medicine, The Johns Hopkins University School of Medicine, Baltimore, MD, USA

Department of Biomedical and Neuromotor Sciences (DIBINEM), University of Bologna, Bologna, Italy

D. Kent Arrell, PhD Center for Regenerative Medicine and Marriott Heart Disease Research Program, Division of Cardiovascular Diseases, Department of Medicine, Department of Molecular Pharmacology and Experimental Therapeutics, and Department of Medical Genetics, Mayo Clinic, Rochester, MN, USA

Jennifer A. Beckman, MSN, BSN Division of Cardiology, Department of Internal Medicine, University of Washington Medical Center, Seattle, WA, USA

Francesca Brambilla, PhD Department of Proteomics and Metabolomics, Institute for Biomedical Technologies-National Research Council, Milan, Segrate, Italy

Emilio Camafeita, PhD Cardiovascular Proteomic Lab and Proteomic Unit, Vascular Pathophysiology Area, Centro Nacional De Investigaciones Cardiovasculares, Madrid, Spain

Jenice Cheah Department of Neurobiology, Physiology, and Behavior, University of California, Davis, Davis, CA, USA

Heaseung Sophia Chung, MS Department of Biological Chemistry, Johns Hopkins University School of Medicine, Baltimore, MD, USA

Luis Miguel Blanco-Colio, PhD Vascular Research Lab, FIIS-Fundacion Jimenez Diaz, Autonoma University of Madrid, Madrid, Spain

Jake Cosme, HBSc Department of Physiology, University of Toronto and Toronto General Hospital Research Institute, Toronto, ON, Canada

Gemma Currie, MB, ChB Institute of Cardiovascular and Medical Sciences, BHF Glasgow Cardiovascular Research Centre, University of Glasgow, Glasgow, UK

Lisandra E. de Castro Brás, PhD Department of Physiology, The Brody School of Medicine, East Carolina University, Greenville, NC, USA

Christian Delles, MD, FRCP, FAHA, FBHS Institute of Cardiovascular and Medical Sciences, BHF Glasgow Cardiovascular Research Centre, University of Glasgow, Glasgow, UK

Dario Di Silvestre, PhD Department of Proteomics and Metabolomics, Institute for Biomedical Technologies-National Research Council, Milan, Segrate, Italy

Michael J. Dunn, BSc, PhD Conway Institute, University College Dublin, Dublin, Ireland

Stephen Farris, MD Division of Cardiology, Department of Internal Medicine, University of Washington Medical Center, Seattle, WA, USA

Clifford Folmes, PhD Division of Cardiovascular Diseases, Department of Medicine, Center for Regenerative Medicine, Mayo Clinic, Rochester, MN, USA

D. Brian Foster, PhD Laboratory of Cardiovascular Biochemistry, Division of Cardiology, Department of Medicine, The Johns Hopkins University School of Medicine, Baltimore, MD, USA

Sarah Franklin, PhD Cardiology Division, Department of Internal Medicine, The Nora Eccles Harrison Cardiovascular Research and Training Institute, University of Utah, Salt Lake City, UT, USA

Department of Biochemistry, University of Utah, Salt Lake City, UT, USA

Ying Ge, PhD Cellular and Molecular Pathology Training Program, University of Wisconsin-Madison, Madison, WI, USA

Human Proteomics Program, University of Wisconsin-Madison, Madison, WI, USA

Jennifer E. Gilda, BS Department of Neurobiology, Physiology, and Behavior, University of California, Davis, Davis, CA, USA

Aldrin V. Gomes, PhD Department of Neurobiology, Physiology, and Behavior, University of California, Davis, Davis, CA, USA

Anthony Gramolini, PhD Department of Physiology, University of Toronto and Toronto General Hospital Research Institute, Toronto, ON, Canada

Department of Physiology, Ted Rogers Centre for Heart Research, University of Toronto and University Health Network, Toronto, ON, Canada

Zachery R. Gregorich, BS Department of Cell and Regenerative Biology, University of Wisconsin-Madison, Madison, WI, USA

Molecular Pharmacology Training Program, University of Wisconsin-Madison, Madison, WI, USA

Rebekah L. Gundry, PhD Department of Biochemistry, Medical College of Wisconsin, Milwaukee, WI, USA

Ronald J. Holewinski, PhD The Advanced Clinical Biosystems Research Institute, Heart Institute, Cedars-Sinai Medical Center, Los Angeles, CA, USA

Tuleagh Innes-Gawn Department of Neurobiology, Physiology, and Behavior, University of California, Davis, Davis, CA, USA

Elena Burillo, PhD Vascular Research Lab, FIIS-Fundacion Jimenez Diaz, Autonoma University of Madrid, Madrid, Spain

Shanxi Jiang, BS Department of Anesthesiology, David Geffen School of Medicine and UCLA, Los Angeles, CA, USA

Kai Kammers, PhD Department of Biostatistics, Johns Hopkins Bloomberg School of Public Health, Baltimore, MD, USA

Mark J. Kohr, PhD Department of Environmental Health Sciences, Bloomberg School of Public Health, Johns Hopkins University, Baltimore, MD, USA

Erin M. Kropp Department of Biochemistry, Medical College of Wisconsin, Milwaukee, WI, USA

Nicole M. Lane, BS, BA Cellular and Molecular Pathology Training Program, University of Wisconsin-Madison, Madison, WI, USA

Department of Cell and Regenerative Biology, University of Wisconsin-Madison, Madison, WI, USA

Merry L. Lindsey, PhD Department of Physiology and Biophysics, Mississippi Center for Heart Research, University of Mississippi Medical Center, Jackson, MS, USA

Research Service, G.V. (Sonny) Montgomery Veterans Affairs Medical Center, Jackson, MS, USA

Diego Martínez-López Vascular Research Lab, FIIS-Fundacion Jimenez Diaz, Autonoma University of Madrid, Madrid, Spain

Department of Nephrology, Vascular Pathology, Hypertension, and Diabetes, IIS-Fundación Jiménez Díaz-UAM, Madrid, Spain

Juan Antonio Lopez, PhD Cardiovascular Proteomic Lab and Proteomic Unit, Vascular Pathophysiology Area, Centro Nacional De Investigaciones Cardiovasculares, Madrid, Spain

Claudius Mahr, DO, FACC, FESC Division of Cardiology, Department of Internal Medicine, University of Washington Medical Center, Seattle, WA, USA

Department of Internal Medicine, University of Washington, Seattle, WA, USA

Roxana Martinez-Pinna, PhD Vascular Research Lab, FIIS-Fundacion Jimenez Diaz, Autonoma University of Madrid, Madrid, Spain

Peter Matt, MD, FETCS Aortic Research Group, Division of Cardiac Surgery, University Hospital Basel, University of Basel, Basel, Switzerland

Pierluigi Mauri, PhD Department of Proteomics and Metabolomics, Institute for Biomedical Technologies-National Research Council, Milan, Segrate, Italy

Laboratory of Medical Science, Scuola Superiore Sant'Anna, Institute of Life Sciences, Pisa, Italy

Mickey R. Miller, PhD Department of Cardiology, The Nora Eccles Harrison Cardiovascular Research and Training Institute, University of Utah, Salt Lake City, UT, USA

Anne M. Murphy, MD Division of Cardiology, Department of Pediatrics, Johns Hopkins University School of Medicine, Baltimore, MD, USA

Christopher I. Murray, PhD Department of Anesthesiology, Pharmacology and Therapeutics, University of British Columbia, Vancouver, BC, Canada

Life Sciences Centre, Vancouver, BC, Canada

Kyriakos N. Papanicolaou, PhD Division of Cardiology, Department of Medicine, Johns Hopkins University, Johns Hopkins School of Medicine, Baltimore, MD, USA

Laurie Parker, PhD Department of Biochemistry, Molecular Biology, and Biophysics, University of Minnesota, Minneapolis, MN, USA

Sarah J. Parker, PhD The Advanced Clinical Biosystems Research Institute, Heart Institute, Cedars-Sinai Medical Center, Los Angeles, CA, USA

Genaro Pimienta, PhD, MSc Biomedical Innovation Department, Center for Scientific Research and Higher Education at Ensenada, Ensenada, Baja California, Mexico

Jun Qu, PhD Department of Biochemistry, University at Buffalo, State University of New York, Buffalo, NY, USA

Department of Pharmaceutical Sciences, University at Buffalo, State University of New York, Buffalo, NY, USA

State of New York Center for Excellence in Bioinformatics and Life Sciences, Buffalo, NY, USA

Genaro Antonio Ramirez-Correa, MD, PhD Division of Cardiology, Department of Pediatrics, Johns Hopkins University School of Medicine, Baltimore, MD, USA

Ingo Ruczinski, PhD Department of Biostatistics, Johns Hopkins Bloomberg School of Public Health, Baltimore, MD, USA

Shichen Shen, BS Department of Biochemistry, University at Buffalo, State University of New York, Buffalo, NY, USA

State of New York Center for Excellence in Bioinformatics and Life Sciences, Buffalo, NY, USA

Xiaomeng Shen, BS Department of Biochemistry, University at Buffalo, State University of New York, Buffalo, NY, USA

State of New York Center for Excellence in Bioinformatics and Life Sciences, Buffalo, NY, USA

Stanley V. Smith, PhD Department of Pharmacology and Toxicology, University of Mississippi Medical Center, Jackson, MS, USA

April Stempien-Otero, MD Division of Cardiology, Department of Internal Medicine, University of Washington Medical Center, Seattle, WA, USA

Irina Tchernyshyov, MSc The Advanced Clinical Biosystems Research Institute, Heart Institute, Cedars-Sinai Medical Center, Los Angeles, CA, USA

Andre Terzic, MD, PhD Center for Regenerative Medicine and Marriott Heart Disease Research Program, Division of Cardiovascular Diseases, Department of Medicine, Department of Molecular Pharmacology and Experimental Therapeutics, and Department of Medical Genetics, Mayo Clinic, Rochester, MN, USA

Jennifer E. Van Eyk, PhD, FAHA, FISHR The Advanced Clinical Biosystems Research Institute, Heart Institute, Cedars-Sinai Medical Center, Los Angeles, CA, USA

Department of Medicine, Johns Hopkins University School of Medicine, Baltimore, MD, USA

Jesus Vazquez, PhD Cardiovascular Proteomic Lab and Proteomic Unit, Vascular Pathophysiology Area, Centro Nacional De Investigaciones Cardiovasculares, Madrid, Spain

Vidya Venkatraman, MS The Advanced Clinical Biosystems Research Institute, Heart Institute, Cedars-Sinai Medical Center, Los Angeles, CA, USA

Jose Luis Martin-Ventura, PhD Vascular Research Lab, FIIS-Fundacion Jimenez Diaz, Autonoma University of Madrid, Madrid, Spain

Thomas M. Vondriska, PhD Department of Anesthesiology, David Geffen School of Medicine and UCLA, Los Angeles, CA, USA

Matthew Waas Department of Biochemistry, Medical College of Wisconsin, Milwaukee, WI, USA

Ian Wright, BSc, CBiol. MRSB Advanced Clinical Biosystems Research Institute, Cedars-Sinai Medical Center, Los Angeles, CA, USA

Strategic Innovations LLC, Earleville, MD, USA

Mingguo Xu, MD, PhD Department of Pediatric Cardiology, Shenzhen Children's Hospital, Shenzhen, Guangdong, China

Division of Cardiology, Department of Pediatrics, Johns Hopkins University School of Medicine, Baltimore, MD, USA

Chapter 1
A Historical Perspective on Cardiovascular Proteomics

Giulio Agnetti and Michael J. Dunn

Abstract The "Manual of Cardiovascular Proteomics" is the result of the concerted effort of many experts in the field and it addresses the core technologies and approaches that have been implemented since its birth. Although each chapter can be read or studied independently of the others; depending on the level of interest, the whole manual should provide a detailed overview on what is available to the modern scientist who wants to embark on a cardiovascular proteomic expedition. Chapter 1 provides the historical perspective and describes the landmark discoveries that propelled the field forward, along with considerations on how to chose a specific approach and what the first steps to complete a proteomic experiment successfully should be. We hope that you will enjoy the first edition and are looking forward to your feedback in order to improve future editions.

Keywords Cardiovascular proteomics • History of proteomics • Proteomic discoveries • History of mass spectrometry

Introduction: The Renaissance of Protein Biochemistry

At the end of the last century, in September 1994 to be exact, a new term was publicly announced during a small meeting titled "2D Electrophoresis: From Protein Maps to Genomes" held in the charming Tuscan hills (*Siena, Italy*). At this meeting, the term "Proteome" was defined by Marc Wilkins from Sydney as the "**prote**in complement of the gen**ome**", resulting in the birth of the field of "proteomics". This rather simple

G. Agnetti, PhD, FAHA (✉)
Laboratory of Cardiovascular Biochemistry, Division of Cardiology, Department of Medicine, The Johns Hopkins University School of Medicine, Baltimore, MD, USA

Department of Biomedical and Neuromotor Sciences (DIBINEM), University of Bologna, Bologna, Italy
e-mail: gagnett1@jhmi.edu

M.J. Dunn, BSc, PhD
Conway Institute, University College Dublin, Dublin, Ireland

G. Agnetti et al. (eds.), *Manual of Cardiovascular Proteomics*,
DOI 10.1007/978-3-319-31828-8_1

moment defined a time in history when the emphasis would slowly but steadily move from the role of genes to that of proteins and their modifications as defining traits of the phenotype, both in health and disease. This "Cultural Revolution" was largely made possible by the remarkable technological advances in the field of protein biochemistry that were achieved in the latter part of the twentieth century. In 1975 Pat O'Farrell and other groups optimized the way to separate and detect over 1100 proteins in a single experiment using two-dimensional polyacrylamide gel electrophoresis (2DE) [1]. Combined with the rapid developments in protein and peptide mass spectrometry (MS), and the creation of protein databases which followed, these combined techniques enabled by the late 1990s the separation, quantification and identification of hundreds to thousands of proteins starting from a single biological sample. Indeed, the study co-Authored by Marc Wilkins the year following the 1994 Siena meeting inaugurated a new era in protein biochemistry [2].

As we write, we have recently celebrated the 21st birthday of the proteome and in retrospect, many things have dramatically improved. To mark this occasion, a Special Issue celebrating the 20th birthday of proteomics, dedicated to the memory of Vitaliano Pallini (the former Supervisor of the Functional Proteomics group at the Molecular Biology Department at the University of Siena, Italy), was recently published in the Journal of Proteomics [3]. We direct the interested reader to this special issue for historical references as several scientists that participated in the birth of proteomics contributed with their personal views and memories.

The "new field" of proteomics (i.e. the complex of technology used to investigate the proteome), which could be also termed *advanced protein biochemistry*, is the result of the incredible technological advances that have allowed protein biochemists to study the subjects of their investigation in a more efficient way. However, along with the technicalities, as often happens in scientific history, new concepts and ideas arose thanks to the availability of these new technologies. Much like the invention of the microscope allowed van Leeuwenhoek to see *animalcules*, which led to the discovery of pathogens, proteomic technologies have revealed an almost unimaginable complexity in terms of what are increasingly referred to as *proteoforms* [4]. Therefore, the impact of proteomics is not only technological, but also highly biological in nature. It is likely that the interest in genes and transcripts that captured the attention of the scientific world (and US Congress) in the 1980s, was also fueled by the relative facility by which nucleic acids can be amplified, a luxury that is still missing in the field of proteomics. However, thanks to the steady advances in protein separation technology, and more recently the implementation of novel MS configurations and protocols, the possibility of mapping the proteome, decorated with its countless post-translational modifications (PTMs), now seems within reach. The outstanding protein scientists who endured long periods of anonymity in the genomic era should not be forgotten as it is thanks to them that we are now capable of achieving all the wonderful things that proteomics is making possible. Thanks to the passion, hard work and dedication of these researchers, the variety and complexity of the proteome has finally emerged. This realization is having and will continue to exert a tremendous impact on modern life sciences and biomedicine and it should be born in mind that, as much as proteomics, more so than other global approaches (or –omics), relies heavily on technologies, the driving force which will propel proteomics in the future will be its impact on biological sciences and biomedicine.

In this chapter, we will highlight the scientific landmarks that made the study of the proteome possible, with specific reference to the cardiovascular system. It is our belief that a thorough understanding of where the proteomic world comes from will highlight the path to a prosperous future.

A Brief History of Proteomics

As mentioned in the introductory paragraph, the "proteomic revolution" could not have been possible without the technique, now 41 years old, which was carefully optimized by Pat O'Farrell in 1975 [1]. Although other groups implemented alternative versions of orthogonal gel separation, the 2DE protocol optimized by O'Farrell was arguably superior at the time it was published [5]. Indeed, by combining the best approaches available at the time and with an eye to the future, he was able to separate and visualize over a thousand proteins from *E. Coli*, while other groups could "only" see a few hundred proteins. Part of this success was made possible by the choice of isotopic labeling to boost sensitivity by about three orders of magnitude with respect to other groups who used protein staining (such as Coomassie Brilliant Blue) [5]. This technique, which is still in use today, is based on the combination of two independent electrophoretic separation methods applied to the sample in an orthogonal fashion. The first separation, commonly referred to as the *first dimension*, facilitates the separation of proteins according to their charge properties (isoelectric point, pI) by isoelectric focusing (IEF) under denaturing conditions. This is followed by the *second dimension*, which exploits the denaturing properties of charged detergents (typically sodium dodecyl sulphate, SDS) in order to separate proteins based on their molecular mass (M_r) (Fig. 1.1, see Chaps. 2 and 7 for details). Both electrophoretic steps are carried out using polyacrylamide gels (PAGE) with different properties that were optimized in the course of the following decade. In fact, another landmark discovery that allowed 2DE to become widely used was the advent of immobilized pH gradients (IPG). The idea, mastered by Pier

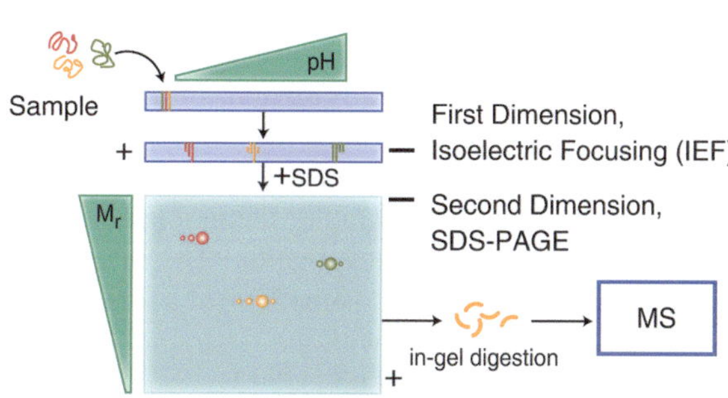

Fig. 1.1 Schematic of 2DE-MS. Proteins are first separated by 2DE according to their pI and molecular mass (M_r). The first separation in 2DE (IEF) is commonly referred to as "first dimension" whereas the second step (SDS-PAGE) is commonly referred to as "second dimension". After in-gel digestion proteins can be identified using MS (see text for details)

Giorgio Righetti and colleagues at the University of Milan (*Italy*) and developed within a large consortium involving several Universities and established scientists as well as an industrial partner (LKB at that time), led to the historical publication by Bjellqvist and colleagues in 1982 [6]. A few years later, based largely on optimization of the IPG 2DE technique in the laboratory of Angelika Gorg, the industrial partner (now Pharmacia) made the technology available at an industrial scale, enabling increased intra- and inter-lab reproducibility of 2DE [7] and the widespread diffusion of 2DE.

Using 2DE technology, Valerie Wasinger and colleagues were able to generate the data disclosed at the'94 Siena meeting and published the following year in the journal Electrophoresis [2]. In what is commonly referred to as the "first proteomic paper", these new technologies were combined with matrix-assisted laser desorption ionization–time of flight MS (MALDI-TOF MS, see Chap. 2), for the rapid identification of proteins using peptide mass fingerprinting (PMF) [8]. This MS-based approach is based on the digestion of the separated proteins with a protease (e.g. trypsin) and was utilized extensively in the early days of proteomics for the rapid identification of proteins. With the more recent popularity of tandem MS (see Chap. 2), allowing accurate determination of the sequence of peptides of the separated proteins, PMF was almost abandoned, only to make a glorious comeback as a rapid way to identify pathogens in biomedical labs in recent years [9] (see Chap. 18). The last key ingredient that was needed to identify proteins efficiently was the development of databases and algorithms to match the obtained MS spectra with protein names. Among the many contributions made by Norman and Leigh Anderson (father and son), the creation of one of the first successful algorithms to analyze 2DE maps [10] and the first online protein database (the Human Protein Index or HPI) [11] are two advances that accelerated the much needed creation of bioinformatic tools to mine the proteome (see Chaps. 12 and 14).

Operator Independence Days in Separation Techniques

Among the limitations of 2DE, the labor intensive and "artisanal" way they are performed paved the way to the success of more automatized techniques. To anybody who has successfully run a 2D gel, it is obvious how much training and effort is required to optimize both the conditions and the manual skills needed to obtain a well-resolved protein profile. With regard to the visual analysis of 2D gels and considering all of the effort involved in generating them, a parallel with visual arts is perhaps not too much a stretch of the imagination. The love-hate relationship of proteomic scientists with 2DE has made it something of a romantic journey, as testified for instance by the severe problems with the separation of membrane proteins, voiced so passionately by Thierry Rabilloud [12, 13].

Although several protein separation techniques have been implemented over the years, liquid chromatography (LC) became extremely popular for several reasons. Firstly its direct coupling with an MS instrument is relatively straightforward, decreasing the chance of contamination and avoiding a "transfer" step. In the origi-

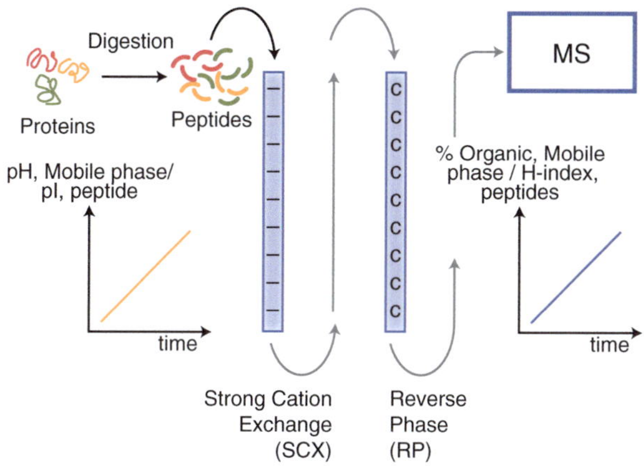

Fig. 1.2 Schematic of LC-MS. In "shotgun" proteomics proteins are first digested into peptides, which are generally more stable and soluble. Peptides can be then separated by liquid chromatographic techniques. Classically, Strong Cation Exchange (SCX) and Reverse Phase (RP) chromatography were combined in series to improve resolution. Liquid chromatography can be physically coupled to MS to identify the resulting peptides (see text and Chap. 4 & 7 for details)

nal iteration of what is referred to today as the "shotgun" approach, John Yates and colleagues also applied an "orthogonal" peptide separation approach by combining strong cation exchange (SCX) and reverse phase (RP) chromatography in sequence [14] (Fig. 1.2, see Chap. 4 & 7). This was in the late'90 and the limitations of shotgun approaches, including limited quantitation capabilities, also became evident in the following years. The use of isotopic metabolic or post- labeling, was successfully applied to improve the quantitation capabilities of LC-MS approaches in the decade that followed [15]. Briefly, labeling proteins with isotopic aminoacids (dividing cells [16]) or tags (post-mitotic primary cultures, tissues, etc. [17]), allowed resolution of peaks originating from the same peptide but deriving from different samples in the MS as a mass shift, in a highly quantitative fashion. Several iterations of these reagents resulted in a technique that is now both robust and commercially available (see Chap. 11) [18].

In a similar fashion, at the end the'90s a labeling approach that improved sensitivity, dynamic range and throughput in 2DE was developed by scientists at Carnegie Mellon in Pittsburgh [19]. This technology, currently commercialized by GE Healthcare, involves the pre-electrophoretic fluorescent labeling of protein samples with *N*-hydroxysuccinimidyl-ester derivatives of fluorescent cyanine (Cy) dyes and is known as two-dimensional difference gel electrophoresis (DIGE). This approach has the advantage that a pair of protein samples can be labeled separately with differently fluorescing Cy3 and Cy5 derivatives. The two samples can be mixed and then separated together on the same 2D gel. The resulting 2D gel is then scanned to acquire the Cy3 and Cy5 images separately using a fluorescent laser scanner. Furthermore, a sample labeled with a third Cy2 dye can be run on each 2D gel and used to normalize the signal among different gels. This DIGE approach dramatically reduces technical variability and exploits the high dynamic range of fluorescent staining for accurate quantitation. The issue of multiple proteins within the same 2DE spot still remains and, therefore, downstream validation of the changes observed in protein levels by other techniques is still essential. In addition, the relatively high price of the dyes and detection systems (a laser scanner able to detect the fluorescent probes) remain limiting factors.

As a result of the enormous advances in MS technology, it is nowadays possible to create an *in silico*, "unbiased" map of the proteome with little sample manipulation prior to MS [20, 21]. The combination of data independent acquisition (DIA) and targeted methods, such as multiple reaction monitoring (MRM [22]), now facilitate the accurate quantitation of peptides in the absence of isotopic labeling and in a proteome-wide fashion (see Chap. 10 on DIA/SWATH). The higher throughput capabilities and automatization of LC *versus* 2DE, combined with the increased capacity of detecting membrane and basic proteins further contributed to the popularity of LC-MS in the following decade and through to the present day. However, one limitation of shotgun approaches that remains to this day is that proteins are first digested into peptides, which are more stable, prior to separation by LC. From the standpoint of a biologist it is easy to understand how, by operating at the peptide level from the outset, important information can be lost. Indeed, so-called "top-down" approaches, which allow the analysis of whole proteins by MS (as opposed to "bottom-down" approaches, such as shotgun proteomics), are becoming increasingly popular (see Chap. 8 and [23]). This is also a consequence of the enormous amount of data and information generated by a single shotgun experiment, which can be challenging to store and analyze. The divide between bottom-up and top-down approaches is a blurred one and in fact some argue that 1D (SDS-PAGE or IEF) and 2DE techniques may be right in the middle, offering the opportunity to separate, or at least reduce the complexity of protein samples to a few intact proteins per spot. The capacity to see gross changes in molecular weight (e.g. degradation or proteolysis), or isoelectric point (e.g. phosphorylation) prior to digestion, combined with the natural propensity of the human brain to recognize patterns, suggest that 2DE may still have a role to play as a balance between top-down and bottom-up approaches, while celebrating its forty-first birthday [24].

Over a Century of Mass Spectrometry

Despite MS having a dominant role in proteomics these days, it is not a new technique. Indeed the first reports of the use of a very distant cousin of a modern MS date back to the work of Joseph John Thomson sometime around the turn of the twentieth century. Unbeknown to this young theoretical physicist at Cambridge University, MS would turn into, arguably, the most revolutionary technology in biomedicine over a century later. This embryonic MS, then called a parabola spectrograph, was initially used to investigate the very intimate components of matter, for which Thomson received a Nobel Prize in 1906 for "discovering the electron" [25].

It would take seven decades for MS to be able to be used to sequence proteins thanks to the work and dedication of Klaus Biemann and colleagues, right around the time when protein biochemists were optimizing 2DE [26]. Indeed, the possibility of analyzing large organic molecules using MS was limited by the ability to convert them into ions that would be able to “fly” in the instrument, a requirement for MS analysis. MS was initially applied to smaller molecules such as metabolites or pharmaceuticals, which could be ionized without being fragmented during the process. The way that an MS instrument works can be equated to that of a very precise molecular scale, as we will see in details in the following Chapter. To ionize large organic molecules represented a major challenge in MS history, because the kinetic energy that needs to be transferred to the peptides for them to enter into a gas phase as ions and be separated using magnetic selectors, would cause their fragmentation.

It took the work of several groups and two Nobel prizes to overcome this limitation in the late ‘80s. Two different approaches were pursued starting from either solid (crystallized) or liquid samples. Both proved to be effective in the end and generated the two sources that are still in use today: electrospray ionization (ESI), and matrix-assisted laser desorption ionization (MALDI). As we will learn in the next Chapter an MS instrument can be divided in three sections: a source, which converts peptides into ions in a gas phase; an analyzer (or multiple analyzers in series, as for tandem MS); and a detector to “count” the peptides ions and/or their fragments, separated according to their mass (mass over charge or *m/z* to be precise) by the analyzer/s (Fig. 1.3).

The Nobel prizes were ultimately awarded to John Fenn (ESI) and Koichi Tanaka (MALDI) in 2002, for “their development of soft desorption ionization methods for mass spectrometric analyses of biological macromolecuels [27]”. A little known fact about MALDI, is that although the Nobel prize was awarded to Dr. Tanaka and his group for their pioneering work on ionizing organic macromolecules from solid phase, the MALDI source, as it is currently used, was developed by two German scientists, Michael Karas and Franz Hillenkamp, who also named the technique [25]. This is one of the first examples of MS applied to the cardiovascular realm as the two scientists had a specific interest in mapping Ca^{2+} stores in

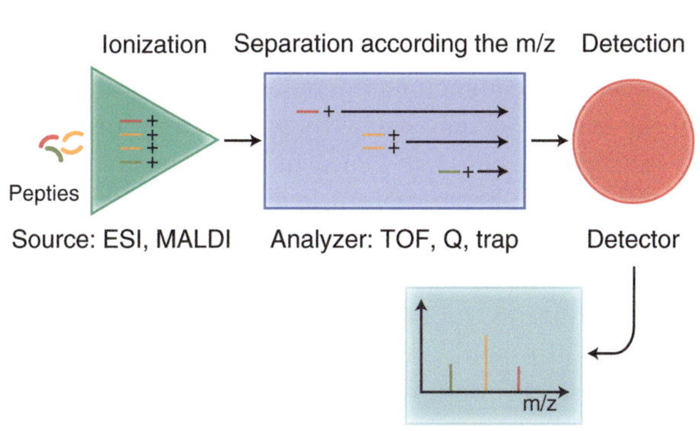

Fig. 1.3 Anatomy of a mass spectrometer. A mass spectrometer is best described by its “anatomical” components: a source, one or more analyzers and a detector (see text and Chap. 2 for details)

cardiac cells [25]. These "soft" ionization methods were first applied to nucleic acids. As we will see in the next paragraph, along with the creation of genomic database, nucleic acid research indirectly contributed to the development of proteomics. Lastly, as we will see in Chapter 2, several advances in MS, including the creation of new analyzers (such as orbital traps [28]) and improved fragmentation approaches [29], have remarkably enhanced the capability of modern instruments to the point that several thousands of proteins and their modifications can be accurately resolved nowadays.

Genomic and Proteomic Databases: From Genes to Proteins, and Back

As mentioned, one of the technologies that was developed during the genomic era and which served the proteomic cause very well was the creation of protein databases. In the first proteomic studies, proteins were identified using chemical Edman degradation and/or genetic tools which allowed scientists to assign a peptide to a protein, by translating it into a genetic sequence and matching it to those available in a particular genome. Since the very beginning of protein sequencing by MS, this new approach helped to identify open reading frames (ORFs) or sequenced genes thus perfecting their publication and annotation [30].

As mentioned above, the creation of protein databases to mine protein sequences was the result of the renaissance mind of the Andersons who created the first online protein database (HPI) [11] and the many implementations which followed. The sequencing of the genome of many organisms, including the human genome, allowed proteomic scientists to build protein databases predicted on the basis of the genetic info and ORFs. On this issue it is interesting to see that proteomic studies helped to find limitations in genomic databases [31]. The quest to match peptide spectra was greatly facilitated by the creation of *search and retrieval systems* (SRSs), algorithms, such as MASCOT [32], capable of scanning through a large database and returning the likelihood of a true-positive match between the experimental mass spectrum and a protein sequence listed *in silico*. As can be seen the sequencing of genomes not only allowed proteomics to flourish but for the former to "pay back" by pinpointing limitations in the algorithms used to predict and annotate genes in the respective databases. Indeed, despite genetic information being there it is important to confirm that it is relevant to the phenotype, or expressed. Even when that is the case, the predictive value of genes is highly limited by mRNA stability, alternative splicing and post-translational modifications [33]. Large-scale proteomic studies are inconceivable without the aid of bioinformatics and this finally allowed us to break free from the "one gene to one protein" dogma [34]. It is our hope that, as our technologies and bioinformatic tools perform more accurately and errors are corrected, the detailed picture of the molecular phenotype (aka the proteome) will finally reach a sublime resolution.

Proteomics with a Heart

We already mentioned how, thanks to the work of Michael Karas and Franz Hillenkamp in Frankfurt, cardiovascular research was involved with cutting-edge proteomic research early on in the history of proteomics. There are many unanswered questions in the cardiovascular realm many of which have a deep impact on public health due to the widespread incidence of a multiplicity of cardiovascular diseases [18]. Proteomics would be a natural tool to investigate mechanisms, generate new hypotheses and test the predictive value of novel biomarkers. We are lucky to have participated in the development of cardiovascular proteomics. Indeed, the first studies on the heart using 2DE anticipated the 1995 publication by Valerie Wasinger and colleagues of a much more unpleasant proteome [2]. Mike Dunn and Peter Jungblut were among the first scientists utilizing these methods to study the human heart, and published the first human heart 2D maps the same year the proteome was "born" [35, 36]. Shortly after, the same groups utilized MS to identify numerous myocardial proteins from 2DE [8, 37]. In their seminal report from 1998, Mike Dunn and colleagues compared cardiac biopsies from dilated cardiomyopathy (DCM) and ischemic heart disease (IHD) patients and found several significant changes that were also monitored at the isolated cell level for fibroblasts, mesothelial and endothelial cells, and cardiac myocytes [38]. To our knowledge this is the first report of a proteomic study addressing cardiovascular disease (Fig. 1.4). Interestingly, the changes in desmin and the chaperone alpha-B-crystallin first reported over 25 years ago are still the object of an intense investigation by several scientists (including our group [39]). Specifically, the idea that desmin can form preamyloid oligomers in the heart, similar to those found in Alzheimer's and Parkinson's diseases as well as several other proteinopathies, suggest that protein misfolding could play a major role in many diseases affecting the majority of the population in westernized societies, including cardiovascular disease. The role of

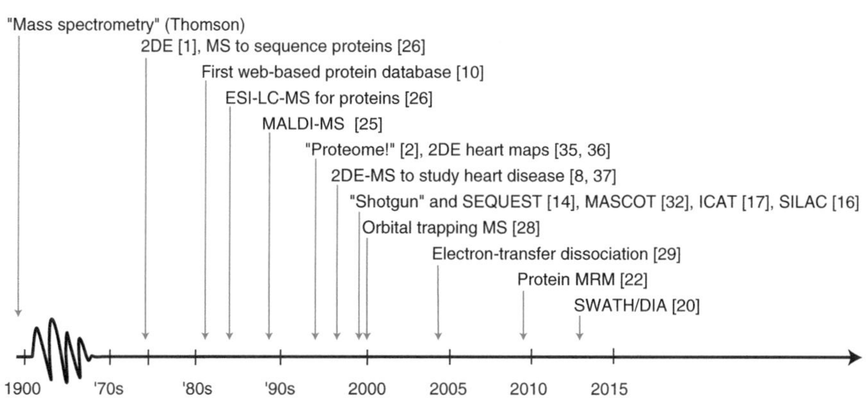

Fig. 1.4 A Timeline for the history of cardiovascular proteomics. See text for details

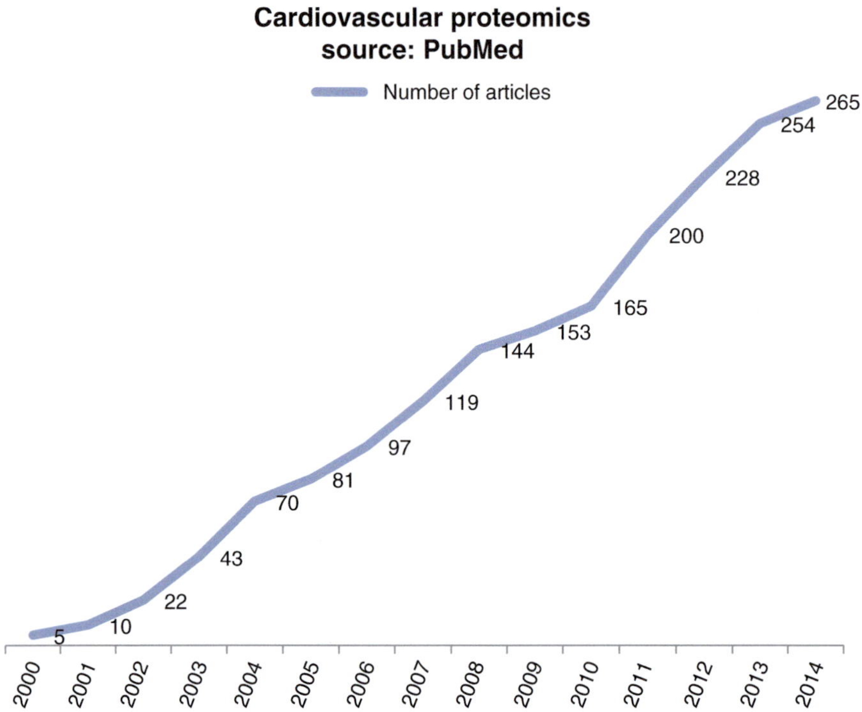

Fig. 1.5 Cardiovascular proteomics articles in the last 15 years. Plot displaying the number of articles identified by the search term "cardiovascular proteomics" per year since 2000 (Source, PubMed)

posttranslational modifications in this process is also rapidly emerging [40]. Therefore, proteomics may also help to generate new views and cutting edge idea in cardiovascular research.

With the new century, proteomics started to feature rapidly in the cardiovascular field. More pioneers of the caliber of Jennifer Van Eyk and Peipei Ping, started exploiting these emerging technologies to address a number of different aspects of cardiovascular disease spanning from signaling [36]. to biomarker discovery [41]. Thanks to the pioneering work of these scientists and others the field grew at an almost exponential pace, scoring a total number of 2011 "cardiovascular proteomics" publications at the time of writing. Although this does not include all of the work done in "specialty areas" and is limited by the searched terms, it does provide a reliable estimate of the upward trend of the field for the last 15 years (Fig. 1.5). The term was used in only five articles in 2000, and in 265 articles in 2014. The appeal of using chromatography over 2DE gels has also rapidly increased and several comparative studies describing the complementary use of the different approaches to achieve more extensive coverage have been published in the last decade. As proteomics evolves, and that is happening very quickly, we are sure that this approach will be increasingly incorporated into studies addressing the main unsolved questions concerning prevention, diagnosis and cure of cardiovascular disease.

Historical Crossroads: Pick Your New Technologies Wisely

Once proteomics became available to cardiovascular scientists at the beginning of this century, an immediate challenge came with it: what technology should they use to address a specific biomedical question? This is true for both basic and translational studies. The number of technological platforms also exploded. To keep up with a continuously evolving field is not an easy task, especially when it is as technologically based as proteomics. In our opinion, one of the responsibilities of the proteomic community is to explain these aspects to the ever-increasing number of collaborators in a way that is unbiased and comprehensible. Untimely, there is a tendency in some scientific environments to claim potential benefits even if they are only on the horizon of a particular technology. The increasing pressure to obtain research funding at a time when the economy is "breathing slowly" may accentuate this tendency.

There are several examples of why it is best to under-promise and over-deliver than *vice versa*. Perhaps one of such example that should be born in mind when establishing a collaborative effort or investing in a new technology is the experience with SELDI (surface-enhanced laser desorption ionization). This is a modified and patented version of MALDI that became quite popular in the early days of proteomics. Despite the promise of accelerating the discovery of new biomarkers, when it was first released, the technology was based on a low resolution MALDI-TOF instrument with a proprietary technology to fractionate the samples directly onto the MALDI (or SELDI) plate, the support that allows samples to enter the instrument. Despite the initial "hype" that seems to accompany most new technologies, the use of SELDI has now rapidly declined, probably because the method did not meet initial expectations, possibly due to the high impact of matrix effects, limited reproducibility and limited identification capabilities [42].

This is just one of the many examples that are a feature of the history of proteomics, and probably of many other fields. Another lesson learned is that the SELDI instrument was originally packaged in a "black box" type of format, claiming a simple sample in/data out operation. This is a very important lesson for new and old proteomic scientists alike. The field is complicated, and as much as there is a purpose in trying to simplify technologies in order to allow their broader use, certain aspects simply cannot be simplified. When a technology does not deliver what it had promised, the impact on its future tends to be more dramatic, possibly because there is a collective memory in the scientific community, which tends to adjust to new concepts with a certain delay. It is also possible that this fact alone may limit the chances of "redemption" for technologies that are initially advertised to address more than can be expected from them. The "market" of proteomic technologies is a bit like the stock market in that it is extremely volatile, and at times expensive. For proteomic beginners it may be wise to rely on the advice of more experienced scientists in the field, and since we all have our passions and beliefs, perhaps diversifying the portfolio of "proteomic brokers" could also prove beneficial. Last but not least, proteomic scientists need to make themselves understandable to the general scientific community, avoiding technical jargon when possible and sharing their unbiased knowledge. This book represents an effort in this very direction.

What Do We Leave Behind as We Move Forward: Bringing the Fun Back into Science

This is a tremendously exciting moment in science, including the cardiovascular field. The capabilities offered by the latest generation of proteomic technologies are almost infinite. We can see most of the proteome, quantify it and finally make sense of it. Thinking back on how hard previous generations of scientists had to work in order to optimize these technologies should make us feel extremely lucky. It took decades to couple an LC with an MS, and decades to analyze peptides and proteins using it. It took decades to sequence the genomes that are available today, create the corresponding protein databases and the algorithms to exploit them to generate proteomic data. It almost appears that everything had to be optimized for the current generation of proteomic scientists to "have fun with it". Proteomics can be painful, as science at large has the tendency to be sometimes. However, it has the potential and capability to bring very diverse expertise around the same table: statisticians, engineers, biochemists and physicians can find some exciting, common ground to move science and medicine forward (hopefully, as is the case with the present book). With this also comes the responsibility to actually change medicine and translate the impact of all these rich technological gifts back into some good for human health. There is an increasing demand for technical expertise and scientists that can translate the difficult languages of math, engineering, protein biochemistry and medicine. It takes time and dedication to have enough knowledge of these languages to be able to contribute something to the table. However, it is not inconceivable that proteomics will become a curriculum in many undergraduate and graduate programs in its own right, as happened previously for biotechnology, molecular biology, etc.. In all, these opportunities for collective growth lie just ahead of us waiting to be picked like ripe fruit.

With such opportunities in prospect, perhaps the best way to move forward is to honor those that took us this far by pursuing the best science possible. Of course quality and integrity are two pillars of science, or at least should be. As we have learned in this chapter, time will tell if what we have fulfilled these two core principles. Moreover, in an era where communication and data access is so rapid it may become increasingly easier to get distracted and lose focus of the scientific goals. In addition, when being exposed to technologies, there is also a risk of falling in love with them. In summary, with all these novelties comes the risk of forgetting the knowledge generated by other means and during earlier times, when perhaps it took longer to separate proteins or to analyze them, but this time could be used to learn about the functional aspects of our discoveries, in-depth. Many proteomic scientists in the cardiovascular field had to write scientific articles explaining why a list of proteins, many of which we had not heard about before, change in a coordinated fashion. What has been published about the role of these mysterious proteins and the biological effect of their known post-translational modifications that we see changing in a particular disease state are very important aspects of our research. To us, another good way to honor the work of the fathers of proteomics would simply be to "do our homework", try to put our observations in the context of what is already known without the need to rewrite science. Lastly, as we stand on the shoulders of the scientists who made us who we are, we honor their legacy by learning from their mistakes and endeavoring to match their successes.

Funding: American Heart Association 12SDG9210000 and 16IRG27240002; NIH P01 HL107153; RFO University of Bologna to GA

References

1. O'Farrell PH. High resolution two-dimensional electrophoresis of proteins. J Biol Chem. 1975;250:4007–21.
2. Wasinger VC, Cordwell SJ, Cerpa-Poljak A, Yan JX, Gooley AA, Wilkins MR, Duncan MW, Harris R, Williams KL, Humphery-Smith I. Progress with gene-product mapping of the Mollicutes: mycoplasma genitalium. Electrophoresis. 1995;16:1090–4.
3. Bini L, Calvete JJ, Turck N, Hochstrasser D, Sanchez J. Special Issue: "20 years of Proteomics" in memory of Vitaliano Pallini. J Proteomics. 2014;107:1–146.
4. Smith, L.M., Kelleher, N.L., and Consortium for Top Down Proteomics. Proteoform: a single term describing protein complexity. Nat Methods. 2013;10:186–7.
5. Righetti PG. The Monkey King: a personal view of the long journey towards a proteomic Nirvana. J Proteomics. 2014;107:39–49.
6. Bjellqvist B, Ek K, Righetti PG, Gianazza E, Gorg A, Westermeier R, Postel W. Isoelectric focusing in immobilized pH gradients: principle, methodology and some applications. J Biochem Biophys Methods. 1982;6:317–39.
7. Gorg A, Drews O, Luck C, Weiland F, Weiss W. 2-DE with IPGs. Electrophoresis. 2009;30 Suppl 1:S122–32.
8. Sutton CW, Pemberton KS, Cottrell JS, Corbett JM, Wheeler CH, Dunn MJ, Pappin DJ. Identification of myocardial proteins from two-dimensional gels by peptide mass fingerprinting. Electrophoresis. 1995;16:308–16.
9. Randell P. It's a MALDI but it's a goodie: MALDI-TOF mass spectrometry for microbial identification. Thorax. 2014;69:776–8.
10. Anderson NL, Taylor J, Scandora AE, Coulter BP, Anderson NG. The TYCHO system for computer analysis of two-dimensional gel electrophoresis patterns. Clin Chem. 1981;27:1807–20.
11. Anderson NG, Anderson L. The Human Protein Index. Clin Chem. 1982;28:739–48.
12. Santoni V, Molloy M, Rabilloud T. Membrane proteins and proteomics: un amour impossible? Electrophoresis. 2000;21:1054–70.
13. Rabilloud T. Membrane proteins and proteomics: love is possible, but so difficult. Electrophoresis. 2009;30 Suppl 1:S174–80.
14. Yates 3rd JR, Carmack E, Carmack E, Hays L, Link AJ, Eng JK. Automated protein identification using microcolumn liquid chromatography-tandem mass spectrometry. Methods Mol Biol. 1999;112:553–69.
15. Patterson SD, Aebersold RH. Proteomics: the first decade and beyond. Nat Genet. 2003;33(Suppl):311–23.
16. Oda Y, Huang K, Cross FR, Cowburn D, Chait BT. Accurate quantitation of protein expression and site-specific phosphorylation. Proc Natl Acad Sci U S A. 1999;96:6591–6.
17. Gygi SP, Rist B, Gerber SA, Turecek F, Gelb MH, Aebersold R. Quantitative analysis of complex protein mixtures using isotope-coded affinity tags. Nat Biotechnol. 1999;17:994–9.
18. Agnetti G, Husberg C, Van Eyk JE. Divide and conquer: the application of organelle proteomics to heart failure. Circ Res. 2011;108:512–26.
19. Unlu M, Morgan ME, Minden JS. Difference gel electrophoresis: a single gel method for detecting changes in protein extracts. Electrophoresis. 1997;18:2071–7.

20. Gillet LC, Navarro P, Tate S, Rost H, Selevsek N, Reiter L, Bonner R, Aebersold R. Targeted data extraction of the MS/MS spectra generated by data-independent acquisition: a new concept for consistent and accurate proteome analysis. Mol Cell Proteomics. 2012;11:O111.016717.
21. Guo T, Kouvonen P, Koh CC, Gillet LC, Wolski WE, Rost HL, Rosenberger G, Collins BC, Blum LC, Gillessen S, Joerger M, Jochum W, Aebersold R. Rapid mass spectrometric conversion of tissue biopsy samples into permanent quantitative digital proteome maps. Nat Med. 2015;21:407–13.
22. Kuhn E, Wu J, Karl J, Liao H, Zolg W, Guild B. Quantification of C-reactive protein in the serum of patients with rheumatoid arthritis using multiple reaction monitoring mass spectrometry and 13C-labeled peptide standards. Proteomics. 2004;4:1175–86.
23. Peng Y, Ayaz-Guner S, Yu D, Ge Y. Top-down mass spectrometry of cardiac myofilament proteins in health and disease. Proteomics Clin Appl. 2014;8:554–68.
24. Oliveira BM, Coorssen JR, Martins-de-Souza D. 2DE: the phoenix of proteomics. J Proteomics. 2014;104:140–50.
25. Griffiths J. A brief history of mass spectrometry. Anal Chem. 2008;80:5678–83.
26. Hudson G, Biemann K. Mass spectrometric sequencing of proteins. The structure of subunit I of monellin. Biochem Biophys Res Commun. 1976;71:212–20.
27. Tabet JC, Rebuffat S. Nobel Prize 2002 for chemistry: mass spectrometry and nuclear magnetic resonance. Med Sci (Paris). 2003;19:865–72.
28. Makarov A. Electrostatic axially harmonic orbital trapping: a high-performance technique of mass analysis. Anal Chem. 2000;72:1156–62.
29. Syka JE, Coon JJ, Schroeder MJ, Shabanowitz J, Hunt DF. Peptide and protein sequence analysis by electron transfer dissociation mass spectrometry. Proc Natl Acad Sci U S A. 2004;101:9528–33.
30. Biemann K. Laying the groundwork for proteomics: mass spectrometry from 1958 to 1988. J Proteomics. 2014;107:62–70.
31. Pandey A, Mann M. Proteomics to study genes and genomes. Nature. 2000;405:837–46.
32. Perkins DN, Pappin DJ, Creasy DM, Cottrell JS. Probability-based protein identification by searching sequence databases using mass spectrometry data. Electrophoresis. 1999;20:3551–67.
33. Rabilloud T, Hochstrasser D, Simpson RJ. Is a gene-centric human proteome project the best way for proteomics to serve biology? Proteomics. 2010;10:3067–72.
34. Gygi SP, Rochon Y, Franza BR, Aebersold R. Correlation between protein and mRNA abundance in yeast. Mol Cell Biol. 1999;19:1720–30.
35. Corbett JM, Wheeler CH, Baker CS, Yacoub MH, Dunn MJ. The human myocardial two-dimensional gel protein database: update 1994. Electrophoresis. 1994;15:1459–65.
36. Jungblut P, Otto A, Zeindl-Eberhart E, Plessner KP, Knecht M, Regitz-Zagrosek V, Fleck E, Wittmann-Liebold B. Protein composition of the human heart: the construction of a myocardial two-dimensional electrophoresis database. Electrophoresis. 1994;15:685–707.
37. Thiede B, Otto A, Zimny-Arndt U, Muller EC, Jungblut P. Identification of human myocardial proteins separated by two-dimensional electrophoresis with matrix-assisted laser desorption/ionization mass spectrometry. Electrophoresis. 1996;17:588–99.
38. Corbett JM, Why HJ, Wheeler CH, Richardson PJ, Archard LC, Yacoub MH, Dunn MJ. Cardiac protein abnormalities in dilated cardiomyopathy detected by two-dimensional polyacrylamide gel electrophoresis. Electrophoresis. 1998;19:2031–42.
39. Agnetti G, Halperin VL, Kirk JA, Chakir K, Guo Y, Lund L, Nicolini F, Gherli T, Guarnieri C, Caldarera CM, Tomaselli GF, Kass DA, Van Eyk JE. Desmin modifications associate with amyloid-like oligomers deposition in heart failure. Cardiovasc Res. 2014;102:24–34.
40. Del Monte F, Agnetti G. Protein post-translational modifications and misfolding: new concepts in heart failure. Proteomics Clin Appl. 2014;8:534–42.
41. Arrell DK, Neverova I, Van Eyk JE. Cardiovascular proteomics: evolution and potential. Circ Res. 2001;88:763–73.
42. Mischak H, Vlahou A, Ioannidis JP. Technical aspects and inter-laboratory variability in native peptide profiling: the CE-MS experience. Clin Biochem. 2013;46:432–43.

Chapter 2
Basic Concepts in Mass Spectrometry and Protein Quantitation

Jake Cosme*, Irina Tchernyshyov*, and Anthony Gramolini

Abstract Mass spectrometry (MS) has provided some paradigm-shifting technology to the field of cardiac biology. Recent advances in MS have made protein identification into a high throughput analytic tool and improved accuracy and sensitivity of protein quantitation. Many of the tools available to scientists trying to answer fundamental questions of basic heart function and mechanisms of disease are quite robust and versatile. MS-based cardiac proteomic approaches have developed to such an extent that a researcher can design experiments to answer clear hypotheses, but also studies can also be 'hypothesis-generating', ultimately leading to deeper analyses and considerations. Here, we will outline the basic concepts of MS in an effort to explain the potential of this technology in investigating cardiac based research questions. Principles of how current instrumentation functions and how data is acquired will be introduced. Protein quantitation in MS is available in many varieties and applications; this chapter will outline current available technologies in protein quantitation such as isotope-labeled and label-free approaches. With the introductory knowledge of MS and protein quantitation, we will examine some key cardiac proteomics studies and discuss how these principles have been applied to answer specific research questions.

*JC and IT are contributed equally

J. Cosme, HBSc
Department of Physiology, University of Toronto and Toronto General Hospital Research Institute, Toronto, ON M5G 1M1, Canada

I. Tchernyshyov, MSc
The Advanced Clinical Biosystems Research Institute, Heart Institute, Cedars Sinai Medical Center, Los Angeles, CA 90048, USA

A. Gramolini, PhD (✉)
Department of Physiology, University of Toronto and Toronto General Hospital Research Institute, Toronto, ON M5G 1M1, Canada

Department of Physiology, Ted Rogers Centre for Heart Research, University of Toronto and University Health Network, 661 University Ave. 14th Floor, Rm 1437, Toronto, ON M5G 1M1, Canada
e-mail: anthony.gramolini@utoronto.ca

G. Agnetti et al. (eds.), *Manual of Cardiovascular Proteomics*,
DOI 10.1007/978-3-319-31828-8_2

Keywords Mass spectrometry • Peptide analysis • Peptide ionization • Peptide fragmentation • Quantative proteomics

Introduction to Mass Spectrometry

Proteins are end-point effectors in cell signaling pathways controlling the majority of cellular functions. Identification of basic components of signaling networks and their relative abundance provides insights into biological understanding of the molecular dynamics of a particular system in health and disease. Mass spectrometry has established itself as a key platform for protein analysis and characterization. In addition to being able to identify and quantify thousands of proteins in one experiment, mass spectrometry can provide detailed structural information as well, such as identifying post-translational modifications. Mass spectrometers can generate information about molecular mass of a polypeptide. In most cases, the molecular mass alone is not sufficient to establish the identity of the peptide or protein unambiguously or to help distinguish isomers (peptides which have the same amino acid composition but different amino acid sequence). To obtain detailed information about amino acid composition of a protein, including those amino acids that were covalently post-translationally modified, an approach called "tandem mass spectrometry" is used. To perform tandem mass spectrometry, the ion of interest is broken into smaller fragments and molecular masses of the resulting products are determined by another round of MS. The fragmentation pattern specifically reflects the structure of the original ion and is used to reconstruct its identity. Two conceptually very different strategies are used to identify and characterize proteins: "top-down" and "bottom-up". "Top-down" MS approach refers to analysis of the intact full-length proteins. This method has a potential advantage for providing the most detailed and complete data about protein variants and isoforms. However, while there is some notable success in using top-down approach, due to inherent difficulties with intact protein separations, ionization and fragmentation, top-down analysis still remains a state of the art and prerogative of only a few expert labs [1]. Some of the challenges of top-down proteomics can be addressed by breaking down proteins into much smaller peptide molecules by means of proteolysis. This approach is commonly referred to as "bottom-up" proteomics. There are a variety of enzymatic or chemical methods that can be used for proteolysis. For these reasons, which will be discussed and highlighted later in this chapter, sequence specific serine protease trypsin is most commonly used in proteomics experiments [2]. Proteolytic digestion generates highly complex samples containing of thousands of peptides. Once cleaved, peptides then are subjected to some form of liquid chromatography (LC), which is commonly used to separate peptides before introduction into a mass spectrometer. This method has been successfully applied to proteomes of various complexity, from cataloging proteins, which make-up cells and tissues [3], to identifying changes in extracellular matrix in a context of ischemia-reperfusion injury [4], to a single protein signaling proteome such as phosphoproteome of cardiac myosin binding protein-C (cMyBP-C) [5]. In this chapter, we will outline methods and principles of peptide analysis by mass spectrometry such as peptide ionization and peptide fragmentation, describe how different mass spectrometers function, and how to obtain quantitative data from a proteomics-based experiment.

How a Mass Spectrometer Works

Ions of different mass-to-charge ratio (m/z), when placed in electrostatic, magnetic or electro-magnetic field, exhibit different patterns of motion. At the basis of 'Mass Spectrometry' is the ability to separate ions by mass-to-charge ratio, from which molecular mass of the analyte can be calculated. Due to recent technological advances, numerous types of mass spectrometers became available to the research community from various vendors. The diverse equipment employ different physical principles and modes of operation, capable of different performance standards and can be used to address different analytical questions. However, regardless of instrument configuration, the fundamental components of a mass spectrometer are the ionization source, the mass analyzer, and the detector. The ion source is the compartment where sample molecules are converted into ions and introduced into other sections of the instrument, namely, the analyzer and the detector. A mass analyzer is the part of the instrument in which ions are separated based on their m/z values. When ion arrives at the detector, its energy is converted into electrical signal, which is transmitted to a computer and recorded in form of a spectrum. A mass spectrum is a two-dimensional plot, where a specific m/z on the x-axis is plotted against the relative abundance of the ion on y-axis. The specific type of the detector is usually supplied to match the analyzer. On the other hand, different ionization methods and analyzers offer great flexibility to the end user and need to be understood in greater detail. In order to better separate ions it is important to avoid accidental collisions with air along the ion pass which can disturb an ion's trajectory. For that reason, the analyzer, detector, and in some instrument configurations, an ion source, are maintained under high vacuum.

Methods of Ionization

For an analyte to be measured in a mass spectrometer, it needs to be ionized and to be brought into a gas phase. Energy needs to be transferred to molecules in condensed (solid) phase to make them volatile. It has proven to be challenging to ionize biological molecules (i.e. peptides/proteins) without destruction or significant degradation. In the late 1980's two "soft ionization" methods were introduced: Matrix Assisted Laser Desorption-Ionization (MALDI) [6] and Electro-Spray Ionization (ESI) [7]. To date, these methods remain most widely used in 'Proteomics'. ESI operates from solution and is easily coupled to many LC separation platforms. The peptide ions are preformed in the liquid phase during chromatographic separation (Fig. 2.1a). While ESI can be used to generate both positive and negative ions, peptides and proteins are almost exclusively analyzed as positive ions. Peptide ions leave the LC column

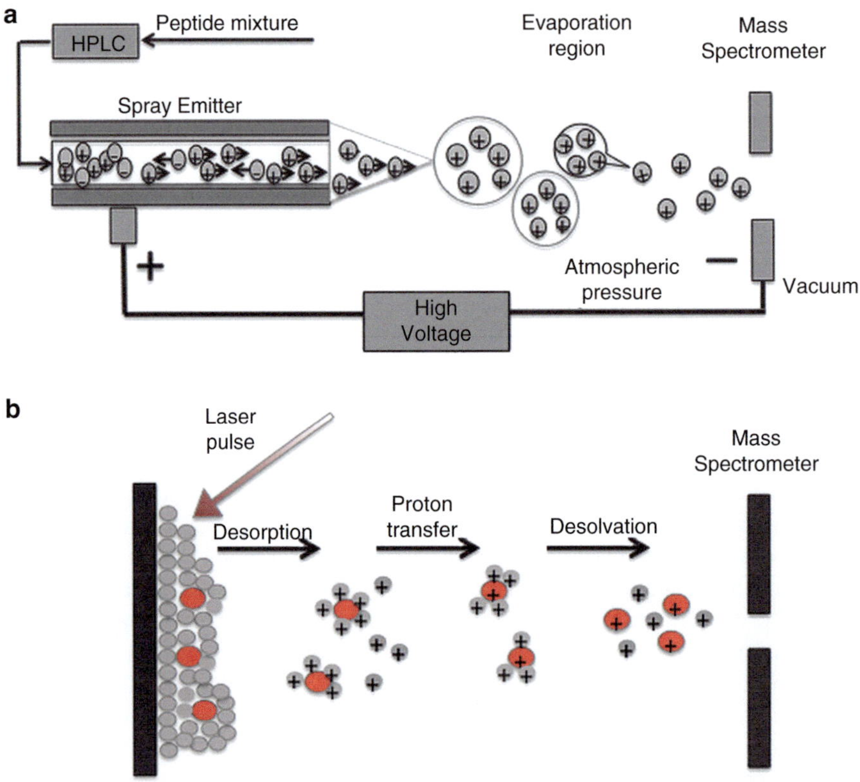

Fig. 2.1 Peptide ionization methods. (**a**) Electrospray ionization source coupled to High Performance Liquid Chromatography (*HPLC*) column. Peptide mixture, generated by proteolytic digestion, is separated by liquid chromatography, ionized by protonation of basic amino-acid residues and sprayed from the emitter towards the MS. As solvent evaporates from the surface of the droplet, ions are forced closer together until charge repulsion causes peptide ions leave the droplet. High voltage potential pulls the ions into the high vacuum region of the mass spectrometer. (**b**) Diagram of Matrix Assisted Desorption Ionization (MALDI). For MALDI peptides, peptides (*red circles*) are embedded in the dense mesh of matrix molecules (*gray circles*). Intense heat of the laser pulse causes matrix to sublime. Evaporating matrix carries non-volatile peptide molecules into a gas-phase, where peptides get ionized via proton transfer from matrix

through a spray emitter capillary. A high voltage power supply is connected between the spray emitter and an inlet into a mass spectrometer. Under these high voltage conditions the liquid leaving a capillary assumes a characteristic shape known as Taylor cone. Positively charged ions accumulate near the surface at the end of the emitter tip, and eventually explode in a mist of fine droplets. Each droplet consists of peptide ions and solvent molecules. As a droplet moves to the negatively charged inlet into the MS, the solvent gradually evaporates. As this happens the droplet becomes smaller and smaller, pushing the charged peptide ions closer and closer together, creating charge repulsion. At some point these repulsion forces overcome the surface tension and the ion leaves the droplet, temporarily decreasing charge repulsion. Released ions are drawn into the high vacuum region of the mass spectrometer.

The predominant mechanism of positive ionization is by transfer of protons from the solvent at low pH. The typical sites of proton attachment are amino acids with the basic side-chains (Arg, Lys, His and N-terminal amine). The charge state of the peptide closely correlates with the amount of proton attachment sites. As mentioned earlier, proteins are typically cleaved using the digestion enzyme trypsin. Trypsin cleaves very specifically at peptide bonds following Arginine or Lysine, leaving each generated peptide to have at least two proton attachment sites and minimum charge +2. An important feature of ESI is that the efficiency of ionization depends on a ratio of sample ions to solvent molecules in a solution (concentration) leaving the ionization capillary, not the total amount of sample. Consequently, sensitivity of the analysis increases when the flow rate decreases. The largest gain in sensitivity is demonstrated at nanoliter flow rates (nanoESI) [8, 9]. As a result, nano ESI has been the most widely used mode of sample introduction in bottom-up proteomics.

In MALDI (Fig. 2.1b), energy required for desorption of sample is generated and transferred from matrix molecules. Prior to analysis, dissolved sample is mixed with saturated solution of small organic compound (matrix) in large molar excess (1:1000) and co-crystallized on a metal sample plate, such that each molecule of analyte is completely surrounded by matrix. This embedding reduces interactions between analyte molecules and makes them more volatile. Short laser pulses are used to very rapidly heat matrix crystals, forcing them to expand and evaporate. The sample molecules are then carried into the gas phase by expanding matrix cloud. The compounds used as matrices are usually relatively more volatile and require less energy to ionize than analyte molecules, whilst preserving analyte molecules from extensive fragmentation. While the precise mechanism of ion formation is still debated, it is believed that peptide and protein sample molecules ionized by the proton transfer from multiple collisions with matrix ions in the vapor phase. Unlike electrospray, MALDI generates singly charged ions. Once in a gas phase, ions are accelerated by an electrostatic field toward the analyzer. Since the laser pulses are very short, ions are released in clusters. For that reason MALDI is usually used in combination with pulse type analyzers. (Time-of–flight (TOFs), which will be reviewed later in this chapter). All the peptides present on the sample plate ionize and enter mass spectrometer at the same time. As such, MALDI has been most useful for analysis of relatively simple samples such as gel bands or gel spots from one- or two-dimensional gel electrophoresis, which will be discussed later on.

Methods of Fragmentation

Tandem mass spectrometry (MS/MS) is performed in two stages. During the first stage (MS1 or survey scan), the ion of interest (the precursor, or parent ion) is measured, isolated from other ions and subjected to fragmentation (MS2). M/z ratios of the resulting fragment ions (the product or daughter ions) are determined at the second stage (Fig. 2.2). Fragmentation pattern is used to gain information about peptide sequence. Collision-induced dissociation (CID), higher-energy collision-induced dissociation

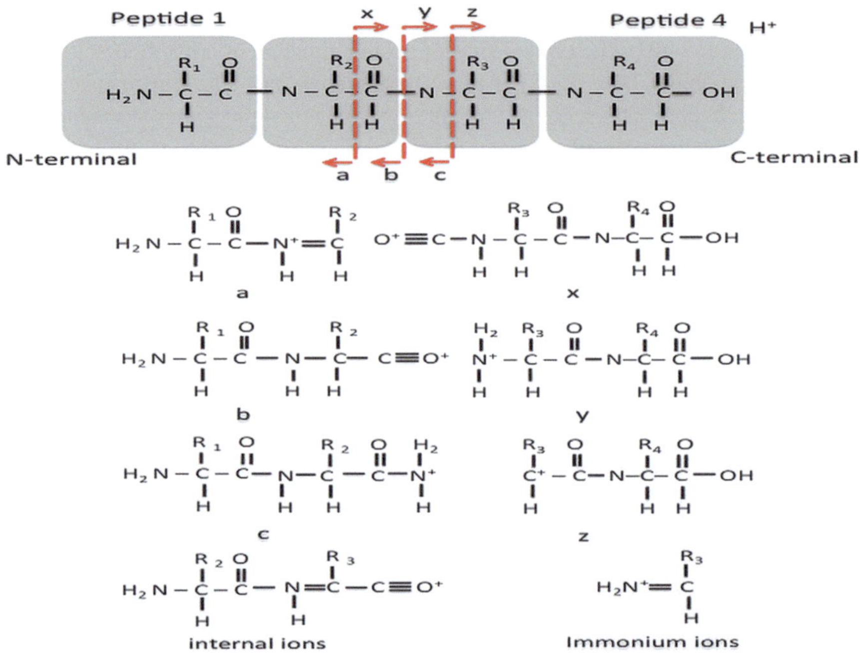

Fig. 2.2 Common peptide fragment ions. Shown is a four amino acid peptide. R- represents amino acid side chains. The cleavage in a peptide backbone can occur in one of the three types of bonds, CH–CO, CO–NH or NH–CH; as a result, six different types of fragments can be produced. Fragment ions, which retain the amino terminus are labeled as a, b, or c ions and those retaining the carboxyl terminus, x, y, or z (This notation was introduced by Roepstorrf and Fohlmann [11] and adopted in the current form by Biemann [12]. Only those fragments that carry at least one charge can be detected. Structures of immonium and internal ions are also shown)

(HCD), and electron transfer disassociation (ETD), are some of the most widely used methods for study of peptides and proteins and will be covered in this section.

In the process leading to fragmentation (activation), the internal energy of an ion increases, chemical bonds start to break, and eventually the ion falls apart. Ion activation methods differ in the amount of energy transferred to the ion and in the means of transfer [10]. Different activation methods produce different ions and provide complementary information about the peptide sequence. Types of ions produced for peptide fragmentation are described in (Fig. 2.3). The cleavage in a peptide chain can occur in one of the three types of bonds, CH–CO, CO–NH or NH–CH; as a result, six different types of fragments can potentially be produced. Fragment ions, which retain the amino terminus are labeled as a, b, or c ions and those retaining the carboxyl terminus, x, y, or z. This notation was introduced by Roepstorrf and Fohlmann [11] and adopted in the current form by Biemann [12]. Only those fragments that carry at least one charge can be detected. Peptide sequencing made is possible by the fact that fragmentation of peptide bonds is random, and various ions differing in mass by one amino acid are produced. Amino acid residues can be determined by mass difference between two successive ions (Fig. 2.2c).

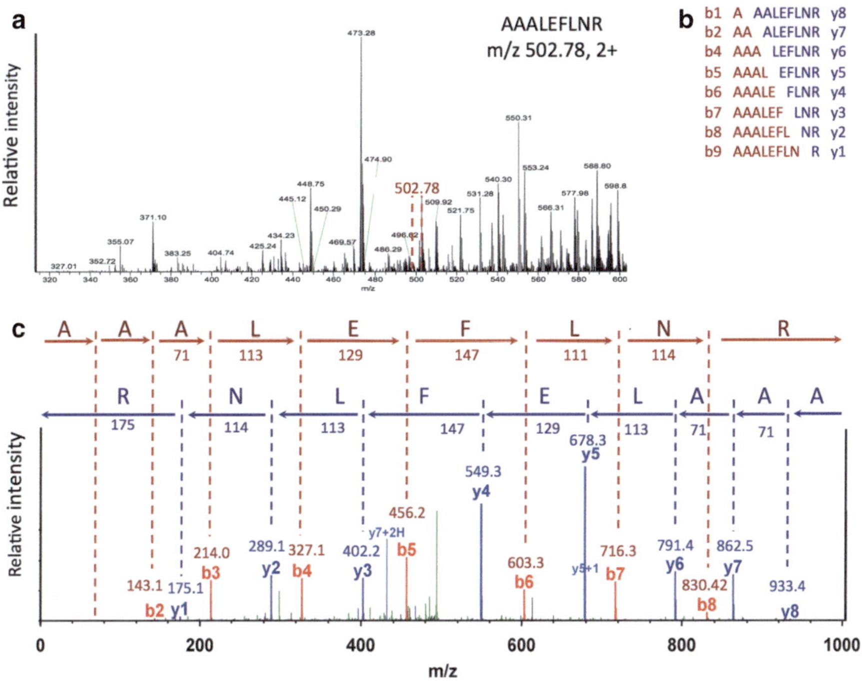

Fig. 2.3 Peptide sequencing by tandem mass spectrometry. (**a**) Peptide ions mass-to-charge ratios are measured by MS1 scan. M/z of a peptide ion is plotted on the x-axis over relative intensity on the y-axis, which indicative of amount of a peptide ion present in the sample. An ion of m/z 502.78 is selected for MS/MS (highlighted in *red*). (**b**) Expected fragmentation pattern of doubly charged *AAALEFLNR* peptide (from panel A) by low-energy CID. Random fragmentation of peptide backbone generates series of b (*red*) – and y (*blue*) – ions. B-ions (and all the other N-terminal fragments) are numbered from the N-terminus. Y-ions (and all the other C-terminal fragments) are numbered from the C-terminus. Subscript indicates the number of amino acid residues in the fragment. (**c**) Experimental MS/MS spectrum of the same peptide is acquired in an ion trap. Measured mass-to-charge ratio of observed fragments is shown on top of each peak. M/z difference between two consecutive ions in the series corresponds to the mass of the specific amino acid residue (Amino acid residue reference data can be found at http://www.matrixscience.com/help/aa_help.html)

The most common mode of fragmentation used for peptide analysis in proteomics is collision-induced dissociation (CID), also called collisional activated dissociation (CAD). As the name implies, the energy required for fragmentation is imparted through collisions of analyte and inert neutral gas. Kinetic energy of moving ions is converted into vibrational energy, leading to the breaking of covalent bonds. The amount of energy transferred in each collision is relatively small, so that typically several hundred collisions occur before an ion dissociates [13]. Low-energy CID favors peptide fragmentation at (–CO-NH), a bond that links two amino-acid residues, which primarily generates ions of b and y types. Different instrument types and configurations allow for the detection of different type of fragment ions. Fragmentation can be performed either in the same device where the ions are trapped or in a separate collision cell. Depending on the type of the analyzer, tandem analysis can either be performed "in space", by combining several analyzers, or "in time", by using the same ion-trapping device to perform different stages of analysis sequentially. When the tandem-in-space configuration is used (Triple Quadrupole, Quadrupole-Time-of-Flight, etc.), the quadrupole collision cell is placed between two analyzers and filled with gas. Ions enter the collision cell in the form of a focused accelerated beam. An ion traveling through the collision chamber experiences numerous impacts with the gas molecules that increase its internal energy and eventually lead to dissociation. Newly formed product ions can continue to undergo activation and dissociation on their way through the collision chamber to the detector. B-ions often do not withstand multiple collisions typical for quadrupoles. When fragmentation is performed in ion traps, the vibrational motion of the pre-selected parent ion population is created by resonant-energy excitation, while the background gas in pumped into a trap. Once an ion is fragmented, it becomes unstable and, after it is expelled from the trap, its fragments are detected. As a result, only one fragmentation event per ion usually happens. Because of that, peptides with liable PTMs (phosphorylation, O-linked glycosylation, citrullination) will fragment preferentially at the PTM site, creating a neutral-loss fragment that dominates the signal. Fragments carrying useful information about amino acid sequence will be relatively few. Although both types of collisions produce b or y-type fragments, ion trap MS/MS spectra have more higher-mass b-type ions. Another characteristic feature of fragmentation spectra generated in ion traps is a reduced stability of product ions in the low-*m/z* range, the so-called "one-third rule". For example, when a precursor ion with $m/z = 900$ is dissociated, fragments with $m/z < 300$ are not consistently observed in the resulting MS2 spectra. This has a very important practical implication for quantitative analysis using isobaric tags (iTRAQ and TMT), as will be discussed later. These limitations of ion traps in tandem analysis have been overcome in a recently introduced beam-type fragmentation on hybrid ion-trap devices; the new technology is known as Higher-energy Collision-induced Dissociation (HCD). HCD differs from low-energy CID in that ions selected for fragmentation are sent to a quadrupole-type collision cell from the ion trap [14]. According to a study by Michalski et al. [15], HCD spectra are more complex than CID showing more low-mass immonium ions and internal fragments. Sequence information is obtained from predominantly y-ions and low-mass b-ions. Higher-mass b-ions are typically underrepresented (Fig. 2.4). Low-energy collisional activation has been a main workhorse in proteomics. It can be done on any commercially available instrument configuration. However, this method has some notable limitations. Because larger polypeptides have a larger number of chemical bonds, more energy needs to be transferred to them to cause fragmentation. The rate of CID reaction with its relatively inefficient energy transfer becomes too slow on a time scale of the mass analyzer detection speed. Also, if peptides have basic amino acids in the middle of a peptide ion that leads to preferential fragmentation at the site of that amino acid, preventing random bond protonation and cleavage. When a PTM is present, it competes with a backbone amide as the preferential site of cleavage, often resulting in premature dissociation of a modifying group and inability to establish site of modification. Alternative fragmentation methods are known as Electron Capture

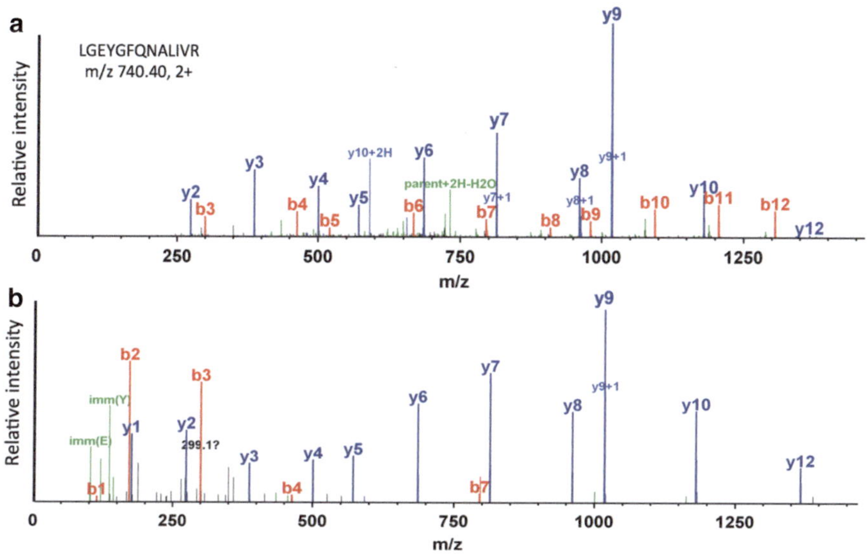

Fig. 2.4 Low-energy CID fragmentation patterns. Shown is MS/MS spectrum of the peptide ion m/z 740.40, 2+. Fragmentation was performed in: (**a**) an ion-trap, (**b**) beam-type collision cell (HCD in this case). b- and y- ion series are indicated. Note the difference between ion-trap and beam-type fragmentation pattern. Ion trap generates more complete b-ion series. On the contrary, only low m/z b-type ions are observed in beam-type spectrum. The manifestation of one-third rule is can be clearly observed in this example. In ion-trap (**a**) no ions are detected below m/z 250. In beam-type (**b**), that region of a spectrum contains low-mass sequence ions (*b2, y1*) and immonium ions

Dissociation (ECD) [16] and its variation Electron Transfer Dissociation (ETD) [17]. ECD involves fragmentation of ions through a reaction of low-energy electrons with positively charged peptides. An extra electron is attached to the site of high proton density, resulting in rapid migration of the hydrogen radical and the dissociation of the ion. For ETD, an extra electron is transferred during a reaction of peptide cation with a radical anion. Subsequent fragmentation has a mechanism similar to that of ECD [18]. Unlike in CID, electron capture results in the cleavage of an –NH-CH- bond, generating mostly c- and z- types of ions. Because activation energy does not get distributed between many bonds, leading to fragmentation of the weakest one, ETD does not cause the loss of PTMs. ETD is most efficient for characterization of peptides with higher charge states (+3 and higher), and for that reason ETD has proven to be indispensable for the analysis of intact proteins [19].

In summary, each activation method has specific strengths and limitations and currently no method is universally applicable to every analytical question. Multiple studies have been performed to compare the activation methods [20]. Combinations of multiple activation methods have been widely used to gain more extensive and often complementary information about the structure of peptides and proteins [21]. In practice, not all of the discussed activation methods have been made commercially available on every instrument configuration (see the next section) and the deciding factor is which instrument is available to the user.

Orbitrap MS

The Orbitrap-class of instruments have become a widely used MS platform since their introduction. Orbitraps share the property of using a Orbitrap mass analyzer to generate m/z signal [22]. Typically they are paired with a linear ion trap and/or quadroupole to store ions for injection to the Orbitrap, as well as to perform MS2 in certain applications. Ions stored in linear ion trap or quadrupole are injected into the Orbitrap via a curved-trap (c-trap) that redirects the ions into the ion trap. Orbitrap mass analyzers generate m/z signal via the harmonic oscillation of ions between a spindle-shaped electrode and the frequency of the ion's oscillation is converted to its m/z signal. Orbitrap MS have a high resolving power with a slightly longer cycle time. Generally, the Orbitrap is paired with a linear ion trap where the Orbitrap performs a high resolution MS1 while the information is generated to perform DDA of MS2 in the linear ion trap that can perform MS2 of multiple ions within one cycle of MS1.

Time-of-Flight MS

TOF is based on the principle that ions exposed to equal high voltage have a m/z-dependent velocity. Smaller ions will arrive at the detector sooner than larger ions and that will be the basis of the m/z measurement generated in TOF MS [23]. Traditional TOF MS is linear but the development of a reflectron which redirects ions to increase the travel time of ions to improve resolution [24]. A standalone TOF MS has applications with MALDI, as ionization of embedded peptides in the matrix are readily available for acceleration in an electrical field for TOF measurements. This approach has been applicable to cardiac research questions, such as the quantification of myosin heavy chain peptides [25]. TOF MS can be used for tandem MS where an in-line CID collision cell is activated for MS2 scans [26]. Quadrupole-Time of Flight (QTOF) MS uses a hybrid MS setup, where a quadrupole MS and TOF MS are paired together [27], similar to how an Orbitrap MS is paired with an ion trap or quadrupole. Ions entering a QTOF MS are guided through quadrupoles via the voltage applied to the quadrupoles. The voltage dictates which m/z range passes through the quadrupoles and towards the TOF analyzer. The quadrupole preceding the TOF MS acts as the collision cell where inert gas is used for CID [28]. TOF analysis in a QTOF MS occurs in the same manner as a standalone TOF MS.

SRM-MS with Triple Quadrupole MS

Selected reaction monitoring MS (SRM-MS) is an approach of targeted MS where proteotypic peptides in a complex mixture are absolutely quantified [29] with spiked-in stable isotope standards. When multiplexing ions for quantification,

SRM-MS is termed multiple reaction monitoring MS (MRM-MS). This is achieved via a triple stage quadrupole MS whereby two stages of mass filtering occur in the first (Q1) and third (Q3) quadrupoles. This setup allows for increased sensitivity and robust detection of low-abundant ions [30]. Previous MRM-MS based studies on cardiovascular disease biomarkers have been assessed in plasma [31, 32].

DDA and DIA (SWATH)

When acquiring MS2 data in a shotgun proteomics experiment, DDA usually picks the most abundant ions in MS1 for fragmentation. DDA inherently biases proteomics studies towards high- and medium- abundant peptides, despite newer instrumentation being able to profile all potential ions after dynamic exclusion, and can lead to issues with reproducibility [33]. Data-independent acquisition (DIA) alternatives to DDA have recently been gaining popularity as instrumentation are able to overcome some of the issues that originally justified the use of DDA. Conceptually, where DDA involves MS2 of top abundant ions identified in MS1, DIA aims to perform MS2 on all ions identified in MS1 survey scans regardless of abundance. Sequential window acquisition of all theoretical fragment ion spectra (SWATH) is one of the emerging DIA technologies used currently. Briefly, All MS1 scans occur within a defined m/z window and cycle time. Within that cycle time, MS2 is performed on all ions identified in MS1 via segmenting the m/z window into smaller 'steps' with narrow m/z windows that when aligned, will cover the entire mass window of MS1 with some overlap. As an example from a proteomics study of degenerative mitral valves, researchers were able to perform SWATH DIA using a 350–1250 m/z mass window in MS1 paired with a 25 Da 'step' totaling 36 windows for MS2 fragmentation [34]. DIA application has been driven by increased use of DIA-compatible platforms and from a technical aspect, is a balance between discovery and targeted approaches.

Methods for Protein Quantitation

Spectral Counting

Spectral counting is a form of quantitation derived from the MS2 spectra generated. The spectra count assigned to a peptide is indicative of how many copies were present in the sample mixture. Spectral counts between experiments can be compared using basic statistics such as t-tests or ANOVA. Additionally, more thorough comparisons can be achieved using analysis such as QSPEC [35] or SAINT [35], where probability-based scoring of spectral counts is employed. QSPEC is designed for binary comparisons where a single experimental condition is compared to a control condition. Spectral counting through SAINT is particularly useful in application of affinity proteomics MS where researchers aim to identify interacting proteins of a candidate protein. Essentially, multiple bait samples together with negative control conditions will be run in an effort to identify putative protein-protein interactions or complexes.

MS1 Peak Integration

When signal is generated from an unfragmented ion during MS1 acquisition, it is presented as an intensity distribution peak over time. This peak is one form of quantitative data, as the peak's area under the curve is a measure of its abundance. Peak integration is useful in multiplexed experiments where ions of peptides from multiple conditions can be measured simultaneously. Skyline [36] is a tool that allows for analysis of MS1 peak integration that will allow for experimental design such as MRM-MS. Briefly, MS1 peak integration uses heavy-labeled peptides spiked in with an unlabeled experimental sample. When MS1 is performed, both labeled and unlabeled peptides will be profiled separately due to m/z shift by the isotope's mass difference. MS1 peaks can then be identified after acquisition and analyzed for abundance. MS2 data can then be used to confirm the identity of the peptide quantified [37]. Multiplexing in MRM-MS is achieved by optimizing gradient conditions of the LC to ensure good separation between the peptides being analyzed. Label-free quantitation (LFQ) is an updated approach of peak integration developed within the MaxQuant analysis pipeline [38]. LFQ in MaxQuant uses ion intensity, retention time values and an m/z range as well to determine a 3D peak to determine an ion's abundance. Briefly, 2D peaks of peptides are identified in LC/MS by ion intensity and retention time. 2D peaks are then aligned based on a threshold m/z window to create 3D peak that can be quantified.

Stable Isotope Labeling by Amino Acids (SILAC)

SILAC is a very useful tool that allows for multiplexing in proteomics. It differs from the previous techniques that can be multiplexed like iTRAQ and TMT in that it does not require a tag and can be used *in vitro* and *in vivo*. SILAC is a metabolic labeling strategy where cells *in vitro* are cultured with heavy isotope-labeled amino acids that are incorporated into the cell's proteome as they synthesize proteins (Fig. 2.5) [39]. Over multiple passages, the cells preferentially contain the heavy amino acid in all their proteins. A simple SILAC workflow is where the one cell culture condition is performed in a heavy-labeled leucine environment. Lysates are prepared and mixed with the control cell culture condition where no SILAC labeling is performed. During MS analysis, identical peptides from each condition will have m/z differing by the mass difference of the heavy and light amino acids in the peptide. Ion intensity curves of MS1 to quantify the abundance of heavy and light

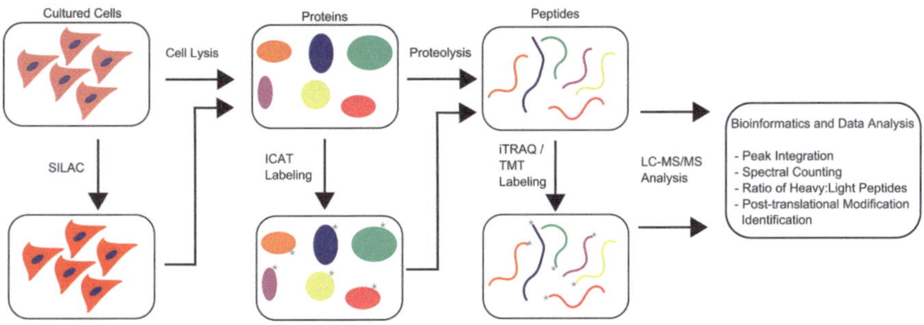

Fig. 2.5 Comparison of labeling strategies for quantitation in proteomics experiments *in vitro*. There are multiple ways where labeling of samples for proteomics can be achieved. At the cell culture level, cells can be labeled metabolically using *SILAC*. If labeling at protein level is desired, the chemical-labeling *ICAT* strategy is applied. After proteolysis, peptides can be labeled chemically using either *iTRAQ* or *TMT* labeling. During *LC-MS/MS* analysis, labeled samples will exist as higher mass-to-charge ions due to the heavy-isotope labeling. Data analysis and bioinformatics are performed using method such as peak integration and spectral counting. With these labeling strategies, the quantitation is based on a ratio of heavy to light peptides. One can also use the workflows to identify and quantify post-translational modifications

samples [40]. SILAC has advantages over chemical labeling strategies such as ICAT or TMT in that the quantitation is dependent on a exogenous chemical reaction to bind the tag to the target protein (ICAT) or peptide (iTRAQ and TMT). It is evident that SILAC labeling is a metabolically labeling strategy that requires living cells or tissues to work, compared to ICAT and isobaric labeling strategies, where chemical labeling is performed on the peptides and proteins. Extrapolating the concept of metabolically labeling, one can imagine heavy-labeling a living system beyond cells *in vitro*. Conceptually, an entire organism can be wholly SILAC-labeled. The SILAC mouse, a mouse with complete integration of heavy lysine was introduced to perform multi-organ proteomics [41]. The use of SILAC mice can be useful in cardiac proteomics as it will allow for labeling of proteins in an *in vivo* environment.

Methods of Labeling Protein Quantitation

Gel-Based Comparisons

Fluorescent two-dimensional gel electrophoresis (2DIGE) is a technique where protein samples are separated on a SDS-PAGE gel prior to MS analysis [42]. This is achieved via separation of proteins based on isoelectric point and molecular weight. In a comparative analysis, proteins of two samples are labeled fluorescently, typically with Cy3 and Cy5 labels, mixed together, and separated on the same two-dimensional gel. Fluorescent intensity is measured for both dyes using a fluorescent scanner. Gel spots identified to have a differential fluorescent intensity are then excised and processed for MS analysis. 2DIGE has benefits of being able to separate internal standards with multiplexed experimental samples on the same gel and allows for consistent method of normalization [43]. 2DIGE has been used in cardiac proteomics to study phosphotase substrate activity [44].

Isotope-Coded Affinity Tag (ICAT) Labeling

ICAT is a labeling strategy in proteomics where proteins are labeled with a H^2-labeled tag (heavy) and is compared in parallel with a H^1-labeled (light) sample. ICAT tags possess three components: a biotin tag, linker, and a thiol reactive group [45]. The thiol reactive group facilitates reduction with the cysteines to attach it to the protein of interest. The linker contains sites where the heavier deuterium replaces the typical hydrogen in the heavy sample. The biotin tag is used in affinity purification so that only ICAT-labeled peptides are analyzed in the MS. When peptides are introduced into the MS for analysis, the MS1 profile should contain two peaks representing each labeled peptide separated by an m/z related to the mass difference of the heavy and light linker regions. Quantitation in ICAT experiments uses the MS1 peak data of the heavy and light peptide and MS2 data is used to identify the peptide in question. ICAT has the benefit of being able to analyze samples in a binary comparison simultaneously, reducing the risk of run-to-run variability.

Isobaric Labeling

Isobaric labeling is a way of quantifying proteins via the addition of tags to the peptides which can be measured against each other in multiplexed samples. The tags all behave the same in the MS except for its fragments, where the location and number of heavy isotope-labeled elements allow for differing m/z in MS data. Isobaric labeling has two popular methods: Isobaric tags for relative and absolute quantitation (iTRAQ) and Tandem mass tag (TMT) [46]. iTRAQ labeling can be multiplexed up to eight samples whereas TMT labeling allows for multiplexing for up to ten samples. This level of multiplexing is an excellent option in time-course proteomics experiments as well as maintaining a robust internal control. iTRAQ labeling uses tags with three components: a reporter group, an amine-specific reactive group, and a balance group. The reporter group contains a combination of C^{13} and N^{15}-labeled elements. To keep the mass of each tag the same, it is balanced by either a combination of C^{13} and O^{18} labeling in the balance region. One can then see that each tag behaves the same, while also having different masses for the reporter regions. Each sample is individually labeled and mixed together for multiplexing the MS analysis. Separation during LC occurs simultaneously, as well as MS1 scans. When MS2 occurs, the reporter region is fragmented and since each sample's tag differs in size, their m/z will be distinct and their relative abundance can be quantified. Typically, iTRAQ reporter region ions range in size from 114 to 117 m/z. TMT tags also contain three components: a reporter group, amine reactive group, and a balance group, and the principle for quantitation is identical to iTRAQ. The reporter group and balance group contain different amounts of C^{13} or N^{15}. The reactive amine group is the structure that will be bind to the tryptic peptide. Each tag has the same number of heavy-isotope labeled elements; the key is that the number in the reporter group differs. Samples of each experimental condition are labeled with a unique TMT tag and mixed together. Since the TMT tags are the same size overall, each condition's peptides elute from the LC simultaneously, as well as MS1 is performed simultaneously. When MS2 fragmentation occurs, the reporter region is fragmented from the labeled peptide and is quantified. Reporter region ions have small m/z and are recorded in the low-mass region of approximately 130 m/z.

Application of MS Technologies to Quantitation in Cardiac Proteomics

Protein quantitation in cardiac proteomics can be achieved with combining the MS technologies with the differing quantitative methods outlined. The combination usually will be dictated by the research being asked and the technology available to the researcher. Moving forward we will focus on the recent studies in cardiac proteomics and how they were able to achieve protein quantitation.

In a study of cardiac extracellular matrix (ECM) proteins [47], researchers were able to identify differentially expressed ECM proteins in the mammalian heart during the course of ischemia/reperfusion injury. Samples were separated in two dimensions, first at the protein via gel electrophoresis, and then at the peptide level via LC. The hybrid MS used for data acquisition was an LTQ Orbitrap instrument. Spectral counting was performed after data transformation. Transforming data using methods such as log transformation and normalization will allow for lower abundant proteins to be assessed. Spectral counting allowed comparison between groups in their time course experiments versus their controls.

Another interesting question that has been investigated using proteomics has been the concept of chamber specificity, in the context of atrial and ventricular differences. It is well-known that atria and ventricle differ at the functional and molecular levels [48], MS-based proteomics were used to identify and quantify the proteomic differences in atria and ventricles [49] at the large scale. The study used a label-free approach to identify the chamber-enriched proteins in the human heart. Data was generated on a new generation QExactive MS platform, which is an Orbitrap-based hybrid MS. Maxquant LFQ data was used as their metric for quantitation. Since a binary comparison was investigated, a *t*-test of LFQ values was used to identify significantly enriched proteins. Data was also integrated with other large-scale data, including microarray data and other proteomic datasets to identify key chamber-enriched proteins.

Concluding Remarks

Having outlined the basics in MS theory, instrumentation, and quantitation, it is quite evident that MS in cardiac proteomics is a very diverse technical field. With interchangeable techniques and multiple methods to answer research questions, there is enormous potential to uncover further understanding of heart function and disease. It is with this diversity that one should be well-versed in the rationale of using one quantitative approach or instrumentation setup over another beyond availability. It will allow researchers to understand the scope and limitations of the data they are generating. MS in general is a very technology-driven field where many instrument or computational shortcomings can be resolved or overcome with only a few iterations of technology development. It is a very exciting future for cardiac proteomics as instrumentation and analysis becomes more accessible to researchers and that the previously difficult hypotheses can be investigated with confidence and accuracy.

References

1. Gregorich ZR, Ge Y. Top-down proteomics in health and disease: challenges and opportunities. Proteomics. 2014;14(10):1195–210.
2. Walmsley SJ, Rudnick PA, Liang Y, Dong Q, Stein SE, Nesvizhskii AI. Comprehensive analysis of protein digestion using six trypsins reveals the origin of trypsin as a significant source of variability in proteomics. J Proteome Res. 2013;12(12):5666–80.
3. Kim MS, et al. A draft map of the human proteome. Nature. 2014;509(7502):575–81.
4. Parker BL, Palmisano G, Edwards AV, White MY, Engholm-Keller K, Lee A, Scott NE, Kolarich D, Hambly BD, Packer NH, Larsen MR, Cordwell SJ. Quantitative N-linked glycoproteomics of myocardial ischemia and reperfusion injury reveals early remodeling in the extracellular environment. Mol Cell Proteomics. 2011;10(8) doi:10.1074/mcp.M110.006833-1-13.
5. Kooij V, Holewinski RJ, Murphy AM, Van Eyk JE. Characterization of the cardiac myosin binding protein-C phosphoproteome in healthy and failing human hearts. J Mol Cell Cardiol. 2013;60:116–20.
6. Karas M, Hillenkamp F. Laser desorption ionization of proteins with molecular masses exceeding 10,000 daltons. Anal Chem. 1988;60:2299–301.
7. Fenn JB, Mann M, Meng CK, Wong SF, Whitehouse CM. Electrospray ionization for mass spectrometry of large biomolecules. Science. 1989;246:64–7.
8. Wilm M, Mann M. Analytical properties of the nanoelectrospray ion source. Anal Chem. 1996;68(1):1–8.
9. Zhou F, Lu Y, Ficarro SB, Webber JT, Marto JA. Nanoflow low pressure high peak capacity single dimension LC-MS/MS platform for high-throughput, in-depth analysis of mammalian proteomes. Anal Chem. 2012;84(11):5133–9.
10. Sleno L, Volmer DA. Ion activation methods for tandem mass spectrometry. J Mass Spectrom. 2004;39:1091–112.

11. Roepstorrf P, Fohlmann J. Proposal for a common nomenclature for sequence ions in mass spectra of peptides. Biomed Mass Spectrom. 1984;11(11):601.
12. Johnson RS, Biemann K. Computer program (SEQPEP) to aid in the interpretation of high-energy collision tandem mass spectra of peptides†. Biomed Environ Mass Spectrom. 1989;18(11):945–57.
13. Hunt DF, Yates JR, Shabanowitz J, Winston S, Hauer CR. Protein sequencing by tandem mass-spectrometry. Proc Natl Acad Sci U S A. 1986;83:6233–7.
14. Olsen JV, Macek B, Lange O, Makarov A, Horning S, Mann M. Higher-energy C-trap dissociation for peptide modification analysis. Nat Methods. 2007;4:709–12.
15. Michalski A, Neuhauser N, Cox J, Mann M. A systematic investigation into the nature of tryptic HCD spectra. J Proteome Res. 2012;11(11):5479–91.
16. Zubarev RA, Kelleher NL, McLafferty FW. Electron capture dissociation of multiply charged protein cations. A nonergodic process. J Am Soc Chem. 1998;120:3265.
17. Syka JE, Coon JJ, Schroeder MJ, Shabanowitz J, et al. Peptide and protein sequence analysis by electron transfer dissociation mass spectrometry. Proc Natl Acad Sci U S A. 2004;101:9528–33.
18. Zhurov KO, Fornelli L, Wodrich MD, Laskay ÜA, Tsybin YO. Principles of electron capture and transfer dissociation mass spectrometry applied to peptide and protein structure analysis. Chem Soc Rev. 2013;42(12):5014–30.
19. Fornelli L, Damoc E, Thomas PM, Kelleher NL, Aizikov K, Denisov E, Makarov A, Tsybin YO. Analysis of intact monoclonal antibody IgG1 by electron transfer dissociation orbitrap FTMS. Mol Cell Proteomics. 2012;11(12):1758–67.
20. Porras-Yakushi T, Sweredoski M, Hess S. ETD outperforms CID and HCD in the analysis of the ubiquitylated proteome. J Am Soc Mass Spectrom. 2015;26(9):1580–7.
21. Zhao P, Viner R, Teo CF, Boons GJ, Horn D, Wells L. Combining high-energy C-trap dissociation and electron transfer dissociation for protein O-GlcNAc modification site assignment. J Proteome Res. 2011;10(9):4088–104.
22. Makarov A, Denisov E, Lange O, Horning S. Dynamic range of mass accuracy in LTQ orbitrap hybrid mass spectrometer. J Am Soc Mass Spectrom. 2006;17(7):977–82.
23. Pitt JJ. Principles and applications of liquid chromatography-mass spectrometry in clinical biochemistry. Clin Biochem Rev. 2009;30(1):19–34.
24. Lewis JK, Wei J, Siuzdak G. Matrix-assisted laser desorption/ionization mass spectrometry in peptide and protein analysis. In: Encyclopedia of analytical chemistry. Chichester: John Wiley & Sons, Ltd; 2006.
25. Helmke SM, Yen C-Y, Cios KJ, Nunley K, Bristow MR, Duncan MW, et al. Simultaneous quantification of human cardiac α- and β-myosin heavy chain proteins by MALDI-TOF mass spectrometry. Anal Chem. 2004;76(6):1683–9.
26. Medzihradszky KF, Campbell JM, Baldwin MA, Falick AM, Juhasz P, Vestal ML, et al. The characteristics of peptide collision-induced dissociation using a high-performance MALDI-TOF/TOF tandem mass spectrometer. Anal Chem. 2000;72(3):552–8.
27. Kristensen DB, Imamura K, Miyamoto Y, Yoshizato K. Mass spectrometric approaches for the characterization of proteins on a hybrid quadrupole time-of-flight (Q-TOF) mass spectrometer. Electrophoresis. 2000;21(2):430–9.
28. Chernushevich IV, Loboda AV, Thomson BA. An introduction to quadrupole–time-of-flight mass spectrometry. J Mass Spectrom. 2001;36(8):849–65.
29. Anderson L, Hunter CL. Quantitative mass spectrometric multiple reaction monitoring assays for major plasma proteins. Mol Cell Proteomics. 2006;5(4):573–88.
30. Keshishian H, Addona T, Burgess M, Kuhn E, Carr SA. Quantitative, multiplexed assays for low abundance proteins in plasma by targeted mass spectrometry and stable isotope dilution. Mol Cell Proteomics. 2007;6(12):2212–29.
31. Keshishian H, Addona T, Burgess M, Mani DR, Shi X, Kuhn E, et al. Quantification of cardiovascular biomarkers in patient plasma by targeted mass spectrometry and stable isotope dilution. Mol Cell Proteomics. 2009;8(10):2339–49.

32. Domanski D, Percy AJ, Yang J, Chambers AG, Hill JS, Freue GV, et al. MRM-based multiplexed quantitation of 67 putative cardiovascular disease biomarkers in human plasma. Proteomics. 2012;12(8):1222–43.
33. Ting YS, Egertson JD, Payne SH, Kim S, MacLean B, Käll L, et al. Peptide-centric proteome analysis: an alternative strategy for the analysis of tandem mass spectrometry data. Mol Cell Proteomics. 2015;14:2301–7.
34. Tan HT, Lim TK, Richards AM, Kofidis T, Teoh KL, Ling LH, et al. Unravelling the proteome of degenerative human mitral valves. Proteomics. 2015;15(17):2934–44.
35. Choi H, Fermin D, Nesvizhskii AI. Significance analysis of spectral count data in label-free shotgun proteomics. Mol Cell Proteomics. 2008;7(12):2373–85.
36. Choi H, Larsen B, Lin Z-Y, Breitkreutz A, Mellacheruvu D, Fermin D, et al. SAINT: probabilistic scoring of affinity purification - mass spectrometry data. Nat Methods. 2011;8(1):70–3.
37. MacLean B, Tomazela DM, Shulman N, Chambers M, Finney GL, Frewen B, et al. Skyline: an open source document editor for creating and analyzing targeted proteomics experiments. Bioinformatics. 2010;26(7):966–8.
38. Schilling B, Rardin MJ, MacLean BX, Zawadzka AM, Frewen BE, Cusack MP, et al. Platform-independent and label-free quantitation of proteomic data using MS1 extracted ion chromatograms in skyline: application to protein acetylation and phosphorylation. Mol Cell Proteomics MCP. 2012;11(5):202–14.
39. Cox J, Mann M. MaxQuant enables high peptide identification rates, individualized ppb-range mass accuracies and proteome-wide protein quantification. Nat Biotechnol. 2008;26(12):1367–72.
40. Ong S-E, Blagoev B, Kratchmarova I, Kristensen DB, Steen H, Pandey A, et al. Stable isotope labeling by amino acids in cell culture, SILAC, as a simple and accurate approach to expression proteomics. Mol Cell Proteomics. 2002;1(5):376–86.
41. Ong S-E, Mann M. A practical recipe for stable isotope labeling by amino acids in cell culture (SILAC). Nat Protoc. 2007;1(6):2650–60.
42. Krüger M, Moser M, Ussar S, Thievessen I, Luber CA, Forner F, et al. SILAC mouse for quantitative proteomics uncovers Kindlin-3 as an essential factor for red blood cell function. Cell. 2008;134(2):353–64.
43. Marouga R, David S, Hawkins E. The development of the DIGE system: 2D fluorescence difference gel analysis technology. Anal Bioanal Chem. 2005;382(3):669–78.
44. Husberg C, Agnetti G, Holewinski RJ, Christensen G, Van Eyk JE. Dephosphorylation of cardiac proteins in vitro – a matter of phosphatase specificity. Proteomics. 2012;12(7):973–8.
45. Shiio Y, Aebersold R. Quantitative proteome analysis using isotope-coded affinity tags and mass spectrometry. Nat Protoc. 2006;1(1):139–45.
46. Rauniyar N, Yates JR. Isobaric labeling-based relative quantification in shotgun proteomics. J Proteome Res. 2014;13(12):5293–309.
47. Barallobre-Barreiro J, Didangelos A, Schoendube FA, Drozdov I, Yin X, Fernandez-Caggiano M, et al. Proteomics analysis of cardiac extracellular matrix remodeling in a porcine model of ischemia/reperfusion injury/clinical perspective. Circulation. 2012;125(6):789–802.
48. Hanna N, Cardin S, Leung T-K, Nattel S. Differences in atrial versus ventricular remodeling in dogs with ventricular tachypacing induced congestive heart failure. Cardiovasc Res. 2004;63(2):236–44.
49. Lu ZQ, Sinha A, Sharma P, Kislinger T, Gramolini AO. Proteomic analysis of human fetal atria and ventricle. J Proteome Res. 2014;13(12):5869–78.

Chapter 3
How to Design a Cardiovascular Proteomics Experiment

Merry L. Lindsey, Aldrin V. Gomes, Stanley V. Smith, and Lisandra E. de Castro Brás

Abstract Proteomics provides the opportunity for the large-scale examination of proteins in fluids or tissue samples, which has significant implications for cardiovascular experimental, translational, and clinical research. Proteomic strategies are useful as a means to provide catalogues of proteins that are present in a sample at the time of evaluation, as a way to quantify complex mixtures and identify proteins differentially expressed among groups (whether it is total amounts or differences in post-translational modifications), and to analyze protein complexes and protein-protein interactions. A crucial component of a successful proteomics examination is the experimental design. In this book chapter, we will discuss how to design a cardiovascular proteomics experiment for optimal success and provide example workflows for different types of experiments with strategies to minimize potential pitfalls.

Keywords Proteomics • Cardiovascular • Insolubility • Sample preparation • Study design • Mass spectrometry

---

M.L. Lindsey, PhD
Department of Physiology and Biophysics, Mississippi Center for Heart Research, University of Mississippi Medical Center, Jackson, MS 39216, USA

Research Service, G.V. (Sonny) Montgomery Veterans Affairs Medical Center, Jackson, MS 39216, USA

A.V. Gomes, PhD
Department of Neurobiology, Physiology, and Behavior, University of California, Davis, Davis, CA 95616, USA

S.V. Smith, PhD
Department of Pharmacology and Toxicology, University of Mississippi Medical Center, Jackson, MS 39216, USA

L.E. de Castro Brás, PhD (✉)
Department of Physiology, The Brody School of Medicine, East Carolina University, Greenville, NC 27834, USA
e-mail: decastrobrasl14@ecu.edu

G. Agnetti et al. (eds.), *Manual of Cardiovascular Proteomics*,
DOI 10.1007/978-3-319-31828-8_3

Introduction

The pathophysiology of cardiovascular disease (CVD) is diverse, due to multiple possible aberrations that can occur in a number of cellular processes. Thus, it is vital to understand how changes in molecular pathways, particularly at the protein level, mediate cellular responses and function. Mass spectrometry (MS)-based proteomic strategies have great potential to advance our understanding of CVD mechanisms at the systems biology level. The word "proteome" originates from the term PROTEins expressed by a genOME, and it refers to all proteins present in a sample at the time of evaluation [1].

Proteomics is an advancement over genomics, providing details in an order of magnitude of increased complexity. Each cell has its own proteome, and the proteome (including all of the post-translationally modified forms of each protein) undergoes fluctuations through biochemical interactions with both the genome and the environment [2]. The complexity of individual proteomes makes good experimental design essential for a successful investigation. A researcher that aspires to analyze the proteome of either tissue or cells using MS-based methods will face experimental challenges. One challenge is the dynamic range across the proteome, as the relative abundance of each protein within the same sample can differ greatly. In plasma, for example, the dynamic range from the highest to lowest protein concentration is ten orders of magnitude [3]. For this reason, the identification and quantitation of low abundance proteins remains a challenge. At the same time, the plasma is an easily sampled source, as compared to obtaining biopsies of the left ventricle (LV), and techniques have been developed to harness its complexity.

The LV has its own layer of complexity. While the LV is comprised primarily of cardiomyocytes, which account for 90 % of the LV volume, other cells types of importance in cardiac research include smooth muscle cells, endothelial cells, fibroblasts, and infiltrating leukocytes [4]. Cardiomyocytes are highly metabolic cells, with 30 % of their volume occupied by mitochondria [5]. Therefore, the relative overabundance of mitochondrial proteins in cardiomyocytes masks the relative lower abundant proteins, such as cardiac extracellular matrix (ECM) proteins. Moreover, many myocardial components contain insoluble proteins, which make identification and quantification using cardiovascular proteomics protocols particularly challenging.

A good experimental design for cardiovascular proteomics will include the considerations listed in Table 3.1 and follow a strict workflow to result in robust and reproducible data (Fig. 3.1) [6]. Careful and reproducible sample preparation and enrichment steps are important, particularly if the study aim is to identify lower abundance components. A successful cardiac study, either at the organ, systemic, sub-cellular, or molecular level, can provide researchers with temporal cellular protein inventories, making it ideal for documenting protein changes between physiological and pathological conditions and through different stages of disease, as well as in response to varying treatment regimens. Here, we describe how proteomic experiments can be modeled to provide experimental designs tailored to cardiovascular studies.

Table 3.1 Major considerations for experimental design [6]

1. Subject selection criteria
2. Sample type and collection
3. Sample storage conditions (including duration)
4. Data acquisition methods
5. Data analysis
6. Results documentation
7. Replication in independent cohorts

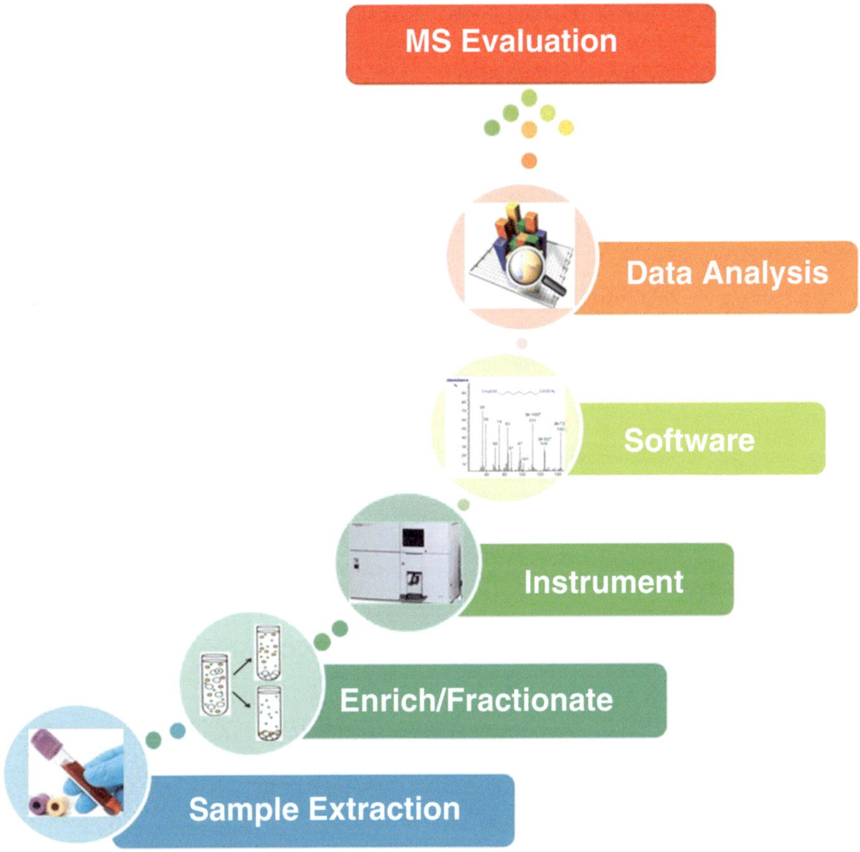

Fig. 3.1 Mass spectrometry workflow

Overall Considerations

Current MS-based proteomic approaches can be used to investigate not only type and levels of proteins, but also signaling pathways, protein-protein interactions, post-translational modifications (PTMs), organelle location, and even protein substrates of enzymes [7]. The type of data needed – qualitative or quantitative – and the biological question asked will dictate which method is the most adequate for a project. Proteomic analysis can be costly, laborious, and time consuming; thus, an effective study design and appropriate approach are critical. Of course, the instrumentation available is also a big factor to consider, and many times the approach can be tailored to the MS instrument to provide the needed answer(s). Table 3.2 describes common approaches used in cardiovascular research that target a particular question.

What Type of Data Do You Want to Obtain?

When studying CVD, three types of biological specimens are frequently used: fluids (eg, serum, plasma, saliva, or urine), cells, and myocardium. When initiating a cardiac study using proteomic-based approaches, the type of sample will have a

Table 3.2 Types of cardiovascular disease and mass spectrometry platforms

	Sample preparation	MS approach	Sample	References
Myocardial infarction	Immunoaffinity enrichment	SELDI-TOF	Human plasma	[8]
	Isobaric mass tag	LC-MS/MS	Rat effluents	[9]
	MW fractionation	HPLC-MS	Mouse LV	[10]
Dilated cardiomyopathy	Myocyte organelle isolation	Gel-free shotgun	Mouse LV	[11]
	None	MudPIT	Human LV	[12]
Hypertrophic cardiomyopathy	Organelle fractionation	2-DE MALDI-TOF	Mouse LV	[13]
	Immunoaffinity depletion	iTRAQ 4-plex labeling	Human plasma	[14]
Heart failure	Porous silicone affinity chip	MALDI-TOF	Human plasma	[15]
	None	2-DIGE MS	Human LV	[16]
	None	2-DE MS	Rat LV	[17]

MS mass spectrometry, *SELDI-TOF* surface-enhanced laser desorption/ionization time-of-flight, *LC* liquid chromatography, *HPLC* high performance LC, *MW* molecular weight, *LV* left ventricle, *MudPIT* multidimensional protein identification technology, *2-DE MALDI* two-dimensional electrophoresis matrix assisted laser desorption/ionization, *iTRAQ* isobaric tag for relative and absolute quantitation, *DIGE* difference in gel electrophoresis great impact on the method selected. For example, the study of cardiac cells involves considerations such as labeled or label-free methods and whether the cells will be cultured or used immediately after isolation, while the study of tissue may include fractionation and/or enrichment steps to reduce sample complexity. Additionally, it is important to consider the type of data expected: qualitative (identification and semi-quantification), quantitative (targeted proteomics), interactions (protein-nucleic acid or protein-protein), identification of protein modifications, or localization.

Qualitative or Quantitative

Global qualitative proteomics involves large-scale protein profiling of a proteome or subcellular proteome and offers the simultaneous detection of hundreds to thousands of proteins in a single experiment with rough quantitative estimates of relative concentration [18]. Bottom-up MS strategies, in which peptide detection is used to infer protein presence, are commonly used in cardiovascular proteomics for large-scale or high-throughput analysis [7]. This method is suitable for studies aimed at identifying the composition of a certain proteome, while at the same time provides semi-quantitative data that can be further validated with other methodologies. Nonetheless, without careful sample preparation, qualitative MS will mostly identify peptides present at high relative abundance, while information regarding low abundance proteins is commonly not obtained [15]. Label-free shotgun MS was used to investigate the proteomic changes in transverse aortic constriction (TAC)-induced heart failure. The use of this global approach resulted in the identification of 538 proteins that were significantly changed after TAC and mapped to 53 pathways, including actin cytoskeleton, mitochondrial function, and the citrate cycle [19]. Huang and colleagues utilized bottom-up MS analysis of enriched phosphopeptides to identify phosphorylation sites in connexin-43 following *in vitro* phosphorylation by CaMKII [20]. They reported 16 CaMKII phosphorylation sites, which represented the first study of CaMKII-mediated phosphorylation of connexin-43 [20].

Quantitative/targeted approaches are often the only way to detect very rare proteins; they are faster and require much less sample than bottom-up analysis. Quantitative approaches make use of stable heavy isotopes (^{2}H, ^{13}C, ^{15}N, and ^{18}O) [21], chemical labeling (isotopic or isobaric tags) [22, 23], enzymatic labeling (the mass tag is introduced in the peptide chain by performing proteolytic digestion in the presence of heavy water) [24], and metabolic labeling (introduced to the whole cell or organism) [25], although label-free methods are also used [15]. Pan et al. developed a new quantitative MS approach to map global Cys-redoxomic in cardiomyocytes under hypoxia using iodoacetyl-based Cys-reactive isobaric tags coupled with LC-MS/MS [26]. In one single analysis, they reported over 260 Cys sites showing significant differences in multiplexed redox modifications from hypoxic cardiomyocytes [26].

Protein-Protein Interactions

Techniques have become sufficiently robust that experiments can reliably determine whether protein-protein interactions occur throughout CVD stages. Identification of protein-protein interactions requires the purification of intact signaling complexes from fluids, cells, or tissue lysates via affinity purification (AP) prior to digestion and MS analysis [27]. Ping et al. studied the role of protein kinase C (PKC) in protecting the myocardium against ischemia/reperfusion (I/R) injury using bottom-up AP-MS to identify PKC signaling complexes in the heart [28]. They identified 36 proteins, including structural, signaling, and stress-activated proteins, which were physically associated with PKC in the mouse myocardium.

PTMs

Two dimensional electrophoresis (2-DE) gel based approaches are commonly used for cardiovascular sample preparation in studies that aim to identify PTMs. PTMs can cause changes in the isoelectric point and/or the molecular weight of the modified protein, which are readily detected by 2-DE. In addition, top-down MS approaches place intact proteins under examination, which decreases sample complexity, and consequently enables a more complete characterization of protein isoforms and PTMs than the bottom-up approach [29]. Top-down proteomics provides the direct study of PTMs by displaying mass discrepancy between the measured mass and the DNA sequence predicted value. However, the physicochemical diversity of intact proteins makes large-scale separation challenging and, thus, traditional top-down studies are primarily recommended for the analysis of a single or small number of proteins. Jia and colleagues used both bottom-up and orthogonal acceleration time-of-flight (TOF), for measurement of intact proteins, to determine the phosphorylation states of human and murine cardiac myosin binding protein-C (cMyBP-C) [30]. Their strategy identified four PKA phosphorylation sites in both the human and murine isoforms, with the characterization of novel sites (Human: Ser^{311} and Mouse: Ser^{307}) providing advances for the complete characterization of the phosphorylation state of cMyBP-C [30]. The 2-DE gel approach has been used to identify candidate *in vivo* substrates of matrix metalloproteinase (MMP)-7 and MMP-9, by focusing on proteins present at lower than expected molecular weights or differentially expressed between wild type and MMP null mice [31, 32].

While the accurate measurements of intact proteins readily provide protein masses that indicate PTMs, it can miss some PTMs due to sub-stoichiometric amounts. The bottom-up MS method can provide good sequence coverage and precise locations of PTM sites. Thus, the combination of these two methods can deliver both maximum coverage and precise PTM identification.

Your Biological Question Will Define Your Approach

Untargeted Approaches

2-DE Gels

As mentioned above, one strength of using 2-DE approaches is the identification of PTMs. This strategy remains a popular proteomics technique and has played a pivotal role in providing insights into the physiology of the heart [33, 34] as well as elucidating markers of disease [35–37]. The major limitation of 2-DE is its limited dynamic range [38]. 2-DE has an estimated maximum dynamic range of 10^4 magnitude [39], while the dynamic range of protein abundance is estimated at 10^6 for cells and tissues and 10^{12} for plasma [3, 40]. Nonetheless, coupling 2-DE with subcellular fractionation methods can reduce sample complexity and increase the range of protein abundance [41]. Research groups have successfully employed 2-DE approaches in the proteomic analysis of biomarkers of cardiomyopathy [9], in cardiac aging [42], in the study of mechanisms underlying cardiomyopathy [43, 44], and diabetic cardiomyopathy [45].

Gel-Free

Gel-free approaches overcome problems of reduced enzyme accessibility to the protein (improves peptide digestion) and inefficient capture of large peptides from the gel leading to decreased protein coverage, and avoids the need to identify potentially hundreds of individual spots [46, 47]. Gel-free MS couples high-efficiency liquid chromatography (LC)-based separation procedures with MS or MS/MS, allowing for very large-scale "shotgun" sequencing of complex mixtures [48].

Label-Free

Even though labeling techniques offer several advantages, one big limitation is the number of samples and groups that can be compared. Thus, label-free methodologies hold great interest for the cardiovascular community. The simplest approach uses the number of peptide fragmentation events (spectral counts) as an estimate of the amount of protein to provide a semi-quantitative measurement [49, 50]. In addition, label-free methods based on the use of ion currents have been reported to provide a level of accuracy comparable to labeling approaches [51].

Stable Isotope

A common metabolic labeling method is Stable Isotope Labeling by Amino acids in Cell culture (SILAC) [25]. This technique uses essential amino acids, such as arginine and lysine, in light or heavy forms to the two cell populations (e.g. treated versus untreated) that are incorporated into each protein after cellular proliferation, resulting in a well-defined mass difference [52]. Another popular technique is ^{15}N-labeling, the metabolic incorporation of inexpensive labeled ammonium salts, that allows the complete labeling of all amino acids in expressed proteins; however, it entails a more difficult detection of the peptide pairs, since the mass difference depends on the amino acid composition [52].

Targeted Approaches

Absolute QUAntification (AQUA)

Targeted liquid chromatography mass spectrometry (LC-MS) approaches allow confirmation and relative quantitation of protein candidates. In addition, with more resource utilization, absolute protein quantitation using stable isotope labeled peptides (Protein AQUA) of candidates is possible [53]. Targeted approaches are typically done either after a discovery experiment has yielded candidates or the candidate pool has been narrowed by some other criteria. Targeted approaches can also be done using entirely *in silico* information, but this method is a tedious, iterative process, and the success rate is variable. The general concept is that, using bottom-up techniques, proteolytic fragments (trypsin is most useful) derived from the target protein are used to identify/quantify the protein of interest. By having several transitions (precursor/fragment ion pairs) for each protein, coupled with the added uniqueness of the fragment retention times via LC, high specificity and sensitivity can be obtained for each protein that is targeted [54]. For targeted discovery approaches, specific peptides or transitions (depending on the type of LC-MS system you are using) represent the information dependent trigger for subsequent experiments involving fragmentation via collision-induced dissociation (CID; also known as collision-activated dissociation, CAD) for sequencing and PTM detection that yield database searchable information for identification [55]. Given certain assumptions or using certain methodologies, intensities from this type of data can be used for quantitation. This combination of sequence information and quantitation is quite useful for streamlining your candidate list and/or for obtaining an overview of candidates and relative amounts rather quickly. For scenarios where more sensitivity is required and there has already been significant narrowing of the candidate list, quantitative proteomics experiments are very viable. The specific peptides and transitions selected and ion counts or intensities are used to indicate relative amounts. Again, coupled with good LC, this approach yields very high sensitivity (picogram or lower amounts) and excellent specificity. If other sources of information (empirical data either collected or derived from any of a number of growing

MS/MS databases such as http://www.mrmatlas.org/index.php) indicate that there are known stable peptides that are LC-MS friendly, heavy-labeled versions of the peptides can be spiked into samples for AQUA [56]. The heavy peptides are monitored the same way as the native peptide with corresponding adjustments for the m/z shift (typically 6–10 Da) of the heavy labeled peptide [57]. By spiking a known concentration of the heavy peptide into your sample and comparison to the intensity of the native peptide, absolute quantitation is obtained. The elution profile, ionization properties and other features are considered identical to the native peptide. This approach is inherently much more costly. Relative quantitation approaches are becoming more acceptable as the instrumentation has become more sensitive and reproducible. Often with relative quantitation, a quick turnaround of useful data is achieved.

Single Reaction Monitoring (SRM) or Multiple-Reaction Monitoring (MRM)

In an SRM or MRM experiment often the core of a targeted experiment, transitions (precursor/fragment combinations) derived from the target protein or proteins, is looped together in the MS method coupled with LC [58]. Tandem MS, either space- (Orbitrap instruments) or time-distinct (Quadrupoles and Ion traps), instruments are used. Specificity and sensitivity are obtained as rather than passing all ions of a particular m/z and measuring intensities, an SRM experiment looks at only the fragment ions after selection in a multistage mass spectrometer [59]. MRM is the application of SRM to multiple product ions. Ion selection occurs at the front-end, these ions enter a CID chamber where they are fragmented; they are then scanned out of a second quadrupole or trap [60]. By looping SRMs, 50–100 different transitions are readily measurable. These can represent different protein candidates or be specific for certain PTM states, if care is made to ensure the PTMs are preserved during the experiment. Typically for each protein candidate, one or two qualifier transitions are chosen along with one quantifier. The former helps to ensure that the correct peptide has been identified or selected while the latter is typically the most intense transition that has been demonstrated to be stable and appropriately concentration dependent over the experimental range. Practically, this sets the limits to about 20–30 different proteins that can be followed after thorough vetting [59]. If more are to be followed, scheduled approaches can be used to maximize instrument time by utilizing precise knowledge of LC elution times of peptides. Using scheduling, measurable transitions can be increased to 250 or so, therefore increasing the number of candidates that can be followed.

Limitations and Pitfalls to Consider

A major assumption of targeted approaches is that with exhaustive proteolytic digestion, the amount of each peptide reflects the amount of the intact protein due to the one to one concordance. The assumptions involved, under general standardized digestion conditions, are reasonable. Unique features of the proteins or peptides derived from the protein may prove problematic in terms of digestion or recovery. Examples are integral membrane or insoluble proteins that are not readily digestible; RNA or DNA binding proteins that might be resistant to digestion and/or recovery; and sticky proteins or peptides that are lost on surfaces. Data must be verified and shown to be reproducible, and detection limits and coefficients of variation should be considered. If there is concern about digestion efficiency and peptide/protein concordance, a targeted method for a housekeeping protein for which there is a commercially available antibody may be useful. Immunoblotting or enzyme-linked immunosorbent assay (ELISA) can then be used to validate the relative concentration and confirm consistency with the targeted approach used.

Study Design (Fig. 3.2)

Before Starting

Since good sample quality is the foundation for analytical success, the importance of generating highly reproducible and thorough sample preparation cannot be overstated. Prior to initializing your sample preparation you need to consider: (a) controls (positive, negative, and standards) and (b) sample variability. The first needs to be representative of the population being studied, comprising age-, sex-, and genotype-matched samples, while the latter will include a power analysis so that you can calculate the optimal number of samples to ensure an adequate power to detect statistical significance. Another critical factor is the source of your samples. The use of circulating blood as sample source for CVD diagnosis is very attractive since it can be collected without any interventional procedures. As such, the majority of proteomic CVD biomarker discovery studies have been done on circulating biomarkers and not on myocardial samples. Although advantageous, the use of plasma or serum comes with serious drawbacks, since the discovered biomarkers may reflect the disease state of any organ in the body. On the other hand, access to human myocardial samples is invasive and therefore rare, which makes the use of animal models essential for the study of CVD.

Sample Extraction

For accurate and reproducible proteomic analyses, sample extraction and preparation is of the utmost importance. Sample extraction must be reproducible, avoid sample alteration, and be compatible with subsequent methods of protein separation and identification [61]. Here we point out important factors to consider when using proteomics analysis of both plasma/serum and myocardium.

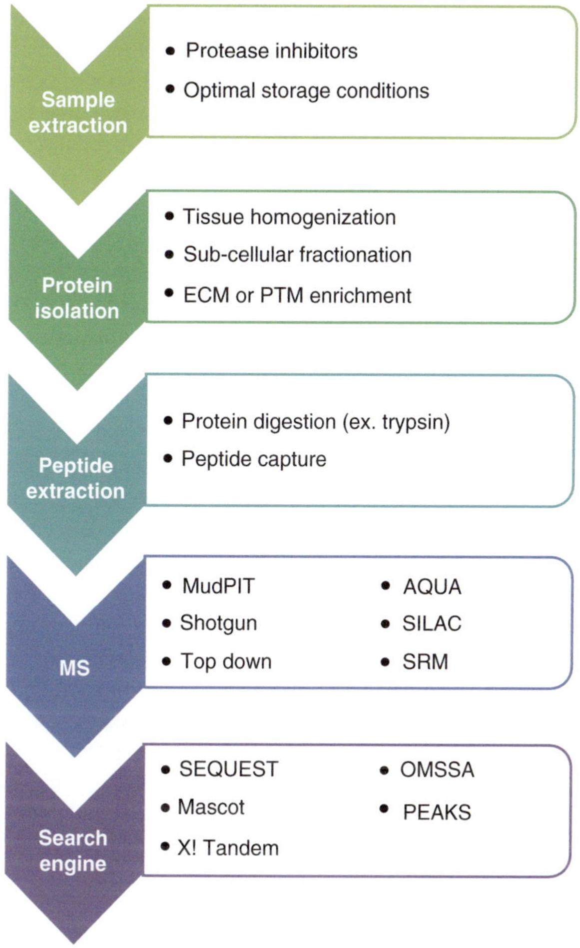

Fig. 3.2 Study design for mass spectrometry (*MS*) based approaches

For proteomic analysis of plasma and serum samples it is important to consider the following points: (1) The type of blood container (e.g. if using serum there is no need for anticoagulants). (2) The lag-time prior to sample processing. Hu and colleagues reported that the time before sample processing greatly affects sample proteome profile, in cases rendering it impossible to distinguish between diseased and healthy controls [62]. (3) The presence of proteases that if not inhibited, and with several cycles of freezing/thawing, will degrade the plasma proteome. Marshall et al. found that the detected changes in the protein profiles of serum of myocardial infarction (MI) patients reflected the balance of protease and aminopeptidase activities *ex vivo* rather than a result of the disease processes [63]. (4) The circulating protein of interest may have poor signal-to-noise ratio, which may lead to poor predictive value and limited clinical use of the identified protein. To circumvent this problem, you can sample the blood at a site more proximal to the diseased tissue, where the concentration of the protein of interest may be highest [64]. (5) The presence of a wide concentration range and number of proteins. In human plasma, more than 10,000 proteins have been identified that range from fM levels to over 600 μM [65]. Highly abundant proteins, such as serum albumin, complement factors, and immunoglobulins, will mask quantification of less abundant proteins. Therefore, methods that remove these proteins or that enrich for your protein of interest may be necessary prior to sample analysis.

Heart tissue collection presents different challenges. First, human biopsies are rare and harder to acquire. Nevertheless, atrial appendage samples can be obtained during surgery for the treatment of atrial fibrillation [66]. The setback with such samples is that underlying pathologies may differ between patients, and additionally, healthy controls are more difficult to acquire. Animal models of CVD are widely used and provide a consistent source for evaluation. The protocol for myocardial tissue isolation will depend on your biological question. As with serum and plasma, myocardial tissue should be stored with protease inhibitors and frozen immediately after extraction. This step will block or inactivate endogenous proteolytic enzymes that are released from subcellular compartments during cells lysis that would otherwise degrade the proteins of interest. [67] Other key issues include the increase of detection sensitivity by reducing complexity of samples and preparation of a protein sample that is biologically relevant. Additionally for the study of PTMs, prevention of in vitro artificial PTM reactions is essential, as well as the use of specific inhibitors to limit de-phosphorylation, such as phosphatase inhibitors [68].

In summary, a set of pre-analytical variables will influence the outcome of the final analysis and the lack of standardized protocols during sample collection and preparation may irreversibly affect the sample analysis and provide bias that distorts the results.

Protein Preparation/Separation Technique (Gel-Free and Gel-Based)

As in any MS-based study, the choice of reagent that is used during tissue homogenization is critical as it needs to be compatible with the downstream analysis. Soluble samples normally do not require the use of a detergent to aid homogenization and work well with gel-free techniques. One attractive advantage of gel-free approaches is the reduced sample handling and thus decreased potential for sample loss. On the other hand, insoluble proteins need further processing, including the use of detergents that have a large effect on protein coverage, and may benefit from further separation by electrophoresis [69]. A gel-based separation is powerful in that it can also provide insight into differential protein regulation in the disease state. Nonetheless, this technique comes with some drawbacks, including loss of protein during gel digestion and peptide capture, bias towards the detection of higher abundance proteins, and lower effectiveness at detecting hydrophobic proteins and proteins with extreme isoelectric points and molecular weights [18, 70].

Protein Fractionation (Directly Proportional to Sample Complexity)

To reduce sample complexity and uncover changes in lower abundance proteins, methods for subcellular fractionation can be applied. These methods include differential centrifugation, flow cytometry, AP isolation, membrane protein enrichment strategies, or density gradient isolation of organelles such as the nucleus or mitochondria. For example, in 1998 a study aimed to identify the human cardiac mitochondrial proteome using 2-DE analysis coupled with MALDI identified merely 46 proteins [71]. Ten years later, another group used sample pre-fractionation to isolate the cardiac mitochondria followed by 1-DE coupled to LC/MS/MS which resulted in the identification of 940 distinct mitochondrial proteins [72]. This example demonstrates the power of fractionation to overcome the issue of complexity. As mentioned above, plasma is a complex proteome with few highly abundant proteins accounting for 90 % of the total protein concentration [73]. Serum albumin is one such abundant protein and thus a leading candidate for selective removal prior to proteomic analysis of less abundant proteins in plasma. Various albumin-depletion methods are available, including immunoaffinity commercial kits, ligand chromatography, and isoelectric trapping [74–76]. As a note of caution, some of these kits merely dilute the total protein concentration of the sample, such that volume loading would show reduced albumin; this, however, also lowers the concentration of low abundance proteins, which makes these kits not useful. In addition, the use of depletion methods may result in the inadvertent removal of low abundance proteins like cytokines, lipoproteins, and peptide hormones [77]. New plasma fractionation methods have been developed that do not involve sample depletion and can be used to measure markers that detect MI or predict outcomes following MI [78].

Fractionating by molecular weight size, by pH, or by solubility are common approaches [78–80]. Similarly in tissue, sub-cellular fractionation selects for proteins of interest and reduces the large dynamic range of proteins found in whole homogenates. Decellularization of LV and aorta has been employed to remove highly abundant mitochondrial and nuclear proteins and enrich for extracellular matrix proteins [79, 80]. Warren and colleagues used a series of differential centrifugations that yielded nuclear, mitochondrial, cytoplasmic, microsomal, and sarcomeric-enriched fractions of infarcted and non-infarcted tissue, in a rodent model of MI [81]. They compared non-fractionated samples and the sarcomeric fraction and found that the sub-cellular enrichment improved protein identification by 4-fold [81].

Mass Spectrometers

Improvements in mass spectrometers over the last decade, as well as lower costs for these machines has allowed more research laboratories and shared facilities to provide more intensive proteomic studies. A summary of several commonly used mass spectrometers is shown in Table 3.3.

Table 3.3 Common mass spectrometry instruments used for peptides and protein identification

Type of mass spectrometer	Configurations	Advantages	Disadvantages
Hybrid quadrupole-orbitrap	Q Exactive (thermo scientific)	High-resolution, accurate-mass detection <1 ppm mass accuracy	Limited mass range (mass range to 6000 m/z)
	Orbitrap fusion lumos tribrid MS (thermo scientific)	High-resolution, accurate-mass detection, optimized for detection of low level PTMs Intact protein characterization	Cost
Triple quad	EVOQ elite triple quadrupole (bruker)	Great for multiple reaction monitoring (MRM) assays	Lower resolution Lower scanning speed
	6495 Triple quadrupole LC/MS (agilent)	Biomarker validation, price, tolerant of high pressure	Lower resolution Lower scanning speed
	TSQ Quantiva (thermo Scientific) 5500/6500 QTRAP (Sciex)	Ultrafast selected-reaction monitoring (SRM) of 500 SRMs High scanning speeds Excellent for MRMs	Lower resolution Lower scanning speed Lower resolution
Quadrupole time-of-flight (Q-TOF)	TripleTOF 5600+ system (Sciex)	MS/MS^{ALL} with SWATH acquisition – enables comprehensive MS/MS quantitation, fast scanning	Cost
	6545 Q-TOF LC/MS (agilent)	High resolving power and sensitivity	Cost
	maXis II (bruker)	High resolving power and sensitivity	Cost

Protein Identification

Several protein sequence databases are commonly used for peptide and protein identification. One of the most common databases utilized is the UniProt database, consisting of Swiss-Prot and its supplement, TrEMBL [82]. Other popular databases include the Entrez Protein database (http://www.ncbi.nlm.nih.gov/Class/MLACourse/Original8Hour/Entrez/) and the reference sequence database (RefSeq, http://www.ncbi.nlm.nih.gov/refseq/) both from the US National Center for Biotechnology Information (NCBI). The International Protein Index (IPI) database was one of the most popular databases used for proteomic investigations, but this database is no longer updated (as of 2/8/13). However, some recent publications still utilize this database [83].

Each database differs in terms of the number of protein sequences available, the number of redundant sequences, and the degree of sequence annotation. The most complete database is the Entrez Protein database; however this database contains many redundant sequences and is not well annotated. The best database containing annotated, non-redundant, and experimentally validated protein sequences is the

Swiss-Prot database. Because the Swiss-Prot database is non-redundant, it only contains a single consensus sequence for each protein such that known variants of a protein are all contained in a single entry. A good compromise is the UniProt database, which contains a mix of annotated and non-annotated protein sequences. Uniprot databases are available for many different species. Using these species specific databases from UniProt allows only proteins from that species to be analyzed, reducing false positives. For example when analyzing MS data from human and mouse samples, the Uniprot Human Proteome (release 03.04.2013, contained 87,656 protein entries) and the Uniprot Murine Proteome (release version 3.68 contained 51,389 entries) are popular.

Expressed sequence tag (EST) databases and genomic databases translated into protein sequences could also be useful for identification of peptide sequences as it sometimes allows for the identification of peptides that are absent in other protein sequence databases. Genomic data from vertebrate and non-vertebrate species could be obtained from the Ensembl and Ensembl Genomes projects respectively. Since peptide sequences obtained from genomic databases often contain DNA sequences translated in all six frames, the size of the database files makes searches computer intensive, and these databases should be used with caution because of potential sequencing errors. These large databases are best utilized when high quality MS/MS spectra is available that could not be identified by searching against a protein sequence database such as UniProt. The large size of the EST databases also limits its use but is a great resource to determine peptides in organisms not well represented in protein databases.

Starting with a relatively small database such as Swiss-Prot, in general the score thresholds are higher as the size of the database increases, resulting in some weaker peptide sequence matches not being identified. It is also possible to remove known contaminants (such as certain keratins and albumin) from the database. Irrespective of the database used, the MS data should be analysed using a target-decoy database to determine the false discovery rate (FDR) of the samples investigated [84]. A false discovery rate of 1 % for both peptide and protein matches is considered the gold standard.

Data Interpretation, Quantification, and Analysis

Many different commercial software programs are available to analyse MS results. Over the last decade the protein identification algorithms used to develop these programs have significantly improved (reviewed in [85]) but many challenges still remain. The main challenge is determining the sequence of the large number of unidentified peptides in typical proteomic experiments. Some of these unidentified peptides is due to unexpected and novel PTMs. Common search engines include Mascot (Matrix Science, Boston, MA), SEQUEST (Thermo-Fisher Scientific, San Jose, CA), OMSSA (NCBI [86]), PEAKS (Bioinformatics Solutions Inc., Waterloo, Canada), and X! Tandem (http://www.thegpm.org/).

Software packages such as Proteome Discoverer (Thermo-Fisher Scientific, Waltham, MA), Scaffold (Proteome Software, Portland, OR), and Progenesis LC-MS (Nonlinear Dynamics, Newcastle, UK) are also available. Proteome Discoverer can utilize several search engines such as SEQUEST, calculate the FDR, and quantify the peptides detected. While some commercial software already have programs that can calculate the FDR, standalone programs that determine the FDR are widely available.

The rapid development of tagged and label-free quantitative proteomic techniques has allowed the determination of protein expression levels in complex samples such as myocardial protein extracts. [87] Peptide labeling allows different samples to be labeled, mixed together, and then subjected to MS. Several well established labeling protocols are widely utilized including Tandem Mass Tags (TMT), Isobaric Tags for Relative and Absolute Quantification (iTRAQ) (each peptide is labeled with a different amine-specific isobaric tag), and SILAC. The main disadvantages of these methods include the high cost of the reagents, the increased complexity of sample preparation, incomplete labeling, and the requirement for more advanced analysis software. The two most commonly used label-free quantification methods include spectral count-based LC-MS/MS and peak intensity-based comparative LC-MS. The main disadvantage of label-free experiments relative to labeled peptides is the need for highly reproducible LC and MS to prevent run-to-run variations in performance of the LC and MS. Label-free quantification using the spectral count method correlates well with quantification by isotope-labeling [88]. Hence, either labeled or label-free methods for quantification of cardiovascular samples should provide adequate results.

An evaluation of the relative quantification of peptides by spectral counts using five different statistical tests showed that the Student's *t*-test was the best statistical test when at least three replicates are available [89]. The Fisher's exact test, AC test, and goodness-of-fit test (G-test) were found to be good statistical tests when the number of replications is two or less. Important parameters including the precursor ion mass tolerance and product ion mass tolerance depend on the instrument being utilized. A common problem in not identifying peptide matches is the use of small mass tolerance ranges. The mass accuracy is more important than the precision. Typical analysis utilizes up to a maximum of two missed protease (most commonly trypsin) cleavage sites and oxidation of methionine residues as dynamic modification. Since many different post-translational modifications of amino acids can occur, including some due to artefacts of sample handling, inclusion of known modifications such as alkylated cysteine as fixed modifications, as well as other modifications expected in some peptides as variable modifications, is important for peptide identification. Unimod (http://www.unimod.org) is a comprehensive and commonly used database which focuses on protein modifications relevant to MS. It is important to limit the number of variable modifications selected, as higher numbers of variable modifications increase the number of peptides incorrectly identified and significantly increase computational time.

Tweaks and Tricks

As one works more with proteomics techniques, these experiences provide insight that makes future experiments easier to perform. Below, we provide insights into a few commonly encountered situations and how to optimize your success with them.

Improved Methods of Digestion for Poorly Solubilized Proteins

The extraction of myocardial membrane and ECM proteins remains a challenge [90]. ECM components are organized into a complex fibrillar, 3-D matrix that cannot easily be solubilized or dissociated into component units [79]. Similarly, protein extraction from cell membranes is challenging due to the hydrophobic nature and poor solubility of these proteins. Recently the development of protocols using differential solubility-based protein fractionation has advanced the study of insoluble proteins [79, 80].

Protein digestion is a critical step in sample preparation for bottom-up and middle-down MS. Optimal protein digestion will enhance sequence coverage and peptide identification. Trypsin is the most widely used enzyme in MS, cleaving at the C-terminus of lysine and arginine with both high efficiency and specificity [91]. However, the use of trypsin only, or any other single protease, does not provide complete proteomic coverage. Additionally, the proximity of proteolytic sites to PTMs can alter protease efficiency at these sites [92]. Chymotrypsin and the endoproteinases Lys-C, Asp-N, and Glu-C have recently gained importance due to advancements in MS techniques. Chymotrypsin cleaves most frequently on the carboxyl side of the hydrophobic aromatic amino acids phenylalanine, tryptophan, and tyrosine and at a lower rate the carboxyl side of leucine. In contrast, endoproteinase Lys-C cleaves at the carboxyl side of lysine only. Asp-N preferentially cleaves proteins at the N-terminus of aspartic and cysteic acid, and Glu-C cleaves at the C-terminus of glutamic and aspartic residues [93, 94]. Due to their specific cleavage sites, the use of these proteinases individually or in combination create unique peptide fragments, offer increased protein coverage, and promote high confidence data [95, 96]. Franklin et al. used a combination of trypsin and chymotrypsin digestions and reported a cardiac nuclear proteome of 1048 proteins [41].

High Throughput Proteomics

In a way this term is an oxymoron, in that it is difficult to be all-inclusive and rapid. If the output measurements are numerous, such as during shotgun proteomics, the best solution is to keep the input numbers low. For example, limiting the number of comparison groups to ≤4 and increasing the number of biological replicates in each group improves the power of the analysis and allows the dataset to be manageable. Additional groups, such as further controls, can be added at the immunoblotting or ELISA validation stage. If the inputs are numerous, consider using a targeted MS approach (e.g., SRM or MRM). Consideration should be given during the design stage for the number of inputs and expected outputs, to keep the experiment feasible and optimize the change of finding results to answer your question.

Interacting with a MS Core: How to Achieve Success

While the best approach is to invest the time and effort to learn to use proteomic technologies first hand, this may not always be possible. Collaborations between your laboratory and a proteomics core can be fruitful, and we offer here some guidance on how to maximize the experience.

Effective communication is key. Setting up at least one face-to-face meeting before the experiment is designed is critical. There needs to be direct and open communication between the individual preparing the samples and the individual running them on the mass spectrometer. Continued communication throughout the analysis is also essential for ensuring the best experimental results. Working with a core is a relationship, not a one day affair. In addition to the exact experiment to be run, it is also important to discuss project deadlines and terms of the collaboration agreement, including whether the core director will be included as an author and how the core will be acknowledged.

Before the experiment is performed, it is important to define approach. The approach used depends on what you want to know. For example, it does not make sense to use a shotgun approach to identify 1000 altered proteins in a group analysis, if all you want to know is whether a particular protein of interest is differentially expressed. For this, a targeted proteomics approach is more suited to the question being asked. Taking the time to contemplate your question early on will strengthen the results obtained. For example, focusing on answering a biological question, rather than posing an open end let us see what is in the mix question, will increase the chance of obtaining mechanistic rather than observational results. Using proteomics to reveal functions rather than mere identifications will provide a more sophisticated level of results.

In designing the experiment, it is important to consider how the samples are prepared, how many samples you have, and what controls are needed. The core needs to trust that the samples provided are the highest quality possible, and providing them with information on any anomalies is important. Testing the core to see if they can provide you with a certain expected result is not a good strategy, if a long-term relationship and trust are wanted. Once these are decided, the mass spectrometry method and instrument used can be decided. Once results are obtained, methods to validate the results and approaches to present the results can be determined. There are several issues to consider while preparing the results for publication, so it is a good idea to start organizing before you write and continue as the experiment is designed and results are acquired.

Many journals, particularly proteomics journals, have a set of standards for publication. The Human Proteome Organization (HUPO) established a Proteomics Standards Initiative (PSI) working group that developed a set of guidelines known as The Minimum Information About a Proteomics Experiment (MIAPE) [97]. With the increased use of mass spectrometric applications in proteomics research, it became clear that it was necessary to establish a set of guidelines for data analysis and criteria for data presentation. This guideline defines community standards for data representation in proteomics to facilitate data comparison, exchange and verification. HUPO-PSI developed guidance modules for reporting the use of proteomic approaches to generate MIAPE-compliant reports [97]. Most proteomics journals have adopted MIAPE reporting; thus, keeping detailed experimental records throughout the study is critical. MIAPE reporting denotes detailed recording of the experimental design and MS acquisition parameters either in the Methods or Supplementary sections. For example, the method (s) used to generate peak lists from raw MS and MS/MS data should be described, and information on the programs used for database searching and the selected search parameters should also be defined [68]. For quantitative MS approaches, the software for data extraction and statistical analysis should be detailed and the results reported. Additionally, any extra handling of data such as outlier removal, data normalization, and characterization of PTMs has to be described.

Many journals also require that MS results be uploaded in a depository such as ProteomExchange (http://www.proteomexchange.org/), which has been established to provide a single point of submission to proteomics repositories [98–100]. This streamlines the process by providing one interface and prevents user confusion about which repository to submit. Once submitted to the ProteomExchange entry point, the data can be automatically distributed to all other repositories. When depositing results, each submission is required to have three components: (1) mass spectrometer output files, (2) study metadata, and (3) peptide/protein identifications. All submissions will include all three components and will be made to the PRoteomics IDEntifications database (PRIDE, http://www.ebi.ac.uk/pride) repository using data sufficiency guidelines established by PRIDE. Some software programs, such as Scaffold, provide an easy template for uploading results into PRIDE. Once the results are uploaded, a PX accession number is provided, which can be used by reviewers and readers to assess MS results.

Concluding Remarks

We present here some guidelines on how to design a proteomics experiment that will provide best case results. Thinking and planning from the start will prevent issues that would become apparent in hindsight. Understanding basic concepts involved in mass spectrometry analysis will provide a proactive understanding of what the expected results will be and how these results will inform us about the basic mechanisms under evaluation. While using institutional cores or outside vendors to provide total support for the mass spectrometry component of the study may be a useful option under certain conditions, investing the time to learn firsthand how to successfully complete a proteomics experiment will pay dividends. Whichever strategy is used, a successful proteomics experiment begins with formulating the right question and optimally preparing the sample for analysis.

References

1. Wilkins MR, Williams KL. The extracellular matrix of the Dictyostelium discoideum slug. Experientia. 1995;51(12):1189–96. Epub 1995/12/18.
2. Abhilash M. Applications of proteomics. Internet J Genomics Proteomics. 2008;4(1):3-10.
3. Anderson NL, Anderson NG. The human plasma proteome: history, character, and diagnostic prospects. Mol Cell Proteomics. 2002;1(11):845–67. Epub 2002/12/19.
4. Doevendans PA, Daemen MJ, de Muinck ED, Smits JF. Cardiovascular phenotyping in mice. Cardiovasc Res. 1998;39(1):34–49. Epub 1998/10/09.
5. Piquereau J, Caffin F, Novotova M, Lemaire C, Veksler V, Garnier A, et al. Mitochondrial dynamics in the adult cardiomyocytes: which roles for a highly specialized cell? Front physiol. 2013;4:102. Epub 2013/05/16.
6. Lindsey ML, Mayr M, Gomes AV, Delles C, Arrell DK, Murphy AM, et al. Transformative impact of proteomics on cardiovascular health and disease: a scientific statement from the American Heart Association. Circulation. 2015;132(9):852–72.
7. Shen X, Young R, Canty JM, Qu J. Quantitative proteomics in cardiovascular research: global and targeted strategies. Proteomics Clin Appl. 2014;8(7–8):488–505. Epub 2014/06/13.
8. Peronnet E, Becquart L, Poirier F, Cubizolles M, Choquet-Kastylevsky G, Jolivet-Reynaud C. SELDI-TOF MS analysis of the Cardiac Troponin I forms present in plasma from patients with myocardial infarction. Proteomics. 2006;6(23):6288–99. Epub 2006/10/31.
9. Koomen JM, Wilson CR, Guthrie P, Androlewicz MJ, Kobayashi R, Taegtmeyer H. Proteome analysis of isolated perfused organ effluent as a novel model for protein biomarker discovery. J Proteome Res. 2006;5(1):177–82. Epub 2006/01/07.
10. Esposito G, Schiattarella GG, Perrino C, Cattaneo F, Pironti G, Franzone A, et al. Dermcidin: a skeletal muscle myokine modulating cardiomyocyte survival and infarct size after coronary artery ligation. Cardiovasc Res. 2015;107(4):431–41. Epub 2015/06/24.
11. Gramolini AO, Kislinger T, Alikhani-Koopaei R, Fong V, Thompson NJ, Isserlin R, et al. Comparative proteomics profiling of a phospholamban mutant mouse model of dilated cardiomyopathy reveals progressive intracellular stress responses. Mol Cell Proteomics. 2008;7(3):519–33. Epub 2007/12/07.
12. Lionetti V, Matteucci M, Ribezzo M, Di Silvestre D, Brambilla F, Agostini S, et al. Regional mapping of myocardial hibernation phenotype in idiopathic end-stage dilated cardiomyopathy. J Cell Mol Med. 2014;18(3):396–414. Epub 2014/01/22.
13. Buscemi N, Murray C, Doherty-Kirby A, Lajoie G, Sussman MA, Van Eyk JE. Myocardial subproteomic analysis of a constitutively active Rac1-expressing transgenic mouse with lethal myocardial hypertrophy. Am J Physiol Heart Circ Physiol. 2005;289(6):2325–33. Epub 2005/09/13.
14. Keshishian H, Burgess MW, Gillette MA, Mertins P, Clauser KR, Mani DR, et al. Multiplexed, quantitative workflow for sensitive biomarker discovery in plasma yields novel candidates for early myocardial injury. Mol Cell Proteomics. 2015;14:2375–93. Epub 2015/03/01.

15. Chen YQ, Bi F, Wang SQ, Xiao SJ, Liu JN. Porous silicon affinity chips for biomarker detection by MALDI-TOF-MS. J Chromatogr B Analyt Technol Biomed Life Sci. 2008;875(2):502–8. Epub 2008/10/25.
16. Wei YJ, Huang YX, Shen Y, Cui CJ, Zhang XL, Zhang H, et al. Proteomic analysis reveals significant elevation of heat shock protein 70 in patients with chronic heart failure due to arrhythmogenic right ventricular cardiomyopathy. Mol Cell Biochem. 2009;332(1–2):103–11. Epub 2009/06/23.
17. Cieniewski-Bernard C, Mulder P, Henry JP, Drobecq H, Dubois E, Pottiez G, et al. Proteomic analysis of left ventricular remodeling in an experimental model of heart failure. J Proteome Res. 2008;7(11):5004–16. Epub 2008/10/17.
18. Edwards AV, White MY, Cordwell SJ. The role of proteomics in clinical cardiovascular biomarker discovery. Mol Cell Proteomics. 2008;7(10):1824–37. Epub 2008/08/01.
19. Dai DF, Hsieh EJ, Chen T, Menendez LG, Basisty NB, Tsai L, et al. Global proteomics and pathway analysis of pressure-overload-induced heart failure and its attenuation by mitochondrial-targeted peptides. Circ Heart Fail. 2013;6(5):1067–76. Epub 2013/08/13.
20. Huang RY, Laing JG, Kanter EM, Berthoud VM, Bao M, Rohrs HW, et al. Identification of CaMKII phosphorylation sites in Connexin43 by high-resolution mass spectrometry. J Proteome Res. 2011;10(3):1098–109. Epub 2010/12/17.
21. Heck AJ, Krijgsveld J. Mass spectrometry-based quantitative proteomics. Expert Rev Proteomics. 2004;1(3):317–26. Epub 2005/06/22.
22. Ross PL, Huang YN, Marchese JN, Williamson B, Parker K, Hattan S, et al. Multiplexed protein quantitation in Saccharomyces cerevisiae using amine-reactive isobaric tagging reagents. Mol Cell Proteomics. 2004;3(12):1154–69. Epub 2004/09/24.
23. Gygi SP, Rist B, Gerber SA, Turecek F, Gelb MH, Aebersold R. Quantitative analysis of complex protein mixtures using isotope-coded affinity tags. Nat Biotechnol. 1999;17(10):994–9. Epub 1999/10/03.
24. Mirgorodskaya OA, Kozmin YP, Titov MI, Korner R, Sonksen CP, Roepstorff P. Quantitation of peptides and proteins by matrix-assisted laser desorption/ionization mass spectrometry using (18)O-labeled internal standards. Rapid Commun Mass Spectrom. 2000;14(14):1226–32. Epub 2000/08/05.
25. Holper S, Ruhs A, Kruger M. Stable isotope labeling for proteomic analysis of tissues in mouse. Methods Mol Biol. 2014;1188:95–106. Epub 2014/07/26.
26. Pan KT, Chen YY, Pu TH, Chao YS, Yang CY, Bomgarden RD, et al. Mass spectrometry-based quantitative proteomics for dissecting multiplexed redox cysteine modifications in nitric oxide-protected cardiomyocyte under hypoxia. Antioxid Redox Signal. 2014;20(9):1365–81. Epub 2013/10/25.
27. Gingras AC, Gstaiger M, Raught B, Aebersold R. Analysis of protein complexes using mass spectrometry. Nat Rev Mol Cell Biol. 2007;8(8):645–54.
28. Ping P, Zhang J, Pierce Jr WM, Bolli R. Functional proteomic analysis of protein kinase C epsilon signaling complexes in the normal heart and during cardioprotection. Circ Res. 2001;88(1):59–62.
29. Gregorich ZR, Chang YH, Ge Y. Proteomics in heart failure: top-down or bottom-up? Pflugers Arch Eur J physiol. 2014;466(6):1199–209.
30. Jia W, Shaffer JF, Harris SP, Leary JA. Identification of novel protein kinase A phosphorylation sites in the M-domain of human and murine cardiac myosin binding protein-C using mass spectrometry analysis. J Proteome Res. 2010;9(4):1843–53.
31. Chiao YA, Zamilpa R, Lopez EF, Dai Q, Escobar GP, Hakala K, et al. In vivo matrix metalloproteinase-7 substrates identified in the left ventricle post-myocardial infarction using proteomics. J Proteome Res. 2010;9(5):2649–57. Epub 2010/03/18.
32. Zamilpa R, Lopez EF, Chiao YA, Dai Q, Escobar GP, Hakala K, et al. Proteomic analysis identifies in vivo candidate matrix metalloproteinase-9 substrates in the left ventricle post-myocardial infarction. Proteomics. 2010;10(11):2214–23. Epub 2010/04/01.

33. Westbrook JA, Wheeler JX, Wait R, Welson SY, Dunn MJ. The human heart proteome: two-dimensional maps using narrow-range immobilised pH gradients. Electrophoresis. 2006;27(8):1547–55.
34. Aye TT, Scholten A, Taouatas N, Varro A, Van Veen TA, Vos MA, et al. Proteome-wide protein concentrations in the human heart. Mol Biosyst. 2010;6(10):1917–27.
35. Jacquet S, Yin X, Sicard P, Clark J, Kanaganayagam GS, Mayr M, et al. Identification of cardiac myosin-binding protein C as a candidate biomarker of myocardial infarction by proteomics analysis. Mol Cell Proteomics MCP. 2009;8(12):2687–99.
36. Lam L, Tsoutsman T, Arthur J, Semsarian C. Differential protein expression profiling of myocardial tissue in a mouse model of hypertrophic cardiomyopathy. J Mol Cell Cardiol. 2010;48(5):1014–22.
37. Bansal A, Dai Q, Chiao YA, Hakala KW, Zhang JQ, Weintraub ST, et al. Proteomic analysis reveals late exercise effects on cardiac remodeling following myocardial infarction. J Proteomics. 2010;73(10):2041–9.
38. Sharma P, Cosme J, Gramolini AO. Recent advances in cardiovascular proteomics. J Proteomics. 2013;81:3–14.
39. Rabilloud T. Two-dimensional gel electrophoresis in proteomics: old, old fashioned, but it still climbs up the mountains. Proteomics. 2002;2(1):3–10.
40. McGregor E, Dunn MJ. Proteomics of the heart: unraveling disease. Circ Res. 2006;98(3):309–21.
41. Franklin S, Zhang MJ, Chen H, Paulsson AK, Mitchell-Jordan SA, Li Y, et al. Specialized compartments of cardiac nuclei exhibit distinct proteomic anatomy. Mol Cell Proteomics. 2011;10(1):M110 000703.
42. Dai Q, Escobar GP, Hakala KW, Lambert JM, Weintraub ST, Lindsey ML. The left ventricle proteome differentiates middle-aged and old left ventricles in mice. J Proteome Res. 2008;7(2):756–65. Epub 2008/01/02.
43. Lang SC, Elsasser A, Scheler C, Vetter S, Tiefenbacher CP, Kubler W, et al. Myocardial preconditioning and remote renal preconditioning--identifying a protective factor using proteomic methods? Basic Res Cardiol. 2006;101(2):149–58. Epub 2005/11/12.
44. Faber MJ, Dalinghaus M, Lankhuizen IM, Bezstarosti K, Dekkers DH, Duncker DJ, et al. Proteomic changes in the pressure overloaded right ventricle after 6 weeks in young rats: correlations with the degree of hypertrophy. Proteomics. 2005;5(10):2519–30. Epub 2005/05/25.
45. Shen X, Zheng S, Thongboonkerd V, Xu M, Pierce Jr WM, Klein JB, et al. Cardiac mitochondrial damage and biogenesis in a chronic model of type 1 diabetes. Am J Physiol Endocrinol Metab. 2004;287(5):E896–905. Epub 2004/07/29.
46. Stastna M, Van Eyk JE. Optimized method for identification of the proteomes secreted by cardiac cells. Methods Mol Biol. 2013;1005:225–35. Epub 2013/04/23.
47. Gramolini AO, Kislinger T, Liu P, MacLennan DH, Emili A. Analyzing the cardiac muscle proteome by liquid chromatography-mass spectrometry-based expression proteomics. Methods Mol Biol. 2007;357:15–31. Epub 2006/12/19.
48. Kislinger T, Gramolini AO, MacLennan DH, Emili A. Multidimensional protein identification technology (MudPIT): technical overview of a profiling method optimized for the comprehensive proteomic investigation of normal and diseased heart tissue. J Am Soc Mass Spectrom. 2005;16(8):1207–20. Epub 2005/06/28.
49. Liu H, Sadygov RG, Yates 3rd JR. A model for random sampling and estimation of relative protein abundance in shotgun proteomics. Anal Chem. 2004;76(14):4193–201. Epub 2004/07/16.
50. de Castro Bras LE, Cates CA, DeLeon-Pennell KY, Ma Y, Iyer RP, Halade GV, et al. Citrate synthase is a novel in vivo matrix metalloproteinase-9 substrate that regulates mitochondrial function in the postmyocardial infarction left ventricle. Antioxid Redox Signal. 2014;21(14):1974–85. Epub 2014/01/03.
51. Casado P, Cutillas PRA, Casado P, Cutillas PR. A self-validating quantitative mass spectrometry method for assessing the accuracy of high-content phosphoproteomic experiments. Mol Cell Proteomics. 2011;10(1):M110 003079.

52. Cappadona S, Baker PR, Cutillas PR, Heck AJ, van Breukelen B. Current challenges in software solutions for mass spectrometry-based quantitative proteomics. Amino Acids. 2012;43(3):1087–108. Epub 2012/07/24.
53. Gerber SA, Rush J, Stemman O, Kirschner MW, Gygi SP. Absolute quantification of proteins and phosphoproteins from cell lysates by tandem MS. Proc Natl Acad Sci U S A. 2003;100(12):6940–5. Epub 2003/05/29.
54. Pitt JJ. Principles and applications of liquid chromatography-mass spectrometry in clinical biochemistry. Clin Biochem Rev. 2009;30(1):19–34.
55. Jones AW, Cooper HJ. Dissociation techniques in mass spectrometry-based proteomics. Analyst. 2011;136(17):3419–29. Epub 2011/06/24.
56. Kim JS, Fillmore TL, Liu T, Robinson E, Hossain M, Champion BL, et al. 18O-labeled proteome reference as global internal standards for targeted quantification by selected reaction monitoring-mass spectrometry. Mol Cell Proteomics. 2011;10(12):M110 007302.
57. Chahrour O, Cobice D, Malone J. Stable isotope labelling methods in mass spectrometry-based quantitative proteomics. J Pharm Biomed Anal. 2015;113:2–20. Epub 2015/05/10.
58. de Hoffmann E. Tandem mass spectrometry: a primer. J Mass Spectrom. 1996;31(2):129–37.
59. Lesur A, Domon B. Advances in high-resolution accurate mass spectrometry application to targeted proteomics. Proteomics. 2015;15(5–6):880–90. Epub 2014/12/30.
60. Pisitkun T, Hoffert JD, Yu MJ, Knepper MA. Tandem mass spectrometry in physiology. Physiology. 2007;22:390–400. Epub 2007/12/13.
61. Arrell DK, Neverova I, Van Eyk JE. Cardiovascular proteomics: evolution and potential. Circ Res. 2001;88(8):763–73.
62. Hu J, Coombes KR, Morris JS, Baggerly KA. The importance of experimental design in proteomic mass spectrometry experiments: some cautionary tales. Brief Funct Genomic Proteomic. 2005;3(4):322–31. Epub 2005/04/09.
63. Marshall J, Kupchak P, Zhu W, Yantha J, Vrees T, Furesz S, et al. Processing of serum proteins underlies the mass spectral fingerprinting of myocardial infarction. J Proteome Res. 2003;2(4):361–72. Epub 2003/08/27.
64. Jaumdally R, Varma C, Macfadyen RJ, Lip GY. Coronary sinus blood sampling: an insight into local cardiac pathophysiology and treatment? Eur Heart J. 2007;28(8):929–40. Epub 2007/03/31.
65. Beck HC, Overgaard M, Melholt RL. Plasma proteomics to identify biomarkers – application to cardiovascular diseases. Transl Proteomics. 2015;7:40–8.
66. Gillinov AM. Advances in surgical treatment of atrial fibrillation. Stroke J Cereb Circulation. 2007;38(2 Suppl):618–23. Epub 2007/01/31.
67. Havanapan PO, Thongboonkerd V. Are protease inhibitors required for gel-based proteomics of kidney and urine? J Proteome Res. 2009;8(6):3109–17. Epub 2009/04/10.
68. Zhao Y, Jensen ON. Modification-specific proteomics: strategies for characterization of post-translational modifications using enrichment techniques. Proteomics. 2009;9(20):4632–41. Epub 2009/09/11.
69. Stanley BA, Neverova I, Brown HA, Van Eyk JE. Optimizing protein solubility for two-dimensional gel electrophoresis analysis of human myocardium. Proteomics. 2003;3(6):815–20. Epub 2003/07/02.
70. Chugh S, Suen C, Gramolini A. Proteomics and mass spectrometry: what have we learned about the heart? Curr Cardiol Rev. 2010;6(2):124–33. Epub 2011/05/03.
71. Rabilloud T, Kieffer S, Procaccio V, Louwagie M, Courchesne PL, Patterson SD, et al. Two-dimensional electrophoresis of human placental mitochondria and protein identification by mass spectrometry: toward a human mitochondrial proteome. Electrophoresis. 1998;19(6):1006–14.
72. Zhang J, Li X, Mueller M, Wang Y, Zong C, Deng N, et al. Systematic characterization of the murine mitochondrial proteome using functionally validated cardiac mitochondria. Proteomics. 2008;8(8):1564–75.

73. Nissum M, Kuhfuss S, Hauptmann M, Obermaier C, Sukop U, Wildgruber R, et al. Two-dimensional separation of human plasma proteins using iterative free-flow electrophoresis. Proteomics. 2007;7(23):4218–27. Epub 2007/11/02.
74. Travis J, Bowen J, Tewksbury D, Johnson D, Pannell R. Isolation of albumin from whole human plasma and fractionation of albumin-depleted plasma. Biochem J. 1976;157(2):301–6. Epub 1976/08/01.
75. Sato AK, Sexton DJ, Morganelli LA, Cohen EH, Wu QL, Conley GP, et al. Development of mammalian serum albumin affinity purification media by peptide phage display. Biotechnol Prog. 2002;18(2):182–92. Epub 2002/04/06.
76. Rothemund DL, Locke VL, Liew A, Thomas TM, Wasinger V, Rylatt DB. Depletion of the highly abundant protein albumin from human plasma using the Gradiflow. Proteomics. 2003;3(3):279–87. Epub 2003/03/11.
77. Chandramouli K, Qian PY. Proteomics: challenges, techniques and possibilities to overcome biological sample complexity. Hum Genomics Proteomics. 2009(1):1–22. Epub 2009/01/01.
78. de Castro Bras LE, Deleon KY, Ma Y, Dai Q, Hakala K, Weintraub ST, et al. Plasma fractionation enriches post-myocardial infarction samples prior to proteomics analysis. Int J Proteomics. 2012;2012:397103. Epub 2012/07/11.
79. de Castro Brás LE, Ramirez TA, DeLeon-Pennell KY, Chiao YA, Ma Y, Dai Q, et al. Texas 3-step decellularization protocol: looking at the cardiac extracellular matrix. J Proteomics. 2013;86:43–52. Epub 2013/05/18.
80. Didangelos A, Yin X, Mandal K, Baumert M, Jahangiri M, Mayr M. Proteomics characterization of extracellular space components in the human aorta. Mol Cell Proteomics. 2010;9(9):2048–62. Epub 2010/06/17.
81. Warren CM, Geenen DL, Helseth Jr DL, Xu H, Solaro RJ. Sub-proteomic fractionation, iTRAQ, and OFFGEL-LC-MS/MS approaches to cardiac proteomics. J Proteomics. 2010;73(8):1551–61. Epub 2010/04/17.
82. Consortium TU. UniProt: a hub for protein information. Nucleic Acids Res. 2015;43(D1):D204–12.
83. Zheng P, Xiong Q, Wu Y, Chen Y, Chen Z, Fleming J, et al. Quantitative proteomics analysis reveals novel insights into mechanisms of action of long noncoding RNA hox transcript antisense intergenic RNA (HOTAIR) in HeLa cells. Mol Cell Proteomics. 2015;14(6):1447–63. Epub 2015/03/13.
84. Elias JE, Gygi SP. Target-decoy search strategy for increased confidence in large-scale protein identifications by mass spectrometry. Nat Methods. 2007;4(3):207–14. Epub 2007/03/01.
85. Nesvizhskii AI. A survey of computational methods and error rate estimation procedures for peptide and protein identification in shotgun proteomics. J Proteomics. 2010;73(11):2092–123. Epub 2010/09/08.
86. Geer LY, Markey SP, Kowalak JA, Wagner L, Xu M, Maynard DM, et al. Open mass spectrometry search algorithm. J Proteome Res. 2004;3(5):958–64. Epub 2004/10/12.
87. Martins-Marques T, Anjo SI, Pereira P, Manadas B, Girao H. Interacting network of the gap junction protein connexin43 is modulated by ischemia and reperfusion in the heart. Mol Cell Proteomics. 2015;14(11):3040–55. Epub 2015/09/01.
88. Zhu W, Smith JW, Huang CM. Mass spectrometry-based label-free quantitative proteomics. J Biomed Biotechnol. 2010;2010:840518. Epub 2009/11/17.
89. Zhang B, VerBerkmoes NC, Langston MA, Uberbacher E, Hettich RL, Samatova NF. Detecting differential and correlated protein expression in label-free shotgun proteomics. J Proteome Res. 2006;5(11):2909–18. Epub 2006/11/04.
90. Lindsey ML, Mayr M, Gomes AV, Delles C, Arrell DK, Murphy AM, et al. Transformative impact of proteomics on cardiovascular health and disease: a scientific statement from the American Heart Association. Circulation. 2015;132(9):852–72. Epub 2015/07/22.
91. Olsen JV, Ong SE, Mann M. Trypsin cleaves exclusively C-terminal to arginine and lysine residues. Mol Cell Proteomics. 2004;3(6):608–14. Epub 2004/03/23.

92. Winter D, Kugelstadt D, Seidler J, Kappes B, Lehmann WD. Protein phosphorylation influences proteolytic cleavage and kinase substrate properties exemplified by analysis of in vitro phosphorylated Plasmodium falciparum glideosome-associated protein 45 by nano-ultra performance liquid chromatography-tandem mass spectrometry. Anal Biochem. 2009;393(1):41–7. Epub 2009/06/25.
93. Ingrosso D, Fowler AV, Bleibaum J, Clarke S. Specificity of endoproteinase Asp-N (Pseudomonas fragi): cleavage at glutamyl residues in two proteins. Biochem Biophys Res Commun. 1989;162(3):1528–34. Epub 1989/08/15.
94. Drapeau GR. The primary structure of staphylococcal protease. Can J Biochem. 1978;56(6):534–44. Epub 1978/06/01.
95. Biringer RG, Amato H, Harrington MG, Fonteh AN, Riggins JN, Huhmer AF. Enhanced sequence coverage of proteins in human cerebrospinal fluid using multiple enzymatic digestion and linear ion trap LC-MS/MS. Brief Funct Genomic Proteomic. 2006;5(2):144–53. Epub 2006/06/15.
96. Meyer B, Papasotiriou DG, Karas M. 100% protein sequence coverage: a modern form of surrealism in proteomics. Amino Acids. 2011;41(2):291–310. Epub 2010/07/14.
97. Taylor CF, Paton NW, Lilley KS, Binz PA, Julian Jr RK, Jones AR, et al. The minimum information about a proteomics experiment (MIAPE). Nat Biotechnol. 2007;25(8):887–93. Epub 2007/08/10.
98. Vizcaino JA, Cote RG, Csordas A, Dianes JA, Fabregat A, Foster JM, et al. The PRoteomics IDEntifications (PRIDE) database and associated tools: status in 2013. Nucleic Acids Res. 2013;41(Database issue):D1063–9. Epub 2012/12/04.
99. Jones P, Martens L. Using the PRIDE proteomics identifications database for knowledge discovery and data analysis. Methods Mol Biol. 2010;604:297–307. Epub 2009/12/17.
100. Martens, L., Jones, P. and Côté, R. 2008. Using the Proteomics Identifications Database (PRIDE). Current Protocols in Bioinformatics. 21:13.8:13.8.1–13.8.19. Epub 2008/04/23.

Chapter 4
Organelle, Protein and Peptide Fractionation in Cardiovascular Proteomics

Mickey R. Miller, Sarah Franklin, and D. Brian Foster

Abstract Proteomics experiments are as diverse as the scientists who perform them. Goals range from the desire to understand a subcellular structure or an individual protein in greater depth, to identification of novel protein-protein interactions. Or perhaps, the goal is to obtain a global protein abundance profile from an animal model of cardiovascular disease or from patient biopsies. Regardless of scale or objective, inevitably, the tools of organelle isolation, protein purification or peptide fractionation will play an integral role. In this chapter, we survey both time-honored and state-of-the-art fractionation techniques, with an emphasis on underlying physical and chemical principles.

Keywords Organelle • Protein • Peptide • Centrifugation • Precipitation • Extraction • Chromatography • Gel Electrophoresis • Capillary Electrophoresis • Orthogonality • Co-isolation Interference

---

M.R. Miller, PhD
Department of Cardiology, Nora Eccles Harrison Cardiovascular Research and Training Institute, University of Utah, Salt Lake City, UT, USA

S. Franklin, PhD (✉)
Cardiology Division, Department of Internal Medicine,
The Nora Eccles Harrison Cardiovascular Research and Training Institute,
University of Utah, 95 South 2000 East, Room 101, Salt Lake City, UT 84112, USA

Department of Biochemistry, University of Utah, Salt Lake City, UT, USA
e-mail: franklin@cvrti.utah.edu; http://www.cvrti.utah.edu/~franklin/Site/index.html

D.B. Foster, PhD (✉)
Laboratory of Cardiovascular Biochemistry, Division of Cardiology, Department of Medicine,
The Johns Hopkins University School of Medicine, Baltimore, MD, USA
e-mail: dbrianfoster@jhmi.edu; http://web1.johnshopkins.edu/fosterlab

G. Agnetti et al. (eds.), *Manual of Cardiovascular Proteomics*,
DOI 10.1007/978-3-319-31828-8_4

Introduction

This may come as a surprise to students or postdoctoral fellows in the midst of inculcating themselves in cardiovascular proteomics, but it wasn't that long ago that the tools we take for granted today, including the use of genetically engineered proteins, fast & sensitive mass spectrometers and informative bioinformatic pathway analyses, were either novelties or fanciful dreams. Even the term "proteome" is only about 20 years old [1]. Back in the day, we called ourselves protein biochemists and we studied only one, or at most, a few proteins at a time! Reductionism ruled. Fast-forward to today. We now have the technology to understand proteins in the context of broader networks and systems. The number of post-translational protein modifications (PTMs) has skyrocketed, as has the permutation of potential effects on protein and protein network function. There has scarcely been a more exciting time to study the proteins of the cardiovascular system. The central theme of this chapter is that though technologies for the separation and analyses of proteins may change, the underlying principles have not. We begin by summarizing a few best practices in sample homogenization, then survey the most widely used methods of organellar, protein and peptide separation, from the tried-and-true to the cutting edge. By calling attention to first principles, our goal is to inform experimental design and to foster an appreciation for how emerging technologies are building on older ones.

Sample Homogenization

Homogenization of tissue or lysis of cells is the first step in organelle and/or protein isolation. Priorities at this stage include preservation of the organelles, the proteins and, if applicable, their post translational modifications. Accomplishing this requires attention to a few best practices that can be boiled down to buffer considerations and homogenization method. Here we summarize a few basic tenets drawn from an old standard of protein knowledge, Methods in Enzymology, Guide to Protein Purification [2], now in its second edition [3], and many articles therein, see [4–6] in particular.

Buffers are often tailored to preserve the organelle of interest. Luckily, for each organelle there is a rich history of successful isolation and no shortage of protocols. A typical buffer is designed to approximate, albeit crudely, the physiological milieu in terms of pH, ionic strength and osmolality. Maintaining pH in the range of 7–8 preserves most enzymatic activity (if desired), whereas preserving the ionic strength and osmolality minimize soluble protein aggregation and maintain organelle integrity. Other buffer components, such as the presence of divalent cation cofactors or detergents, vary depending on the nature of the organelle and whether it is essential to preserve biological activity of the proteins. For example, isolation of mitochondria requires an iso-osmotic buffer to prevent their rupture. Alternately, when isolating myofilaments, it is often desirable to employ a buffer that dissolves membranous organelles away with detergents.

While preserving organelle integrity is a priority, it also essential to minimize any biological activity that might compromise protein integrity (e.g. proteolysis) or otherwise confound the intended proteomic studies (e.g. in vitro kinase/phosphatase activity). An excellent start is to use ice-cold buffers and refrigerated lab equipment to reduce overall enzymatic activity. Proteolysis in the early stages of isolation can subsequently be minimized by inclusion of protease inhibitors that target serine-, thiol-, metallo- and acid-activated-proteases. Depending on the goal of the study, it may be necessary to preserve the posttranslational profile of the organelle. Fortunately, many of the enzymes that add and remove PTMs can be inhibited. For example perturbation of protein phosphorylation can be minimized by inclusion of phosphatase inhibitors such β-glycerophosphate, orthovanadate and sodium fluoride, while kinase activity is minimized by inclusion of Mg^{2+}-and Ca^{2+}-chelators, such as ethylenediaminetetraacetic acid (EDTA).

Once the appropriate buffer has been prepared and chilled, we consider the actual homogenization procedure itself. If it is too harsh, organelles may be ruptured or sheared and proteins denatured. If homogenization is too mild, overall yield may be compromised. Here, the choice of homogenization technique is often tailored to the tissue under study. Some tissues are amenable to gentle disruption by hand with either glass or Teflon Potter-Elvehjem homogenizers, which resemble tight-fitting mortar & pestles. Others, like the heart, are more fibrous and can be difficult to homogenize by hand, in which case, it may be desirable to use a mechanized homogenizer that resembles a narrow-bore immersion hand blender. Interestingly, as "omic" technologies have emerged, so too have high-throughput homogenization tools including methods based on cavitation with ball-bearings that enable homogenization of many samples simultaneously (e.g. the "Bullet Blender"). These advances in automated homogenization, are important as they may substantially minimize sample-to-sample variability that can confound proteomics experiments.

Cells, particularly from cultures, may be disrupted in a variety of ways. Again shearing may be used, as in the case of a tight fitting homogenizer, or alternatively by passing cells through a narrow aperture at high pressure using a French press or a simple syringe needle. Cavitation of cells with high frequency sound waves, known as sonication, is also effective. Finally cells may also be disrupted by treatment with a mild detergent. As with tissues, care must be taken not to over-homogenize to preserve the integrity of the organelles.

Regardless of the buffer or homogenization method used, optimal organellar fractionation is only obtained from fresh, rather than frozen, samples. This is not to suggest that the frozen tissue is not amenable to proteomic analysis, only that subcellular fractionation is best reserved for fresh tissue/cells. In short, freezing and thawing compromises both protein and organelle integrity; proteins often denature and organelles may fuse and/or rupture.

Finally, for global-scale peptide-centric proteomics experiments in which organelle or protein separation are not a priority, buffer considerations switch from preserving biological function to eradicating it as quickly as possible. Instead, a premium is placed on preserving the proteins and their PTMs, which is accomplished by denaturing pernicious activity. Proteins can be denatured by including high concentrations of chaotropes such as urea and guanidine-HCl, or detergents such as sodium dodecyl sulfate. Denaturation also results from extreme changes in pH or by boiling proteins. Though all of these methods can be used to abolish unwanted enzymatic activity, each has unique shortcomings with respect to its capacity to preserve PTMs or induce *in vitro* artifacts and must be implemented mindfully.

Separation of Particles

Why Do We Separate Organelles, Proteins and Peptides?

Though the specific reasons for a particular separation are innumerable, the basic motivation is often to understand or characterize organelles, proteins and peptides in greater detail. We separate particles of interest, or sets of particles, from the rest. In the case of organelle separation, it is often desirable to study proteins and processes in the context of a cell's componentry or subcellular localization. For instance, isolation of mitochondria or myofilaments for functional and proteomic characterization are critical to advancing our understanding of heart disease. From the proteomic standpoint, we understand that mass spectrometers can't identify all the proteins and PTMs in a sample, therefore we subfractionate subcellular components to minimize contaminants and maximize discovery within a particular domain.

Alternatively, individual protein purification remains a critical aspect of proteomics. In this case, the goal is to study a particular protein quite intensively. The task then, is to separate it from thousands of other proteins in the cell, while preserving native binding partners and PTMs. In some cases, proteins can be genetically tagged to facilitate protein isolation. In others, tagging may be deleterious to function, alter protein folding or subcellular localization, in which case isolating native protein is still desirable. This is only possible with a detailed understanding of all the available tools of protein separation.

Or perhaps the goal of the experiment is global characterization of nearly all proteins in the cell in a case/control design. In that case, one might dispense with organellar and possibly protein separation. However, there are technical reasons why maximizing protein identification would still require extensive separation at the peptide level. Briefly, in the parlance of proteomics, we seek to maximize proteome depth, or coverage. The challenge is that peptides (and therefore proteins) of low abundance are often difficult to detect. Whereas low abundance DNA or RNA species can be identified within a complex DNA or RNA sample through amplification by polymerase chain reaction, no such amplification step exists for the identification of a low abundance protein within a complex protein mixture. In fact, mass spectral identification tends to be biased *against* the identification of rare proteins. One reason is that abundant proteins yield more peptides after proteolysis, which in the mass spectrometer, yield high parent-ion currents in MS^1. In workflows that employ data-dependent data acquisition (DDA), these parent-ions are sampled preferentially for fragmentation in MS^2. In essence, lower abundance proteins are under-sampled or masked by the more abundant proteins in the sample [7]. In light of this, the key to maximizing peptide identification on a given instrument is to decrease sample complexity by optimizing peptide separation.

First Principles of Particle Separation

Whether isolating intact organelles, purifying a protein, or spreading peptides over multiple fractions prior to analysis, the goal is the same – separation, separation, separation. In the interests of brevity and clarity, organelles, proteins, and peptides will, hereinafter, be described collectively as particles in a solvent. Particle separation methods can be broadly classified into two types, rate and equilibrium methods. Rate methods are based on the principle that particles moving with different velocities can be separated over time. In contrast, equilibrium methods manipulate the interaction between a particle and its environment to affect its end- position relative to other particles, irrespective of rate.

Directed Movement of Particles

In the modern biomedical research laboratory, there are a few main ways to get particles to move in a directed fashion (i.e. vectorially). One method is to apply tremendous forces, often many times the force of gravity, to move a particle through a solvent, as in the case of centrifugation. A second way is to elicit flow of the solvent that contains freely tumbling particles, as in the case of chromatography. A third way it to apply an electric field to drive charged particles, which is the very definition of electrophoresis. Separation technologies have evolved to alter the migration of particles by taking advantage of their distinctive combinations of properties, which include their size, shape, polarity, charge and biological function.

Diffusion: Undirected Particle Movement and Its Effect on Separation

The extent to which one can separate populations of particles from each other is termed resolution. If particles have been completely separated we say they have been resolved or that resolution between populations is high. If particle populations overlap substantially, the resolution is low. Resolution depends on several factors that include intrinsic capabilities of the method and the capacity to modify the properties of the particle to maximum advantage. These factors are explored in greater detail later. There is, however, one particle property that often compromises resolution if left unchecked – diffusion. Diffusion, the random movement of particles, is what prevents particles with identical properties, from migrating vectorially at a uniform rate. Instead, a population of identical particles travels at an average velocity, and diffusion contributes to the dispersion of the individual particle velocities. Large spreads on velocity or the equilibrium position of particles increase the likelihood of overlap between populations with marginally different properties. As you will see in the following sections, limiting particle diffusion to maximize resolution is a recurring theme across the methods of centrifugation and chromatography and electrophoresis.

Centrifugation: The Workhorse of Organellar Separation

Fractionation by Centrifugation

At its root, centrifugation is a procedure by which particles are separated by imposing large forces on a sample, by spinning it quickly, in a tube or bucket, around a fixed axis. Reports of the use of hand-driven centrifuges date back to the mid-fifteenth century when it was used to separate milk from cream [8]. Today modern laboratory centrifuges come in many forms, ranging from compact tabletop models to floor-standing ultracentrifuges capable of spinning samples at tens of thousands of revolutions per min (RPM). The relationship between the speed of rotation, or angular velocity (ω), and relative centrifugal force (RCF) is given by:

$$RCF = \frac{angular\ acceleration}{gravitational\ acceleration} = \frac{\omega^2 r}{g} \tag{4.1}$$

where

ω, the angular velocity in radians/s and
r, the radius of rotation, i.e. distance from the axis
g is earth's gravitational acceleration

Since angular velocity is most commonly understood in the laboratory in terms of RPM we can re-write this as:

$$RCF = 1.118 r_{mm} \left(\frac{RPM}{1000} \right)^2 \tag{4.2}$$

where

r_{mm} is radius of rotation in mm

Centrifugation separates particles in the sample on the basis of size, shape and/or density by varying the speed of the centrifuge rotor, run time, or the density of the medium through which they travel. The utility of two common organellar fractionation strategies, differential velocity sedimentation and density gradient centrifugation, are outlined briefly here.

Differential Velocity Centrifugation

Differential velocity centrifugation (DVC), not surprisingly, is a strategy that capitalizes on the fact that applying high forces to particles causes them to migrate at differing speeds, largely depending on their size, or technically, their mass. Centrifugation causes particles that migrate quickly, to move to the bottom of a tube or bucket, where they form a semisolid pellet. This is called sedimentation. Lighter, slower moving particles are largely confined to the supernatant. Decanting the supernatant yields a pellet that is enriched and a supernatant that is depleted of a given particle.

DVC therefore depends on the sedimentation rate of particles. The speed of a particle is determined by the balance of forces imposed on it, which include the centrifugal force and the opposing effects of buoyancy and drag. Expressed formally, we see,

$$\frac{dr}{dt} = \frac{M_W\left(1-\bar{v}\rho\right)}{N_A f}\omega^2 r \tag{4.3}$$

where

M_W is the molecular weight of the particle
$1-\bar{v}\rho$ term accounts for buoyancy where
$\bar{v}$ is the partial specific volume of the particle, the inverse of particle density and
ρ is the density of the solvent
f is the frictional coefficient, or drag term, which reflects particle shape and solvent viscosity
N_A is Avogadro's number

Breaking the equation down, we can see that the sedimentation rate is a function of angular acceleration ($\omega^2 r$), which is determined by the experimentalist, and a host of particle properties intrinsic to the particle itself in a given solvent. These intrinsic properties define a sedimentation propensity or index (s) so,

$$\frac{dr}{dt} = s\omega^2 r \tag{4.4}$$

where

$$s = \frac{M_W\left(1-\bar{v}\rho\right)}{N_A f} \tag{4.5}$$

In other words, the sedimentation coefficient, s, is simply a function of mass, shape and density of the particle, as well as the viscosity and density of the solvent. What

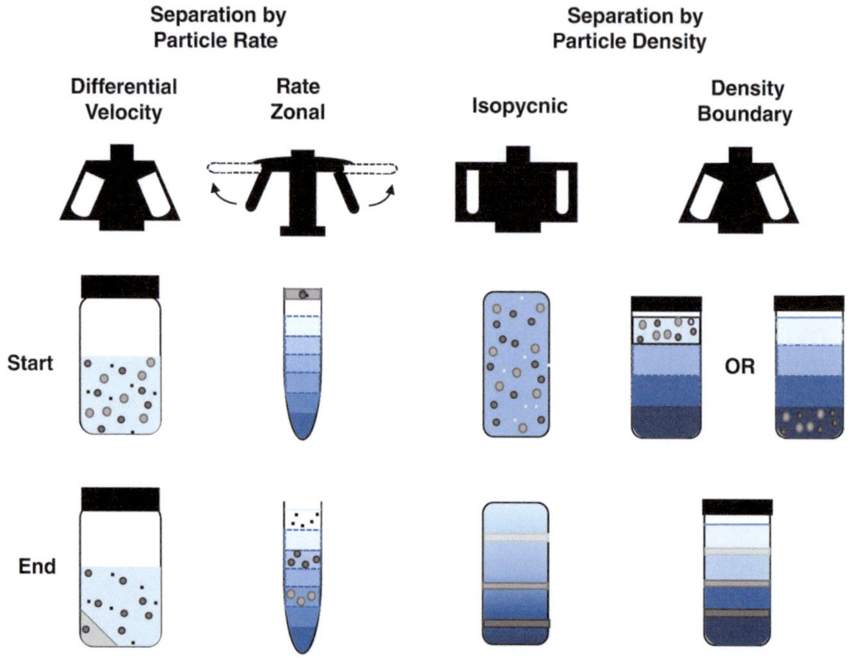

Fig. 4.1 Commonly-used centrifugation methods

this means is that heavier particles will sediment faster than lighter ones. If two particles have comparable mass, a compact shape will sediment faster than an extended one, due to reduced drag. Likewise, between comparable masses, the particle with lower buoyancy will sediment faster than a high buoyancy particle (e.g. high lipid content).

In the context of subcellular fractionation, we can see that differential velocity centrifugation is a procedure that capitalizes on the fact that organelles have different sedimentation coefficients. For instance nuclei are much larger and heavier than mitochondria. Mitochondria, in turn, are larger and less buoyant than plasma membranes. In practical terms, this means that organelles with high sedimentation coefficients can be harvested at lower centrifugation speeds. By conducting successive rounds of centrifugation and progressively higher and judiciously chosen speeds, it is possible separate heavy organelles from light ones. Figure 4.1 outlines the separation of organelles obtained in a sequential centrifugation with increasing centrifugation speeds and times. An initial low-speed centrifugation will separate nuclei from the homogenate. The post-nuclear supernatant (PNS) can then be centrifuged at higher speeds and for longer periods of time to further separate organelles from the sample. The number of centrifugation steps required for a given study will depend on the subcellular compartment(s) desired.

It should be noted that while the differential velocity centrifugation method is easy to perform and organellar enrichment can be substantial, the purity of the organelles is still somewhat crude for a couple of reasons. Firstly, all the particles are fairly homogeneously distributed in the tube or bucket before initiating centrifugation. The second reason is that the sedimentations coefficients for each organelle are not discrete, but cover a range. Therefore, organelles with overlapping sedimentation coefficients will co-fractionate in a given solvent. Each of these challenges can be addressed by additional centrifugation strategies.

Density-Gradient Centrifugation

Rate Zonal Gradient Density Centrifugation

One reason the differential velocity centrifugation strategy affords only crude organellar enrichment, is that prior to centrifugation, biological samples are homogeneously dispersed. Upon centrifugation small particles near the bottom of the tube will pellet as fast as large particles near the top of the tube, as they have a shorter distance to travel. Rate zonal centrifugation solves this problem by changing the density and viscosity of the solvent. First, consider a centrifuge tube filled near the top with a solvent whose density is greater than a typical tissue homogenate (e.g. 10 % w/v sucrose). The homogenate can then be applied in a thin layer near the top of the tube. As a result, all particles in the sample begin their sedimentation from roughly the same starting point. Larger particles can now outpace smaller ones to the bottom of the tube. Rate zonal centrifugation is optimally performed in swinging bucket rotors. A tube that is parallel to the centrifugal force maximizes migration distance and therefore resolution between particles. To summarize, the judicious choice of centrifuge rotor, speed and run time allows organelles of differing sedimentation coefficients to be distributed from the top of the tube to the bottom. In practical terms, however, using a single-density solvent is suboptimal, since the effects of particle diffusion dampen resolution over the course of centrifugation. This problem is alleviated either by using multiple solvents of differing density and viscosity (a discontinuous gradient; see Fig. 4.2) or a single continuous gradient. A typical procedure would be to construct a discontinuous gradient with solvents varying from 10 to 40 % sucrose. It should be emphasized that as long as the density of the solvent does not exceed the density of the particle or organelle, fractionation proceeds primarily on the basis of particle size, not its density. The density (and viscosity) gradient simply increases resolution by limiting diffusion. A corollary of this is that if centrifugation time were extended significantly, all organelles would pellet.

Isopycnic (Equilibrium) Gradient Density Centrifugation

The second challenge that confounds standard DVC is that organelles come in a range of sizes and, therefore, sedimentation coefficients. Organelles with overlapping coefficients can never be fully resolved by rate methods alone. This is where

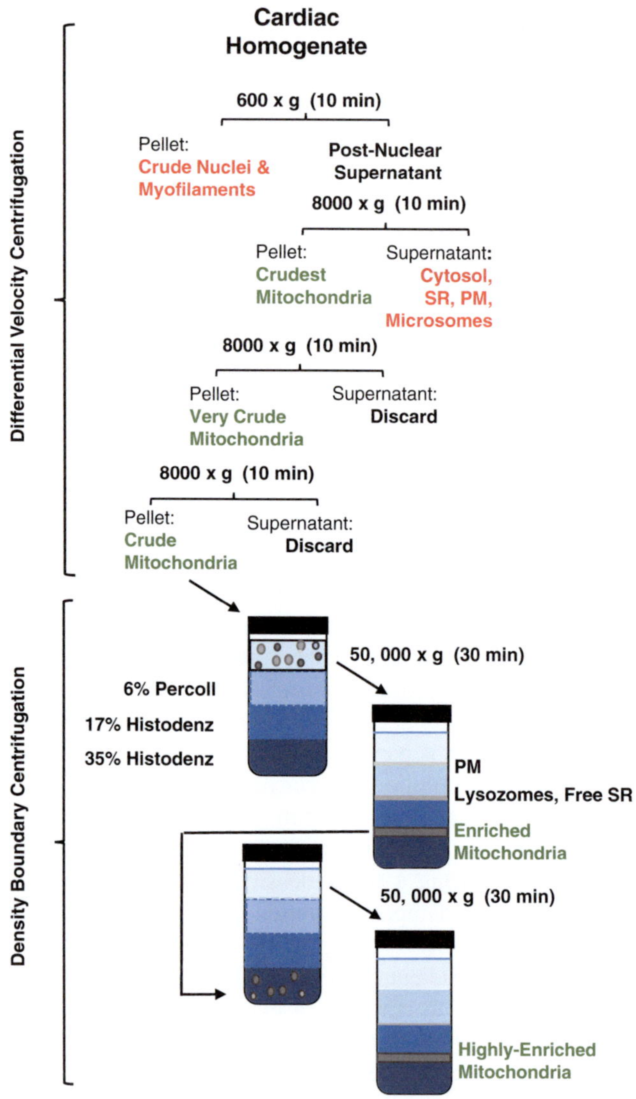

Fig. 4.2 A mitochondrial isolation protocol illustrates the combined use of rate and equilibrium centrifugation. This strategy was used by Foster et al. [9] to obtain highly-enriched mitochondria whose further subfractionation led to identification of a cytoprotective mitochondrial potassium channel, mitoROMK (It is based on the work of Taylor et al. [10] and Storrie and Madden [11]). Differential velocity centrifugation (DVC) and density boundary centrifugation (DBC) were used to isolate highly-enriched mitochondria from nearly 1 kg of bovine heart tissue. Following removal of myofilament and nuclei by low-speed centrifugation, multiple rounds of DVC at 8000 × g minimize contamination of particles with non-overlapping sedimentation coefficients (e.g. sarcoplasmic reticulum, SR; plasma membrane, PM). Remaining free contamination can only be removed by methods that are independent of sedimentation rate. DBC is an equilibrium method that separates particles by density. Electron microscopic examination of preparations after two rounds of DBC reveals that mitochondria are essentially pure. The only (rarely) observable contamination consists of membranes that remain tethered to the mitochondrial outer membrane

equilibrium methods distinguish themselves. Remember that an equilibrium method depends only on the final spatial resolution of particles, not the rate they travel. The key principle of equilibrium methods is that that one can manipulate the particle/solvent interaction to achieve spatial resolution. Isopycnic centrifugation exemplifies this. Specifically, in isopycnic centrifugation, the density of the solvent is manipulated relative to the density of the particle. In a continuous high-density gradient, there is a final position where the solvent density equals the density of the particle (the isopycnic point). Solvents commonly used in isopycnic centrifugation share the intriguing property that they can form self-generated gradients upon centrifugation. Examples would include Cesium chloride solution which is used for preparation of high-purity DNA, or Percoll is often used for isolation of high purity mitochondria. When using a self-generating gradient protocol, the sample is mixed with the solvent to form a homogeneous solution and upon centrifugation migrates to the equilibrium point as the density gradient is established. Centrifugation proceeds until this equilibrium is reached, which often takes many hours. Indeed, as particles near their isopycnic points, they migrate very slowly. For this reason, isopycnic centrifugation is best performed in rotors with vertical or near-vertical tubes to minimize the distance the particles must travel to reach equilibrium.

Boundary Gradient Density Centrifugation

Boundary centrifugation is a variant of equilibrium centrifugation performed with a discontinuous density (ρ) gradient and is illustrated in Fig. 4.2. The principle is that organelles will migrate to the interface of two solvents where

$$\rho_{solvent1} < \rho_{organelle} < \rho_{solvent2}.$$

This method is more easily tailored to organelle fractionation than isopycnic method. Take, as a specific example, the work of Storrie & Madden, who optimized discontinuous gradient to resolve nuclei, mitochondria, endoplasmic reticulum and lysosomes [11]. They found that mitochondria have a density greater than 37 % w/v sucrose but less than 50 % w/v sucrose, and therefore migrate to the interface or boundary if the two layers. Initial contaminating endoplasmic reticulum has a density less than 37 % sucrose but above greater than 10 % and will migrate to the 10/37 boundary. This method is a highly effective way of increasing the resolution of organelles of different densities and the resulting enrichment can be very high. It is noteworthy that newcomers to centrifugation frequently confuse this method with rate zonal centrifugation. This is understandable since discontinuous density gradients are used in both cases. However, rate zonal centrifugation uses the gradient to control the sedimentation velocity and diffusion. In boundary centrifugation, however, the density gradient is chosen to resolve final position of particles in the centrifuge tube. This is keenly illustrated by the fact that it can be performed by mixing the sample with *the densest solvent at the bottom of the centrifuge tube* before layering the rest of the discontinuous gradient. Upon centrifugation, the particles float to their final position.

Evaluating Organelle Enrichment

Knowing the degree of enrichment or purity of an organelle preparation is key to understanding what conclusions may safely be drawn from its use. Purity can often be assessed qualitatively by electron microscopy and visually inspecting a broad field [12]. A nother time-honored quantitative approach is to track the fold-enrichment of protein known to be found exclusively in the organelle of interest. This may be done by measuring enzymatic activity per mg of total protein from successive steps of the enrichment protocol. Alternatively immunoblotting may be used as a proxy for enzymatic activity provided that equal amounts of protein from each step of the protocol are analyzed. In addition to enrichment of specific markers, it is also desirable to track the depletion of contaminants similarly by enzymatic or immunoblot analysis. Using mitochondria as example, it would be desirable to track the fold-enrichment of an enzyme like citrate synthase, as well as depletion of cytosolic, plasma membrane, sarcoplasmic reticulum and nuclear markers [13].

Choosing an Organellar Enrichment Strategy

Centrifugation is by far the most commonly-used tool for organellar separation. Procedures may yield either crude or highly-enriched organelles, depending on the type of centrifugation used. Which procedures are right for you depends on the goal of the experiment. This is because there is an inherent trade-off between organelle purity and isolation time. A DVC strategy may yield a cruder product, but it can be accomplished quickly, which minimizes preparation-induced changes to the proteome that accrue with time, particularly the loss of post-translational modifications. Therefore, if the motivation behind organelle fractionation is simply to reduce sample complexity, crude preparations may be sufficient. On the other hand, if the goal is to ascribe a subcellular location to the proteins that are ultimately identified, high organelle purity is important. However, high-purity protocols are more complicated and often involve both DVC and either rate-zonal or density boundary centrifugation as illustrated in Fig. 4.2. Isolation can take several hours, over which time the proteome may be affected.

Protein and Peptide Fractionation

Just like organelles, proteins and peptides have distinct physical properties that can be harnessed for their separation or isolation. These properties include charge, size & shape, hydrophobicity and biological activity. Before the advent of genetically-engineered high affinity tags, co-opting the properties of the native protein was often the only way to purify it to homogeneity. Today, knowledge of protein biochemistry is still relevant, whether the goal is protein purification for intensive study, or simply to minimize sample complexity by maximizing protein separation for broader-scale proteomic investigation. In this section, we introduce the primary methods by which proteins and peptides may be separated. These include precipitation, extraction, centrifugation, chromatography, and electrophoresis.

Selective Protein Precipitation

Precipitation is one of the oldest tools in protein separation. It capitalizes on the fact that soluble proteins may be selectively rendered insoluble by altering the ionic strength or pH of the solution [14]. It is essentially a phase separation technique where proteins undergo a transition from the soluble to insoluble phase and phases are then separated by centrifugation. Here, we focus on selective precipitation from which proteins can be easily resolubilized and which preserve biological activity. Techniques such as acetone, ethanol and trichloroacetic acid precipitation, in contrast, are used to precipitate proteins indiscriminately for the purpose of removing non-protein impurities, rather than for selective enrichment.

Ammonium Sulfate Precipitation

In the 1880s Franz Hofmeister analyzed how adding salts to solutions of hen egg white lysozyme caused the protein to precipitate [15]. Anions and cations were ranked in decreasing order of their propensity to precipitate protein. The studies were extended by others, but the ranking is still known as the Hofmeister series.

Anions: $PO_4^{3-} > SO_4^{2-} > CH_3COO^- > Cl^- > Br^- > NO_3^- > ClO_4^- > I- > SCN^-$

Cations $NH_4^+ > Rb^+ > K^+ > Na^+ > Cs^+ > Li^+ > Mg^+ > Ca^{2+} > Ba^{2+}$

From that series, one salt in particular, ammonium sulfate ($(NH_4)_2SO_4$ or AS), emerged as a particularly useful tool for selective protein precipitation, a frequent first step in protein purification of native proteins. Each protein is soluble because of how it orders water molecules immediately around it - its hydration shell. The protein-H_2O interaction is mediated most strongly by charged (i.e. ionizable) amino acids of the protein through ion-dipole interactions. Selective precipitation methods work by interfering with a protein's hydration shell. One way to alter the protein-water interaction is to dramatically alter the ionic strength of a solution. Briefly, ions in solution order the water around them just as proteins do. In other words, ions such as NH_4^+ and SO_4^{-2} interact with water more strongly than do neighboring water molecules. At sufficiently high ionic strength, ions can compete with proteins to order the water, effectively stripping the hydration shell. A compromised hydration shell favors both non-specific ionic and hydrophobic protein interactions, which lead to formation of high molecular weight aggregates. The aggregates, in turn, can be separated from soluble proteins by centrifugation. AS is uniquely suited to protein precipitation because of its high solubility and ionic strength. The amount of AS added to a solution is reported as the fraction present in a saturated solution, which, at a sensible working temperature near 0 °C, is 70.6 g/100 g water. For instance, a protocol may recommend adding AS slowly to 10 % saturation, stirring the solution for a few minutes, then centrifuging it. In this way, one may fractionate the proteome according to protein insolubility by iterative addition of AS and centrifugation. Most proteins precipitate between 5 and 70 % saturation.

Isoelectric Point Precipitation

A protein's isoelectric point is defined as the pH at which the protein has no net charge, and is determined by the relative number of basic and acidic amino acids. A protein particularly rich in glutamic and aspartic acid will have a low isoelectric point (e.g. Acidic leucine-rich nuclear phosphoprotein 32, pI 3.98), whereas proteins rich in lysine and arginine will have high isoelectric points (e.g. Histone H1, pI 10.8). In practical terms, this means that the pH of the solution can be altered to render groups of proteins neutral. When a protein is neutral, the ion-water interactions that form the hydration shell are minimized. At the isoelectric point, a protein is at its most hydrophobic and therefore thermodynamically driven to interact with other proteins, aggregate and precipitate. As in AS precipitation, pH may be altered, and solutions centrifuged, iteratively, to separate proteins of differing isoelectric points. Bulk isoelectric precipitation of proteins from solution is rarely used in proteomics today, however capitalizing on isoelectric point of proteins remains relevant, as it is foundational to the practice of isoelectric focusing in 2-dimensional gel electrophoresis discussed later.

Differential Protein Extraction

Protein extraction methods, like precipitation methods, work by manipulating how proteins partition between soluble and insoluble phases. Common extractions like detergent or acid extraction work by selectively increasing protein solubility. In this section, we also include methanol/chloroform extraction as a liquid extraction method that partitions the major classes of biological molecules on the basis of their solubility in solvents of differing polarity.

Detergent Extraction

Hydrophobic membrane proteins are largely surrounded by lipids to prevent exposure to the local aqueous environment that would otherwise render the proteins insoluble. The major challenge of successfully fractionating these proteins is maintaining their solubility during extraction. Chaotropic agents such as detergents are capable of fractionating and extracting membrane proteins while maintaining their solubility. Detergents are amphipathic molecules that contain a hydrophilic polar head at the end of a long hydrophobic tail. In aqueous solutions, the hydrophobic tails of several detergent molecules interact with each other while their polar heads form hydrogen bonds with water molecules to form organized structures known as micelles. The lowest concentration at which individual detergent molecules aggregate to form micelles defines the detergent's critical micelle concentration (CMC). A detergent's CMC is also the value above which membrane proteins can be extracted. Furthermore, the number of detergent monomers that aggregate to form micelles also affects the detergent's efficiency in extracting proteins. As each detergent varies in its CMC value and aggregation number, the efficiency in which each detergent is able to extract soluble proteins also varies. Therefore, proteins can be extracted and separated based on their solubility in detergents of increasing solubilization efficiency [16].

Figure 4.3 illustrates the fractionation that occurs upon sequential extractions using detergents of increasing extraction efficiency. Starting with a weaker detergent like digitonin, the cells are washed to extract proteins that are most soluble. Following centrifugation, the supernatant contains a highly enriched fraction of cytoplasmic proteins. The pellet is resuspended in a stronger detergent solution to extract proteins that are less soluble. The process continues until the final extraction with sodium dodecyl sulfate (SDS) resulting in the solubilization of the remaining proteins of the sample. Differential detergent centrifugation typically yields four distinct subproteomes enriched in (1) cytosolic, (2) membrane, (3) nuclear, and (4) cytoskeletal proteins [13]. Unfortunately, due to the wide variety of proteins and their interactions, this approach results in only a modest specificity of the subcellular compartments extracted. One exception, however, is the high specificity in the extraction of proteins of lipid rafts [16].

pH-Based Extraction

The IN sequence protein extraction method is a sequential extraction method based on the solubility of proteins at various pH levels. Proteins in a sample are first extracted at neutral pH (7.4) in which only the most soluble proteins remain in solution resulting in a cytoplasmic-enriched extract. The insoluble portion of the first extraction is then homogenized in an acidic solution in which more proteins, particularly myofilament proteins are solubilized and extracted. The remaining insoluble pellet contains mostly membrane proteins. This method is useful in that it can serve to deplete cardiac samples of highly abundant myofilament proteins to investigate the lower abundant cytosolic proteins in samples. IN sequence extraction also provides extracts compatible for further separation by 2D-PAGE or high-performance liquid chromatography (HPLC) [17, 18]. Acid extraction of chromatin fractions provides great enrichment of histone proteins as it isolates proteins tightly bound to DNA [19–22].

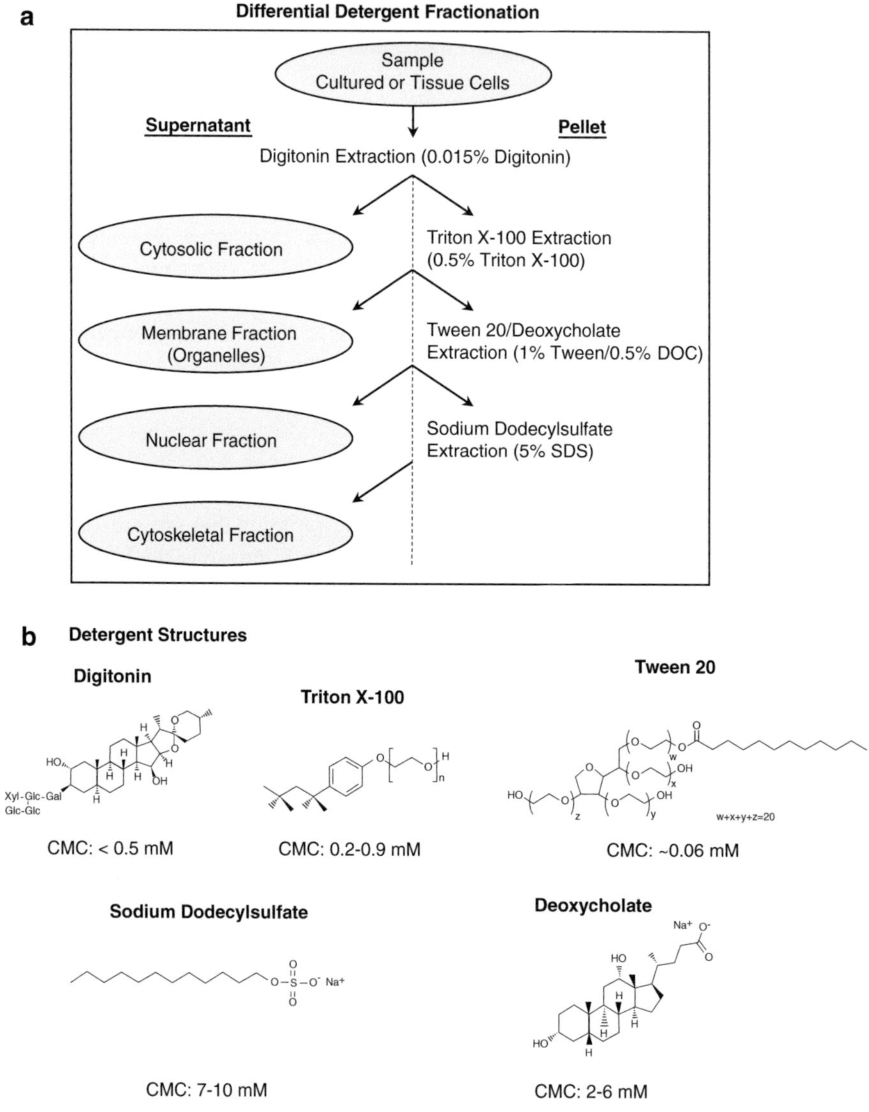

Fig. 4.3 Differential detergent extraction. (**a**) A non-ultracentrifugation-based workflow diagram for organelle enrichment via differential detergent extraction. Sequential extraction using detergents of increasing extraction efficacy yields distinct subproteomes. (**b**) The chemical structures of several commonly used detergents for differential detergent extraction with their respective critical micelle concentrations (CMCs). Solubilization efficiency generally increases with detergents of increasing CMC

Methanol/Chloroform Extraction

Proteomic cognoscenti would argue that the methanol/chloroform extraction method has little place in a discussion of protein fractionation. The method *is* broadly used in proteomics, however, as a means of near-quantitative protein precipitation and contaminant removal [23], though proteins are not resolved from each other. The method entails adding 4 volumes of methanol to 1 volume of aqueous biological sample and mixing thoroughly. Chloroform is subsequently added and also mixed thoroughly. Finally, 3 volumes of high purity water are added, mixed thoroughly and centrifuged. Since chloroform is immiscible in methanol/water, after centrifugation, the tube contains two solvent phases. The upper phase is the polar phase which contains methanol/water. The bottom, denser, phase consists of chloroform. The fractionation principle in liquid extraction is that solutes are extracted into solvents with like properties. The pertinent property, in this case, is polarity or hydrophilicity. Polar hydrophilic substances including metabolites and salts are extracted into the upper polar solvent phase. Hydrophobic, low-polarity species such as long chain aliphatic lipids are extracted into the bottom low polarity chloroform phase. Proteins, however, contain amino acids that are both hydrophilic (ionic or polar) and hydrophobic (apolar) and therefore migrate to the interface of the two solvent phases. The upper layer can be collected at this point. Following removal of the upper layer, lipids and proteins can be retrieved by adding 3 volumes of methanol to the residual sample, vortexing and centrifuging. Since chloroform is miscible in methanol, centrifugation reveals a single liquid phase and a pellet. The liquid phase containing methanol chloroform and lipids can be removed. The protein pellet can then be resolubilized for downstream proteomic applications.

Liquid Chromatographic Methods

The essence of liquid chromatography is rooted in extraction. However, rather than partitioning proteins in the liquid phase as in section "Methanol/Chloroform Extraction", proteins are extracted by their interaction with a solid phase in a process called solid-phase extraction. Chromatography is a widely used laboratory implementation of solid phase extraction. The distinction is that chromatography allows for movement of liquid solvent through a stationary, yet porous solid phase, or matrix. This affords continuous partitioning of the solutions that flow through the matrix. In practical terms, the matrix is packed into a long column and equilibrated by pumping buffer through it before protein solutions are applied. How proteins partition themselves between the mobile solvent and stationary matrix, is determined by the physical/chemical properties of each protein, the matrix and the composition of solvent buffer used. In this way, the proteins can be separated from each other. We begin by briefly introducing the physical and chemical properties of commonly used matrices and the principles underlying protein separation in each case.

Ion Exchange Chromatography (IEC)

Proteins contain both positive and negative charges at near-physiological pH, both of which can be used to advantage in ion exchange chromatography [24, 25]. Matrices can be derivitized to bear either negatively or positively charged ligands. When a positively charged matrix is used to adsorb proteins through their negatively charged amino acids, this is called anion-exchange chromatography. Adsorption through positive amino acids to a negatively charged matrix is called cation-exchange chromatography. Both anion- and cation- exchange matrices come in "weak" and "strong" forms. A weak anion exchange resin is one whose ligand is partially ionized at the working pH, as in the case of diethylaminoethyl (DEAE)-Sepharose. A strong anion exchanger is fully ionized at the working pH, such as quaternary ammonium (Q)-Sepharose. Similarly, carboxymethyl (CM) and sulfopropyl (SP) ligands are examples of weak and strong cation exchangers, respectively.

A typical DEAE anion exchange chromatography experiment is depicted schematically in Fig. 4.4. Assume the column is equilibrated in low salt buffer around pH 7.5, prior to application of protein sample soluble in the same buffer. As a general rule proteins whose pI <7.5 would bear a degree of negative charge and will adsorb the positively charged DEAE matrix via ionic interactions. Proteins whose pI >7.5 would bear differing degrees of positive charge and would be less likely to adsorb DEAE. These non-adsorbed proteins can be washed away with application of more equilibration buffer. Elution proceeds when a linear salt gradient is applied. Proteins with little negative charge are easily displaced by increasing Cl^- in the salt gradient and elute at a lower retention volume or retention time. Proteins with greater negative charge will adsorb more strongly to the column and desorption requires higher salt concentrations. These proteins elute at higher retention volumes or retension times.

Hydrophobic Interaction Chromatography (HIC)

Given the right matrix and buffer conditions, proteins can also be coaxed to adsorb to the stationary phase through its hydrophobic residues [26, 27]. Matrices are derivatized with apolar cyclic compounds or short alkyl chains to yield hydrophobic resins, as in the case of Phenyl-Sepharose and Octyl-Sepharose respectively. Proteins are applied to the column equilibrated in high salt (NaCl or KCl) buffers. This is important since hydrophobic interactions, in contrast to ionic interactions, are favored by high salt. Unbound protein is washed through with high salt buffer prior to elution of bound proteins with either a linear or stepwise gradient with low salt buffer (See Fig. 4.4). Hydrophobic interactions also differ from ionic interactions in that they are also temperature dependent. Hydrophobic interactions are stronger at typical lab room temperatures (20–25 °C) than at 4 °C. A classic example is the room temperature isolation of EF-Hand-containing Ca^{2+}-binding proteins including calmodulin [28] and caltropin [29], whose retention on Phenyl-Sepharose is dramatically reduced at 4 °C [28]. Finally, because binding is favored by high salt, HIC is well suited to follow IEC or AS precipitation in a multi-step protein purification strategy.

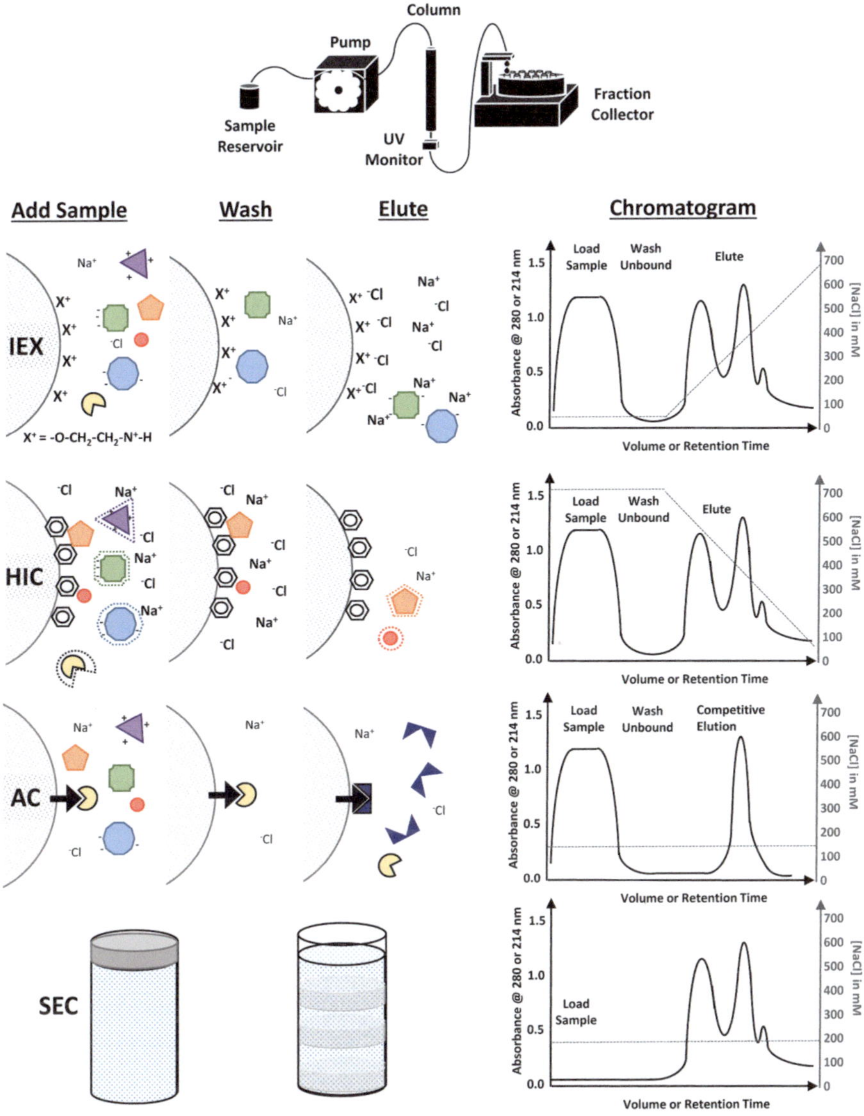

Fig. 4.4 Chromatographic methods. (**a**) A simplified schematic of a typical low-pressure chromatographic apparatus is shown in silhouette. (**b**) The principles of four types of chromatography are depicted. Proteins are introduced to the column in column equilibration buffer and subsequently washed with more buffer; non-adsorbed particles are removed. Elution conditions vary with the type of chromatography. In IEX, adsorbed proteins are often eluted with a linear gradient of increasing salt (NaCl, or KCl) concentration. In HIC the equilibration buffer is typically a high salt buffer and elution proceeds with a linear gradient of decreasing salt. For affinity chromatography, competitive elution depends on the concentration of a competing ligand that will either bind the matrix or the adsorbed ligand. In SEC, where proteins diffuse within, but do not adsorb to, the matrix, elution is a function of buffer volume applied to the column. In chromatography, protein profiles are typically monitored spectrophotometrically, in real-time, by absorbance at 280 nm

Size-Exclusion Chromatography

Size exclusion chromatography (SEC) is singular among chromatography techniques in the sense that the separation of proteins does not result from adsorption to the matrix. Instead, the matrix consists of beads of porous gel; protein separation results from the diffusion of protein into and out of the pores in the gel [30, 31]. SEC, also known as gel filtration or gel permeation chromatography, is one of few laboratory methods in which particle diffusion can be harnessed to advantage. In practice, a small volume of protein solution is applied to a long (1 m) equilibrated column, buffer flow is immediately resumed and proteins proceed through the resin. The principle behind protein separation is that larger proteins cannot easily penetrate the pores and are partially or completely excluded from the gel beads. As a result, they remain mobile and elute from the column in a relatively small volume. Smaller proteins are free to diffuse into the pores, which retards their progress through the column, and more buffer is required to elute them. Matrices can be manufactured to have different pore sizes, to maximize separation over different size ranges.

It is noteworthy that SEC does not fractionate the basis of molecular weight *per se*, but rather its effective size in solution, called the Stokes' radius. This is the same property that factors into centrifugal separation, which, as discussed previously, depends greatly on the shape of the molecule. In SEC, the Stokes' radius has opposite effect on the speed of migration. In contrast to centrifugation, where solvent friction is significant, in SEC, it is minimal, as proteins tumble with the mobile buffer phase. What matters most, is the extent to which a protein can diffuse into the gel pores. Therefore, molecules with compact structures penetrate the gel beads more easily than oblong structures that cut a wide swath as they flow.

Affinity Chromatography

Affinity chromatography (AC) is one of the most powerful tools for protein purification. Chosen properly, it can often reduce the isolation of a particular protein to a single step. Rather than partitioning proteins by general physical/chemical properties, adsorption is accomplished by harnessing a protein's biological activity or binding proclivity. To do this, the matrix is derivitized or processed to incorporate ligands that bind a particular protein or group of proteins of interest. Examples include enzyme inhibitors or substrate analogs, antibodies, hormones, and specific DNA sequences, to name only a few [32, 33]. In a typical experiment, a protein solution is applied to the conjugated matrix and buffer is used to wash away unbound protein. Buffer composition will depend on whether the goal of the experiment is maximal protein purity or preserving any potential interacting proteins. Elution of the targeted proteins is elicited by interfering with the protein-ligand interaction in one of a few ways. A common method is low pH elution, often with the amino acid, glycine, at pH 2.5. If the protein is eluted into fractions containing a neutralizing buffer such as ammonium bicarbonate, biological activity can be preserved. If activity is dispensable, strong detergents, like SDS, are used to elute proteins by denaturing them.

From a proteomic perspective, however, both low-pH and SDS elution may be suboptimal, as they also elute proteins that adsorb non-specifically to the matrix rather than the ligand. Contamination from nonspecific adsorption is not trivial, and represents a confounding factor in affinity chromatography, particularly in microscale experiments. Therefore, whenever possible, it is preferable to perform a specific elution with a competitive inhibitor (CI) of the protein-ligand interaction. Remember that the bound protein is actually in equilibrium between the stationary and mobile phases, determined by its association and dissociation rate constants, k_{on} and k_{off} respectively. Competitive elution works by shifting the equilibrium. Specifically, in the presence of a competitive inhibitor, when a protein briefly dissociates from the matrix ligand, the inhibitor will bind and prevent its reassociation. Whether the inhibitor is chosen to bind the protein or the ligand is immaterial, as long as the protein-ligand interaction is interrupted. The result is displacement of the protein of interest from the column. The two factors that contribute most to the success of competitive elution strategies are the concentration of CI used, and its affinity for either the ligand or protein, which ideally should be greater than that of the protein-ligand interaction.

While drugs, inhibitors and analogs are often among the most specific ligands, their conjugation to matrices can be challenging. To facilitate the process, manufacturers have developed matrices of varying chemistry to permit conjugation to disparate functional groups on the amino acids of proteins (e.g. N-hydroxysuccinamide (NHS)-Sepharose, cyanogen bromide-activated Sepharose), or small organic compounds (e.g. Epoxide-activated (EAH Sepharose)). Developed in the days before protein engineering, these resins remain relevant today, particularly for the purification of native proteins. Yet there is little question that in research labs today, affinity methods are most commonly used in the context of immunoaffinity purification (IP) and purification of proteins bearing engineered epitope tags.

Affinity purification is widely used for large-scale purification of recombinant fusion proteins and epitope- tagged proteins from expression systems such as *E. coli* or baculovirus. Proteins may be engineered to include protein or peptide sequences whose activity or binding properties are easily harnessed. Prominent historical examples include engineering of glutathione S-transferase (GST) fusion proteins and Histidine (His)-tagged protein. GST-bearing proteins bind to glutathione agarose, can be washed extensively to remove contaminants, and eluted competitively with reduced glutathione. His-tagged proteins bind to columns charged with nickel. The general principle is that of biological metal chelation [34]. Poly-His tracts were engineered to bind nickel preferentially over iron, magnesium and calcium. Upon sample binding, the column can be washed with high salt concentrations to minimize nonspecific binding prior to competitive elution with imidazole, a histidine analog.

The chromatography column, while still favored for large scale affinity purification, is often eschewed for microscale experiments in favor of batch incubation in a microfuge tube, as it requires less matrix and experiments can be easily parallelized. Briefly, the affinity resin is incubated with homogenate/lysate for a specified time often at 4°. The resin is then pelleted by low speed centrifugation. The supernatant, containing unbound proteins, is removed and the matrix is washed by dispersing it in fresh buffer. A second round of centrifugation pellets the washed matrix. Washing is repeated several times. Finally, the matrix is dispersed in a small volume of elution buffer and centrifuged anew. The supernatant contains the eluted protein. This step is usually repeated to maximize yield.

Microscale affinity methods are exemplified by immunoaffinity purification (IAP). Many protocols exist for the conjugation of antibodies to commercially-available matrices. They can be conjugated directly, via traditional NHS or CNBr chemistries, though these methods offer little control over the crosslinking site on the antibody and the orientation of the antibody relative to the mobile phase. Other procedures use matrices that have been pre-conjugated with high-affinity antibody-binding proteins such as Protein A or Protein G. Incubating antibody with Protein A/G-matrices results in stereospecific binding that orients the antibodies with their antigen-binding interfaces facing away from the matrix and toward the mobile phase. The antibody is then crosslinked to Protein A/G, with agents like disuccinimidyl suberate, to immobilize the antibody and limit antibody contamination in subsequent IAP experiments. Working with engineered proteins bearing tags such as FLAG, c-myc, or hemagglutinin, offers the advantage that conjugated antibody-matrices are commercially available, as are peptides that can be used for competitive elution. Finally, there are several commercially-available epitope tags that do not require the use of antibodies. Examples include streptavidin-binding peptide, calmodulin-binding peptide, and maltose-binding protein. All are amenable to competitive elution.

Finally, the advent of tandem affinity protein purification (TAP) [35, 36] was a watershed advance for the identification of protein-protein interactions. The method now has many variants, but, at its root, it entails engineering two high-affinity tags, in series, at one of the termini of a protein of interest. Sequential affinity purification steps result in highly-purified tagged protein along with stably-bound protein partners. Over the years, considerable effort has been invested to optimize the design of the affinity tags to maximize yield of novel specific interactions [37, 38].

Factors that Affect Chromatographic Separation

The term that describes the extent to which proteins can separated from each other is called *resolution*. When proteins have been completely separated, we say that they have been resolved. The resolution of a column purification procedure depends on a couple of factors – retention, efficiency and selectivity (Fig. 4.5).

Column efficiency refers its capacity to produced tight symmetrical peaks. This is determined by several factors, including matrix bead size, uniformity of pore size, column packing, column length and flow rate. Of these parameters, the ones that have the greatest impact on efficiency are bead size and uniformity. Reducing bead size increases the functional surface area of the matrix which impacts protein transfer from mobile to immobile phases. However, it does so at the expense of increasing resistance to buffer flow, incurring the need for high pressure pumps. Fast protein liquid chromatography

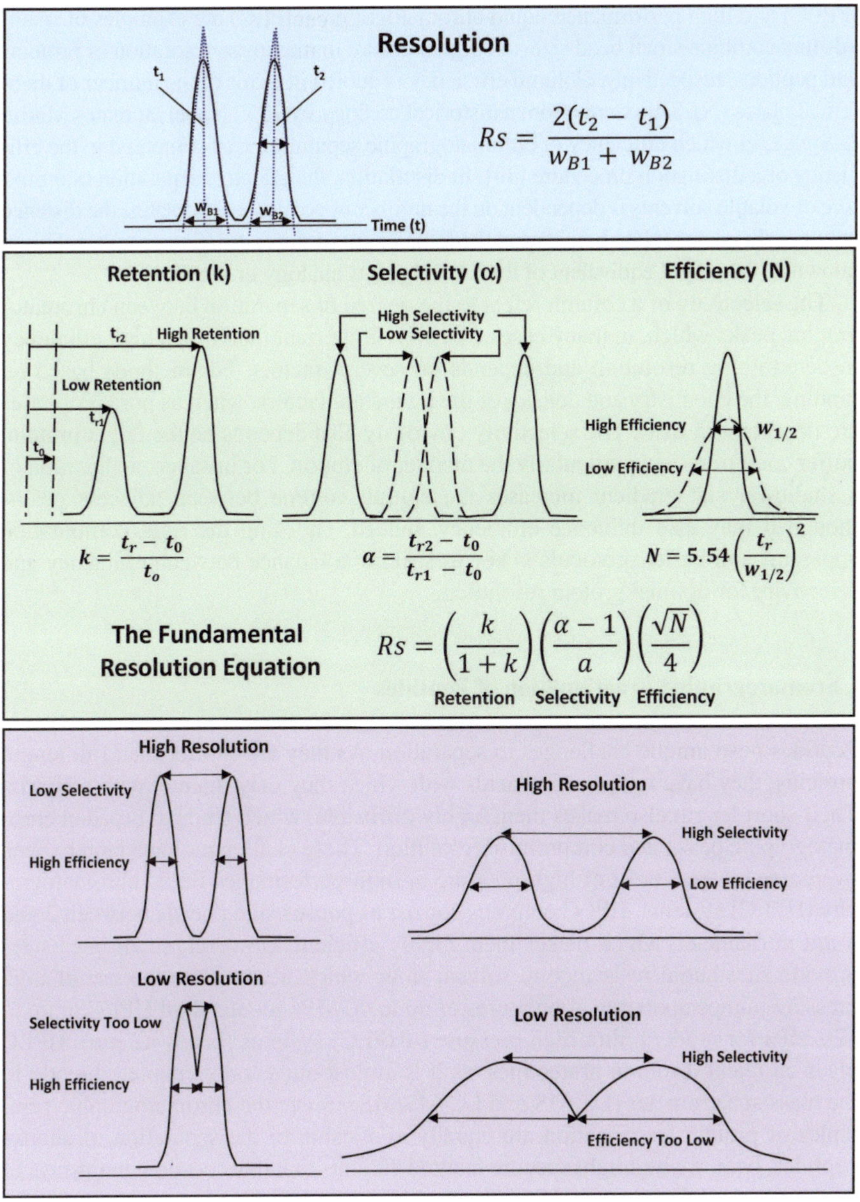

Fig. 4.5 Chromatographic resolution. (*Panel 1*) Resolution describes the extent to which particle populations or chromatographic peaks (monitored spectrophotometrically) have been separated. (*Panel 2*) Chromatographic resolution is dependent on three factors: retention, selectivity and efficiency. Retention is a function of a particle's propensity to adsorb to the column. Selectivity describes the distance between adjacent chromatographic peaks. Efficiency refers to the capacity of the column to generate narrow chromatographic peaks. (*Panel 3*) High resolution can be achieved with modest selectivity if column efficiency is high. Alternatively, resolution of proteins can also be achieved with only modest column efficiency if selectivity is sufficiently high

(FPLC) and high performance liquid chromatography (HPLC) are examples of methods that combine small bead size and high pressure to maximize resolution of proteins and peptides, respectively. Column efficiency is quoted in terms of the number of theoretical plates (N). This stems from a historical analogy by 1952 Nobel laureates Martin & Synge, in which efficiency of chromatographic separation was compared to the efficiency of a distillation procedure [39]. In distillation, the effective separation of a mixture of volatile solvents is dependent on the number of condensers or plates, the distance between the plates (plate height) and the total length of the distillation chamber. This is known as the height equivalent of theoretical plates analogy or HETP.

The selectivity of a column refers to the degree of separation between chromatographic peaks which, in many cases, is a more important factor than high efficiency in determining resolution and depends on several factors. For methods based on binding, the chemistry and density of the ligand are factors, whereas pore properties are important in SEC. The selectivity obviously also depends on the target protein, buffer conditions and particularly the manner of elution. For instance, application of a shallower salt gradient increases the elution volume between adjacent peaks, though it may also influence efficiency. Indeed, choosing the right combination apparatus and buffer protocols is key to striking a balance between efficiency and selectivity for optimal protein resolution.

Chromatographic Fractionation of Peptides

Peptides pose unique challenges to separation. As they are shorter than full length proteins, they have fewer amino acids with which they may interact with a matrix. Their short length also makes them highly diffusible, which tends to broaden chromatographic peaks and compromise resolution. These challenges have largely been overcome by the advent of high-pressure or high-performance liquid chromatography (HPLC) systems. HPLC columns consist of porous silica beads between 2 and 5 μm in diameter, which makes them highly efficient. However, small bead sizes provide substantial resistance to solvent flow, which necessitates the use of high pressure pumps operating at pressures of up to 60 MPa for standard HPLC or up to 120 MPa for modern ultra-high pressure (uHPLC) systems (beads <2 μm). HPLC plays an integral role in proteomics, as it is almost universally coupled directly to the mass spectrometer (LC-MS and LC-MS/MS). Since the chromatographic principles of protein fractionation are equally applicable to the separation of shorter peptides, here, we highlight specific method variants that have become mainstays of proteomic workflows.

Low-pH Reversed-Phase HPLC

Reversed phase HPLC (RP-HPLC) is the workhorse chromatographic method used in proteomics, owing to its high efficiency and mass-spectrometry compatible solvent system, usually acidified water and acetonitrile. The term, reversed-phase, refers to the fact that this method employs a stationary phase (matrix) that is substantially less polar than the solvent running through it. The matrix consists of silica covered in alkyl chains, of which the most commonly-used consists of 18 carbons (C18). RP-HPLC is most commonly performed at low pH, using solvent acidified with strong acids such as trifluoroacetic acid (TFA) or formic acid (FA). These acids enhances the hydrophobicity of peptides and, therefore, their adsorption to alkyl matrices, in two ways. First, at pH 2.5–2.8, the carboxylic acid sidechains of aspartic acid and glutamic acid residues are predominantly protonated. Secondly, the conjugate bases (e.g. trifluoroacetate or formate) serve as hydrophobic anions that bind, and effectively neutralize, the remaining positively charged residues: arginine, lysine and histidine [40]. For this reason TFA and FA are called ion pairing agents. FA (0.1 % vol/vol) is favored when RP-HPLC is coupled to mass spectrometry, as TFA tends to suppress electrospray-ionization of peptides [41]. So peptides bind RP columns on the basis of hydrophobicity, enhanced by ion pairing agents. However, in contrast to HIC methods that manipulate salt or detergent concentrations to elute proteins, peptides are eluted from RP-HPLC columns by decreasing the polarity of the solvent, commonly with a linear gradient of increasing acetonitrile.

Two-Dimensional Liquid Chromatography Fractionation Strategies

Two-dimensional liquid chromatography (2D-LC) is one of the most broadly adopted workflows for global-scale peptide fractionation in proteomics. It entails fractionation by successive rounds of HPLC, where the eluted fractions of the first column are subsequently subjected to RP-HPLC-MS/MS [42]. 2D-LC offers the advantage that fractionation in the first dimension reduces sample complexity, which favors greater resolution in the second. Low sample complexity and high resolution reduce the kind of peptide co-isolation in the mass spectrometer that compromises identification of low abundance peptides. For 2D-LC strategies to be effective, the elution behavior (i.e. retention times), of peptides from the two columns should be poorly correlated [43]. Chromatographic methods whose elution is completely uncorrelated are said to be orthogonal. Maximizing orthogonality (i.e. minimizing elution correlation) of peptide separation is the holy grail of 2D-LC strategies. Weapons in the arsenal include the use of different types of chromatography, conducting chromatography at different pH, or pooling eluted fractions in creative ways.

Increasing Orthogonality Through Different Types of Chromatography

The work of Wolters et al. marked a seminal advance in 2D-LC as applied to proteomics. The authors called their approach Multi-Dimensional Protein Identification technology or MUDPIT [42]. The intriguing aspect of their study is that it employed tandem online configuration of SCX beads and reversed phase C18 beads. Peptides were loaded on to SCX resin at low pH and eluted on to the RP column with a stepwise salt gradient of ammonium acetate which was washed through the system before application of an acetonitrile gradient to elute peptides from RP into the mass spectrometer. This was repeated in multiple cycles following successive ammonium acetate steps. The online configuration was elegant, but more importantly, the SCX fractionation was orthogonal, leading to identification of many more peptides than by RP-HPLC alone. It has, since, been widely adopted for 2DLC. Many prefer to perform SCX offline (uncoupled from MS) as it permits greater flexibility to optimize SCX resolution (e.g. linear gradients of NaCl or KCl) without worrying about the MS-compatibility of the buffer components, since peptides may be easily desalted prior to RP-HPLC-MS/MS.

The first **hydrophilic interaction liquid chromatography** (**HILIC**) methods were published in the 1970s but its use, particularly in proteomic and metabolomic studies has skyrocketed over the last 10 years. It is a variant of aqueous normal phase (ANP) chromatography, and thus a conceptual mirror image of RP chromatography. Recall that in RP chromatography, apolar alkyl groups on the matrix partition peptides on the basis of hydrophobicity. Gradients of increasing apolar solvent are used to elute progressively more hydrophobic peptides. In contrast, ANP and HILIC use polar matrices that partition more polar molecules. ANP is commonly performed with underivitized silica columns and is used to separate both hydrophilic and hydrophobic substances. HILIC matrices such as polysulfoethyl A and TSKgel Amide 80 are optimal only for polar molecules. In HILIC, peptides are loaded in apolar solvent such as acetonitrile and eluted with a linear gradient of increasing aqueous solvent. Separation by hydrophilicity has proven highly effective for many omic studies. It is particularly well-suited to metabolomic studies since many small organic metabolites are polar. In proteomic 2D-LC workflows, the method is highly orthogonal to RP-HPLC, both in principle and in practice. In fact, HILIC conducted at pH 4.5 exhibits greater orthogonality than does SCX conducted at pH 3.25 (see [44, 45]). HILIC is also effective for capturing and resolving peptides bearing hydrophilic PTMs, such as phosphorylation and glycosylation. Such peptides generally do not bind well to RP columns and often elute in the void volume or very early in acetonitrile gradients.

Increasing Orthogonality by Changing Buffer pH

Basic reversed-phase HPLC (bRP-HPLC), as its name would suggest, is simply a variant of RP-HPLC conducted at a higher pH [46]. The column is equilibrated in 10 mM triethylammonium bicarbonate (TEAB) pH 8.5, and a typical elution gradient might range between 5 and 45 % acetonitrile in TEAB. This method has been made possible with the development of a new generation of silica matrices that are stable when pH >8 [47]. The method is orthogonal to low-pH RP-HPLC in the second dimension of 2D-LC because changing the pH alters the charge on the ionizable amino acids and, therefore, the overall polarity of the peptides. In turn, this affects the strength with which the peptides adsorb to the C18 matrix, the amount of apolar solvent required for elution, and ultimately the retention time of the peptides.

Increasing Orthogonality Through Creative Pooling

One of the challenges of a 2D-LC strategy is that if many fractions are collected over the course of elution in the first dimension, this increases the number of RP-HPLC runs to be performed in the second dimension. Therefore, any 2D LC strategy is a trade-off between peptide separation in the first dimension and the increased time and cost of performing the subsequent LC-MS/MS. It is therefore often desirable to reduce the number of fractions while preserving orthogonality. Pooling adjacent fractions from the first dimension makes little sense as one could simply have collected fewer fractions of greater volume. Moreover, adjacent fractions contain peptides with similar properties that may not be resolved in the 2nd dimension. Concatenated pooling is simply the act of pooling fractions at intervals rather than sequentially (e.g. pooling elution fractions 1, 16, 31 then 2, 17, 32 then 3, 18, 33 etc.). This reduces the number of fractions while ensuring the peptide composition is chemically diverse, since the original fractions were collected at different retention times. This chemical diversity yields better peptide resolution upon subsequent RP-HPLC than one would obtain from sequential collection of the same number of fractions [48, 49].

Rate Zonal Centrifugation Revisited

As we have seen, centrifugation may be the workhorse for organellar isolation, but it can be equally valuable for protein fractionation. Remember that the primary factor determining particle fractionation in rate-zonal centrifugation is molecular size. Proteins in our cells rarely work in isolation. Often they operate as part of elaborate structures – scaffolds and machines – that carry out the business of the cell in processes ranging from replication, transcription and translation to protein quality control and energy production, and specifically in the case of the heart, muscle contraction. These protein macrocomplexes are eminently suited to resolution by rate-zonal centrifugation using a discontinuous sucrose gradient. Typical sucrose gradients range from 5 to 40 % w/v. A classic example of the utility of this method is the fractionation of mitochondrial respiratory complexes [50]. Following centrifugation, the proteins in each sucrose layer can be captured by piercing a hole in the bottom of the centrifuge tube (polyallomer) and collecting drops from each layer in separate microfuge tubes.

Electrophoresis

Electrophoresis simply refers to the movement of particles in an electric field. How fast each particle moves is determined by the balance of forces acting on it. The electrical driving force F_{el} is

$$F_{el} = qE \tag{4.6}$$

where

q is the charge on the particle
E is the electric field strength

As we discussed in the case of centrifugation, when a particle moves through a solution, there is a counteracting frictional force (F_f), or drag, that is a function of its size (r), shape, viscosity of the medium (η) and the speed at which the particle is moving (v). This is known as Stokes' Law and is given by

$$F_f = 6\pi\eta rv \tag{4.7}$$

In a constant electric field, the velocity of a given particle is constant, therefore $F_{el} = F_f$, so

$$qE = 6\pi\eta rv \tag{4.8}$$

Then the velocity of the particle is

$$v = \frac{qE}{6\pi\eta r} \tag{4.9}$$

Therefore, in any given electric field, the intrinsic electrophoretic mobility of the particle, μ_{el} is

$$\mu_{el} = \frac{v}{E} = \frac{q}{6\pi\eta r} \tag{4.10}$$

Electrophoretic mobility, μ_{el}, is to electrophoresis what the sedimentation coefficient is to centrifugation. Simply put, the intrinsic mobility of a particle in an electric field is determined by the ratio of charge to Stokes radius (size & shape). For comparable masses, the particle with greater net charge will move faster than lesser charged ones. Larger molecules will migrate more slowly than small ones. Shape also matters here, as it does in centrifugation and SEC. Oblong molecules will migrate slower than spherical particles of the same mass owing to increased drag.

Gel Electrophoresis

Arne Tiselius is considered the father of particle electrophoresis. It was conducted in the liquid phase in a U-shaped tube with electrodes immersed at both ends. Application of an electric field caused charged particles in the liquid to migrate to either electrode according to the principles summarized in section on "Electrophoresis". Tiselius was awarded a Nobel Prize in 1948 for his work on the electrophoresis of colloids (e.g. [51]). Yet despite this major advance in protein separation, liquid electrophoresis was fraught by a host of challenges, according to fellow Nobel laureate, Oliver Smithies [52, 53]. First, despite its utility, a drawback of liquid electrophoresis was, as noted for centrifugation and chromatography, that resolution of migrating particles is compromised by the zone-broadening effects of diffusion. Fortunately, as with the other methods, this can be partially alleviated, by increasing the frictional force opposing the particle movement. One solution was to use a mixed phase separation in which proteins were electrophoresed through saturated filter paper [54]. This was subsequently refined by using sheets of starch granules [55]. However, the advent of starch gels, in 1955 by Oliver Smithies, is broadly recognized as a watershed for the electrophoretic resolution of proteins [56]. Gels are porous mesh-like stationary phases through which particles are driven by the electric field. They increase resolution, not only by dampening diffusion, but also act as a molecular sieve, allowing smaller particles to penetrate easily and migrate quickly, whereas larger molecules are retarded by the meshwork. Over the years, gels have been refined to increase the uniformity of the meshwork and their durability, as reflected by the use of agarose gels and polyacrylamide gels [57, 58].

Today, polyacrylamide gels are the most commonly used medium for electrophoretic movement of proteins. They are formed when a solution of acrylamide and N, N′-bis-methylene-acrylamide (bis-acrylamide) is mixed with compounds, ammonium persulfate (APS) and tetramethylethylenediamine (TEMED). TEMED catalyzes free-radical formation from APS. The free radicals react with acrylamide monomers to initiate polymerization. Linear acrylamide chains grow and are punctuated by incorporation of bis-acrylamide, which serves to connect or crosslink adjacent acrylamide polymers. Polymerization, therefore, creates a porous polyacrylamide meshwork whose density is proportional to the concentration of acrylamide. The value of using polyacrylamide lies in the ease with which the one can manipulate the density of the polymer network by controlling the extent of crosslinking. A dense meshwork effectively limits band broadening and is, therefore, highly suited to analysis of small proteins, say 3–20 kDa. However, it is only poorly penetrable by proteins with higher molecular weights (e.g. >100 kDa). A low density meshwork allows large proteins to penetrate the gel more easily, at the expense of modest diffusion control for small proteins, and even loss of resolution if they migrate with the solvent front.

Denaturing Gel Electrophoresis

Sodium Dodecyl Sulfate Polyacrylamide Gel Electrophoresis (SDS-PAGE)

Polyacrylamide gels were already well-established for protein electrophoresis in the early/mid-60s [57, 58] when researchers first used the denaturing detergent, sodium dodecyl sulfate (SDS) in the preparation of proteins for gel electrophoresis [59]. Its early use was motivated by a simple desire to dissociate tightly bound proteins such as viral coat proteins, to make them more amenable to electrophoresis. It was only later observed that inclusion of SDS also renders protein migration on polyacrylamide gels inversely proportional to the log of their molecular weight, not their charge/size ratios [60]. The reasons are two-fold. First, dodecyl sulfate binds to both hydrophobic and basic residues on proteins in a way that is remarkably uniform. In fact, SDS binds the average protein at a ratio of 1.4 g SDS/1 g protein. This confers a uniform negative charge to the protein, eliminating charge as a variable in electrophoretic mobility. Secondly, SDS destroys the secondary and tertiary structure of proteins, effectively stringing out denatured proteins as they migrate under voltage. This nullifies the impact of protein shape on migration, leaving size as the only determinant of mobility in SDS gels. Some proteins do migrate anomalously in SDS gels, however. Very basic proteins (high pI), tend to bind more SDS and their migration is retarded relative to other proteins of the same size. A classic example would be cardiac troponin I, whose molecular weight is 24 kDa but migrates closer to 30 kDa in a Tris-glycine buffered gel.

Electrophoresis entails applying an electric field to a set of protein samples. Clearly, factors like the ionic strength and pH of the buffer system are critical parameters. Prior to the early 1970s, buffer systems varied, but one that gained widespread acceptance was the gel and buffer system of Laemmli [61]. The Laemmli buffer system improved gel resolution of proteins in two ways. It was the first system to incorporate SDS in both the gels and running buffer, ensuring complete protein denaturation throughout electrophoresis. Second, it was one of few "discontinuous" buffer systems that addressed a major barrier to resolution – that when protein samples are loaded into their reservoirs and the voltage is applied, *not all proteins enter the gel at the same time*. The Laemmli system is depicted in Fig. 4.6. The system is discontinuous in that it uses two abutted gels, each with its own buffer. The upper stacking gel serves as a staging area where proteins enter from the sample reservoir and are focused into a narrow band prior to entering a separating gel. Briefly, samples are prepared in a buffer that contains Tris-HCl pH 6.8, 2 % w/v SDS, 10 % (w/v) glycerol, a reducing agent (e.g. dithiothreitol) to fully denature proteins and cleave covalent disulfide bonds. Samples are then loaded into the sample reservoir and the cathode chamber is filled with gel running buffer containing 25 mM Tris 192 mM glycine, 0.2 % SDS w/v. Applying voltage (e.g. 150 V) drives the proteins into a high porosity stacking gel (pH 6.8). As the proteins proceed through the stacking gel, they are focused into a single tight band as they are sandwiched between chloride ions (from the sample and stacking gel buffers) and glycine ions (from the running buffer). The sandwiching effect is exacerbated as the proteins reach the interface of the stacking and separating gels. The proteins, therefore, enter the separating gel more or less at the same time, and are subsequently resolved by molecular weight.

Today, SDS-polyacrylamide gels are available commercially that fit just about any protein separation application, from highly crosslinked gels to resolve smaller proteins to low acrylamide gels to accommodate high molecular weight proteins. Moreover, there are even gradient gels that allow optimal resolution of both high and low molecular weight proteins on the same gel. While Laemmli gels remain popular, the high pH of the separating gel buffer causes progressive hydrolysis of polyacrylamide and limits shelf life. Newer formulations, such as the now widely-used Bis-Tris gels are prepared at lower pH (6.4) to extend shelf life. The running buffers also contain different ions (e.g. morpholinopropanesulfonic acid, MOPS; Methylenesulfononic acid, MES) to replace glycine in the running buffer.

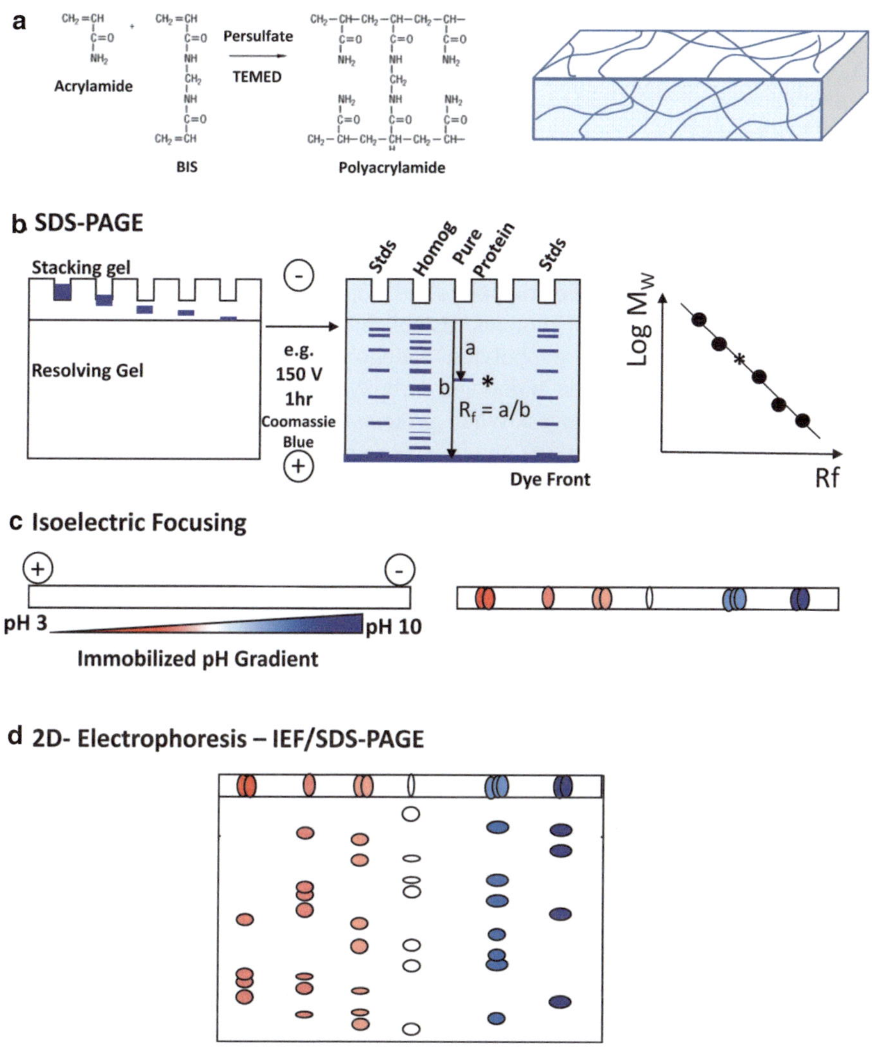

Fig. 4.6 Gel electrophoresis of proteins. (**a**) Polyacrylamide gels are preferred for most protein separations. Free radical-induced polymerization of acrylamide monomers in the presence of bis-acrylamide yields a porous a gel whose meshwork acts as a molecular sieve for the separation of proteins. (**b**) Stacking and resolving proteins by molecular weight. In a discontinuous gel & buffer system, under constant voltage, the stacking gel serves as a staging zone where proteins from the sample reservoir are compacted within the first few minutes of electrophoresis such that proteins in a sample may enter the resolving gel at the same time. The resolving gel separates proteins as they migrate toward the anode. The migration distance of proteins after a given time, relative to the solvent front (R_f), is inversely proportional to the log of their molecular weight. (**c**) Isoelectric focusing. A schematic depiction of an IPG strip and its orientation relative to the anode and cathode. Homogeneous protein samples are applied to the strip and under high voltage will migrate and become a focused band of protein where pI = pH. (**d**) Following IEF, and equilibration in appropriate buffer, the IPG strip may be placed atop an SDS gel which allows proteins to be resolved by mass. The IEF/SDS-PAGE 2D gel offers high resolving power

Isoelectric Focusing

As previously mentioned, proteins, absent reagents like SDS, migrate through liquid or gel in an electric field on the basis of mass and charge. Remember from section "Isoelectric Point Precipitation". that the charge on a protein can be changed by altering the pH of the solution. Moreover, there is a pH at which the protein has no net charge, called the isoelectric point. For this discussion, the salient point is that the isoelectric point is also the pH at which a protein is immobile in an electric field. If one could construct a gel bearing a pH gradient, proteins would migrate, upon electrophoresis, to their isoelectric points and stop. This is the principle underlying isoelectric focusing (IEF) of proteins. Establishing such pH gradients is possible through the use of carrier ampholytes. Ampholytes are small molecules (300–600 Da) that contain aliphatic amino and carboxylic acid groups and exist as zwitterions across broad pH ranges. When proteins and ampholytes are mixed and subjected to an electric field, the ampholytes segregate quickly according to their charge, and in so doing, establish a pH gradient. Proteins, less mobile owing to their size, then slowly migrate in the electric field until they reach their isoelectric point. The voltage and time required to optimally focus protein is somewhat dependent on the length of the IEF gel used, but >5000 V-h was used in work by O'Farrell [62].

Gel-based IEF is primarily an analytical tool. Historically, IEF using carrier ampholytes suffered from a few complications. The pH gradients tended to be unstable after about 3 h, owing to the poorly understood phenomenon called cathodic drift. Cathodic drift causes acidification of the gel near the anode, deterioration of the pH gradient in the neutral region and loss of bands at the basic end of the gel. Moreover, proteins could interact with the ampholytes, with the potential to interfere with the mobility of each. Finally, because ampholytes were actually complex mixtures of many different compounds, batch-to batch variability could be an issue. Many of these challenges were addressed by the development of gel strips in which the ampholytes are fixed, so-called immobilized pH gradients (IPG) [63–65]. IPG-based IEF is now the preferred method used in 2D gel electrophoresis discussed shortly hereafter.

IEF can also be performed in the liquid phase. Solution IEF has long been a valuable tool in preparative (large-scale) protein purification strategies as a complement to column chromatography. Agilent's Offgel fractionation system is a modern day variant of traditional liquid phase IEF.

Native Gel Electrophoresis (NGE)

Notwithstanding the power of denaturing gel electrophoresis, there are times when it is desirable to resolve proteins by the mass of their quaternary or macromolecular structure. Doing so requires that denaturing agents, such as SDS, be omitted from sample preparation, gels and running buffers. Yet apart from denaturation, SDS serves other functions in electrophoresis. It's a solubilizing agent for hydrophobic proteins and its protein-binding properties ensure particle velocity varies inversely with log MW. Maximizing solubility and ensuring separation by molecular weight while minimizing denaturation are the major challenges to successful native-gel electrophoresis. Early efforts to harness the power of the Laemmli buffer system by replacing SDS with milder non-denaturing detergents, such as Triton X-100, met with limited success, particularly when applied to hydrophobic membrane protein complexes.

The key innovation, developed by Schägger and von Jagow [66], was to address issues of solubility and charge separately. Sample preparation consists of using the detergent, lauryl maltoside and aminocaproic acid to maximize protein solubility of hydrophobic complexes. Variable protein charge was minimized by masking complexes with Coomassie blue G, a dye which, like SDS, binds proteins at a reasonably uniform ratio. There are two variants on the method, one that uses Coomassie Blue in the cathode buffer (Blue-Native PAGE) and one in which it is omitted (Native PAGE). Native gels commercially available today are still based on the methods of Schägger and von Jagow [66], with minor modifications.

2D Gel Electrophoresis– Orthogonal Resolution of Proteins

Protein samples, such as crude tissue homogenates and cell lysates, contain thousands of proteins of differing abundance. Combining gel methods sequentially affords the opportunity to resolve substantially more proteins than any single method in isolation. The concept of differential or orthogonal electrophoresis was first applied in the 50s [67, 68]. Markham resolved serum and urine proteins using a two dimensional paper electrophoresis technique in which the pH of the running buffers differed between the first and second dimensions [67]. Smithies and Poulik used a combination of paper electrophoresis and starch gel electrophoresis to resolve serum globins [68]. These studies demonstrated the power of using methods with uncorrelated protein migration profiles. In the case of the serum globins, the 2D method resolved 20 proteins instead of 5 by starch gel electrophoresis alone [52, 68, 69]. Today, orthogonal protein separation by 2D gel electrophoresis is one of the foundational technologies of proteomics. The most widely used gel combinations are IEF/SDS-PAGE and Native PAGE/SDS-PAGE.

2D Gels: IEF/SDS-PAGE

SDS-PAGE and denaturing IEF are two powerful electrophoresis techniques, each capable of resolving hundreds of proteins. Combining IEF and SDS-PAGE dates back to the work of O'Farrell [62]. Using ^{14}C-labeling he was able to detect 1100 protein spots across a broad abundance range. Seasoned proteomic practitioners will still appreciate the quality of those first 2D gel images, in light of some of the challenges associated with carrier ampholyte-based IEF discussed in section "Isoelectric Focusing". The 1980s saw the emergence of the immobilized pH gradient (IPG) [63–65, 70], which exhibited little cathodic drift while exhibiting higher resolution and greater sample loading capacity.

In practical terms, classic 2D gel electrophoresis is defined by separation, in the first dimension, by IEF to separate proteins according to their isoelectric points, followed by SDS-PAGE, in the second dimension, to resolve proteins by mass. The principal steps include sample preparation, hydration of the IPG-gel strip, performing IEF, equilibrating the IEF gel strip in SDS buffer before applying the strip atop an SDS polyacrylamide gel for the second dimension of electrophoresis. Obtaining high-quality well-resolved 2D gels remains an art form that requires painstaking commitment to best practices at each step of the process. Newcomers to 2D gels would be well-served by familiarizing themselves with the work of Görg, Righetti, Rabilloud and others [71, 72].

Expertise aside, a major factor that determines the resolving power of 2D gels is their size. Long IPG strips provide greater spatial resolution for IEF, which is a near-equilibrium separation method. SDS-PAGE, on the other hand, is a rate-based method in which protein separation increases with longer run times that are, likewise, afforded by larger gels. Today, large format 2D gels can resolve between 2000 and 5000 protein spots, depending on the method of protein visualization [73].

One of the great strengths of the classic 2D gel is that the combination of IEF and SDS PAGE is well-suited to resolve multiple post translationally modified variants of a given protein, called proteoforms. Many PTMs have relatively low mass and can be difficult to distinguish by SDS-PAGE. Yet if a modification either confers or ablates a charge on a protein, it can often be resolved by IEF. An example of the former would be phosphorylation of serines, threonines or tyrosines whereas lysine acetylation and methylation would exemplify the latter. Visualized on a 2D gel, the major proteoforms of a protein appear as closely-spaced horizontally-aligned spots, like beads on a string.

2D-gels have a long and dignified history in cardiovascular proteomics. They were, and to a degree still are, a workhorse platform for the study of the proteomics of heart failure, genetic cardiomyopathy, ischemia reperfusion injury and preconditioning. Seminal contributions from the labs of cardiovascular proteomics luminaries including Drs. Michael Dunn [74–76], and Jennifer Van Eyk [17, 77], to name only a few, all harnessed the resolving power of 2D gels. Today, the prominence of 2D gels has waned a bit as the rapid pace of mass spectrometer development has made chromatography-based (or gel-free) workflows a prime choice for global scale cardiac proteomics. Nevertheless, their resolving power, particularly with respect to proteoforms, will ensure their utility in the proteomic arsenal for years to come.

2D Gels: Native PAGE/SDS PAGE

Another 2D gel format that has proven quite useful combines native and denaturing gel electrophoresis. If the power of classic IEF/SDS-PAGE lies in its resolution, the main advantage of native/denaturing PAGE is that it preserves native macromolecular structure in the first dimension. The procedure entails performing native gels according to Schägger and von Jagow [66], excising the gel lane longitudinally and equilibrating it in SDS-containing buffer before setting it atop a denaturing gel. It has gained prominence for the study of high molecular weight protein complexes, particularly the four mitochondrial respiratory chain complexes and ATP synthase [66, 78].

Diagonal Gel Electrophoresis –Orthogonal Resolution of Selected Proteins

Over the years, clever 2D gel strategies have proven useful, even when the two gel dimensions differ in a specific variable, such that they are orthogonal for only a select group of proteins, In this case, most proteins migrate on the diagonal while proteins affected by the perturbed variable migrate off the diagonal. The classic application of diagonal gels is for mapping of intra and intermolecular disulfide bonds (e.g. [78–80]). In this case, SDS-PAGE is performed in both dimensions. The variable is the absence or presence of a reducing agent, typically DTT for Laemmli gels. Proteins are run, un-reduced, in the first dimension. The gel lane is then excised longitudinally and equilibrated in buffer containing DTT, to reduce disulfide bonds, prior to running the second dimension. The migration of non-disulfide bonded proteins is unaffected by the change in buffer conditions and therefore appear, upon staining, as a diagonal line of proteins in the 2D gel. However, proteins joined by disulfide bonds in the first dimension migrate at a molecular weight that reflects the sum of the masses of the tethered proteins. In the second dimension, with disulfide bonds cleaved, the proteins run at their individual (lower) molecular weights, off the diagonal axis.

3D- and 4D-Gel Electrophoresis: Ultimate Orthogonal Protein Separation

From the methods outlined in section "2D Gel Electrophoresis– Orthogonal Resolution of Proteins", we can begin to appreciate that gel electrophoresis of proteins comes in two principal classes: native and denaturing. Among the denaturing methods, this chapter has only scratched the surface. We have introduced SDS-PAGE and IEF, but this constitutes only a subset of useful methods (e.g. urea gels, urea-glycerol, Phos-Tag gels) and subtle variants that fulfill niches in protein biochemistry. We have also seen that combining gel methods, to confer further protein separation, is often as simple as excising a gel lane and equilibrating it with a buffer to be used in the subsequent dimension. It should not be surprising then, that three or four rounds of electrophoresis can be linked in succession. A prime example can be found again in the work of Schägger and colleagues, who performed two rounds of BN-PAGE, with and without the detergent dodecyl-maltoside, followed by SDS-Urea-PAGE and, in turn, by SDS-Tricine PAGE, to identify two novel proteins associated with the mitochondrial ATP synthase complex [81]. This serves to illustrate that the number of possible higher-order electrophoresis strategies is high and may be tailored tractably for specific protein projects.

Liquid Phase Electrophoresis Revisited: Everything Old Is New Again

As previously mentioned, early liquid phase electrophoresis suffered the drawback of protein diffusion. Though diffusion can be a problem for any separation technique, the problem is compounded in electrophoresis because applying high voltages to the aqueous solutions generates heat, which in turn, increases particle movement. The technical term for this phenomenon is convection. Advances in liquid electrophoresis have largely been driven by the need to limit convection.

Free-Flow Electrophoresis

Free-flow Electrophoresis (FFE) is a gel-free fractionation method that overcomes the resolution-compromising effects of convection by making particles literally go-with-the-flow of the running buffer. Specifically, in FFE, laminar buffer flow runs perpendicular to the electric field. Protein sample is applied at one end of a long flow chamber and particles stream down the chamber as they are subjected to the electric field. The separation principal is determined by the choice of buffer composition. If the pH is constant, it is a zone electrophoresis method that sorts particles by charge to mass ratio. If a pH gradient perpendicular to flow is used, particles stream toward their isoelectric point. In contemporary implementations of FFE, samples are collected into tubes at the end of the chamber and distributed to the wells of a 96-well plate. The elegance of the method belies its age, as it dates back to the 60s, and was once widely used for cell sorting prior to modern flow cytometry. The method has been resurrected and automated for the proteomic era not only for protein fractionation, but isolation of organelles [82–84].

Capillary Electrophoresis

Capillary electrophoresis emerged in 1967 as another solution to the convection problem. Reducing the scale of the liquid electrophoresis apparatus by driving proteins through narrow bore capillaries (300 μm), with high surface-to-volume ratios, reduced heat generation while accelerating heat dissipation. In fact, as capillary diameter drops, higher voltages can be applied without overheating. Today, voltages of 30 kV, applied to a capillary with an internal diameter of <100 μm, yield highly efficient separation.

Interestingly, the use of such high voltages and small diameter tubes leads to new physical phenomena that have an impact on particle mobility. Electrical forces are no longer the sole determinants of particle movement; bulk flow of the liquid phase emerges as a new factor. The phenomenon is called electroosmotic flow (EOF) and is caused by the behavior of molecules at the interface of the solution and the glass capillary. Briefly, glass is simply fused silica, which contains silanol (SiOH) groups at its surface that are deprotonated (negative) at pH >4. Positive ions in the running buffer, therefore, migrate to the capillary wall and form a double layer. One layer (the Stern Layer) consists of positive ions bound to the SiO^- in the glass wall. The second, is a loose mobile layer of charge. When high voltage is applied, the mobile cations in the second layer migrate toward the cathodic end of the capillary. The moving layer interacts with ions and polar groups in the bulk aqueous buffer layer, pulling the solution along. Formally, electroosmotic mobility (μ_{EOF}) is defined as

$$\mu_{EOF} = \frac{\epsilon}{4\pi\eta} E\zeta \qquad (4.11)$$

where

ε is the dielectric constant of the solution,
η is the viscosity of the solution,
E is the field strength, and
ζ is the zeta potential i.e. the potential between the charged double layer, at the capillary interface, and the bulk solution in the capillary

Modern capillary electrophoresis is routinely performed capillaries from 25 to 75 μm in diameter. Peptide resolution and detection is optimized by combining and tweaking both electrophoretic and electroosmotic particle migration.

A schematic representation of a typical capillary electrophoresis system is depicted in Fig. 4.7. Fundamentally, it consists of a high-voltage power supply, two electrodes, a capillary tube, a way of introducing the sample, and a detector. Prior to sample application, buffer is flushed through the capillary. Samples are often loaded electrokinetically by applying a voltage across the capillary where one end is immersed in the sample. After sample loading, both ends of the capillary are immersed in the running buffer and voltage can be applied for particle separation. Particles migrate along the capillary at rates governed by a combination of electrophoretic and electroosmotic mobility. Common detector systems include on-column monitoring using spectrophotometry (i.e. U/V absorbance) or laser-induced fluorescence (LIF).

CE, like other separation technologies has many variants that often differ simply with respect to the buffer employed. The type discussed above is an example of liquid capillary zone electrophoresis (CZE), though gels can also be used (capillary gel electrophoresis, CGE). It should perhaps not be surprising that CE lends itself to isoelectric focusing (CIEF) and even the kind of particle stacking observed in Laemmli gels, through a process called capillary isotachophoresis (CITP). Finally, the advantages of electroosmotic flow can be combined with stationary matrices to perform capillary electrochromatography (CEC). CE has also been used to resolve hydrophobic particles in a process known as micellar electrokinetic chromatography (MEKC).

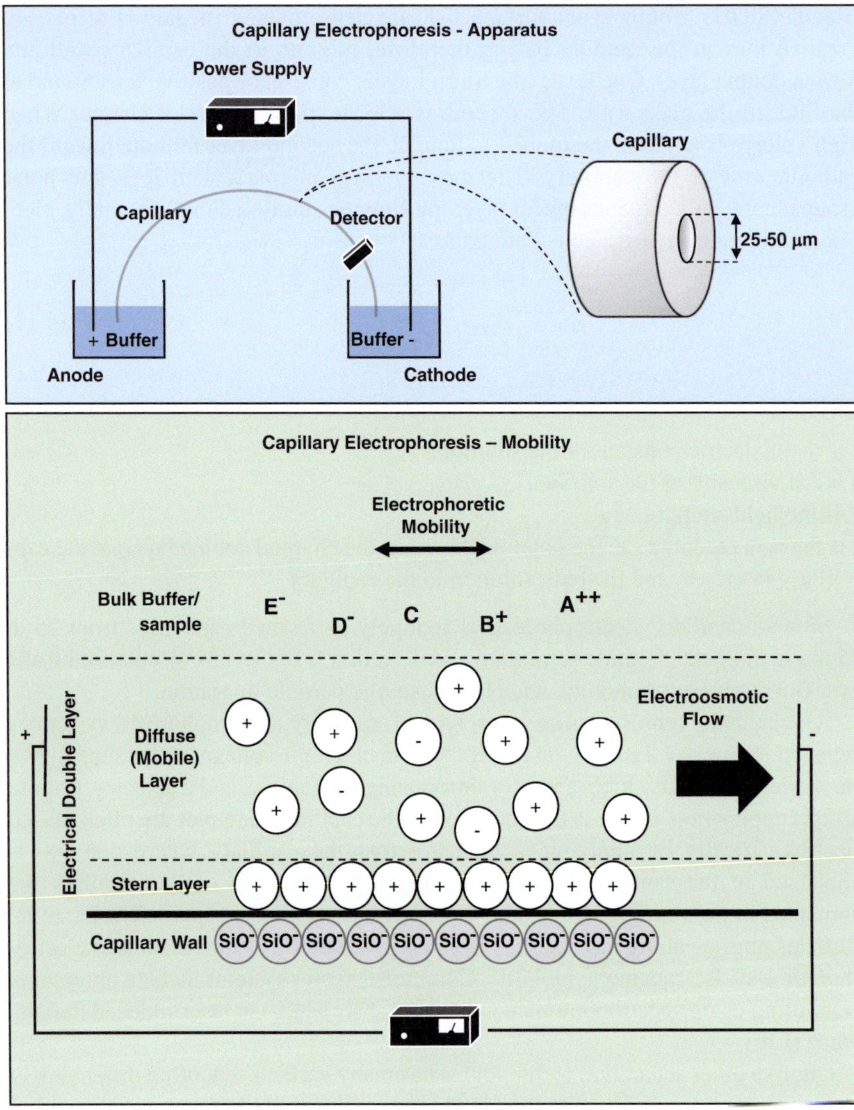

Fig. 4.7 Capillary electrophoresis. (*Upper Panel*) Schematic representation of a capillary electrophoresis apparatus. (*Bottom Panel*) Particle mobility in capillary electrophoresis is determined by the electrophoretic mobility of the particle, determined by ratio of charge to size, as well as electroosmotic flow. Electroosmotic flow is dependent on the chemistry of the capillary wall and the ionic composition of the buffer used. Together, they set the zeta potential and the composition of the diffuse mobile ion layer that migrates in response to applied high voltage

CE vs. HPLC: Will CE Replace RP-HPLC as a Front End for MS/MS?

CE has the potential to make a large impact in the field of proteomics. A prime reason is that the efficiency of CE separation is very high. In the parlance of HETP (see section "Factors that Affect Chromatographic Separation"), CE is capable ultra-high resolution owing to an efficiency that routinely exceeds 100,000 theoretical plates. The efficiency stems from a few factors. Firstly, buffer flow from EOF in CE is more uniform than in HPLC. For instance, high pressure causes flow to be faster near the center of a column than it is at its sides. Moreover, in HPLC, the liquid phase flows through small silica particles which generates turbulence, or eddy flows, which are absent in liquid phase CE. With fewer flow imperfections, longitudinal diffusion is minimized. A second advantage of CE is that, as a liquid phase separation technique, it is largely immune from adsorptive losses that occur in HPLC, which limit the MS sensitivity of peptide detection below 100 ng. Finally, the apparatus is relatively simple compared with HPLC, and highly amenable to parallel analyses. Together, these strengths have made CE arrays a workhorse technology for the analysis of DNA fragments.

So why is inline CE-MS/MS not the *de facto* standard for proteomic workflows? After all, CE was rapidly adopted for DNA analysis. In fact, CE has been used for offline protein and peptide separation, detected by UV absorption, for years. Moreover, CE has also been used for protein separation prior to matrix-assisted laser desorptive ionization (MALDI)-MS. Nevertheless, coupling CE directly to MS, using electrospray ionization, has historically proven highly challenging [85–87]. A second issue is that the ultra-high resolution of CE leads to narrow peptide peaks whose identification require very high MS^2 sampling rates. Mass spectrometers with duty cycles fast enough to take full advantage of CE have come to market only recently. These factors likely explain why it has taken so long for CE-MS/MS to emerge as a robust tool for analysis of complex proteomes. For now, CE-MS/MS-based proteomic profiling is still performed, principally, by the pioneers of the methods [88–90]. The early results are tantalizing, however, as hundreds of peptides can be detected from as little as a single nanogram of protein. This kind of success will surely spur commercial investment.

Putting It All Together: Selected Examples of Subcellular Fractionation and Subproteomic Analysis in Cardiac Research

Several cardiovascular proteomics studies of various subcellular compartments and/or organelles have vastly deepened our understanding of the proteins involved under both physiological and pathological conditions in the heart. In each study, at least one fractionation technique was applied – most studies incorporated multiple fractionation techniques – to investigate the subproteome of interest. We will briefly highlight some of the research that has contributed to our understanding of several subproteomes in cardiac cells.

The Cardiac Sarcomere Subproteome

Sarcomeric proteins are the major molecular components that drive cardiomyocyte function. Initial mapping of cardiac sarcoplasmic reticulum proteins was accomplished through a combination of subcellular fractionation and specialized detergent extraction [91]. The subproteomes of cardiac myofilaments have also been the subject of several studies. One such study used IN sequence extraction of rabbit ventricular myocytes followed by 2D-PAGE and MS to identify a novel phosphorylation of myosin light chain 1 [17]. In another study, differential detergent fractionation allowed for the isolation of cardiac myofilaments from rat ventricular myocytes in a recent proteomics study that revealed the dynamic interplay of kinases and phosphatases in regulating the phosphorylation of cardiac myofilaments [92].

The Cardiac Nuclear Subproteome

Diseased cardiomyocytes undergo significant changes in cellular plasticity through global changes in gene expression. These changes must be preceded by alterations in chromatin structure. In an attempt to map the entire cardiac nuclear proteome, cardiac nuclei were fractionated three different ways to give data for acid-extracted, chromatin-bound, and nucleoplasmic proteins thus dividing the total protein content of the nucleus but increasing the total number of proteins identified in the cardiac nucleus [93]. Acid extraction of cardiomyocyte nuclear fractions led to a proteomics analysis that revealed differential expression of core histones and histone variants as well as the chromatin-associated protein HMGB2 suggesting important roles in regulation of gene expression in diseased cardiomyocytes [20]. A follow-up study in which nuclear proteins were extracted by detergent extraction complemented the previous study by showing that the chromatin-associated protein, nucleolin, was also differentially expressed in diseased cardiomyocytes adding another regulator of gene expression in cardiac cells [21].

Histone PTMs and their role in changes in gene expression by altering chromatin structure is a relatively new focus in proteomics. While several proteomics studies have examined the enzymes that generate or remove histone PTMs in cardiac cells [reviewed in 94–96], very little has been studied in regards to the PTMs themselves. One recent study examined the cardiac acetyl-lysine proteome of fractions obtained by differential centrifugation followed by MS mapped the cardiac acetyl-lysine proteome to both histone and non-histone proteins of the guinea pig heart [97]. With a growing understanding of the importance of a histone code in gene regulation, it is expected that more proteomics studies of the histone proteome in the heart will be forthcoming.

The Cardiac Membrane Subproteome

Proteins in the plasma membrane are the starting point of signaling pathways that determine a cell's response to external stimuli. In general, the hydrophobicity of membrane-bound proteins has presented a challenge in studying the membrane subproteome. However, fractionation with a non-ionic detergent resulted in the successful extraction of hydrophobic cardiac membrane-bound proteins that were identified by MS [98]. Also, fractions obtained by discontinuous sucrose density gradient centrifugation from failing and non-failing human heart tissue were delipidated and analyzed by 2D-PAGE and MS to successfully characterize the membrane microdomain subproteome [99].

The Cardiac Mitochondria Subproteome

The mitochondria is the center for energy metabolism, and alterations in myocardial energy metabolism have been linked to heart disease. An early study of the mitochondria purified from bovine heart by a combination of detergent extraction and sucrose gradient centrifugation followed by 2D-PAGE provided an early map of the cardiac mitochondrial subproteome [50]. Later studies isolated mitochondria from cardiac tissue by a combination of differential centrifugation and Percoll or metrizamide gradient fractionation followed by MS analysis to characterize the proteome in cardiac mitochondria both under normal conditions and under stress [10, 100, 101]. Another significant study used the IN sequence extraction method to examine the protective effects of preconditioning in the cardiac mitochondrial subproteome and identified, for the first time, a subset of enzymes post-translationally modified by phosphorylation in cardiac mitochondria [77].

Concluding Remarks: Frontiers in Fractionation

The purpose of this chapter has been two-fold. Firstly, we hoped to introduce newcomers to the array of separation technologies at their disposal for the separation of organelles, proteins and peptides. Secondly, we wished to provide an overview of the basic principles that underlie these methods. Though established long ago, they are not only relevant to modern proteomics, but key to troubleshooting existing fractionation protocols and designing new separation methods.

Today, the two main drivers of new separation technologies are the quests to maximize resolution and to lower detection limits. Advances promise unparalleled proteome depth. To that end, after years of toil, capillary electrophoresis as a front end separation method for tandem mass spectrometry may almost be ready for prime-time. Indeed, cutting edge separation, coupled with advances in MS instrumentation, have put the prospect of single-cell proteomics on the table. For instance, peptide detection and identification from a nanogram of protein by CZE is notable, since this represents the protein content of about ten eukaryotic cells. Moreover, mass cytometry has enabled signal transduction cascades to be queried on a cell-by-cell basis [102]. Either of these accomplishments would have been considered fanciful only 10 years ago. Over the next 10 years, advances in microfluidic workflows, including inline single-cell/organelle lysis & proteolysis will, no doubt, extend these gains. Stay tuned.

Funding MRM was supported by an NIH T32 grant, 5T32HL007576-30. SF was supported by an NIH/NHLBI R01 HL130424 and the Nora Eccles Treadwell Foundation. DBF was supported by an American Heart Association National Scientist Development Grant 12SDG12060056, NIH/NHLBI R21HL108052 and the Zegar Family Foundation

References

1. Wasinger VC, Cordwell SJ, Cerpa-Poljak A, Yan JX, Gooley AA, Wilkins MR, Duncan MW, Harris R, Williams KL, Humphery-Smith I. Progress with gene-product mapping of the Mollicutes: Mycoplasma genitalium. Electrophoresis. 1995;16:1090–4.
2. Deutscher MP. Guide to protein purification. Methods Enzymol. vol. 182. Academic: San Diego; 1990. p. 894.
3. Deutscher MP, Burgess RR. Guide to protein purification. Methods Enzymol. vol. 463. Academic: San Diego; 2009. p. 915.
4. Linn S. Strategies and considerations for protein purifications. In: Deutscher MP, editors. Methods Enzymol. Academic: San Diego; 1990. p. 9–15.
5. Deutscher MP. Maintaining protein stability. Methods Enzymol. 1990;182:83–9.
6. Stoll VS, Blanchard JS. Buffers: principles and practice. In: Deutscher MP, editor. Methods Enzymol. Academic: San Diego; 1990. p. 24–38.
7. Liu H, Sadygov RG, Yates JR. A model for random sampling and estimation of relative protein abundance in shotgun proteomics. Anal Chem. 2004;76:4193–201.
8. Buie J. Evolution of the lab centrifuge. Lab Manager. May 7, 2010.
9. Foster DB, Ho AS, Rucker J, Garlid AO, Chen L, Sidor A, Garlid KD, O'Rourke B. Mitochondrial ROMK channel is a molecular component of mitoK(ATP). Circ Res. 2012;111:446–54.
10. Taylor SW, Fahy E, Zhang B, Glenn GM, Warnock DE, Wiley S, Murphy AN, Gaucher SP, Capaldi RA, Gibson BW, Ghosh SS. Characterization of the human heart mitochondrial proteome. Nat Biotechnol. 2003;21:281–6.
11. Storrie B, Madden EA. Isolation of subcellular organelles. In: Deutscher MP, editor. Methods Enzymol. Academic: San Diego; 1990. p. 203–25.
12. Agnetti G, Husberg C, Van Eyk JE. Divide and conquer: the application of organelle proteomics to heart failure. Circ Res. 2011;108:512–26.
13. Michelsen U, von Hagen J. Isolation of subcellular organelles and structures. Methods Enzymol. 2009;463:305–28.
14. Englard S, Seifter S. Precipitation techniques. In: Deutscher MP, editor. Methods Enzymol. Academic: San Diego; 1990. p. 285–300.
15. Hofmeister F. The Original German. II. Arch Exp Pathol Pharmacol. 1888;24: 247–260.
16. Rockstroh M, Muller SA, Jende C, Kerzhner A, von Bergen M, Tomm JM. Cell fractionation – an important tool for compartment proteomics. J Integr OMICS. 2011;1:135–43.
17. Arrell DK, Neverova I, Fraser H, Marban E, Van Eyk JE. Proteomic analysis of pharmacologically preconditioned cardiomyocytes reveals novel phosphorylation of myosin light chain 1. Circ Res. 2001;89:480–7.
18. Kane LA, Neverova I, Van Eyk JE. Subfractionation of heart tissue: the "in sequence" myofilament protein extraction of myocardial tissue. Methods Mol Biol. 2007;357:87–90.
19. Shechter D, Dormann HL, Allis CD, Hake SB. Extraction, purification and analysis of histones. Nat Protoc. 2007;2:1445–57.
20. Franklin S, Chen H, Mitchell-Jordan S, Ren S, Wang Y, Vondriska TM. Quantitative analysis of the chromatin proteome in disease reveals remodeling principles and identifies high mobility group protein B2 as a regulator of hypertrophic growth. Mol Cell Proteomics MCP. 2012;11(M111):014258.
21. Monte E, Mouillesseaux K, Chen H, Kimball T, Ren S, Wang Y, Chen JN, Vondriska TM, Franklin S. Systems proteomics of cardiac chromatin identifies nucleolin as a regulator of

growth and cellular plasticity in cardiomyocytes. Am J Physiol Heart Circ Physiol. 2013;305:H1624–38.
22. Chen H, Monte E, Vondriska TM, Franklin S. Systems proteomics of healthy and diseased chromatin. Methods Mol Biol. 2013;1005:77–93.
23. Wessel D, Flugge UI. A method for the quantitative recovery of protein in dilute solution in the presence of detergents and lipids. Anal Biochem. 1984;138:141–3.
24. Ion exchange chromatography and chromatofocusing principles & methods handbook. GE Healthcare. http://www.gelifesciences.com/file_source/GELS/Service%20and%20Support/Documents%20and%20Downloads/Handbooks/pdfs/Ion%20Exchange%20Chromatography.pdf.
25. Rossomando EF. Ion-exchange chromatography. In: Deutscher MP, editor. Methods Enzymol. Academic: San Diego; 1990. p. 309–17.
26. Kennedy RM. Hydrophobic chromatography. In: Deutscher MP, editor. Methods Enzymol. Academic: San Diego; 1990. p. 339–43.
27. Hydrophobic interaction and reversed phase chromatography: principles & methods. GE Healthcare. https://webcache.googleusercontent.com/search?q=cache:125NuYEM0i4J:https://proteins.gelifesciences.com/~/media/protein-purification-ib/documents/selection-guide-pdfs/hydrophobic_interaction_chromatographyhic.pdf%3Fla%3Den+&cd=3&hl=en&ct=clnk&gl=us.
28. Gopalakrishna R, Anderson WB. Ca2+-induced hydrophobic site on calmodulin: application for purification of calmodulin by phenyl-Sepharose affinity chromatography. Biochem Biophys Res Commun. 1982;104:830–6.
29. Mani RS, Kay CM. Purification and characterization of a novel 12,000-Da calcium binding protein from smooth muscle. Arch Biochem Biophys. 1992;296:442–9.
30. Stellwagen E. Gel filtration. In: Deutscher MP, editor. Methods Enzymol. Academic: San Diego; 1990. p. 317–28
31. Gel filtration: principles & methods. Protein Sample Preparation. http://www.gelifesciences.com/file_source/GELS/Service%20and%20Support/Documents%20and%20Downloads/Handbooks/Protein_sample_preparation_handbook.pdf.
32. Affinity chromatography: principles & methods. GE Healthcare. http://www.gelifesciences.com/file_source/GELS/Service%20and%20Support/Documents%20and%20Downloads/Handbooks/Affinity_chromatography_handbook.pdf.
33. Ostrove S. Affinity chromatography: general methods. In: Deutscher MP, editor. Methods Enzymol. Academic: San Diego; 1990. p. 357–71.
34. Porath J. Immobilized metal ion affinity chromatography. Protein Expr Purif. 1992;3:263–81.
35. Rigaut G, Shevchenko A, Rutz B, Wilm M, Mann M, Seraphin B. A generic protein purification method for protein complex characterization and proteome exploration. Nat Biotechnol. 1999;17:1030–2.
36. Puig O, Caspary F, Rigaut G, Rutz B, Bouveret E, Bragado-Nilsson E, Wilm M, Seraphin B. The tandem affinity purification (TAP) method: a general procedure of protein complex purification. Methods (San Diego Calif). 2001;24:218–29.
37. Li Y. Commonly used tag combinations for tandem affinity purification. Biotechnol Appl Biochem. 2010;55:73–83.
38. Xu X, Song Y, Li Y, Chang J, Zhang H, An L. The tandem affinity purification method: an efficient system for protein complex purification and protein interaction identification. Protein Expr Purif. 2010;72:149–56.
39. Martin AJP, Synge RLM. A new form of chromatogram employing two liquid phases. Biochem J. 1941;35:1358–68.
40. Guo DC, Mant CT, Hodges RS. Effects of ion-pairing reagents on the prediction of peptide retention in reversed-phase high-performance liquid chromatography. J Chromatogr. 1987;386:205–22.
41. Mallet CR, Lu Z, Mazzeo JR. A study of ion suppression effects in electrospray ionization from mobile phase additives and solid-phase extracts. Rapid Commun Mass Spectrom RCM. 2004;18:49–58.
42. Wolters DA, Washburn MP, Yates 3rd JR. An automated multidimensional protein identification technology for shotgun proteomics. Anal Chem. 2001;73:5683–90.

43. Giddings JC. Concepts and comparisons in multidimensional separation. J High Resolut Chromatogr. 1987;10:319–23.
44. Boersema PJ, Mohammed S, Heck AJ. Hydrophilic interaction liquid chromatography (HILIC) in proteomics. Anal Bioanal Chem. 2008;391:151–9.
45. Gilar M, Olivova P, Daly AE, Gebler JC. Orthogonality of separation in two-dimensional liquid chromatography. Anal Chem. 2005;77:6426–34.
46. Gilar M, Olivova P, Daly AE, Gebler JC. Two-dimensional separation of peptides using RP-RP-HPLC system with different pH in first and second separation dimensions. J Sep Sci. 2005;28:1694–703.
47. Wyndham KD, O'Gara JE, Walter TH, Glose KH, Lawrence NL, Alden BA, Izzo GS, Hudalla CJ, Iraneta PC. Characterization and evaluation of C18 HPLC stationary phases based on ethyl-bridged hybrid organic/inorganic particles. Anal Chem. 2003;75:6781–8.
48. Wang Y, Yang F, Gritsenko MA, Wang Y, Clauss T, Liu T, Shen Y, Monroe ME, Lopez-Ferrer D, Reno T, Moore RJ, Klemke RL, Camp 2nd DG, Smith RD. Reversed-phase chromatography with multiple fraction concatenation strategy for proteome profiling of human MCF10A cells. Proteomics. 2011;11:2019–26.
49. Yang F, Shen Y, Camp 2nd DG, Smith RD. High-pH reversed-phase chromatography with fraction concatenation for 2D proteomic analysis. Expert Rev Proteomics. 2012;9:129–34.
50. Hanson BJ, Schulenberg B, Patton WF, Capaldi RA. A novel subfractionation approach for mitochondrial proteins: a three-dimensional mitochondrial proteome map. Electrophoresis. 2001;22:950–9.
51. Tiselius A. Electrophoresis of serumglobulin I. Biochem J. 1937;31:313–7.
52. Smithies O. How it all began: a personal history of gel electrophoresis. Methods Mol Biol. 2012;869:1–21.
53. Smithies O. Early days of gel electrophoresis. Genetics. 1995;139:1–4.
54. Kunkel HG, Tiselius A. Electrophoresis of proteins on filter paper. J Gen Physiol. 1951;35:89–118.
55. Kunkel HG, Slater RJ. Zone electrophoresis in a starch supporting medium. Proc Soc Exp Biol Med Soc Exp Biol Med (New York NY). 1952;80:42–4.
56. Smithies O. Zone electrophoresis in starch gels: group variations in the serum proteins of normal human adults. Biochem J. 1955;61:629–41.
57. Raymond S, Weintraub L. Acrylamide gel as a supporting medium for zone electrophoresis. Science. 1959;130:711.
58. Devillez EJ. Preparative acrylamide electrophoresis with the conventional vertical starch gel apparatus. Anal Biochem. 1964;9:485–6.
59. Pederson T. Turning a PAGE: the overnight sensation of SDS-polyacrylamide gel electrophoresis. FASEB J. 2008;22:949–53.
60. Shapiro AL, Vinuela E, Maizel Jr JV. Molecular weight estimation of polypeptide chains by electrophoresis in SDS-polyacrylamide gels. Biochem Biophys Res Commun. 1967;28: 815–20.
61. Laemmli UK. Cleavage of structural proteins during the assembly of the head of bacteriophage T4. Nature. 1970;227:680–5.
62. O'Farrell PH. High resolution two-dimensional electrophoresis of proteins. J Biol Chem. 1975;250:4007–21.
63. Gelfi C, Righetti PG. Preparative isoelectric focusing in immobilized pH gradients. II. A case report. J Biochem Biophys Methods. 1983;8:157–72.
64. Gianazza E, Dossi G, Celentano F, Righetti PG. Isoelectric focusing in immobilized pH gradients: generation and optimization of wide pH intervals with two-chamber mixers. J Biochem Biophys Methods. 1983;8:109–33.
65. Westermeier R, Postel W, Weser J, Gorg A. High-resolution two-dimensional electrophoresis with isoelectric focusing in immobilized pH gradients. J Biochem Biophys Methods. 1983;8:321–30.
66. Schägger H, von Jagow G. Blue native electrophoresis for isolation of membrane protein complexes in enzymatically active form. Anal Biochem. 1991;199:223–31.
67. Markham RL. A modified method of two-dimensional zone electrophoresis applied to mucoproteins in serum and urine. Nature. 1956;177:125–6.

68. Smithies O, Poulik MD. Two-dimensional electrophoresis of serum proteins. Nature. 1956;177:1033.
69. Poulik MD, Smithies O. Comparison and combination of the starch-gel and filter-paper electrophoretic methods applied to human sera: two-dimensional electrophoresis. Biochem J. 1958;68:636–43.
70. Bjellqvist B, Ek K, Righetti PG, Gianazza E, Gorg A, Westermeier R, Postel W. Isoelectric focusing in immobilized pH gradients: principle, methodology and some applications. J Biochem Biophys Methods. 1982;6:317–39.
71. Gorg A, Obermaier C, Boguth G, Harder A, Scheibe B, Wildgruber R, Weiss W. The current state of two-dimensional electrophoresis with immobilized pH gradients. Electrophoresis. 2000;21:1037–53.
72. Rabilloud T, Chevallet M, Luche S, Lelong C. Two-dimensional gel electrophoresis in proteomics: past, present and future. J Proteomics. 2010;73:2064–77.
73. Weiss W, Gorg A. High-resolution two-dimensional electrophoresis. Methods Mol Biol. 2009;564:13–32.
74. Heinke MY, Wheeler CH, Chang D, Einstein R, Drake-Holland A, Dunn MJ, dos Remedios CG. Protein changes observed in pacing-induced heart failure using two-dimensional electrophoresis. Electrophoresis. 1998;19:2021–30.
75. Weekes J, Wheeler CH, Yan JX, Weil J, Eschenhagen T, Scholtysik G, Dunn MJ. Bovine dilated cardiomyopathy: proteomic analysis of an animal model of human dilated cardiomyopathy. Electrophoresis. 1999;20:898–906.
76. Jiang L, Tsubakihara M, Heinke MY, Yao M, Dunn MJ, Phillips W, dos Remedios CG, Nosworthy NJ. Heart failure and apoptosis: electrophoretic methods support data from micro- and macro-arrays. A critical review of genomics and proteomics. Proteomics. 2001;1:1481–8.
77. Arrell DK, Elliott ST, Kane LA, Guo Y, Ko YH, Pedersen PL, Robinson J, Murata M, Murphy AM, Marban E, Van Eyk JE. Proteomic analysis of pharmacological preconditioning: novel protein targets converge to mitochondrial metabolism pathways. Circ Res. 2006;99:706–14.
78. Wang SB, Foster DB, Rucker J, O'Rourke B, Kass DA, Van Eyk JE. Redox regulation of mitochondrial ATP synthase: implications for cardiac resynchronization therapy. Circ Res. 2011;109:750–7.
79. Brown JR, Kauffman DL, Hartley BS. The primary structure of porcine pancreatic elastase. The N-terminus and disulphide bridges. Biochem J. 1967;103:497–507.
80. Brennan JP, Wait R, Begum S, Bell JR, Dunn MJ, Eaton P. Detection and mapping of widespread intermolecular protein disulfide formation during cardiac oxidative stress using proteomics with diagonal electrophoresis. J Biol Chem. 2004;279:41352–60.
81. Meyer B, Wittig I, Trifilieff E, Karas M, Schagger H. Identification of two proteins associated with mammalian ATP synthase. Mol Cell Proteomics MCP. 2007;6:1690–9.
82. Satori CP, Kostal V, Arriaga EA. Review on recent advances in the analysis of isolated organelles. Anal Chim Acta. 2012;753:8–18.
83. Zischka H, Braun RJ, Marantidis EP, Büringer D, Bornhövd C, Hauck SM, Demmer O, Gloeckner CJ, Reichert AS, Madeo F, Ueffing M. Differential analysis of Saccharomyces cerevisiae mitochondria by free flow electrophoresis. Mol Cell Proteomics. 2006;5:2185–200.
84. Islinger M, Li KW, Loos M, Liebler S, Angermüller S, Eckerskorn C, Weber G, Abdolzade A, Völkl A. Peroxisomes from the heavy mitochondrial fraction: isolation by zonal free flow electrophoresis and quantitative mass spectrometrical characterization. J Proteome Res. 2010;9:113–24.
85. Sun L, Zhu G, Yan X, Champion MM, Dovichi NJ. Capillary zone electrophoresis for analysis of complex proteomes using an electrokinetically pumped sheath flow nanospray interface. Proteomics. 2014;14:622–8.
86. Fonslow BR, Yates 3rd JR. Capillary electrophoresis applied to proteomic analysis. J Sep Sci. 2009;32:1175–88.
87. Issaq HJ, Janini GM, Chan KC, Veenstra TD. Sheathless electrospray ionization interfaces for capillary electrophoresis-mass spectrometric detection advantages and limitations. J Chromatogr A. 2004;1053:37–42.

88. Sun L, Zhu G, Yan X, Zhang Z, Wojcik R, Champion MM, Dovichi NJ. Capillary zone electrophoresis for bottom-up analysis of complex proteomes. Proteomics. 2016;16(2):188–96.
89. Wang Y, Fonslow BR, Wong CC, Nakorchevsky A, Yates 3rd JR. Improving the comprehensiveness and sensitivity of sheathless capillary electrophoresis-tandem mass spectrometry for proteomic analysis. Anal Chem. 2012;84:8505–13.
90. Li Y, Compton PD, Tran JC, Ntai I, Kelleher NL. Optimizing capillary electrophoresis for top-down proteomics of 30-80 kDa proteins. Proteomics. 2014;14:1158–64.
91. Macri J, McGee B, Thomas JN, Du P, Stevenson TI, Kilby GW, Rapundalo ST. Cardiac sarcoplasmic reticulum and sarcolemmal proteins separated by two-dimensional electrophoresis: surfactant effects on membrane solubilization. Electrophoresis. 2000;21:1685–93.
92. Yin X, Cuello F, Mayr U, Hao Z, Hornshaw M, Ehler E, Avkiran M, Mayr M. Proteomics analysis of the cardiac myofilament subproteome reveals dynamic alterations in phosphatase subunit distribution. Mol Cell Proteomics MCP. 2010;9:497–509.
93. Franklin S, Zhang MJ, Chen H, Paulsson AK, Mitchell-Jordan SA, Li Y, Ping P, Vondriska TM. Specialized compartments of cardiac nuclei exhibit distinct proteomic anatomy. Mol Cell Proteomics MCP. 2011;10(M110):000703.
94. Dirkx E, da Costa Martins PA, De Windt LJ. Regulation of fetal gene expression in heart failure. Biochim Biophys Acta. 2013;1832:2414–24.
95. Karbassi E, Vondriska TM. How the proteome packages the genome for cardiovascular development. Proteomics. 2014;14:2115–26.
96. Rosa-Garrido M, Karbassi E, Monte E, Vondriska TM. Regulation of chromatin structure in the cardiovascular system. Circ J Off J Jpn Circ Soc. 2013;77:1389–98.
97. Foster DB, Liu T, Rucker J, O'Meally RN, Devine LR, Cole RN, O'Rourke B. The cardiac acetyl-lysine proteome. PLoS One. 2013;8:e67513.
98. Donoghue PM, Hughes C, Vissers JP, Langridge JI, Dunn MJ. Nonionic detergent phase extraction for the proteomic analysis of heart membrane proteins using label-free LC-MS. Proteomics. 2008;8:3895–905.
99. Banfi C, Brioschi M, Wait R, Begum S, Gianazza E, Fratto P, Polvani G, Vitali E, Parolari A, Mussoni L, Tremoli E. Proteomic analysis of membrane microdomains derived from both failing and non-failing human hearts. Proteomics. 2006;6:1976–88.
100. Zhang J, Liem DA, Mueller M, Wang Y, Zong C, Deng N, Vondriska TM, Korge P, Drews O, Maclellan WR, Honda H, Weiss JN, Apweiler R, Ping P. Altered proteome biology of cardiac mitochondria under stress conditions. J Proteome Res. 2008;7:2204–14.
101. Zhang J, Li X, Mueller M, Wang Y, Zong C, Deng N, Vondriska TM, Liem DA, Yang JI, Korge P, Honda H, Weiss JN, Apweiler R, Ping P. Systematic characterization of the murine mitochondrial proteome using functionally validated cardiac mitochondria. Proteomics. 2008;8:1564–75.
102. Bendall SC, Simonds EF, Qiu P, Amir el AD, Krutzik PO, Finck R, Bruggner RV, Melamed R, Trejo A, Ornatsky OI, Balderas RS, Plevritis SK, Sachs K, Pe'er D, Tanner SD, Nolan GP. Single-cell mass cytometry of differential immune and drug responses across a human hematopoietic continuum. Science. 2011;332:687–96.

Online Resources

www.chromacademy.com: an online e-learning resource for analytical scientists, centered on chromatography, which includes instructional video.

http://chemwiki.ucdavis.edu/: a chemistry-centered resource with core modules covering analytical, biological, theoretical, inorganic, organic and physical chemistry. The topics covered in the chapter as well as an overview of mass spectrometry can be found in the analytical chemistry module under "Instrumental Analysis".

Chapter 5
Vascular Proteomics

Jose Luis Martin-Ventura, Elena Burillo, Roxana Martinez-Pinna, Diego Martinez-Lopez, Luis Miguel Blanco-Colio, Emilio Camafeita, Juan Antonio Lopez, and Jesus Vazquez

Abstract Atherothrombosis remains one of the main causes of morbidity and mortality in the western countries. The evolution from the initial steps to clinical events in atherothrombosis is a continuum of integrated and increasingly complex biological processes. To understand a process of such complexity, there is a need of developing new techniques, which could help to identify novel mediators of atherothrombosis that could become potential diagnostic, prognostic and/or therapeutic targets. In this chapter, we have described vascular samples including cells, tissues and blood (circulating cells, lipoproteins and extracellular vesicles) used in proteomic studies, as well as potential challenges and limitations. Following this, we made a succinct description of proteomics technologies (gel-based and MS-based approaches) applied to vascular samples and we included some examples to highlight what proteomics have added to previous techniques and/or concepts. On the whole, proteomics, in combination with other complementary approaches, is emerging as a very powerful tool that is expected to improve the diagnosis and treatment of patients at high risk of suffering a cardiovascular event.

Keywords Vascular • Atherothrombosis • Atherosclerosis • Abdominal aortic aneurysm • Biomarkers • Quantitative proteomics • Mass spectrometry • Diagnosis • Prognosis • Therapy

J.L. Martin-Ventura, PhD (✉) • E. Burillo, PhD • R. Martinez-Pinna, PhD
D. Martinez-Lopez • L.M. Blanco-Colio, PhD
Vascular Research Lab, FIIS-Fundacion Jimenez Diaz,
Autonoma University of Madrid, Madrid, Spain
e-mail: JLMartin@fjd.es; eburilloi@gmail.com; rmpinna@gmail.com; diegomartilo@hotmail.com

E. Camafeita, PhD • J.A. Lopez, PhD • J. Vazquez, PhD
Cardiovascular Proteomic Lab and Proteomic Unit, Vascular Pathophysiology Area, Centro Nacional De Investigaciones Cardiovasculares, Madrid, Spain

G. Agnetti et al. (eds.), *Manual of Cardiovascular Proteomics*,
DOI 10.1007/978-3-319-31828-8_5

Introduction

Atherothrombosis remains one of the main causes of morbidity and mortality in the western countries. Atherothrombosis is a systemic disease affecting different vascular territories, mainly coronary, carotid, abdominal aorta and iliofemoral arteries. Chronic vascular remodeling is a very complex process involving different mechanisms, such as lipid deposition, oxidative stress, inflammatory cell recruitment, proliferation, proteolysis, apoptosis, neovascularization, calcification and fibrosis. These processes could lead to clinical complications due to arterial wall rupture, involving medial and adventitial rupture in abdominal aortic aneurysm (AAA) and intimal cap rupture in complicated atherothrombotic plaques [1, 2] (Figs. 5.1 and 5.2).

It is well established that atherothrombosis is linked to lipid/lipoprotein deposition in the arterial wall, readily observed from the initial stages of the disease. Consistently, increased systemic low-density lipoprotein (LDL) and reduced high-density lipoprotein (HDL) cholesterol levels are known risk factors for atherothrombosis. LDL retention in atherosclerotic plaques has been related to their modification (mainly by oxidation) and their uptake by scavenger receptors of phagocytes, leading to foam cell formation. However, oxidative stress could also produce modifications in molecules involved in other processes associated to vessel wall remodeling (e.g. nitric oxide- related endothelial dysfunction). Oxidative stress is also involved in AAA progression associated to the presence of red blood cells (RBCs) and neutrophils in the intraluminal thrombus of human AAA, as well as macrophages in the wall [2]. The pathological vascular wall remodeling progress to more advanced stages, characterized by vascular smooth muscle cell (VSMC) proliferation and migration, leukocyte infiltration and differentiation, angiogenesis, as well as adventitial immune inflammation and fibrosis. Moreover, proteolysis of extracellular matrix (ECM) proteins like collagen or elastin, along with VSMC death, is involved in both plaque desestabilization and AAA dilatation, potentially favoring their rupture.

The evolution from the initial steps to clinical events in atherothrombosis is a continuum of integrated and increasingly complex biological processes. To understand a process of such complexity, there is a need of developing novel techniques, which could help to identify novel mediators of atherothrombosis that could become potential therapeutic targets. For these purposes, two main strategies are being developed: imaging and biomarker discovery. Different imaging modalities (including optical imaging, magnetic resonance imaging, and nuclear imaging) are giving clues of the importance of specific type of cells in the development of atherothrombosis [1, 3]. In addition, identification of novel biomarkers by high-throughput techniques (e.g. proteomics) could afford the identification of novel mediators involved in disease diagnosis, progression and/or treatment [4]. In this respect, most patients with atherothrombosis are

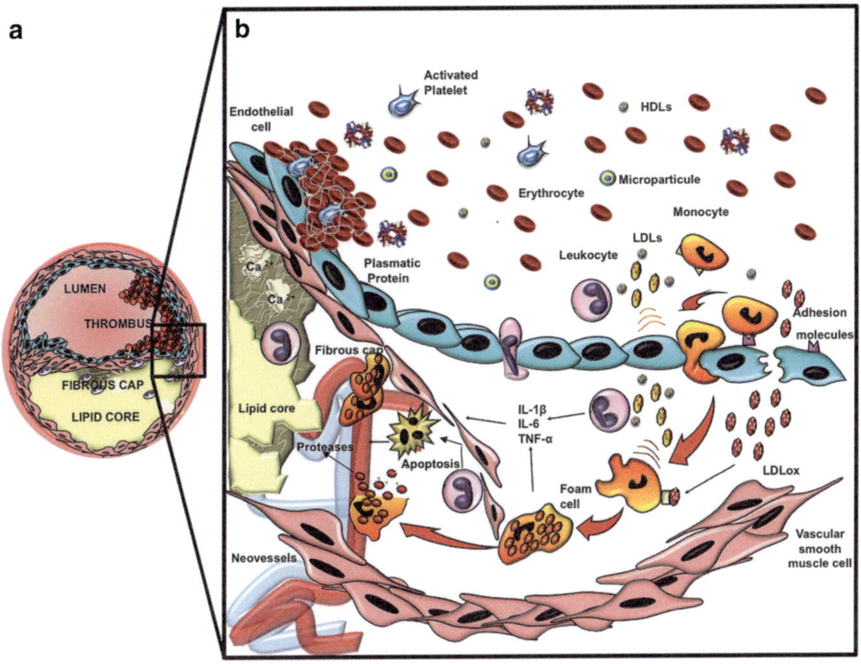

Fig. 5.1 Pathogenic mechanisms involved in atherosclerosis. (**a**) Schematic representation of the different regions of human atherosclerotic plaques. (**b**) The initial event in the atherosclerotic process is endothelial injury and the formation of fatty streaks originated by trapping of lipoproteins (*LDL*) and the appearance of leukocyte adhesion molecules on the endothelial cells, triggering leukocyte infiltration. Leukocytes migrate into the subendothelial space and differentiate into macrophages. Uptake of LDL (modified as oxidized LDL) via scavenger receptors leads to foam cell formation. Interactions between macrophage foam cells and T lymphocytes establish a chronic inflammatory process, with the secretion of cytokines (such as TNF-α) and other proinflammatory mediators, which promote the proliferation and the change of the phenotype of VSMCs. These cells synthesize extracellular matrix proteins that lead to the development of the fibrous cap. The more advanced stages of atherosclerosis are characterized by the presence of a lipid-rich necrotic core, with presence of neovessels, calcification and intraplaque hemorraghes. Necrosis or apoptosis of macrophages and SMCs result in the formation of a necrotic core and accumulation of extracellular cholesterol. Red Blood Cells (RBCs), coming from leaky neovesssels, participate in both the oxidative process and cholesterol accumulation. Finally, rupture of atherosclerotic plaques lead to the formation of a thrombus, giving rise to the most harmful clinical complications (myocardial infarction, stroke)

asymptomatic, so early detection could be key for cardiovascular (CV) prevention. Even when patients are diagnosed, identification of prognostic biomarkers could help to prevent a future CV event. Moreover, we should keep in mind that AAA evolution is not linear and thus, the follow-up of these patients based only on AAA size is not sufficient. Finally, biomarkers could also help to tailor therapy.

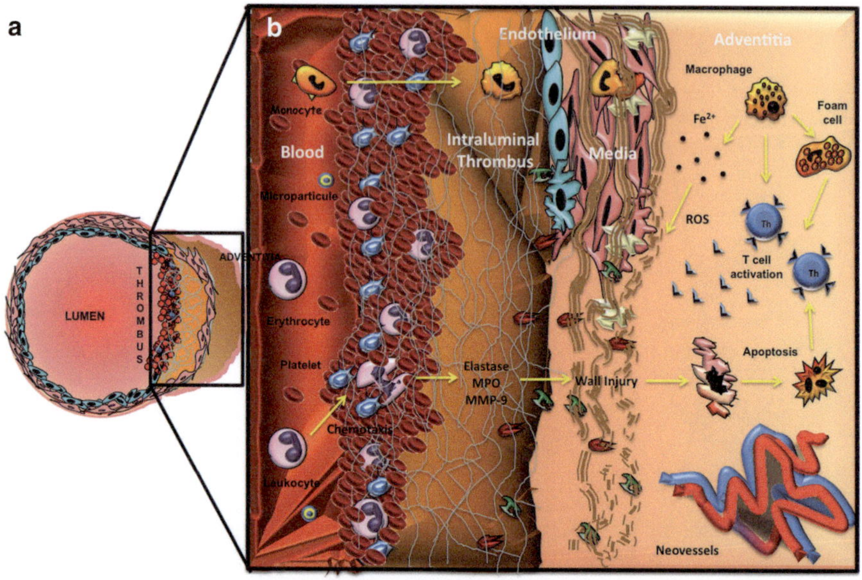

Fig. 5.2 Pathogenic mechanisms involved in abdominal aortic aneurysms. (**a**) Schematic representation of the different regions of human AAA. (**b**) AAA is characterized by the presence of intraluminal thrombus, which exerts its pathogenic effect through platelet activation, fibrin formation and trapping of erythrocytes and neutrophils, leading to oxidative and proteolytic injury of the arterial wall. These injuries of the wall favours the depletion of vascular smooth muscle cells (VSMC) in the media and an adventitial response characterized by the presence of immune-inflammatory cells, iron deposits, cholesterol crystals, fibrosis and neoangiogenesis

Samples Analyzed in Vascular Diseases: Challenges and Limitations

Vascular Sampling

The sampling is a critical step in differential proteomics. Samples can be obtained from humans (patients with atherothrombosis and controls) or from experimental models. Experimental models of hypercholesterolemia (e.g. ApoE$^{-/-}$ mice, LDL receptor$^{-/-}$ mice) are widely used in the study of the mechanisms of atherosclerosis. A common experimental model used in AAA is the elastase perfusion model, which in the case of rats, is also able to develop an intraluminal thrombus [5]. Another experimental model reflecting mainly adventitial inflammatory processes is the AngII model in ApoE$^{-/-}$ mice. However, there are several differences between these animal models that should be taken into account when interpreting the results obtained (e.g. time necessary to develop the disease, mechanisms of thrombosis, etc) [1, 6].

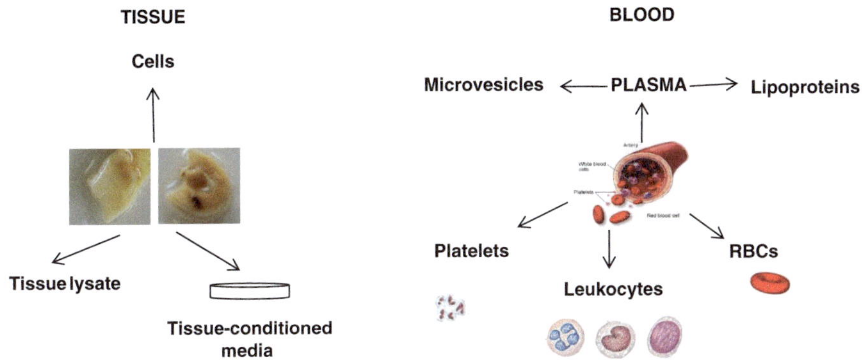

Fig. 5.3 Samples analyzed in proteomic studies. Two main sources of samples to analyze potential mechanisms/mediators involved in vascular diseases are tissues or blood. Tissues can be obtained from surgery and then, resident cells can be isolated from these tissues (primary cultures, lasser microdissection). On the other hand, different types of samples can be obtained from blood: plasma (containing lipoproteins and microvesicles, which can be isolated by serial centrifugation) and circulating cells (isolated by different protocols in order to obtain the highest purification)

Sampling is directly related to the question raised and should be addressed before starting any kind of sample collection and proteomic analysis. For example, if the objective of the study is to understand the mechanisms of atherothrombosis, analysis of cell or tissue extracts will be preferred. In contrast, if the aim of the study is to discover circulating biomarkers that could be present in plasma or serum, either tissue/cell secretome or plasma/serum should be directly investigated (Fig. 5.3).

Vascular Cells and Tissues

Proteomic analysis can be performed on vascular cells. However, primary culture of vascular cells may produce changes in phenotype due to cell extraction and culture. For example, proteases that can be used to dissociate vascular smooth muscle cells (VSMCs) from an arterial wall will disrupt all cell-cell and cell-extracellular matrix interactions, leading to the induction of expression of adhesion molecules and neosynthesis of ECM when the cells are plated in culture dishes. Exploring the proteome of cultured cells will probably provide information on the intrinsic pathological potential of the cells and their capacity to respond to a particular stimulus rather than on the original proteome expressed within the arterial wall before culture-induced phenotype adaptation.

Proteomic analysis can also be performed directly on tissue samples. However, this approach is difficult since atherosclerotic plaques/AAA are very heterogeneous (in lipid content, presence of intraplaque hemorrhage, fibrosis, neovascularization, calcification…). Lipids, which are very abundant in the case of atherothrombotic tissues, are normally present in the protein extracts, potentially interfering with the analysis. Moreover, differential proteomics is based on the comparison between at least two conditions and this implies that proteomes from pathological arteries must be compared with those of the corresponding healthy arteries, often difficult to obtain in similar conditions. In this respect, some high-abundant plasmatic proteins (e.g. albumin) are present in the pathological tissue in higher amount that in healthy arteries and can interfere with the analysis. The retention of those high-abundant molecules from the circulation could be due, at least in part, through their attachment to the ECM (mainly to proteoglycans). ECM is modified during vascular pathological remodeling in parallel to the change in the VSMC phenotype (from a contractile to a synthetic phenotype). Due to the interest of ECM modification during atherothrombosis, Mayr's group set up specific conditions to study the ECM proteome in normal or vascular pathological conditions [7, 8].

Tissues can be homogenized and subsequently analyzed; however, the information provided by this approach will likely be limited to abundant constitutive proteins. Tissues could also be laser-microdissected, which would allow to analyze specific areas of the plaque (such as the intima, as described in de la Cuesta et al. [9]) or different cells in AAA (such as the subpopulation of macrophages, as described in Boytard et al. [10]), providing valuable spatial information. Another way to understand the topology of the vascular tissue and the cells involved is to dissect the different parts of the tissues and study them comparatively. For example, advanced atherosclerotic plaques can be divided into culprit area and the adjacent non-complicated area, with the possibility to analyze the contribution of activated VSMC, inflammatory cells and RBCs in the pathological mechanisms associated to plaque instability (e.g. proteolysis, apoptosis) [11]. Similarly, AAA tissue can be divided into thrombus and wall, and more specifically, the wall can be dissected into media and adventitia [12]. The mechanisms and cells involved in AAA media and adventitia are different (e.g. VSMC apoptosis in the media vs immune inflammatory reaction in the adventitia). However, careful dissection of the tissue must be performed by a trained vascular biology/surgeon. As commented before, all these procedures must be done rapidly, including fast freezing steps. However, another possibility is to analyze plaques/AAA that have been kept in paraffin. For this purpose, Van Eyk's group described an improved method of protein extraction from aortas by modifying both the temperature and the pressure of the protocol. Moreover, they observed that for long-term storage, aorta was preserved better by paraffin-embedding after fixation in formalin [13]. This is also an important aspect that should be considered for biobanking, even more when, in some cases, there can be reduced availability of tissues in the future due to changes in surgery strategies (e.g. use of endovascular prosthesis instead of open surgery for AAA treatment).

For both tissues and cells, proteomic analysis can also be performed on conditioned medium. Focusing on conditioned medium will allow one to target more specifically secreted/released proteins provided that culture conditions do not induce necrosis or cell lysis. This approach can be used to discover circulating biomarkers [14]. Moreover, plaque or AAA conditioned culture medium could offer substantial advantages in the analytical proteomics of biological fluids, as they provide a source of intact or, more importantly, degraded tissue proteins released by the tissue that could serve as potential circulating biomarkers of atherothrombotic disease.

Blood Collection

The blood compartment is easily accessible to proteomic analysis in order to discover biomarkers that could be used in diagnosis, prognosis or to evaluate the efficiency of a treatment in a particular disease. Interestingly, there are some conditions where only blood can be obtained. For example, as no surgery is indicated for AAA patients with aortic diameter between 3 and 5 cm (or in patients with diagnosed carotid stenosis <70 %), human samples from these patients could only be obtained from blood, either circulating cells or plasma, in order to analyze mechanisms involved in AAA/atherosclerosis evolution towards surgery or occasionally rupture.

Blood may reflect directly or indirectly a cardiovascular pathological state and the proteome of circulating cells or of plasma may be modified accordingly. Blood sampling must be carried out when it is most appropriate. For example, if the objective is to discover markers of coronary atherosclerosis, the blood should not be drawn a few hours after revascularization since stenting likely induces significant changes in the blood proteome that could mask markers of atherosclerosis found in basal conditions. Ideally, blood should be obtained at a similar time of the day since it has been demonstrated that circulating proteins could be modified due to circadian rhythms [15]. Others factors that could influence the proteomic analysis of blood components are the presence of hemolysis (e.g. RBC lysis could lead to oxygen radicals release), as well as the conditions when the blood was taken (e.g. if blood is taken after anaesthesia, after fasting conditions, etc).

Serum Vs Plasma

Plasma is obtained by collecting blood in an anticoagulant solution and subsequent centrifugation. In contrast, serum is obtained after coagulation, a process that involves the activation of proteases in cascade leading to the formation of a clot containing activated leukocytes and platelets. These activated blood cells can release many proteins and proteases [16], which, in addition to coagulating proteases, will drastically affect the serum proteome. This must be taken into account in order to normalize blood sampling (type of anticoagulant, time of clotting for serum, centrifugation speed…).

A major challenge for analyzing plasma proteome is the high dynamic range of protein concentration. Analysis of crude plasma does not provide information on thousands of proteins and peptides that are masked by high- or moderately-abundant proteins. Several approaches can be used to eliminate abundant protein interfering with proteomic analysis (e.g. immuno-affinity columns, normally followed by chromatographic fractionation). More recently, serial/tandem depletion (using the same type of columns) or ultradepletion methods have demonstrated an increase in the number of proteins depleted [17, 18]. However, highly abundant protein depletion (e.g. albumin) may involve non-specific loss of other low-abundance proteins, which may reflect important pathophysiologic pathways, such as cytokines, known to participate in atherothrombosis.

Circulating Cells

Due to the important role that circulating cells plays in atherothrombosis (neutrophils, monocytes, RBCs, platelets), addressing the differential protein profile of these cells could afford valuable information of this systemic pathological setting. However, when isolating circulating cells, it must be taken into account that many cells, such as polymorphonuclear neutrophils or platelets, may be activated during the isolation process (centrifugation, temperature, pipetting…). For example, if the idea is to discover platelet biomarkers, it is known that preactivation of platelets could influence the levels of some biomarkers [19]. Similarly, specific isolation of circulating cells by using magnetic beads coupled to antibodies could led to activation of some circulating cells. The purity of the preparation is also critical as the proteome of contaminating cells may interfere significantly with the proteome of interest (e.g. reticulocytes in RBCs preparations [20]). Likewise, contamination of plasma proteins could also happen (mainly in platelets [21]). For that reasons, standarization of protocols for isolation of cells is key to be able to compare the results obtained in different studies. Another issue that is relevant in the case of RBCs is the large dynamic range for cellular protein concentration, due to the high concentration of cytosolic Hb (97 %). One alternative is analyzing the membrane fraction, which is not devoid of some technical difficulties (e.g. proteins from other compartments) [22].

Lipoproteins/Extracellular Vesicles

Another way to circumvent the dynamic problem of plasma protein concentration is to analyse plasma subproteomes. Regarding atherothrombosis, and due to the key role of lipid homeostasis, lipoproteins can be an interesting target. Among lipoproteins, high-density lipoprotein (HDL) is the more protein-rich particle. It is generally accepted that some of the beneficial role of HDL in atherothrombosis could go beyond their cholesterol efflux properties, being related to its protein cargo. Lipoproteins can be isolated by different procedures, although the more common is based on ultracentrifugation. However, this procedure is known to increase the contamination of high-abundance plasmatic proteins (e.g. albumin) and weakly-bound proteins could be released due to the high stress conditions applied to the particles. In the case of HDL, there exist other isolation methods, like ApoA-1 affinity columns or gel filtration chromatrography, used alone or in combination, although it is not clear which of those isolated HDL particles resemble better the in vivo (physiological) situation. Another grade of complexity comes from the fact that there are different subpopulations of LDL and HDL particles. In this regard, previous studies have shown that proteins distribute throughout the HDL family in distinct patterns according to particle density, size and ionic character [23, 24].

Microvesicles (MV), including microparticles and exosomes are membranous vesicles released by cells into extracellular fluids, thereby mediating intercellular communication in physiological and pathological processes, among them atherothrombosis [25]. Given the fact that MV proteins represent less than 0.07 % of total protein plasma content, obtaining highly purified microvesicle fractions may become difficult. In addition, despite the differential centrifugation speeds required for each type of MVs (exosomes vs microparticles), cosedimentation is a common feature [26]. Moreover, lipoprotein contamination (mainly very low density lipoproteins) is usually present in MVs. In addition, characterization of MV is neither a completely solved question, and different methods are usually combined to assure a highly enriched MV sample (electron microscopy, nanotracking analysis, flow-cytometry).

Proteomics Approaches Applied to the Study of Atherothrombosis

Gel-Based Approaches for Quantitative Proteomics

A large number of vascular proteomic investigations have employed two-dimensional gel electrophoresis (2DE) to separate proteins, combined with quantitative computer analysis to detect differentially expressed proteins and MS technologies to identify proteins of interest [27]. In 2DE, proteins are firstly separated in the first dimension by isoelectric focusing based on their charge, while in the second dimension, proteins are separated according to their molecular mass by SDS-polyacrylamide gel electrophoresis (SDS-PAGE). The 2DE technique provides a comprehensive protein map where protein abundance changes, isoforms, or posttranslationally modified proteoforms can be detected [28]. The 2D differential in gel electrophoresis (DIGE) technique was proposed to circumvent the high variability associated to conventional 2DE by running different samples on the same gel [29]. In DIGE, each sample is labeled with a different fluorophore; then the samples are mixed and run on the same 2D gel, and relative protein abundance is assessed based on the use of different fluorescence excitation wavelengths [30]. Following sample separation, protein spots are visualized using standard protein staining methods; these stains must be sensitive, reproducible and compatible with protein identification by MS. Once the spots of interest have been selected, they can be excised from the gel to be identified by MS. Both MALDI- and ESI-based MS methods are used to produce peptide mass fingerprints and peptide fragment fingerprints which are matched against the corresponding theoretical fingerprints derived from sequence databases for protein identification.

Successful examples of 2D-DIGE technology applied to cell, tissues or plasma subproteomes in atherothrombosis have been published [9, 31–33]. In one of those studies, the analysis of the intima of coronary plaques by 2D-DIGE allowed the identification of several differentially expressed proteins in atherosclerosis, among them ferritin [9]. Interestingly, Lepedda et al. [31] showed an increase in ferritin expression in unstable vs stable carotid atherosclerotic plaques. The increased levels of ferritin indicate higher iron deposits in plaques, which can be associated to processes of intraplaque hemorraghe, a main determinant of plaque progression

[34]. Interestingly, intraplaque hemorrhage and increased intraplaque vessel formation are independently related to clinical outcome [35]. A recent meta-analysis highlights the relation of iron status with atherothrombosis [36]. Another example of the usefulness of 2D/2D-DIGE is the study of AAA thrombus-conditioned media in order to find AAA biomarkers. Among proteins identified in thrombus-conditioned media, we observed an increase in peroxiredoxin-1 (PRX-1) in the luminal part of the thrombus [32]. PRX-1 was associated to RBCs in the thrombus and RBCs incubated with pro-oxidant molecules in vitro released PRX-1 to the extracellular medium. Finally, PRX-1 levels were significantly increased in the serum of patients with AAA in relation to controls. Interestingly, PRX-1 was associated to both aortic diameter and AAA growth. More importantly, the combination of PRX-1 and aortic size has significantly additive value in predicting growth compared with size alone, the actual parameter used in the follow-up of AAA patients. This data highlight the potential value of biomarkers, in conjunction with imaging techniques, in the prognosis of patients with AAA.

Gel-Free Strategies

While gel-based techniques remained the method of choice for high-resolution proteome analysis for a number of years [37], it suffers from low reproducibility issues and is biased against membrane-associated proteins, low abundance proteins, or proteins with extreme isoelectric point or molecular weight. But the most important drawback is its limited capacity for protein identification, being only able to characterize hundreds of proteins, a number that in most of cases only constitutes a minor fraction of total proteome content of biological samples. More recently, gel-free approaches have been developed based on the proteolytic digestion of whole protein extracts followed by chromatographic separation and on-line MS/MS analysis of the resulting peptides. This technique, termed liquid chromatography coupled to mass spectrometry (LC-MS) allows the "shotgun" identification of thousands of proteins present in cell or tissue protein extracts [38].

Several examples of "shotgun" technology applied to vascular samples have been recently published [12, 39–42]. An interesting example is its application to study proteolysis, which is a main mechanism in pathological vascular remodeling, both in atherosclerosis and AAA. Proteolysis of ECM proteins (mainly collagen) has been linked to plaque rupture of advanced atherosclerotic plaques. Therefore, big efforts have been done in order to address the functional role of different proteases, mainly through animal models [43]. Stegemann et al. globally analyze matrix-metalloproteinase (MMPs) substrates in human vasculature by high-throughput shotgun proteomic analysis [39]. Interestingly, the authors described novel candidates of MMP-3, -9 and -14, including ECM proteins associated with the basement membrane, elastic fibers (emilin-1), and other extracellular proteins (periostin, tenascin-X). Proteolysis can also alter the function of many secreted bioactive molecules. In this respect, we recently described by shotgun proteomics approaches complement retention and proteolysis in conditioned media of intraluminal thrombus of AAA [12]. Since complement proteins are mainly synthesized by the liver, the increased levels observed in the thrombus is due to their retention from plasma. The highly enriched proteolytic conditions present in AAA thrombus induce the cleavage of complement C3 leading to C3a, which favor the recruitment and activation of neutrophils, playing a detrimental role in AAA.

Recently, LC-MS technologies have facilitated sensitive and high-throughput analysis of post-translational modifications (PTMs) in the setting of myocardial injury, where key proteins involved in calcium regulation have been found differentially affected by a variety of PTMs (e.g. phosphorylation and oxidative modifications) [44]. Due to the low stoichiometry of PTMs, global quantitation methods often involve a PTM-enrichment step or a fractionation procedure prior to LC-MS analysis [45].

Despite the development of rigorous statistical analyses to facilitate protein quantification in label-free proteomics experiments, stable isotope labeling (SIL) of peptides allows a very accurate relative quantitation of proteins and is the method of choice to detect changes in protein abundance across several samples.

Labeling Techniques for Quantitative Proteomics

SIL may be achieved by metabolic, chemical and enzymatic methods. Metabolic labeling techniques include stable isotope labeling by amino acids in cell culture (SILAC) and amino acid-coded mass tagging (AACM). These techniques incorporate isotopically labeled amino acids into cell culture media lacking their unlabeled counterparts. Although direct labeling of proteins within cells allows a full control of the experimental conditions, particularly when sample preparation requires considerable manipulation, this method requires a complete metabolic incorporation of labeled amino acids in cell cultures, and is not amenable to study samples that cannot be labeled (i.e. human tissues). In vitro labeling methods are based on a chemical reaction by which an isotope label is incorporated into the peptides produced by protein digestion. Derivatisation of cysteine residues with an isotope coded affinity tag (ICAT) [46, 47], or enzymatic labeling with ^{18}O [48, 49] are examples of this type of strategy.

The employment of isobaric tags for relative and absolute quantification (iTRAQ) reagents [50] is one of the most-widely used chemical labeling approaches. In this method, iTRAQ reagents incorporate isobaric mass labels to peptide N-termini and lysine side chains. The resulting derivatized peptides belonging to different samples are chemically indistinguishable, and having the same total mass are detected together as a single peptide ion. However, during MS/MS fragmentation of the peptide ion, the coding agents are cleaved producing reporter ions that have unit mass differences for the differentially coded peptides and allow relative protein abundance to be determined. The iTRAQ approach has two main advantages: in one hand, the signal intensity of the samples are added up, so that less amount of samples are required to achieve the same sensitivity. In the other hand, it allows the parallel quantitation of four samples at the same time. Recently, a new version of iTRAQ reagents have been developed that allows multiplexed quantitation of 8 samples [51]. The TMT approach is another example of stable isobaric labeling [52]; it has the same chemistry than iTRAQ, and allows protein quantitation in either 6-plex or 10-plex mode, although the later requires a minimum resolution in MS/MS mode that is only affordable in some types of mass spectrometers.

Several papers have resorted to stable isotope labeling in combination with gel-free proteomics approaches for the analysis of vascular samples [53–57]. As part of the Systems Approach to Biomarker Research in CVD Initiative, Yin et al. [54] identified several plasma protein biomarkers of new-onset myocardial infarction and atherosclerotic CVD in a very large CV cohort analyzed by discovery- (135 patients and 135 controls by iTRAQ-LC-MS) and targeted-based MS experiments (336 atherosclerotic CVD case–control pairs by MRM). This interesting work demonstrated that addition of a multiple-marker protein panel associated to risk of myocardial infarction or new-onset of atherosclerotic CVD increased the predictive value compared with a model with clinical risk factors alone. Another recent study has also addressed the proteomic profiling by both discovery and targeted MS analysis of plasma samples from patients undergoing a therapeutic planned myocardial infarction [58]. This is an excellent designed study that through improved methodological advances (multiplexed iTRAQ, intensive depletion of abundant plasma proteins, optimized fractionation methods, along with the use of the latest MS instrumentation) has been able to identify the largest number of proteins to date (around 5000). In this study, the biological variability was controlled by using serial samples from the same person at different time points, which is key on human studies.

Label-Free Shotgun Strategies for Quantitative Proteomics

Label-free shotgun approaches rely on the measurement of peptide MS/MS rates (spectral counts) or chromatographic peptide peak area (peptide-ion intensities). Spectral count-based strategies have been favored by the increase of MS scan speed, while the development of high-resolution instruments (e.g. orbitrap) has promoted peptide-ion intensity-based methods. Despite that label-free strategies have the advantage of a simplified sample preparation, powerful bioinformatic tools are required to extract, normalize and organize the highly complex MS data generated from this kind of experiments. In the last years, several tools have been developed for analyzing spectral count data [59–61] or chromatographic peak areas [62, 63], while other tools are able to retrieve and analyze information from both spectral counts and chromatographic peak areas [64, 65]. However, the choice of appropriate data analysis methods still remains a major challenge in label-free quantitative proteomics [66].

Label-free shotgun strategies have been used in several studies involving vascular samples [8, 44, 67–69]. Didangelos et al. analyzed the changes in ECM and associated proteins involved in the pathological remodeling process in AAA by means of an LC-MS, label-free approach [8]. Interestingly, ECM proteins and their degradation products were selectively extracted from aneurysmal and control aortas using a solubility-based sub-fractionation methodology. In another study [67], the authors address the proteome of circulating leukocytes of patients with coronary occlusion using iTRAQ reagents for the discovery phase followed by validation of some biomarker candidates by label free LC-MS/MS. The authors identified a bactericidal/permeability-increasing protein as a potential biomarker for severe atherosclerotic coronary stenosis.

Selected Reaction Monitoring (SRM) for Quantitative Proteomics

MS-based proteomics is a field of intense research mainly due to the need of high-throughput validation of biomarkers, given the limitations of targeted methods based on antibodies or other affinity capture techniques. Selected (or Multiple) Reaction Monitoring (SRM or MRM) is a highly selective MS scan mode for quantification of low-abundant proteins in protein mixtures and for characterization of post-translational modifications [70]. SRM is a powerful method exploiting the capabilities of triple quadrupole mass spectrometers to filter in specific precursor peptides from a few interesting proteins (e.g. potential biomarkers) in complex protein mixtures. In an SRM assay, the so-called transition filter is monitored with high selectivity: the first quadrupole, which is used as a mass filter, focused in the mass of the precursor peptide of interest, while a specific diagnostic peptide fragment is predefined in the third quadrupole; peptide fragmentation of the precursor peptide is achieved in the second quadrupole. Moreover, the chromatographic retention time can be used to increase the confidence of detection and quantification [71]. The multiplexing ability of SRM enables several precursor-diagnostic transitions to be monitored in a single LC–MS run [20].

Several examples of targeted MS-based proteomics has been recently published in the vascular field [72–74]. HDLs are modified during atherothrombosis leading to “dysfunctional” HDL particles. Oxidative modifications of ApoA1, the main constituent of HDL, have been observed in human atheroma [75]. In an interesting study, increased oxidation of ApoA1 has been quantified by SRM in HDL from patients with CVD and from control subjects [73]. In a recent study, we observed by SRM that systemic levels of hepcidin were increased in AAA patients compared to controls [74]. Hepcidin is a key peptide in the regulation of iron homeostasis as a major determinant of iron retention in phagocytes [76]. Our data derived from SRM experiments let us explain that the localized tissue iron that was overloaded in AAA could be associated to increased hepcidin systemic concentrations.

Concluding Remarks and Perspectives

High-throughput techniques have been used in the search of novel cardiovascular biomarkers. The study of vascular cells or tissues by proteomics approaches is expected to contribute to the identification of novel biomarkers implicated in the pathogenesis of atherothrombosis, as well as of potential therapeutic targets. Blood proteomics may also help unveiling biomarkers that could be used in the diagnosis and prediction of cardiovascular events. While proteomics is playing a major role in the discovery phase of potential biomarkers, validation has traditionally relied on antibody-based methods such as Western Blotting and ELISA. However, MS targeted-methods will become a major contributor for the quantification of these potential biomarkers and their modifications in the validation phase. In any case, even when a biomarker does not meet all the criteria to be translated into the clinical practice (specificity, sensitivity), it could be useful to understand the pathological mechanisms of the disease and may potentially be used as an imaging and therapeutic target. Due to the complex nature of atherothrombosis, it seems clear that a single marker measurement will not be enough to specifically detect a disease alone. In this respect, a multimarker approach has been suggested to provide higher sensitivity and specificity for cardiovascular diseases than is afforded using single markers considered individually [54]. On the whole, proteomics, in combination with other complementary approaches like genomics, metabolomics and imaging techniques, is emerging as a very powerful tool that is expected to improve the diagnosis and treatment of patients at high risk of suffering a cardiovascular event.

References

1. Bentzon JF, Otsuka F, Virmani R, Falk E. Mechanisms of plaque formation and rupture. Circ Res. 2014;114:1852–66.
2. Michel JB, Martin-Ventura JL, Egido J, Sakalihasan N, Treska V, Lindholt J, et al. Novel aspects of the pathogenesis of aneurysms of the abdominal aorta in humans. Cardiovasc Res. 2011;90:18–27.
3. Swirski FK, Nahrendorf M. Imaging macrophage development and fate in atherosclerosis and myocardial infarction. Immunol Cell Biol. 2013;91:297–303.
4. Tuñón J, Martín-Ventura JL, Blanco-Colio LM, Lorenzo O, López JA, Egido J. Proteomic strategies in the search of new biomarkers in atherothrombosis. J Am Coll Cardiol. 2010;55:2009–16.
5. Anidjar S, Salzmann JL, Gentric D, Lagneau P, Camilleri JP, Michel JB. Elastase-induced experimental aneurysms in rats. Circulation. 1990;82:973–81.
6. Zaragoza C, Gomez-Guerrero C, Martin-Ventura JL, Blanco-Colio L, Lavin B, Mallavia B, et al. Animal models of cardiovascular diseases. J Biomed Biotechnol. 2011;2011:497841.
7. Didangelos A, Yin X, Mandal K, Baumert M, Jahangiri M, Mayr M. Proteomics characterization of extracellular space components in the human aorta. Mol Cell Proteomics. 2010;9:2048–62.
8. Didangelos A, Yin X, Mandal K, Sajc A, Smith A, Xu Q, et al. Extracellular matrix composition and remodeling in human abdominal aortic aneurysms: a proteomics approach. Mol Cell Proteomics. 2011;10:M111.008128.
9. de la Cuesta F, Alvarez-Llamas G, Maroto AS, Donado A, Zubiri I, Posada M, et al. A proteomic focus on the alterations occurring at the human atherosclerotic coronary intima. Mol Cell Proteomics. 2011;10:M110.003517.
10. Boytard L, Spear R, Chinetti-Gbaguidi G, Acosta-Martin AE, Vanhoutte J, Lamblin N, et al. Role of proinflammatory CD68(+) mannose receptor(−) macrophages in peroxiredoxin-1 expression and in abdominal aortic aneurysms in humans. Arterioscler Thromb Vasc Biol. 2013;33:431–8.
11. Moreno JA, Ortega-Gómez A, Delbosc S, Beaufort N, Sorbets E, Louedec L, et al. In vitro and in vivo evidence for the role of elastase shedding of CD163 in human atherothrombosis. Eur Heart J. 2012;33:252–63.

12. Martinez-Pinna R, Madrigal-Matute J, Tarin C, Burillo E, Esteban-Salan M, Pastor-Vargas C, et al. Proteomic analysis of intraluminal thrombus highlights complement activation in human abdominal aortic aneurysms. Arterioscler Thromb Vasc Biol. 2013;33:2013–20.
13. Fu Z, Yan K, Rosenberg A, Jin Z, Crain B, Athas G, et al. Improved protein extraction and protein identification from archival formalin-fixed paraffin-embedded human aortas. Proteomics Clin Appl. 2013;7:217–24.
14. Blanco-Colio LM, Martín-Ventura JL, Muñóz-García B, Orbe J, Páramo JA, Michel JB, et al. Identification of soluble tumor necrosis factor-like weak inducer of apoptosis (sTWEAK) as a possible biomarker of subclinical atherosclerosis. Arterioscler Thromb Vasc Biol. 2007;27:916–22.
15. Dominguez-Rodriguez A, Abreu-Gonzalez P, Kaski JC. Inflammatory systemic biomarkers in setting acute coronary syndromes – effects of the diurnal variation. Curr Drug Targets. 2009;10:1001–8.
16. Fontaine V, Jacob MP, Houard X, Rossignol P, Plissonnier D, Angles-Cano E, et al. Involvement of the mural thrombus as a site of protease release and activation in human aortic aneurysms. Am J Pathol. 2002;161:1701–10.
17. Shuford CM, Hawkridge AM, Burnett JC, Muddiman DC. Utilizing spectral counting to quantitatively characterize tandem removal of abundant proteins (TRAP) in human plasma. Anal Chem. 2010;82:10179–85.
18. Tan SH, Mohamedali A, Kapur A, Baker MS. Ultradepletion of human plasma using chicken antibodies: a proof of concept study. J Proteome Res. 2013;12:2399–413.
19. Martin-Ventura JL, Madrigal-Matute J, Martinez-Pinna R, Ramos-Mozo P, Blanco-Colio LM, Moreno JA, et al. Erythrocytes, leukocytes and platelets as a source of oxidative stress in chronic vascular diseases: detoxifying mechanisms and potential therapeutic options. Thromb Haemost. 2012;108:435–42.
20. Barasa B, Slijper M. Challenges for red blood cell biomarker discovery through proteomics. Biochim Biophys Acta. 1844;2014:1003–10.
21. Burkhart JM, Gambaryan S, Watson SP, Jurk K, Walter U, Sickmann A, et al. What can proteomics tell us about platelets? Circ Res. 2014;114:1204–19.
22. Pasini EM, Lutz HU, Mann M, Thomas AW. Red blood cell (RBC) membrane proteomics – part II: comparative proteomics and RBC patho-physiology. J Proteomics. 2010;73:421–35.
23. Asztalos BF, Schaefer EJ. High-density lipoprotein subpopulations in pathologic conditions. Am J Cardiol. 2003;91:12E–7.
24. Gordon SM, Deng J, Tomann AB, Shah AS, Lu LJ, Davidson WS. Multi-dimensional co-separation analysis reveals protein-protein interactions defining plasma lipoprotein subspecies. Mol Cell Proteomics. 2013;12:3123–34.
25. Loyer X, Vion AC, Tedgui A, Boulanger CM. Microvesicles as cell-cell messengers in cardiovascular diseases. Circ Res. 2014;114:345–53.
26. Palmisano G, Jensen SS, Le Bihan MC, Lainé J, McGuire JN, Pociot F, et al. Characterization of membrane-shed microvesicles from cytokine-stimulated β-cells using proteomics strategies. Mol Cell Proteomics. 2012;11:230–43.
27. McGregor E, Dunn MJ. Proteomics of heart disease. Hum Mol Genet. 2003;12 Spec No 2:R135–44.
28. Görg A, Weiss W, Dunn MJ. Current two-dimensional electrophoresis technology for proteomics. Proteomics. 2004;4:3665–85.
29. Unlü M, Morgan M, Minden J. Difference gel electrophoresis: a single gel method for detecting changes in protein extracts. Electrophoresis. 1997;18:2071–7.
30. Alban A, David SO, Bjorkesten L, Andersson C, Sloge E, Lewis S, et al. A novel experimental design for comparative two-dimensional gel analysis: two-dimensional difference gel electrophoresis incorporating a pooled internal standard. Proteomics. 2003;3:36–44.
31. Lepedda AJ, Cigliano A, Cherchi GM, Spirito R, Maggioni M, Carta F, et al. A proteomic approach to differentiate histologically classified stable and unstable plaques from human carotid arteries. Atherosclerosis. 2009;203:112–8.
32. Martinez-Pinna R, Ramos-Mozo P, Madrigal-Matute J, Blanco-Colio LM, Lopez JA, Calvo E, et al. Identification of peroxiredoxin-1 as a novel biomarker of abdominal aortic aneurysm. Arterioscler Thromb Vasc Biol. 2011;31:935–43.

33. Lepedda AJ, Nieddu G, Zinellu E, De Muro P, Piredda F, Guarino A, et al. Proteomic analysis of plasma-purified VLDL, LDL, and HDL fractions from atherosclerotic patients undergoing carotid endarterectomy: identification of serum amyloid A as a potential marker. Oxid Med Cell Longev. 2013;2013:385214.
34. Michel JB, Virmani R, Arbustini E, Pasterkamp G. Intraplaque haemorrhages as the trigger of plaque vulnerability. Eur Heart J. 2011;32:1977–85, 85a, 85b, 85c.
35. Hellings WE, Peeters W, Moll FL, Piers SR, van Setten J, Van der Spek PJ, et al. Composition of carotid atherosclerotic plaque is associated with cardiovascular outcome: a prognostic study. Circulation. 2010;121:1941–50.
36. De Das S, Krishna S, Jethwa A. Iron status and its association with coronary heart disease: systematic review and meta-analysis of prospective studies. Atherosclerosis. 2015;238: 296–303.
37. Hanash SM. Biomedical applications of two-dimensional electrophoresis using immobilized pH gradients: current status. Electrophoresis. 2000;21:1202–9.
38. Link AJ, Eng J, Schieltz DM, Carmack E, Mize GJ, Morris DR, et al. Direct analysis of protein complexes using mass spectrometry. Nat Biotechnol. 1999;17:676–82.
39. Stegemann C, Didangelos A, Barallobre-Barreiro J, Langley SR, Mandal K, Jahangiri M, et al. Proteomic identification of matrix metalloproteinase substrates in the human vasculature. Circ Cardiovasc Genet. 2013;6:106–17.
40. Kaga E, Karademir B, Baykal AT, Ozer NK. Identification of differentially expressed proteins in atherosclerotic aorta and effect of vitamin E. J Proteomics. 2013;92:260–73.
41. Dinardo CL, Venturini G, Zhou EH, Watanabe IS, Campos LC, Dariolli R, et al. Variation of mechanical properties and quantitative proteomics of VSMC along the arterial tree. Am J Physiol Heart Circ Physiol. 2014;306:H505–16.
42. Delporte C, Noyon C, Raynal P, Dufour D, Nève J, Abts F, et al. Advancement in stationary phase for peptide separation helps in protein identification: application to atheroma plaque proteomics using nano-chip liquid chromatography and mass spectrometry. J Chromatogr A. 2015;1385:116–23.
43. Newby AC. Matrix metalloproteinase inhibition therapy for vascular diseases. Vascul Pharmacol. 2012;56:232–44.
44. Smith LE, White MY. The role of post-translational modifications in acute and chronic cardiovascular disease. Proteomics Clin Appl. 2014;8:506–21.
45. Agnetti G, Husberg C, Van Eyk JE. Divide and conquer: the application of organelle proteomics to heart failure. Circ Res. 2011;108:512–26.
46. Gygi SP, Rist B, Gerber SA, Turecek F, Gelb MH, Aebersold R. Quantitative analysis of complex protein mixtures using isotope-coded affinity tags. Nat Biotechnol. 1999;17:994–9.
47. von Haller PD, Yi E, Donohoe S, Vaughn K, Keller A, Nesvizhskii AI, et al. The application of new software tools to quantitative protein profiling via isotope-coded affinity tag (ICAT) and tandem mass spectrometry: II. Evaluation of tandem mass spectrometry methodologies for large-scale protein analysis, and the application of statistical tools for data analysis and interpretation. Mol Cell Proteomics. 2003;2:428–42.
48. Martinez-Acedo P, Núñez E, Gómez FJS, Moreno M, Ramos E, Izquierdo-Álvarez A, et al. A novel strategy for global analysis of the dynamic thiol redox proteome. Mol Cell Proteomics. 2012;11:800–13.
49. Bonzon-Kulichenko E, Perez-Hernandez D, Nunez E, Martinez-Acedo P, Navarro P, Trevisan-Herraz M, et al. A robust method for quantitative high-throughput analysis of proteomes by 18O labeling. Mol Cell Proteomics. 2010;10:M110.003335–M110.
50. Ross PL, Huang YN, Marchese JN, Williamson B, Parker K, Hattan S, et al. Multiplexed protein quantitation in Saccharomyces cerevisiae using amine-reactive isobaric tagging reagents. Mol Cell Proteomics. 2004;3:1154–69.
51. Ow SY, Cardona T, Taton A, Magnuson A, Lindblad P, Stensjo K, et al. Quantitative shotgun proteomics of enriched heterocysts from Nostoc sp. PCC 7120 using 8-plex isobaric peptide tags. J Proteome Res. 2008;7:1615–28

52. Viner RI, Zhang T, Second T, Zabrouskov V. Quantification of post-translationally modified peptides of bovine alpha-crystallin using tandem mass tags and electron transfer dissociation. J Proteomics. 2009;72:874–85.
53. Prentice RL, Paczesny S, Aragaki A, Amon LM, Chen L, Pitteri SJ, et al. Novel proteins associated with risk for coronary heart disease or stroke among postmenopausal women identified by in-depth plasma proteome profiling. Genome Med. 2010;2:48.
54. Yin X, Subramanian S, Hwang SJ, O'Donnell CJ, Fox CS, Courchesne P, et al. Protein biomarkers of new-onset cardiovascular disease: prospective study from the systems approach to biomarker research in cardiovascular disease initiative. Arterioscler Thromb Vasc Biol. 2014;34:939–45.
55. Jorge I, Burillo E, Mesa R, Baila-Rueda L, Moreno M, Trevisan-Herraz M, et al. The human HDL proteome displays high inter-individual variability and is altered dynamically in response to angioplasty-induced atheroma plaque rupture. J Proteomics. 2014;106:61–73.
56. Moxon JV, Liu D, Moran CS, Crossman DJ, Krishna SM, Yonglitthipagon P, et al. Proteomic and genomic analyses suggest the association of apolipoprotein C1 with abdominal aortic aneurysm. Proteomics Clin Appl. 2014;8:762–72.
57. Burillo E, Lindholt JS, Molina-Sanchez P, Jorge I, Martinez-Pinna R, Blanco-Colio LM, et al. ApoA-I/HDL-C levels are inversely associated with abdominal aortic aneurysm progression. Thromb Haemost. 2015;113:1335–46.
58. Keshishian H, Burgess MW, Gillette MA, Mertins P, Clauser KR, Mani DR, et al. Multiplexed, quantitative workflow for sensitive biomarker discovery in plasma yields novel candidates for early myocardial injury. Mol Cell Proteomics. 2015;14:2375–93.
59. Park SK, Venable JD, Xu T, Yates JR. A quantitative analysis software tool for mass spectrometry-based proteomics. Nat Methods. 2008;5:319–22.
60. Searle BC. Scaffold: a bioinformatic tool for validating MS/MS-based proteomic studies. Proteomics. 2010;10:1265–9.
61. Heinecke NL, Pratt BS, Vaisar T, Becker L. PepC: proteomics software for identifying differentially expressed proteins based on spectral counting. Bioinformatics. 2010;26:1574–5.
62. Mueller LN, Rinner O, Schmidt A, Letarte S, Bodenmiller B, Brusniak MY, et al. SuperHirn – a novel tool for high resolution LC-MS-based peptide/protein profiling. Proteomics. 2007;7: 3470–80.
63. Bouyssié D, Gonzalez de Peredo A, Mouton E, Albigot R, Roussel L, Ortega N, et al. Mascot file parsing and quantification (MFPaQ), a new software to parse, validate, and quantify proteomics data generated by ICAT and SILAC mass spectrometric analyses: application to the proteomics study of membrane proteins from primary human endothelial cells. Mol Cell Proteomics. 2007;6:1621–37.
64. Cox J, Mann M. MaxQuant enables high peptide identification rates, individualized p.p.b.-range mass accuracies and proteome-wide protein quantification. Nat Biotechnol. 2008;26:1367–72.
65. Cox J, Hein MY, Luber CA, Paron I, Nagaraj N, Mann M. Accurate proteome-wide label-free quantification by delayed normalization and maximal peptide ratio extraction, termed MaxLFQ. Mol Cell Proteomics. 2014;13:2513–26.
66. Goeminne LJ, Argentini A, Martens L, Clement L. Summarization vs peptide-based models in label-free quantitative proteomics: performance, pitfalls, and data analysis guidelines. J Proteome Res. 2015;14:2457–65.
67. Bleijerveld OB, Wijten P, Cappadona S, McClellan EA, Polat AN, Raijmakers R, et al. Deep proteome profiling of circulating granulocytes reveals bactericidal/permeability-increasing protein as a biomarker for severe atherosclerotic coronary stenosis. J Proteome Res. 2012;11:5235–44.
68. Rocchiccioli S, Pelosi G, Rosini S, Marconi M, Viglione F, Citti L, et al. Secreted proteins from carotid endarterectomy: an untargeted approach to disclose molecular clues of plaque progression. J Transl Med. 2013;11:260.
69. Kessler T, Zhang L, Liu Z, Yin X, Huang Y, Wang Y, et al. ADAMTS-7 inhibits re-endothelialization of injured arteries and promotes vascular remodeling through cleavage of thrombospondin-1. Circulation. 2015;131:1191–201.

70. Blanco-Colio LM, López JA, Martínez-Pinna Albar R, Egido J, Martín-Ventura JL. Vascular proteomics, a translational approach: from traditional to novel proteomic techniques. Expert Rev Proteomics. 2009;6:461–4.
71. Pernemalm M, Lehtiö J. Mass spectrometry-based plasma proteomics: state of the art and future outlook. Expert Rev Proteomics. 2014;11:431–48.
72. Tölle M, Huang T, Schuchardt M, Jankowski V, Prüfer N, Jankowski J, et al. High-density lipoprotein loses its anti-inflammatory capacity by accumulation of pro-inflammatory-serum amyloid A. Cardiovasc Res. 2012;94:154–62.
73. Yassine HN, Jackson AM, Reaven PD, Nedelkov D, Nelson RW, Lau SS, et al. The application of multiple reaction monitoring to assess Apo A-I methionine oxidations in diabetes and cardiovascular disease. J Transl Proteomics. 2014;4–5:18–24.
74. Martinez-Pinna R, Lindholt JS, Madrigal-Matute J, Blanco-Colio LM, Esteban-Salan M, Torres-Fonseca MM, et al. From tissue iron retention to low systemic haemoglobin levels, new pathophysiological biomarkers of human abdominal aortic aneurysm. Thromb Haemost. 2014;112:87–95.
75. Huang Y, Di Donato JA, Levison BS, Schmitt D, Li L, Wu Y, et al. An abundant dysfunctional apolipoprotein A1 in human atheroma. Nat Med. 2014;20:193–203.
76. Theurl I, Theurl M, Seifert M, Mair S, Nairz M, Rumpold H, et al. Autocrine formation of hepcidin induces iron retention in human monocytes. Blood. 2008;111:2392–9.

Chapter 6
Stem Cell Proteomics

Matthew Waas, Stephen Farris, Erin M. Kropp, Jennifer A. Beckman, April Stempien-Otero, Claudius Mahr, and Rebekah L. Gundry

Abstract Recent advances in stem cell and proteomic technology carry tremendous potential to impact our understanding of mechanistic underpinnings and fundamental pathophysiology of cardiovascular disease. In this chapter, we introduce investigators new to these disciplines to the various types of stem cells relevant to cardiovascular biology and how, when combined with state-of-the-art proteomic analyses, they may be exploited for mechanistic and translational studies related to cardiomyopathies, coronary atherosclerotic disease, and heart failure. Although the potential of these emerging technologies is just beginning to be explored, this chapter aims to illustrate how integration of novel stem cell and proteomic technologies is poised to make significant contributions to future advanced therapies and diagnostics in cardiovascular medicine.

Keywords Stem cells • Cardiomyopathy • Coronary atherosclerotic disease • Heart failure • Disease modeling • Cell surface markers • Epigenetics • Cell surface proteomics • Histone proteomics • Top down proteomics

M. Waas • E.M. Kropp • R.L. Gundry, PhD (✉)
Department of Biochemistry, Medical College of Wisconsin,
Milwaukee, WI 53226, USA
e-mail: mwaas@mcw.edu; ekropp@mcw.edu; rgundry@mcw.edu

S. Farris, MD • J.A. Beckman, MSN, BSN • A. Stempien-Otero, MD
Division of Cardiology, Department of Internal Medicine,
University of Washington Medical Center, Seattle, WA 98195, USA
e-mail: sfarris@cardiology.washington.edu; jabeckma@uw.edu; april@uw.edu

C. Mahr, DO, FACC, FESC
Division of Cardiology, Department of Internal Medicine,
University of Washington Medical Center, Seattle, WA 98195, USA

Department of Internal Medicine, University of Washington, Seattle, WA, USA
e-mail: cmahr@uw.edu

G. Agnetti et al. (eds.), *Manual of Cardiovascular Proteomics*,
DOI 10.1007/978-3-319-31828-8_6

Introduction

The fusion of modern mass spectrometry and proteomic strategies with state-of-the-art stem cell technologies provides a powerful platform for addressing outstanding challenges and questions in cardiovascular medicine and cardiac biology. Investigators new to these fields may find it daunting to navigate the complexities, especially as analytical instrumentation and their respective capabilities are rapidly evolving and stem cell culturing and differentiation protocols are continually improving our ability to generate better defined progeny of diverse lineages. Moreover, as these technologies and applications to cardiovascular medicine largely remain in their infancy, specific challenges must be recognized and addressed before the full potential of these technologies can be realized. In this chapter, we merge basic science and clinical perspectives to provide an overview apt for clinicians, students, and scientists alike. We begin by defining various classes of stem cells relevant to cardiovascular biology, subsequently provide clinically-focused examples illustrating the potential impact of various stem cell technologies for mechanistic and translational studies, and conclude with an overview of state-of-the-art proteomic strategies especially relevant for addressing outstanding questions related to stem cell and cardiovascular biology.

Unique Properties of Stem Cells

Stem cells are defined by an ability to self-renew indefinitely and to differentiate into at least one cell type other than itself. Several types of stem cells exist and vary by location and niche, developmental stage when they are present, and ability to form differentiated progeny. Some stem cell lines are representative of early developmental stages and have greater potential to form multiple cell types and organs (*e.g.* embryonic stem cells and neuronal stem cells). Other stem cells are present in developed adults, but are more limited in their differentiation potential and niche. For example, intestinal stem cells are located in intestinal crypts and give rise to four main cell types making up intestinal epithelial lining. For scientists new to this field, the number of stem cell types, culturing strategies, differentiation protocols, and in some cases, controversy surrounding their existence and definition, can appear overwhelming. Moreover, despite increasing media attention, the specific properties of cells optimal for mechanistic and translational studies remain to be fully established. In this section, we briefly outline several stem cell types with particular relevance to the cardiovascular field and describe their origins, differentiation potential, how they are defined, and provide an overview of their relevance to cardiovascular pathophysiology.

Pluripotent Stem Cells

Human pluripotent stem cells (hPSC), which include embryonic and induced pluripotent stem cells, are defined by an ability to differentiate into cell types from each of the three germ layers (*i.e.* endoderm, ectoderm, mesoderm). Human embryonic stem cells (hESC) are derived from the inner cell mass of the blastocyst stage embryo [1]. Human induced pluripotent stem cells (hiPSC), which have functional potential similar to hESC, can be derived from explanted somatic cells (*e.g.* fibroblasts, mononuclear cells) by using any one of a number of strategies (*e.g.* virus, protein, mRNA, chemicals) to overexpress transcription factors responsible for maintenance of pluripotency [2, 3]. While hiPSC were originally generated using retroviral integration of four transcription factors [4, 5], modern protocols do not necessitate viral integration into the genome [6, 7] and are continually evolving to accommodate reprogramming of a variety of somatic cell types (reviewed in Rony et al. [8]). Currently, hESC and hiPSC are widely used *in vitro* to generate various cell types including neural, cardiac, hepatic, and retinal epithelial cells. Of particular relevance to the cardiovascular field, both hiPSC and hESC can be used to generate cardiomyocytes, endothelial cells and vascular smooth muscle *in vitro* for mechanistic and translational studies. Furthermore, hiPSC can be derived from patients to generate cell types containing a specific genotype of interest.

Adult Stem Cells

Found throughout the body, adult stem cells are typically responsible for maintaining differentiated cell populations characterized by high turnover (*e.g.* blood, skin, gastrointestinal tract). These cells are commonly encountered in a specific niche and their differentiation potential ranges from multipotent to unipotent (*i.e.* ability to differentiate into multiple or one cell type, respectively). Of the many adult stem cell types described, two have received significant attention regarding their utility in cardiovascular research and medicine. Bone marrow-derived mononuclear cells (BMMC) are heterogeneous cell populations in the bone marrow defined by morphology (*e.g.* nuclear morphology and cytoplasm without granules). BMMC contain stem cells that give rise to hematopoietic cells and can potentially differentiate into non-hematopoietic cells. Furthermore, they are a viable source for generating patient-derived hiPSC [9]. Mesenchymal stem cells (MSC) are an adult stem cell population found in bone marrow, but have also been isolated from other tissues (*e.g.* adipose) [10]. MSC are defined by their ability to adhere to plastic; expression of cell surface markers CD105, CD73, and CD90; and have the capacity to differentiate into osteoblasts, adipocytes, and chondroblasts [9]. MSC and BMMC are relevant to cardiovascular biology and therapy based on their effects to modulate immune response, cardiomyocyte cell survival and remodeling via direct and indirect interactions. Moreover, they are useful in mechanistic studies of vascular disease and hypertension [11–13].

Other Stem Cell Like Populations

Additional cell types of relevance to the cardiovascular field include progenitor cells, which are similar to stem cells in that they can differentiate into one or more cell types but are incapable of indefinite self-renewal [14]. Several populations of adult resident "cardiac stem cells" and "cardiac progenitor cells" have been described, including KIT+, ISL1+, SCA1+, side-population, and cardiosphere-derived progenitor cells (see recent reviews [15, 16]); however, the definition, existence, prevalence, functional potential, and utility of these populations are not entirely defined and some are fervently debated [17–20]. Vascular wall progenitor cells, including endothelial, smooth muscle, and multipotent vascular progenitor cells, are also highly relevant to cardiovascular biology, and their diverse properties, functions, roles in disease, and therapeutic potential have been the subject of an excellent recent review [21]. Finally, direct reprogramming – the delivery of cardiac transcription factors or miRNA to induce cellular changes in epigenetics and gene regulation – can stimulate terminally differentiated fibroblasts to become induced cardiomyocytes (iCM) [22]. This approach is being pursued both for directly reprogramming mouse cells *ex vivo* and *in vitro* [22, 23] and human cells *ex vivo* [24], and may avoid concerns such as tumorigenicity associated with hPSC derived cardiomyocytes. However, variations in efficiency and reproducibility will need to be addressed (reviewed in [25, 26]).

Stem Cell

A cell that possesses two fundamental properties: (1) indefinite self-renewal – where a cell proceeds through cell division and maintains its undifferentiated state and (2) the ability to differentiate into at least one cell type other than itself. Stem cells are classified according to their differentiation potential.

Totipotent: the ability to differentiate into all extraembryonic and embryonic cells in an organism (e.g. zygote)
Pluripotent: the ability to differentiate into cells from any of the three germ layers (e.g. embryonic stem cell)
Multipotent: the ability to differentiate into multiple, but limited, cell types (e.g. mesenchymal stem cell)
Unipotent: the ability to differentiate into a single cell type (e.g. spermatogenic stem cell)

Human Pluripotent Stem Cell Derivatives for Studies on Cardiovascular Biology and Disease

Of the stem cell types relevant to cardiovascular medicine, the most versatile are hPSC owing to their differentiation potential and utility for modeling early human development. Differentiation strategies for cardiomyocytes (hPSC-CM) have significantly advanced over the past decade (reviewed in [27, 28]). In general, contemporary protocols are capable of reproducibly generating cultures >80 % positive for cardiomyocyte markers and are achieved by modulation of Wnt signaling pathways

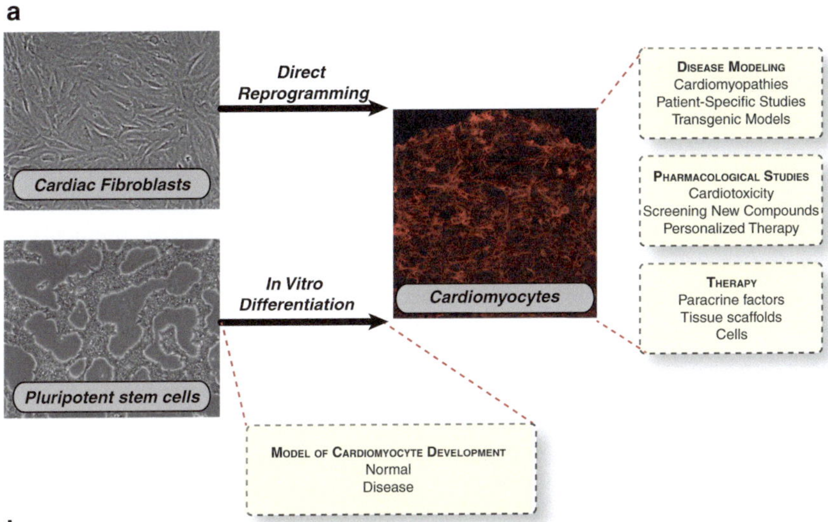

b

CLASS	TECHNIQUES	METRICS
Morphology	Microscopy, Immunocytochemistry	Ultrastructure, cell shape/size, contractility
Transcript Abundance	qRT-PCR, Microarray, RNA-Seq	Reference markers
Protein Abundance	Flow Cytometry, Immunocytochemistry	Reference markers, subpopulation heterogeneity
Functional Tests	Calcium Handling, Electrophysiology	Baseline profiles, response to cardioactive drugs
Bioenergetics	Extracellular flux, Imaging	Mitochondrial organization, Metabolic function

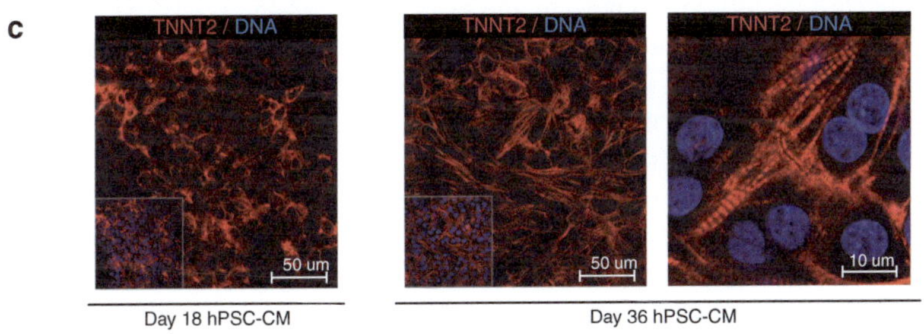

Fig. 6.1 (**a**) Summary of *in vitro* approaches for generating cardiomyocytes and their potential utility in mechanistic and translational studies. (**b**) Common techniques used to characterize hPSC-CM and iCM phenotypes. (**c**) Immunocytochemistry of hPSC-CM on days 18 and 36 of *in vitro* differentiation illustrating transition from a rounded phenotype at early stages to a more elongated phenotype at later stages. *Inset* shows overlay of TNNT2 and DNA stain

on cells cultured in two and three-dimensional platforms [29–31]. Spontaneously contracting cells can be produced in as few as 7–8 days [30] with early stage cells having rounded morphology and later stage cells exhibiting elongated morphology, sarcomeric organization, and potential gap junctions [32, 33] (Fig. 6.1). Despite recent developments in differentiation strategies to produce cells that robustly express cardiomyocyte specific genes, hPSC-CM cultures are heterogeneous with respect to functional subtype identity and fail to generate a phenotype consistent with adult cardiomyocytes [28]. Rather, gene expression and functional phenotypes of hPSC-CM mimic cardiomyocytes from the embryonic or fetal heart as evidenced by the lack of key features such as T-tubule structure, well-formed sarcomeres and gap junctions, and multi-nucleation [28, 34–36]. As a result, strategies to drive maturation towards a phenotype more closely resembling cardiomyocytes of the adult heart are being pursued, which include mechanical and electrical stimulation, manipulation of energy substrates and/or hormones, and co-culturing with vascular constructs [33, 37–42]. However, a single protocol to efficiently produce adult-like cells has not yet been widely adopted. In addition to cardiomyocytes, protocols for generating vascular smooth muscle cells (hPSC-VSMC), endothelial cells (hPSC-EC), and pericytes, have also been described [43–46]. Overall, these protocols have been successful in generating cells expressing appropriate markers, and phenotypic differences between patient and control cells correlate with those found in disease (e.g. Hutchinson-Gilford progeria and supravalvular aortic stenosis [47–49]). Further efforts to define positive and negative markers for hPSC-VSMCs, hPSC-EC, and pericytes will facilitate the use and improvement of differentiation models of these cell types.

Although differentiation protocols will undoubtedly continue to evolve for all hPSC-derivatives, progeny generated with current strategies have already begun to demonstrate potential utility for numerous applications (Fig. 6.1). First, the timing and hierarchy of molecular events during *in vitro* differentiation closely mimic those of human development [50], and thus allow for study of molecular dynamics during very early stages of human development impossible to study *in vivo*. Second, these models generate unlimited numbers of cells that can be used for studying cellular biology in normal and disease states [45, 47–49, 51]. Third, as cardiotoxicity studies traditionally employ either animal models or overexpression systems (*e.g.* CHO cells) to predict potential harmful effects of compounds on cardiac function, current evidence suggests it may be possible to use hPSC-CM for these studies as a more accurate alternative system of assessing potential cardiotoxic effects of compounds in pre-clinical studies [52, 53]. In this way, toxicity studies could expand to routinely include human cells of diverse genetic origins to evaluate drug efficacy and safety according to genotype and promote advancement towards patient-specific medical therapy (*i.e.* "personalized medicine"). Fourth, as modern reagents and protocols enable efficient transfection of hPSC [54–56], the use of gene editing strategies (*e.g.* knockout, knockdown) may provide a pragmatic alternative to animal models for initial inquiries into functional consequences of genetic mutations during early development and studies of novel protein function in a cell type specific manner. Fifth, as discussed in more detail below, patient-specific cells can be used to study cellular phenotypes and molecular mechanisms resulting from specific genetic mutations and it may become possible to use iCM or hPSC-derivatives to repair damaged tissue and improve cardiac function by transplanting cells, tissue constructs, or secreted factors. Independent of whether the goal is revascularization, transplantation of cells or tissues that functionally integrate into the host myocardium, or to provide protective effects against further damage, modern stem cell technologies have the potential to revolutionize therapeutic options for cardiovascular disease.

Translational Applications of Stem Cell Proteomics

Recent advances in stem cell technologies and proteomic technologies carry tremendous potential to impact our understanding of mechanistic underpinnings and fundamental pathophysiology of cardiovascular disease, ultimately leading to improved clinical care. From less common monogenic disorders such as Hypertrophic Cardiomyopathy (HCM) and Congenital Long QT Syndrome (LQTS) to highly prevalent polygenic syndromes such as hypertension (HTN), coronary artery disease (CAD) and heart failure (HF), stem cells and their derivatives are increasingly investigated for their utility in both research and clinical use. Recent reviews have comprehensively summarized ongoing efforts in the cardiovascular field to generate stem cell derivatives apt for pharmacological testing [57, 58], disease modeling [59–62], and the current state of stem cell-based clinical trials [63, 64]. In this section, we provide three clinically-focused examples to demonstrate the applicability of various stem cell technologies to address specific clinical needs in cardiovascular medicine. Within each example, we briefly outline currently outstanding questions that could be addressed using proteomic technologies (Fig. 6.2).

Example 1: Inherited Cardiac Disease

Clinical Overview

HCM is the most common heritable cardiovascular disorder with an estimated prevalence of 1:500 in the U.S [65, 66]. Transmitted in an autosomal dominant manner, HCM is the most frequent cause of sudden death in young adults [66]. Genetic missense or nonsense mutations in cardiomyocyte sarcomere genes, most frequently beta myosin heavy chain or myosin binding protein C, induce cardiomyocyte dysfunction [67]. Sarcomere mutation induces abnormal calcium sensitivity, myofibrillar disarray, hypertrophy, and cardiomyocyte apoptosis. These changes manifest clinically with variable penetrance and severity even within a single family. Classically, patients develop focal hypertrophy, often confined to the interventricular septum. Left ventricular outflow tract obstruction can occur and pathologically elevated filling pressures are prominent in some forms of the disease resulting in exertional chest pain and dyspnea. Cardiac fibrosis and ventricular arrhythmias are also frequent. There are no known therapies to ameliorate either pathologic remodeling or occurrence of arrhythmias.

Another relatively common genetic disorder is Congenital Long QT syndrome (LQTS). Loss-of-function mutations in either the slow rectifying K^+ channel (KCNQ1 which causes LQTS-1) or the fast rectifying K^+ channel (KCNH2 which causes LQTS-2) occur in 50–80 % of cases [68]. These mutations slow repolarization of cardiomyocyte action potentials, pathologically prolonging the QT segment. This delayed repolarization lengthens the QT portion of the electrocardiogram (ECG) and thus predisposes patients to polymorphic ventricular tachycardia and sudden cardiac

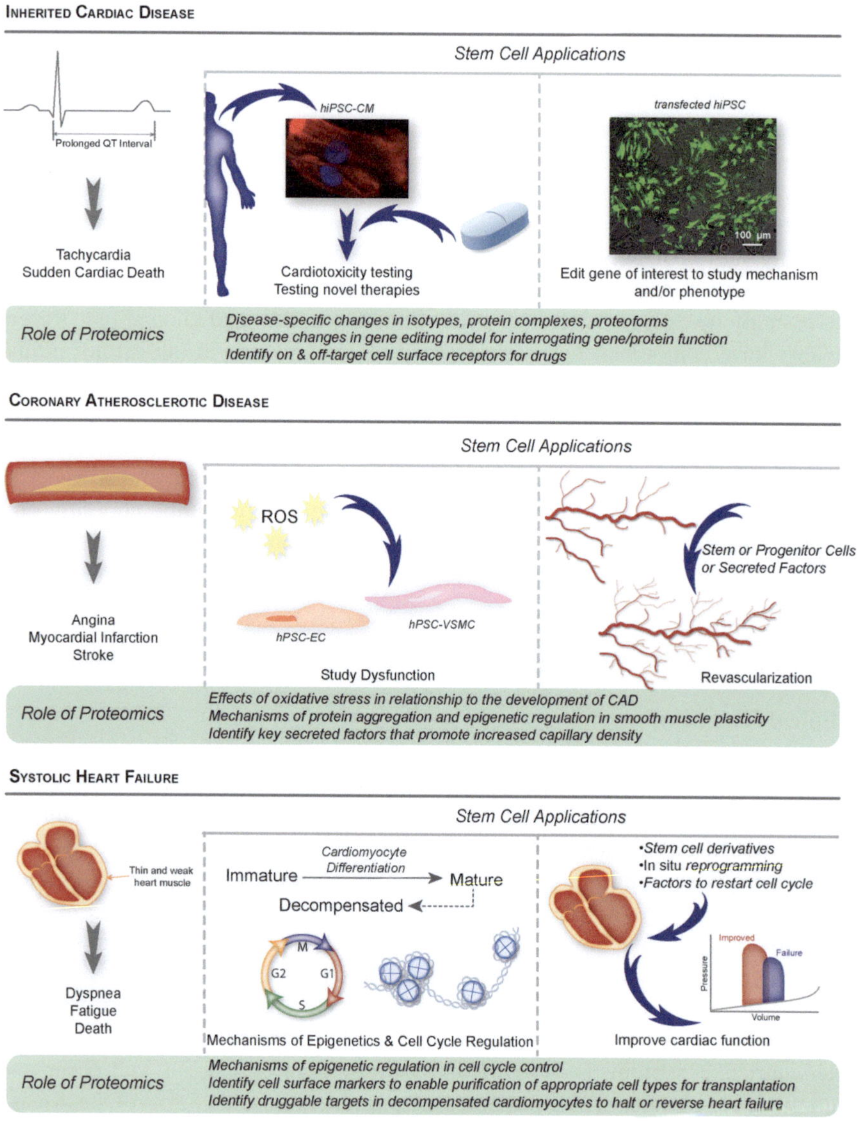

Fig. 6.2 Graphical representations of three major types of cardiovascular disease. For each disease type, examples of how stem cells, stem cell derivatives and proteomic technologies may be applied to address outstanding questions and challenges in research and clinical applications are illustrated

death. Although beta-blockers may decrease risk of arrhythmia, implantable cardiac defibrillators are often indicated. At this time in the U.S., patients are not routinely screened by ECG for LQTS or other cardiomyopathies, with the exception of competitive athletes. Tragically, this disease is often discovered only after exposure to medications with QT prolonging side effects such as fluoroquinolone antibiotics.

Stem Cell Applications

Though HCM and LQT syndromes are highly prevalent, current therapies are often insufficient due to limited knowledge of basic mechanical and electrophysiological properties of adult cardiomyocytes, such as the definition, evaluation and treatment of diastolic dysfunction on the cellular and myofibrillar level, impact of fibroblasts and fibrosis on cardiomyocyte electromechanics, and physiology of ion channel dysfunction and propagation of malignant arrhythmias. Currently, major obstacles to using primary human adult cardiomyocytes for *in vitro* studies include technical challenges associated with isolating live cells with high yield, they fail to undergo cell division, and they are relatively challenging to culture *ex vivo*. Thus, there is understandable enthusiasm for use of patient-specific hiPSC-CM or directly reprogrammed iCM to study inherited diseases as they carry patient-specific gene mutation(s) contributing to variable penetrance and phenotype. To date, hiPSC-CM and hiPSC-EC lines from numerous inherited cardiomyopathies have been developed and shown to recapitulate physiological properties of the disease [45, 62, 69] and reviewed in [53, 70, 71]. Accordingly, this approach to modeling cardiovascular "disease in a dish" may be used for screening potential pharmacological therapies and determining molecular mechanisms of disease similar to seminal studies showing genetic mutations affect localization of KCNQ1 [72] and effects of pharmacotherapies on calcium transients and action potential duration in hPSC-CM containing Timothy syndrome [73]. One example of ongoing efforts in this area is the NIH-sponsored development of a "Biorepository of Human iPSCs for Studying Dilated and Hypertrophic Cardiomyopathy" at Stanford University to generate a bank of 600 hiPSC lines from control and cardiomyopathy patients for use in genetic and drug screening studies and distribution to the scientific community.

Role of Proteomics

Mutations in HCM result in a large range of clinical phenotypes, including heterogeneous chamber morphology, contractile function, and arrhythmic risk. Although many specific genetic mutations have been identified, genotype-phenotype correlation is variable. Of particular interest are mutations in sarcomere proteins myosin heavy chain and troponin carrying increased risk of sudden death [74–78]. Coupling patient specific hiPSC-CM with proteomic approaches could be used to determine if the variable occurrence of sudden cardiac death is related to other important processes such as changes in calcium handling or responsiveness to catecholamines [79, 80]. Proteomic approaches, such as top-down analysis of intact proteins and quantitative analysis of subcellular proteomes, could be utilized to determine how specific gene mutations affect protein post-translational modifications and to determine effects on localization of calcium handling proteins to uncover pathways leading from myofibril disarray to impaired relaxation. These studies would help to elicit a mechanism for the physiological observations such as variations in catecholamine sensitivity and electrophysiological profiles observed in inherited cardiomyopathies and LQTS. Moreover, modern proteomic strategies that allow for identification of ligand-receptor interactions could prove invaluable for determining which mutations are susceptible to off-target drug effects. Importantly, patient-specific hiPSC-CM provide a platform to make these studies feasible, providing sufficient material for proteomic and molecular studies.

Example 2: Coronary Atherosclerotic Disease

Clinical Overview

Coronary atherosclerotic disease (CAD) is a chronic, progressive, complex polygenic disorder with many well-established risk factors including gender, advanced age, hypertension, elevated low density lipoprotein levels, low levels of high density lipoprotein, diabetes and tobacco use [81]. In CAD, pro-atherogenic lipoproteins infiltrate the vessel wall and initiate development of lipid-rich plaques. Retained lipids subsequently induce a positive feedback loop of chronic inflammation, progressive lipid deposition, proteolytic medial degeneration and formation of vulnerable, thin-capped, lipid-rich atheromas. The presentation of CAD is highly variable, ranging from sudden spontaneous plaque rupture and acute myocardial infarction, to sub-clinical ischemia, or progressive angina. Questions remain surrounding pathogenesis of CAD including outcome prediction in high-risk patient cohorts, mechanisms of plaque rupture and reverse cholesterol transport. For example, despite contemporary medical therapy for primary and secondary prevention, many patients experience progressive increase in plaque burden or recurrent myocardial infarction. Thus, improved prediction models and therapies are needed to provide benefit beyond traditional risk models (*e.g.* Framingham [81, 82]) and inflammatory markers (*e.g.* C-reactive protein [83]).

Stem Cell Applications

Percutaneous or surgical revascularization improves myocardial blood flow and reduces mortality in ischemic heart disease. However, despite restoration of epicardial blood flow, many patients continue to experience sub-endocardial tissue ischemia. To address the need for revascularization, several preclinical and early human studies have evaluated the efficacy of increasing myocardial perfusion or microvascular density using endothelial progenitor cells and MSCs (reviewed in [12, 13]). These studies, in addition to our own study on the use of CD34+ and CD34– BMMC [84], largely demonstrate pre-clinical safety and feasibility of such approaches, albeit with modest improvement in capillary density. Interestingly, the most promising effect is a trend toward decreased fibrosis and pathologic remodeling, hallmarks of ischemic heart disease [12, 13] but the mechanisms underlying these changes are unclear and are the subject of ongoing investigations [85]. In addition, as differentiation protocols continue to improve, hPSC-VSMC and hPSC-EC may become useful for the study of vascular disease, including mechanisms of plasticity and studies on the role of endothelial dysfunction, oxidative stress and signaling, autophagy, and accumulation of protein aggregates, which are especially relevant to atherosclerosis (reviewed in [86–89]). Finally, hPSC-EC and hPSC-VSMC could be used for screening pro-angiogenic drugs [90] and co-culturing with macrophages could benefit the study of crosstalk between macrophages and smooth muscle cells [91] to reveal new insights into disease mechanisms and targets for therapy.

Role of Proteomics

In regard to cellular therapies for revascularization, it is yet unknown whether a paracrine mechanism of action is responsible for some of the effects observed in transplantation of BMMC and MSC. Thus, proteomic analyses could be used to identify secreted factors *in vitro* and *in vivo* to reveal key components affecting capillary growth, which could ultimately promote development of more selective and specific molecules than transplanted cells. Strategies to differentiate hPSC into VSMC and EC would benefit from cell-type specific cell surface markers to enhance identification and selection of cells and tracking cells post-injection [92–95] that could be identified using proteomic approaches (discussed below). Finally, mechanisms of protein aggregation, effects of oxidative stress, and epigenetic regulation in smooth muscle plasticity [96] could also be investigated at the proteome level to benefit our understanding of vascular pathology.

Example 3: Systolic Heart Failure

Clinical Overview

The clinical syndrome of heart failure (HF) is defined by impaired myocardial function, ultimately leading to an inability to meet metabolic demand, manifesting as dyspnea, fatigue, fluid retention, and, at worst, impaired end-organ perfusion. In response, catecholamine release and renin-angiotensin-aldosterone are upregulated, further propagating a vicious cycle of neurohormonal activation, fluid retention, myocardial fibrotic remodeling and cardiac dysfunction. In HF with reduced ejection fraction (HFrEF), the left ventricular ejection fraction measures <40 % and systolic dysfunction results from loss of functional cardiomyocytes and alterations of cytoarchitecture and left ventricular geometry (*i.e.* "systolic heart failure"). Targeted neurohormonal therapy with beta-blockers, renin-angiotensin and aldosterone inhibitors has proven effective for all etiologies of HFrEF and indeed reduces cardiac mortality, morbidity and infrequently normalizes cardiac function. However, despite effective therapies, the overall 5-year mortality from HF exceeds 50 % from time of diagnosis.

Stem Cell Applications

The *in vitro* differentiation process of hPSC to hPSC-CM largely recapitulates major events in human embryonic heart development, including the transition from an immature, proliferative cellular phenotype to a maturing, non-proliferative phenotype. Accordingly, with continued time in culture, hPSC-CM follow a metabolic progression from glucose utilization to fatty acid oxidation and demonstrate a shift from expression of troponin isoforms robust in human fetal heart (*TNNI1, TNNI2*) to that expressed in the adult (*TNNI3*) [97, 98]. Altogether, this model may be valuable for mechanistic studies of cardiomyocyte dysfunction in HF. Specifically, using this model to decipher processes involved in cell cycle regulation, including cell cycle exit and reentry, cytokinesis, proliferation, and hypertrophy (reviewed in Ahuja et al. [99]), could reveal new strategies to stimulate cell cycle re-entry in diseased heart. Furthermore, as cardiomyocytes interact with macrophages, fibroblasts and other cell types, *in vitro* co-culture models with hPSC-CM may benefit studies regarding the effects of intercellular interactions on cardiomyocyte function [100, 101]. Additionally, as immature hPSC-CM and cardiomyocytes in the failing heart display several similarities in regard to fetal gene expression, metabolic regulation, and cellular electrophysiology [28, 102–104], early hPSC-CM could serve as a platform to study cellular mechanisms in autophagy, arrhythmias and apoptosis as they relate to impaired cardiomyocyte function [105–107].

Ultimately, patients suffering from systolic HF may benefit from cardiomyocyte replacement therapies aimed at restoring myocardial morphology and function. Phase I/II clinical trials to investigate use of hPSC derivatives in the eye are underway [108], thus enthusiasm regarding the possibility of such approaches for other organ systems is warranted. While numerous studies have tested the potential for various bone marrow derived stem cells and hPSC-CM to remuscularize the heart in rodent and larger non-primate mammalian models of infarction and/or heart failure (reviewed in Chong and Murry [109]), a recent study by Chong et al. [110] provides the most advanced analysis to date in a non-human primate model of myocardial infarction. By surgically transplanting hESC-CM into infarcted myocardium, this study successfully demonstrates important concepts in feasibility of clinical-scale production of hESC-CM and that transplanted cells can engraft, electromechanically couple, and become vascularized within the host myocardium. Importantly, this study also demonstrates the need for further efforts to eliminate potential arrhythmias observed in hearts where hESC-CM engraft. Though promising, several remaining challenges include potential immune-mediated rejection of allogeneic cells, how to standardize implantation protocols, and ensuring proper electromechanical integration into the host myocardium.

Role of Proteomics

Cell cycle processes are regulated by tightly choreographed expression of cell cycle genes responsible for DNA replication, nuclear division and cell division. These genes, in turn, are regulated in part by post-translational modifications of histones (*i.e.* epigenetic regulation). For example, addition of acetyl groups to histones serves to free DNA and prime it for transcription. On the other hand, modifications such as acetyl group removal or addition of methyl groups shield DNA from transcriptional machinery. Using hPSC-CM models to study cell cycle processes, proteomic analysis of histone signatures could potentially identify critical, time-sensitive epigenetic modifications regulating cardiomyocyte exit from the cell cycle and therefore discover therapeutic targets aimed at unlocking the proliferation block in adult cells. With regards to cardiomyocyte replacement therapies, proteomic analyses of the plasma membrane are poised to reveal novel cell surface markers that could provide a non-genetic approach to identifying and selecting subtype-specific cells at the appropriate maturation stage with the objective of generating well-defined cell populations that pose minimal arrhythmia risk.

Overview of Outstanding Questions

Stem cells and their derivatives carry tremendous potential for providing a relevant platform to study disease mechanisms, screen novel pharmacotherapies, and for generating cell and tissue products for regenerative medicine. However, major remaining barriers need to be overcome to realize the full potential of such approaches (reviewed in [111, 112]). As discussed below, proteomic analyses have the potential to address many of these challenges. Whether the application is to model a disease, discover novel pharmacotherapies, graft tissue with the purpose of revascularization, or to increase functional cardiomyocyte mass, the optimal cell type and milieu will vary among applications and remains largely undefined. For all stem cell classes, reproducible strategies to produce and select the appropriate cell type in sufficient scale and purity are imperative. Considering hPSC-CM, future studies will need to address heterogeneity, as cultures contain both non-hPSC-CM as well as atrial, ventricular and pacemaker cells that may span many stages of development [30, 110, 113–122]. *In vitro* differentiation variability and inherent differences among hPSC lines exacerbate problems of heterogeneity, which together limit translational and clinical application. While a number of strategies have been proposed to enrich for hPSC-CM, they either involve a mutagenic insertion of transgenes into the genome [115, 123] or are unable to select for a specific chamber subtype or maturation stage specific hPSC-CM [124–127]. Moreover, given the tumorigenic potential of hPSC, strategies to eliminate remnants of undifferentiated cells will be required for clinical applications [128].

Major Outstanding Questions and Challenges Related to using Stem Cell Derivatives for Mechanistic and Translational Studies

- Which cell type(s) is most appropriate for each application?
- How should cells with the appropriate phenotype be identified and selected?
- For a particular application, are homogeneous populations or cell mixtures most effective?
- In the case of hPSC-derivatives, how can we eliminate tumorigenic cells that may be present in the culture?

Proteomic Approaches for Addressing Current Needs in Stem Cell and Cardiovascular Biology

Currently available mass spectrometry instrumentation, sample preparation strategies, biochemical tools, and bioinformatic workflows can be combined in an infinite number of ways to assess protein content within a sample. Here, we describe three broad types of proteomic strategies especially well-suited to addressing outstanding questions related to applications of stem cells to cardiac biology described above (Fig. 6.3). Certainly, approaches we did not have space to discuss here will also have an impact and readers are directed to other chapters in this volume for additional relevant content [129]. Importantly, although each approach is discussed separately for purposes of clarity, it is possible and often advantageous to integrate multiple strategies into a single study (*e.g.* quantitative top-down analyses of histone proteins).

Cell Surface Proteomics

Relevance

Cell surface proteins are transmembrane, GPI-anchored, and extracellular matrix proteins such as receptors, enzymes, and transporters that participate in inter- and intracellular communication, cellular structure, and adhesion. As such, the cell surface proteome is a rich source of markers, drug targets, and molecules involved in the dynamic interplay between cells and their microenvironment, rendering this class of proteins especially relevant to stem cell biology and cardiovascular applications. First, cell surface proteins may be exploited as accessible markers of live cells for antibody-based cell sorting, termed immunophenotyping. As discussed above, it is not expected that a single hPSC-CM phenotype will be universally applicable for all research and clinical applications (*e.g.* atrial vs. ventricular; proliferative vs. non-proliferative). However, there is no singular metric for evaluating cardiomyocyte maturation stage; rather, extensive morphological, molecular, and functional measurements are required [34, 42] (Fig. 6.1b). Altogether, without appropriate tools to select specific subtypes at specific maturation stages with high purity, it is currently challenging to perform quantitative comparisons to determine which populations are optimal for each application. Thus, similar to how immunophenotyping has enabled selection of clinically relevant subpopulations within the hematopoietic stem cell system [130], we and others [131–134] have proposed that this strategy will be relevant for identification and selection of subtype and maturation stage specific hPSC-derivatives that can be tested for their utility in research and clinical applications. Moreover, development of cell surface marker panels for assessing and selecting maturation stage-specific cells would be valuable for rapid quantitative comparisons of hPSC-CM generated by various strategies designed to accelerate maturation *in vitro*. Although it may eventually be possible to manipulate culture

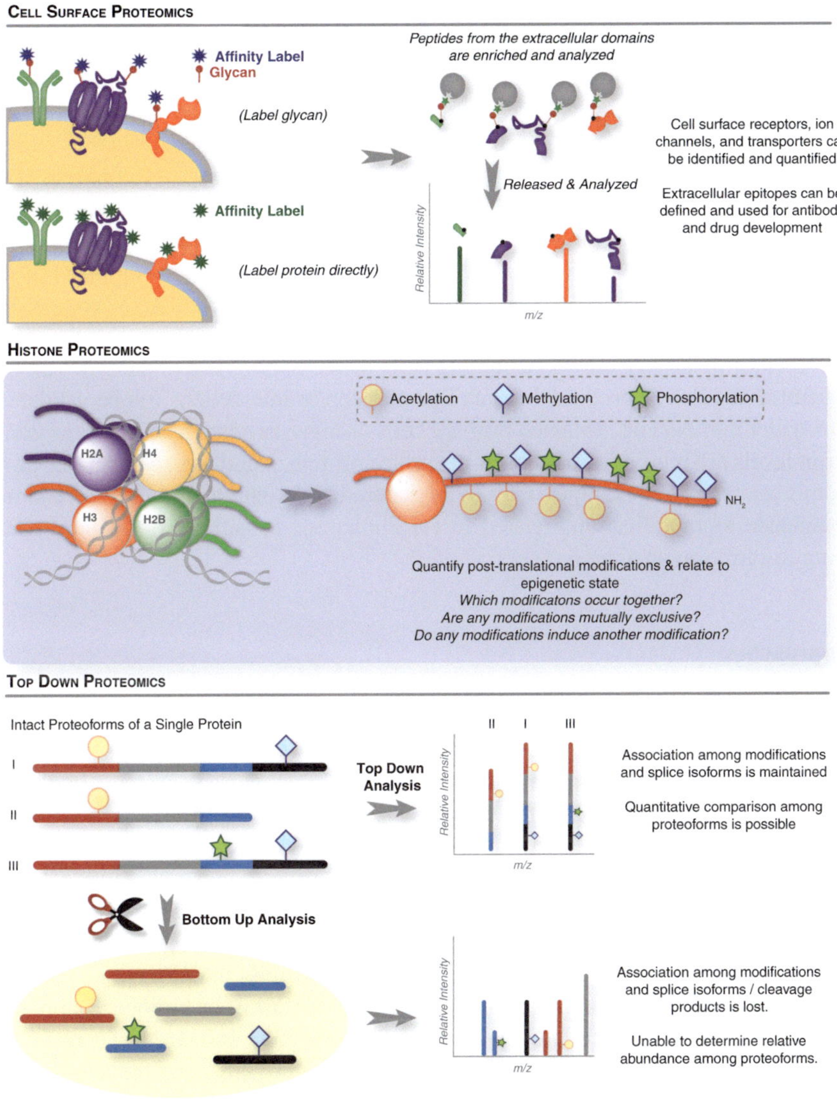

Fig. 6.3 Graphical representations of three major types of proteomic strategies highly relevant for the study of stem cells and their derivatives within the context of cardiovascular biology and medicine. Illustrated are the major technical aspects and features of the approach

conditions to drive differentiation towards pure populations of cells at select endpoints, cell surface markers will remain important quality control measures for non-genetic, live cell analyses. Finally, increased understanding of cell surface receptors present at various maturation stages could identify new targets and facilitate development of novel methods for driving maturation forward, especially once the full cell surface proteome of adult *vs* developmentally immature human cardiomyocytes is defined. Overall, application of proteomic technologies for the identification and quantitation of cell surface markers could have a significant impact on the future utility of stem cell derivatives in cardiovascular science and medicine.

Currently, more than 60 % of drug therapies target plasma membrane proteins [135]. Thus, beyond immunophenotyping, cell surface proteomic techniques can also be used to discover unknown or off-target receptors of common pharmacotherapies to reveal why they have a highly varied response among patients (*e.g.* beta-blockers, angiotensin converting enzyme inhibitors, aldosterone antagonists) and for identifying accessible receptors on live cells for targeted drug delivery. In both of these applications, hPSC-CM can be generated in large numbers to support the receptor discovery efforts, assuming cells of the appropriate phenotype can be generated *in vitro*. Subsequently, such ligand-receptor interactions can be validated on smaller numbers of explanted primary cells or in tissue sections. Finally, as transcript levels are unreliable predictors of surface protein abundance, cell surface proteomic analyses, for example, may more accurately reveal changes in ion channel abundance or modifications related to catecholamine sensitivity or calcium handling in cardiomyopathies.

Approaches

The available reagents and proteomic strategies for studying the cell surface proteome have expanded considerably in the past decade (see reviews [136–138]). It is now possible to identify, quantify, and characterize cell surface proteins in a way that provides experimental evidence of their subcellular localization. Historically, cell surface proteins have been challenging to study due to the limited availability of antibodies, the relatively low abundance of cell surface compared to intracellular proteins, the presence of hydrophobic and heavily glycosylated domains, and the challenges in isolating plasma membrane proteins without contamination from intracellular membranes. As outlined in Table 6.1, affinity enrichment based proteomic approaches are gaining momentum for a vast array of biological applications, including stem cell biology [134, 138–140, 154, 160–163]. There is no singular strategy yet that can identify all cell surface proteins, and the approaches summarized in Table 6.1 are largely complementary. In general, strategies that apply an affinity capture tag to cell surface proteins while the cells are intact, rather than after lysis, offer the highest level of specificity for plasma membrane proteins and can provide experimental evidence of localization, generating a snapshot of the cell surface proteome location at a specific time or condition. Such strategies therefore have the advantage of being independent of database annotations and localization and topology prediction algorithms. Another emerging area is the use of novel linking agents [158] for identifying cell surface receptors for specific ligands of interest whose receptors are currently unknown (*e.g.* biologics, small molecules,

Table 6.1 Mass spectrometry based proteomic strategies for selective identification of cell surface proteins

Cell surface capture	Oligosaccharide on extracellular domain of protein conjugated to biotin *via* hydrazone bond, streptavidin-based enrichment [134, 139, 140]
Advantages	Highly specific for cell surface proteins
Limitations	Limited to N-glycoproteins
Aniline-catalyzed oxime ligation	Oligosaccharide on extracellular domain of protein conjugated to biotin *via* oxime ligation, streptavidin-based enrichment [141, 142]
Advantages	Highly specific for cell surface proteins, oxime ligation may be more stable than hydrazine bond
Limitations	Limited to N-glycoproteins
Amine biotinylation	Free amines conjugated to biotin, streptavidin-based enrichment [143–145]
Advantages	Not limited to glycosylated proteins
Limitations	Published results vary in regard to specificity for plasma membrane proteins
Lectin affinity	Glycoproteins or glycopeptides from cell lysate and/or digests selectively enriched by lectin affinity chromatography [146–149]
Advantages	Tailored enrichment based on lectin, complementary to chemical labeling
Limitations	Limited to glycoproteins, often lower yield compared to chemical labeling strategies
Cationic colloidal silica	Positively charged silica increases density of plasma membrane and facilitates physical separation from other cellular components [150, 151]
Advantages	May be particularly suited to apical vs. basal types of analyses and cell types with membranes more permeable to chemical labeling approaches
Limitations	Often lower yield compared to chemical labeling strategies, specificity for cell surface proteins varies among reports
GPI shaving	Metabolic labeling via azido sugar analog, phospholipase release; alkyne-based enrichment and phospholipase release followed by lectin affinity enrichment; phospholipase D shaving [152, 153]
Advantages	Highly specific for cell surface GPI-anchored proteins
Limitations	Phospholipase efficiency may be limited by modifications; Lectin specificity may limit range of identified proteins
Matrisome	Analysis of insoluble fraction of tissue depleted of other cellular components, coupled with bioinformatic filtering [129, 154, 155]
Advantages	Applicable to analysis of extracellular matrix proteins in tissue
Limitations	Preparation of insoluble fractions can be challenging to reproduce routinely for quantitative analyses
Cell shaving	Proteinase K or thermolysin release of extracellular domain [156, 157]
Advantages	Not limited to glycosylated proteins
Limitations	Mammalian cells may not be stable in extreme pH for proteinase K

(continued)

Table 6.1 (continued)

Ligand-based receptor capture	Tri-functional molecule bound to ligand interacts with oxidized oligosaccharide on extracellular domain of cell surface receptor, streptavidin enrichment [158, 159]
Advantages	Possible to identify direct interactors and near neighbors; not limited by transient interactions or antibody affinity
Limitations	Limited to N-glycosylated receptors

Included are major features of the approach, selected references, notable advantages and limitations

circulating secreted factors). These technologies offer advantages over yeast-two-hybrid and antibody-antigen based immunoprecipitation which may fail to detect transient interactions and require high affinity antibodies, respectively. Finally, relevant to all types of proteomic studies, but particularly those focused on hydrophobic transmembrane proteins, is the development of new mass spectrometry compatible surfactants. Whereas common laboratory detergents are incompatible with mass spectrometry (*e.g.* SDS, Triton X-100, CHAPS, NP-40), the availability and diversity of mass spectrometry compatible surfactants is growing and already have proven advantageous for analysis of membrane proteins [164–167].

Histone Proteomics

Relevance

Post-translational modifications within histones affect the ability of the transcription machinery to access DNA and thus the extent of gene transcription. The list of modified residues and modification types is growing and includes, but is not limited to methylation, acetylation, propionylation, citrullination, deimination, numerous types of acylation, crotonyllysine, glycoxidation, and ubiquitination [168, 169]. These histone modifications are carefully maintained by tightly controlled expression and post-translational modification of the proteins responsible for these marks, including histone acetyl transferases, deacetylases, methyl transferases, and demethyl transferases. The commitment to a specific cell fate involves tight regulation of gene expression and increasing evidence suggests that epigenetic marks regulate cellular differentiation and commitment to various cell types including cardiomyocytes, smooth muscle, and endothelial cells [170–172]. Consistent with these data, epigenetic abnormalities have been correlated to both developmental defects and pathological mechanisms in adult onset cardiomyopathies (reviewed in [173, 174]). Proteomic analyses of histones in hPSC-derived cells in comparison to tissue specimens will be critical to determining how closely epigenetic marks acquired *in vitro* recapitulate those acquired during human development. Once elucidated, tracking histone marks may serve as a quality control metric to indicate when cultured cells have epigenetically achieved a differentiated phenotype. Moreover, due to the power of epigenetic modification to change cell function, studies of pharmacologic mediators of histone modifications are underway for cancer. For cardiac disease in which beneficial effects of therapy are more challenging to quickly assess, proteomic analyses of histones in hPSC-derived progeny could be used as a platform to evaluate efficacy and safety of histone deacetylase inhibitors currently under consideration for treatment of cardiovascular diseases ranging from CAD to HF [173]. Finally, as epigenetic regulation has been linked to arterial calcification [175, 176], histone analyses using hPSC-VSMC to elucidate these mechanisms could provide novel targets for management of CAD.

Approaches

Mass spectrometry is a powerful tool in epigenetic studies as it is possible to identify and quantify multiple modification types and locations within a single experiment and several approaches have been well-described. First, bottom-up discovery-based approaches are suited to the identification of previously unknown modifications as well as quantitative comparisons among samples [150, 177, 178]. Second, if the site and type of modification is known or predicted, targeted mass spectrometry approaches (*e.g.* multiple reaction monitoring, parallel reaction monitoring, see Chaps. 9 and 13), can be used to specifically monitor modifications at specific sites in a quantitative manner [179]. Importantly, these assays can be highly multiplexed to monitor multiple modifications across numerous sites on multiple proteins within a single experiment. Third, application of modern top-down proteomic strategies, discussed in Chap. 8 and briefly below, allow for the stoichiometry of modifications within a histone protein to be measured, relationships that are largely lost when performing proteolytic digestion as in bottom up approaches (Fig. 6.3). Importantly, because the masses of several common modifications are similar (*e.g.* tri-methylation (42.046950) and acetylation (42.010565) are within 0.036385 Da), high resolution and mass accuracy of modern instrumentation are invaluable for the analysis of histone modifications.

Top Down Proteomics

Relevance

Post-translational modifications, splice isoforms, and products of proteolytic cleavage play key roles in a variety of cellular regulatory mechanisms and are increasingly recognized as important determinants and indicators of cardiovascular pathologies [180, 181]. For a single gene, numerous molecular forms (*i.e.* proteoforms) of the protein are possible, arising from any number of combinations of allelic variants, alternative RNA splicing, *in vivo* proteolysis, and post-translational modifications [182]. A key approach to analyzing proteoforms in biology and disease context is termed "top down" proteomics, where intact proteins are fragmented during mass spectrometry analysis to obtain amino acid sequence information [183, 184]. The major distinction between this approach and more traditional "bottom up" approaches is that the protein does not undergo chemical or proteolytic digestion prior to mass spectrometry analysis, but rather the intact protein is measured (Fig. 6.3). This preserves the stoichiometry among post-translational modifications, proteolytic cleavage products, and products of splicing events, among others. Top down proteomics has already made an impact on cardiac biology including the study of troponin I [185] and tropomyosin [186] in diseased tissue and blood, and apolipoprotein proteoforms in lipoprotein particles [187]. Recently, this strategy revealed previously uncharacterized chamber-specific post-translational modifications of cardiomyocyte myofilament proteins under basal conditions [188]. Of relevance to stem cell applications described above, top down analyses could assist in the study of post-translational changes in sarcomere complexes and myosin isotype switching in cardiomyopathy patient-specific hiPSC-CM, mapping epigenetic changes associated with cardiomyocyte cell cycle regulation, and defining secreted factors that play a role in paracrine effects associated with stem cell based therapies.

Proteoform

A "specific molecular form of a protein product arising from a specific gene." Proteoforms arise from combinations of allelic variants, alternative RNA splicing, *in vivo* proteolysis, and post-translational modifications [182].

Approaches

Top down proteomic analyses are becoming more routine in modern mass spectrometry laboratories due to the increasing availability of suitable instrumentation, data analysis platforms [189–193], and requisite sample preparation strategies [193, 194]. While the specific capabilities of available instrumentation and sample and chromatographic quality will ultimately determine the types of analyses possible for a particular study, in general, proteins up to ~50 kDa can be analyzed on commonly available instruments (*e.g.* those with Orbitrap and TOF analyzers) [195, 196], and larger (>100 kDa) proteins are possible with more advanced instrumentation and native mass spectrometry [193, 197]. For more details, readers are directed to Chap. 8 dedicated to this topic.

The Future of Stem Cell Proteomics for Cardiovascular Biology and Disease

Overall, the integration of novel stem cell and proteomic technologies are poised to have a tremendous impact on cardiovascular science. While current evidence suggests a promising outlook, significant technical hurdles remain to be overcome. As experience has taught us to be mindful of the fundamental details within each discipline, from how pluripotent stem cell culture density affects their susceptibility to metabolic inhibition [126] to how the source of trypsin in a proteomics experiment can affect the proteotypic profile [198], we are reminded of the adage "*the devil is in the detail*". For these reasons, synergistic efforts among scientists with expertise in stem cell biology, protein chemistry, mass spectrometry, bioinformatics, physiology, and bioengineering that align with their clinical counterparts are imperative to develop truly innovative strategies that transcend convention and have the potential to fulfill the promise of stem cell-based approaches for expanding our understanding of cardiac biology and developing advanced therapies in cardiovascular medicine.

Acknowledgements This work was supported by American Heart Association 15GRNT24980002 to R.G. and NIH grant R01HL094384 to A.S.O. E.K. is a member of the Medical Scientist Training Program at the Medical College of Wisconsin, which is partially supported by an NIH T32 grant GM080202. S.F. was partially supported by an American College of Cardiology Fellowship.

References

1. Thomson JA, Itskovitz-Eldor J, Shapiro SS, Waknitz MA, Swiergiel JJ, Marshall VS, et al. Embryonic stem cell lines derived from human blastocysts. Science. 1998;282(5391):1145–7.
2. Takahashi K, Yamanaka S. Induction of pluripotent stem cells from mouse embryonic and adult fibroblast cultures by defined factors. Cell. 2006;126(4):663–76. Epub 2006/08/15.
3. Gonzalez F, Boue S, Izpisua Belmonte JC. Methods for making induced pluripotent stem cells: reprogramming a la carte. Nat Rev Genet. 2011;12(4):231–42.
4. Yu J, Vodyanik MA, Smuga-Otto K, Antosiewicz-Bourget J, Frane JL, Tian S, et al. Induced pluripotent stem cell lines derived from human somatic cells. Science. 2007;318(5858):1917–20. Epub 2007/11/22.
5. Takahashi K, Tanabe K, Ohnuki M, Narita M, Ichisaka T, Tomoda K, et al. Induction of pluripotent stem cells from adult human fibroblasts by defined factors. Cell. 2007;131(5):861–72. Epub 2007/11/24.
6. Fusaki N, Ban H, Nishiyama A, Saeki K, Hasegawa M. Efficient induction of transgene-free human pluripotent stem cells using a vector based on Sendai virus, an RNA virus that does not integrate into the host genome. Proc Jpn Acad Ser B Phys Biol Sci. 2009;85(8):348–62.
7. Ban H, Nishishita N, Fusaki N, Tabata T, Saeki K, Shikamura M, et al. Efficient generation of transgene-free human induced pluripotent stem cells (iPSCs) by temperature-sensitive Sendai virus vectors. Proc Natl Acad Sci U S A. 2011;108(34):14234–9.
8. Rony IK, Baten A, Bloomfield JA, Islam ME, Billah MM, Islam KD. Inducing pluripotency in vitro: recent advances and highlights in induced pluripotent stem cells generation and pluripotency reprogramming. Cell Prolif. 2015;48(2):140–56. Epub 2015/02/04.
9. Cuende N, Rico L, Herrera C. Concise review: bone marrow mononuclear cells for the treatment of ischemic syndromes: medicinal product or cell transplantation? Stem Cells Transl Med. 2012;1(5):403–8. Epub 2012/12/01.
10. Dominici M, Le Blanc K, Mueller I, Slaper-Cortenbach I, Marini F, Krause D, et al. Minimal criteria for defining multipotent mesenchymal stromal cells. The International Society for Cellular Therapy position statement. Cytotherapy. 2006;8(4):315–7.
11. Mirotsou M, Jayawardena TM, Schmeckpeper J, Gnecchi M, Dzau VJ. Paracrine mechanisms of stem cell reparative and regenerative actions in the heart. J Mol Cell Cardiol. 2011;50(2):280–9. Epub 2010/08/24.

12. Karantalis V, Hare JM. Use of mesenchymal stem cells for therapy of cardiac disease. Circ Res. 2015;116(8):1413–30. Epub 2015/04/11.
13. Afzal MR, Samanta A, Shah ZI, Jeevanantham V, Abdel-Latif A, Zuba-Surma EK, et al. Adult bone marrow cell therapy for ischemic heart disease: evidence and insights from randomized controlled trials. Circ Res. 2015;117(6):558–75. Epub 2015/07/15.
14. Seaberg RM, van der Kooy D. Stem and progenitor cells: the premature desertion of rigorous definitions. Trends Neurosci. 2003;26(3):125–31. Epub 2003/02/20.
15. Chong JJ, Forte E, Harvey RP. Developmental origins and lineage descendants of endogenous adult cardiac progenitor cells. Stem Cell Res. 2014;13(3 Pt B):592–614. Epub 2014/12/03.
16. Martinez EC, Kofidis T. Adult stem cells for cardiac tissue engineering. J Mol Cell Cardiol. 2011;50(2):312–9. Epub 2010/08/17.
17. Naumova AV, Modo M, Moore A, Frank JA, Murry CE. Response to cardiac regeneration validated. Nat Biotechnol. 2015;33(6):587. Epub 2015/06/10.
18. Malliaras K, Marban E. Cardiac regeneration validated. Nat Biotechnol. 2015;33(6):587. Epub 2015/06/10.
19. Ellison GM, Vicinanza C, Smith AJ, Aquila I, Leone A, Waring CD, et al. Adult c-kit(pos) cardiac stem cells are necessary and sufficient for functional cardiac regeneration and repair. Cell. 2013;154(4):827–42. Epub 2013/08/21.
20. van Berlo JH, Kanisicak O, Maillet M, Vagnozzi RJ, Karch J, Lin SC, et al. c-kit+ cells minimally contribute cardiomyocytes to the heart. Nature. 2014;509(7500):337–41. Epub 2014/05/09.
21. Psaltis PJ, Simari RD. Vascular wall progenitor cells in health and disease. Circ Res. 2015;116(8):1392–412. Epub 2015/04/11.
22. Ieda M, Fu JD, Delgado-Olguin P, Vedantham V, Hayashi Y, Bruneau BG, et al. Direct reprogramming of fibroblasts into functional cardiomyocytes by defined factors. Cell. 2010;142(3):375–86. Epub 2010/08/10.
23. Qian L, Huang Y, Spencer CI, Foley A, Vedantham V, Liu L, et al. In vivo reprogramming of murine cardiac fibroblasts into induced cardiomyocytes. Nature. 2012;485(7400):593–8. Epub 2012/04/24.
24. Nam YJ, Song K, Luo X, Daniel E, Lambeth K, West K, et al. Reprogramming of human fibroblasts toward a cardiac fate. Proc Natl Acad Sci U S A. 2013;110(14):5588–93. Epub 2013/03/15.
25. Yamakawa H, Ieda M. Strategies for heart regeneration: approaches ranging from induced pluripotent stem cells to direct cardiac reprogramming. Int Heart J. 2015;56(1):1–5. Epub 2015/03/07.
26. Sadahiro T, Yamanaka S, Ieda M. Direct cardiac reprogramming: progress and challenges in basic biology and clinical applications. Circ Res. 2015;116(8):1378–91. Epub 2015/04/11.
27. Batalov I, Feinberg AW. Differentiation of cardiomyocytes from human pluripotent stem cells using monolayer culture. Biomark Insights. 2015;10 Suppl 1:71–6. Epub 2015/06/09.
28. Hartman ME, Dai DF, Laflamme MA. Human pluripotent stem cells: prospects and challenges as a source of cardiomyocytes for in vitro modeling and cell-based cardiac repair. Adv Drug Deliv Rev. 2015;96:3–17. Epub 2015/05/20.
29. Cao N, Liang H, Huang J, Wang J, Chen Y, Chen Z, et al. Highly efficient induction and long-term maintenance of multipotent cardiovascular progenitors from human pluripotent stem cells under defined conditions. Cell Res. 2013;23(9):1119–32. Epub 2013/07/31.
30. Burridge PW, Matsa E, Shukla P, Lin ZC, Churko JM, Ebert AD, et al. Chemically defined generation of human cardiomyocytes. Nat Methods. 2014;11(8):855–60. Epub 2014/06/16.
31. Lian X, Zhang J, Azarin SM, Zhu K, Hazeltine LB, Bao X, et al. Directed cardiomyocyte differentiation from human pluripotent stem cells by modulating Wnt/beta-catenin signaling under fully defined conditions. Nat Protoc. 2013;8(1):162–75. Epub 2012/12/22.
32. Hwang HS, Kryshtal DO, Feaster TK, Sanchez-Freire V, Zhang J, Kamp TJ, et al. Comparable calcium handling of human iPSC-derived cardiomyocytes generated by multiple laboratories. J Mol Cell Cardiol. 2015;85:79–88. Epub 2015/05/20.

33. Zhu R, Blazeski A, Poon E, Costa KD, Tung L, Boheler KR. Physical developmental cues for the maturation of human pluripotent stem cell-derived cardiomyocytes. Stem Cell Res Ther. 2014;5(5):117. Epub 2015/02/18.
34. Robertson C, Tran DD, George SC. Concise review: maturation phases of human pluripotent stem cell-derived cardiomyocytes. Stem Cells. 2013;31(5):829–37. Epub 2013/01/29.
35. Vunjak Novakovic G, Eschenhagen T, Mummery C. Myocardial tissue engineering: in vitro models. Cold Spring Harb Perspect Med. 2014;4(3). a014076. Epub 2014/03/05.
36. Moretti A, Laugwitz KL, Dorn T, Sinnecker D, Mummery C. Pluripotent stem cell models of human heart disease. Cold Spring Harb Perspect Med. 2013;3(11). a014027. Epub 2013/11/05.
37. Tulloch NL, Muskheli V, Razumova MV, Korte FS, Regnier M, Hauch KD, et al. Growth of engineered human myocardium with mechanical loading and vascular coculture. Circ Res. 2011;109(1):47–59. Epub 2011/05/21.
38. Muscari C, Giordano E, Bonafe F, Govoni M, Guarnieri C. Strategies affording prevascularized cell-based constructs for myocardial tissue engineering. Stem Cells Int. 2014;2014:434169. Epub 2014/02/11.
39. Ruan JL, Tulloch NL, Saiget M, Paige SL, Razumova MV, Regnier M, et al. Mechanical stress promotes maturation of human myocardium from pluripotent stem cell-derived progenitors. Stem Cells. 2015;33(7):2148–57. Epub 2015/04/14.
40. Wen JY, Wei CY, Shah K, Wong J, Wang C, Chen HS. Maturation-based model of arrhythmogenic right ventricular dysplasia using patient-specific induced pluripotent stem cells. Circ J. 2015;79(7):1402–8. Epub 2015/05/15.
41. Yang X, Rodriguez M, Pabon L, Fischer KA, Reinecke H, Regnier M, et al. Tri-iodo-l-thyronine promotes the maturation of human cardiomyocytes-derived from induced pluripotent stem cells. J Mol Cell Cardiol. 2014;72:296–304. Epub 2014/04/17.
42. Keung W, Boheler KR, Li RA. Developmental cues for the maturation of metabolic, electrophysiological and calcium handling properties of human pluripotent stem cell-derived cardiomyocytes. Stem Cell Res Ther. 2014;5(1):17. Epub 2014/01/29.
43. Cheung C, Bernardo AS, Pedersen RA, Sinha S. Directed differentiation of embryonic origin-specific vascular smooth muscle subtypes from human pluripotent stem cells. Nat Protoc. 2014;9(4):929–38. Epub 2014/03/29.
44. Orlova VV, Drabsch Y, Freund C, Petrus-Reurer S, van den Hil FE, Muenthaisong S, et al. Functionality of endothelial cells and pericytes from human pluripotent stem cells demonstrated in cultured vascular plexus and zebrafish xenografts. Arterioscler Thromb Vasc Biol. 2014;34(1):177–86. Epub 2013/10/26.
45. Theodoris CV, Li M, White MP, Liu L, He D, Pollard KS, et al. Human disease modeling reveals integrated transcriptional and epigenetic mechanisms of NOTCH1 haploinsufficiency. Cell. 2015;160(6):1072–86. Epub 2015/03/15.
46. Dash BC, Jiang Z, Suh C, Qyang Y. Induced pluripotent stem cell-derived vascular smooth muscle cells: methods and application. Biochem J. 2015;465(2):185–94. Epub 2015/01/07.
47. Ge X, Ren Y, Bartulos O, Lee MY, Yue Z, Kim KY, et al. Modeling supravalvular aortic stenosis syndrome with human induced pluripotent stem cells. Circulation. 2012;126(14):1695–704. Epub 2012/08/24.
48. Zhang J, Lian Q, Zhu G, Zhou F, Sui L, Tan C, et al. A human iPSC model of Hutchinson Gilford Progeria reveals vascular smooth muscle and mesenchymal stem cell defects. Cell Stem Cell. 2011;8(1):31–45. Epub 2010/12/28.
49. Liu GH, Barkho BZ, Ruiz S, Diep D, Qu J, Yang SL, et al. Recapitulation of premature ageing with iPSCs from Hutchinson-Gilford progeria syndrome. Nature. 2011;472(7342):221–5. Epub 2011/02/25.
50. Mummery CL, Zhang J, Ng ES, Elliott DA, Elefanty AG, Kamp TJ. Differentiation of human embryonic stem cells and induced pluripotent stem cells to cardiomyocytes: a methods overview. Circ Res. 2012;111(3):344–58. Epub 2012/07/24.
51. Sallam K, Kodo K, Wu JC. Modeling inherited cardiac disorders. Circ J. 2014;78(4):784–94. Epub 2014/03/19.

52. Braam SR, Tertoolen L, van de Stolpe A, Meyer T, Passier R, Mummery CL. Prediction of drug-induced cardiotoxicity using human embryonic stem cell-derived cardiomyocytes. Stem Cell Res. 2010;4(2):107–16. Epub 2009/12/26.
53. Liang P, Lan F, Lee AS, Gong T, Sanchez-Freire V, Wang Y, et al. Drug screening using a library of human induced pluripotent stem cell-derived cardiomyocytes reveals disease-specific patterns of cardiotoxicity. Circulation. 2013;127(16):1677–91. Epub 2013/03/23.
54. Cerbini T, Luo Y, Rao MS, Zou J. Transfection, selection, and colony-picking of human induced pluripotent stem cells TALEN-targeted with a GFP gene into the AAVS1 safe harbor. J Vis Exp. 2015;(96). doi:10.3791/52504. Epub 2015/03/06.
55. Hatada S, Subramanian A, Mandefro B, Ren S, Kim HW, Tang J, et al. Low-dose irradiation enhances gene targeting in human pluripotent stem cells. Stem Cells Transl Med. 2015;4(9):998–1010. Epub 2015/07/18.
56. Merkle FT, Neuhausser WM, Santos D, Valen E, Gagnon JA, Maas K, et al. Efficient CRISPR-Cas9-mediated generation of knockin human pluripotent stem cells lacking undesired mutations at the targeted locus. Cell Rep. 2015;11(6):875–83. Epub 2015/05/06.
57. Mathur A, Loskill P, Shao K, Huebsch N, Hong S, Marcus SG, et al. Human iPSC-based cardiac microphysiological system for drug screening applications. Sci Rep. 2015;5:8883. Epub 2015/03/10.
58. Jahnke HG, Steel D, Fleischer S, Seidel D, Kurz R, Vinz S, et al. A novel 3D label-free monitoring system of hES-derived cardiomyocyte clusters: a step forward to in vitro cardiotoxicity testing. PLoS One. 2013;8(7):e68971. Epub 2013/07/19.
59. Yang C, Al-Aama J, Stojkovic M, Keavney B, Trafford A, Lako M, et al. Concise review: cardiac disease modeling using induced pluripotent stem cells. Stem Cells. 2015;33(9):2643–51. Epub 2015/06/03.
60. Ng KM, Tse HF. Modeling hereditary cardiac disease with patient-specific-induced pluripotent stem cells: opportunities and concerns. J Cardiovasc Pharmacol. 2012;60(4):406–7. Epub 2012/06/02.
61. Sallam K, Li Y, Sager PT, Houser SR, Wu JC. Finding the rhythm of sudden cardiac death: new opportunities using induced pluripotent stem cell-derived cardiomyocytes. Circ Res. 2015;116(12):1989–2004. Epub 2015/06/06.
62. Kamdar F, Klaassen Kamdar A, Koyano-Nakagawa N, Garry MG, Garry DJ. Cardiomyopathy in a dish: using human inducible pluripotent stem cells to model inherited cardiomyopathies. J Card Fail. 2015;21(9):761–70. Epub 2015/05/03.
63. Fisher SA, Brunskill SJ, Doree C, Mathur A, Taggart DP, Martin-Rendon E. Stem cell therapy for chronic ischaemic heart disease and congestive heart failure. Cochrane Database Syst Rev. 2014;(4):CD007888. Epub 2014/04/30.
64. Clifford DM, Fisher SA, Brunskill SJ, Doree C, Mathur A, Watt S, et al. Stem cell treatment for acute myocardial infarction. Cochrane Database Syst Rev. 2012;(2):CD006536. Epub 2012/02/18.
65. Maron BJ, Gardin JM, Flack JM, Gidding SS, Kurosaki TT, Bild DE. Prevalence of hypertrophic cardiomyopathy in a general population of young adults. Echocardiographic analysis of 4111 subjects in the CARDIA Study. Coronary Artery Risk Development in (Young) Adults. Circulation. 1995;92(4):785–9. Epub 1995/08/15.
66. Maron BJ. Hypertrophic cardiomyopathy: a systematic review. JAMA. 2002;287(10):1308–20. Epub 2002/03/12.
67. Gersh BJ, Maron BJ, Bonow RO, Dearani JA, Fifer MA, Link MS, et al. 2011 ACCF/AHA guideline for the diagnosis and treatment of hypertrophic cardiomyopathy: a report of the American College of Cardiology Foundation/American Heart Association Task Force on Practice Guidelines. Circulation. 2011;124(24):e783–831. Epub 2011/11/10.
68. Bezzina CR, Lahrouchi N, Priori SG. Genetics of sudden cardiac death. Circ Res. 2015;116(12):1919–36. Epub 2015/06/06.
69. Wang G, McCain ML, Yang L, He A, Pasqualini FS, Agarwal A, et al. Modeling the mitochondrial cardiomyopathy of Barth syndrome with induced pluripotent stem cell and heart-on-chip technologies. Nat Med. 2014;20(6):616–23. Epub 2014/05/13.

70. Roden DM, Hong CC. Stem cell-derived cardiomyocytes as a tool for studying proarrhythmia: a better canary in the coal mine? Circulation. 2013;127(16):1641–3. Epub 2013/03/23.
71. Tanaka A, Yuasa S, Node K, Fukuda K. Cardiovascular disease modeling using patient-specific induced pluripotent stem cells. Int J Mol Sci. 2015;16(8):18894–922. Epub 2015/08/15.
72. Moretti A, Bellin M, Welling A, Jung CB, Lam JT, Bott-Flugel L, et al. Patient-specific induced pluripotent stem-cell models for long-QT syndrome. N Engl J Med. 2010;363(15): 1397–409. Epub 2010/07/28.
73. Yazawa M, Hsueh B, Jia X, Pasca AM, Bernstein JA, Hallmayer J, et al. Using induced pluripotent stem cells to investigate cardiac phenotypes in Timothy syndrome. Nature. 2011;471(7337):230–4. Epub 2011/02/11.
74. Maron BJ, Maron MS. Hypertrophic cardiomyopathy. Lancet. 2013;381(9862):242–55. Epub 2012/08/10.
75. Fananapazir L. Advances in molecular genetics and management of hypertrophic cardiomyopathy. JAMA. 1999;281(18):1746–52. Epub 1999/05/18.
76. Tardiff JC. Sarcomeric proteins and familial hypertrophic cardiomyopathy: linking mutations in structural proteins to complex cardiovascular phenotypes. Heart Fail Rev. 2005;10(3):237–48. Epub 2006/01/18.
77. Schaub MC, Hefti MA, Zuellig RA, Morano I. Modulation of contractility in human cardiac hypertrophy by myosin essential light chain isoforms. Cardiovasc Res. 1998;37(2):381–404. Epub 1998/06/06.
78. Petzhold D, Lossie J, Keller S, Werner S, Haase H, Morano I. Human essential myosin light chain isoforms revealed distinct myosin binding, sarcomeric sorting, and inotropic activity. Cardiovasc Res. 2011;90(3):513–20. Epub 2011/01/26.
79. Schober T, Huke S, Venkataraman R, Gryshchenko O, Kryshtal D, Hwang HS, et al. Myofilament Ca sensitization increases cytosolic Ca binding affinity, alters intracellular Ca homeostasis, and causes pause-dependent Ca-triggered arrhythmia. Circ Res. 2012;111(2): 170–9. Epub 2012/06/01.
80. Nagata M, Shimizu M, Ino H, Yamaguchi M, Hayashi K, Taki J, et al. Hemodynamic changes and prognosis in patients with hypertrophic cardiomyopathy and abnormal blood pressure responses during exercise. Clin Cardiol. 2003;26(2):71–6. Epub 2003/03/11.
81. Wilson PW. Established risk factors and coronary artery disease: the Framingham Study. Am J Hypertens. 1994;7(7 Pt 2):7S–12. Epub 1994/07/01.
82. Lloyd-Jones DM, Larson MG, Beiser A, Levy D. Lifetime risk of developing coronary heart disease. Lancet. 1999;353(9147):89–92. Epub 1999/02/19.
83. Ridker PM. High-sensitivity C-reactive protein: potential adjunct for global risk assessment in the primary prevention of cardiovascular disease. Circulation. 2001;103(13):1813–8. Epub 2001/04/03.
84. Stempien-Otero A, Helterline D, Plummer T, Farris S, Prouse A, Polissar N, et al. Mechanisms of bone marrow-derived cell therapy in ischemic cardiomyopathy with left ventricular assist device bridge to transplant. J Am Coll Cardiol. 2015;65(14):1424–34. Epub 2015/04/11.
85. Farris S, Stempien-Otero A. Allogeneic precursor cells for systolic heart failure: a need for mechanisms in humans. Circ Res. 2015;117(6):494–7. Epub 2015/09/01.
86. Michel JB, Li Z, Lacolley P. Smooth muscle cells and vascular diseases. Cardiovasc Res. 2012;95(2):135–7. Epub 2012/05/29.
87. Wang F, Lerman A, Herrmann J. Dysfunction of the ubiquitin-proteasome system in atherosclerotic cardiovascular disease. Am J Cardiovasc Dis. 2015;5(1):83–100. Epub 2015/06/13.
88. Willis MS, Townley-Tilson WH, Kang EY, Homeister JW, Patterson C. Sent to destroy: the ubiquitin proteasome system regulates cell signaling and protein quality control in cardiovascular development and disease. Circ Res. 2010;106(3):463–78. Epub 2010/02/20.
89. Bertoluci MC, Ce GV, da Silva AM, Wainstein MV, Boff W, Punales M. Endothelial dysfunction as a predictor of cardiovascular disease in type 1 diabetes. World J Diabetes. 2015;6(5):679–92. Epub 2015/06/13.
90. Katoh M. Therapeutics targeting angiogenesis: genetics and epigenetics, extracellular miRNAs and signaling networks (review). Int J Mol Med. 2013;32(4):763–7. Epub 2013/07/19.

91. Koga J, Aikawa M. Crosstalk between macrophages and smooth muscle cells in atherosclerotic vascular diseases. Vascul Pharmacol. 2012;57(1):24–8. Epub 2012/03/10.
92. Lin CS, Lue TF. Defining vascular stem cells. Stem Cells Dev. 2013;22(7):1018–26. Epub 2013/01/22.
93. Gomez D, Owens GK. Smooth muscle cell phenotypic switching in atherosclerosis. Cardiovasc Res. 2012;95(2):156–64. Epub 2012/03/13.
94. Zhang L, Hoffman JA, Ruoslahti E. Molecular profiling of heart endothelial cells. Circulation. 2005;112(11):1601–11. Epub 2005/09/08.
95. Stroncek JD, Grant BS, Brown MA, Povsic TJ, Truskey GA, Reichert WM. Comparison of endothelial cell phenotypic markers of late-outgrowth endothelial progenitor cells isolated from patients with coronary artery disease and healthy volunteers. Tissue Eng Part A. 2009;15(11):3473–86. Epub 2009/05/14.
96. Liu R, Leslie KL, Martin KA. Epigenetic regulation of smooth muscle cell plasticity. Biochim Biophys Acta. 2015;1849(4):448–53. Epub 2014/06/18.
97. Bhavsar PK, Dhoot GK, Cumming DV, Butler-Browne GS, Yacoub MH, Barton PJ. Developmental expression of troponin I isoforms in fetal human heart. FEBS Lett. 1991;292(1–2):5–8. Epub 1991/11/04.
98. Bedada FB, Chan SS, Metzger SK, Zhang L, Zhang J, Garry DJ, et al. Acquisition of a quantitative, stoichiometrically conserved ratiometric marker of maturation status in stem cell-derived cardiac myocytes. Stem Cell Rep. 2014;3(4):594–605. Epub 2014/11/02.
99. Ahuja P, Sdek P, MacLellan WR. Cardiac myocyte cell cycle control in development, disease, and regeneration. Physiol Rev. 2007;87(2):521–44. Epub 2007/04/13.
100. Deb A. Cell-cell interaction in the heart via Wnt/beta-catenin pathway after cardiac injury. Cardiovasc Res. 2014;102(2):214–23. Epub 2014/03/05.
101. Abramochkin DV, Lozinsky IT, Kamkin A. Influence of mechanical stress on fibroblast-myocyte interactions in mammalian heart. J Mol Cell Cardiol. 2014;70:27–36. Epub 2014/01/07.
102. Dirkx E, da Costa Martins PA, De Windt LJ. Regulation of fetal gene expression in heart failure. Biochim Biophys Acta. 2013;1832(12):2414–24. Epub 2013/09/17.
103. Doenst T, Nguyen TD, Abel ED. Cardiac metabolism in heart failure: implications beyond ATP production. Circ Res. 2013;113(6):709–24. Epub 2013/08/31.
104. Kehat I, Molkentin JD. Molecular pathways underlying cardiac remodeling during pathophysiological stimulation. Circulation. 2010;122(25):2727–35. Epub 2010/12/22.
105. Bayeva M, Sawicki KT, Butler J, Gheorghiade M, Ardehali H. Molecular and cellular basis of viable dysfunctional myocardium. Circ Heart Fail. 2014;7(4):680–91. Epub 2014/07/17.
106. Cung TT, Morel O, Cayla G, Rioufol G, Garcia-Dorado D, Angoulvant D, et al. Cyclosporine before PCI in patients with acute myocardial infarction. N Engl J Med. 2015;373(11):1021–31. Epub 2015/09/01.
107. Yang KC, Kyle JW, Makielski JC, Dudley Jr SC. Mechanisms of sudden cardiac death: oxidants and metabolism. Circ Res. 2015;116(12):1937–55. Epub 2015/06/06.
108. Schwartz SD, Hubschman JP, Heilwell G, Franco-Cardenas V, Pan CK, Ostrick RM, et al. Embryonic stem cell trials for macular degeneration: a preliminary report. Lancet. 2012;379(9817):713–20. Epub 2012/01/28.
109. Chong JJ, Murry CE. Cardiac regeneration using pluripotent stem cells – progression to large animal models. Stem Cell Res. 2014;13(3 Pt B):654–65. Epub 2014/08/05.
110. Chong JJ, Yang X, Don CW, Minami E, Liu YW, Weyers JJ, et al. Human embryonic-stem-cell-derived cardiomyocytes regenerate non-human primate hearts. Nature. 2014;510(7504):273–7. Epub 2014/04/30.
111. Dixit P, Katare R. Challenges in identifying the best source of stem cells for cardiac regeneration therapy. Stem Cell Res Ther. 2015;6:26. Epub 2015/04/18.
112. Rosen MR, Myerburg RJ, Francis DP, Cole GD, Marban E. Translating stem cell research to cardiac disease therapies: pitfalls and prospects for improvement. J Am Coll Cardiol. 2014;64(9):922–37. Epub 2014/08/30.
113. Davis RP, van den Berg CW, Casini S, Braam SR, Mummery CL. Pluripotent stem cell models of cardiac disease and their implication for drug discovery and development. Trends Mol Med. 2011;17(9):475–84. Epub 2011/06/28.

114. Burridge PW, Keller G, Gold JD, Wu JC. Production of de novo cardiomyocytes: human pluripotent stem cell differentiation and direct reprogramming. Cell Stem Cell. 2012;10(1):16–28. Epub 2012/01/10.
115. Elliott DA, Braam SR, Koutsis K, Ng ES, Jenny R, Lagerqvist EL, et al. NKX2-5(eGFP/w) hESCs for isolation of human cardiac progenitors and cardiomyocytes. Nat Methods. 2011;8(12):1037–40. Epub 2011/10/25.
116. Yang L, Soonpaa MH, Adler ED, Roepke TK, Kattman SJ, Kennedy M, et al. Human cardiovascular progenitor cells develop from a KDR+ embryonic-stem-cell-derived population. Nature. 2008;453(7194):524–8. Epub 2008/04/25.
117. Lian X, Hsiao C, Wilson G, Zhu K, Hazeltine LB, Azarin SM, et al. Robust cardiomyocyte differentiation from human pluripotent stem cells via temporal modulation of canonical Wnt signaling. Proc Natl Acad Sci U S A. 2012;109(27):E1848–57.
118. Uosaki H, Fukushima H, Takeuchi A, Matsuoka S, Nakatsuji N, Yamanaka S, et al. Efficient and scalable purification of cardiomyocytes from human embryonic and induced pluripotent stem cells by VCAM1 surface expression. PLoS One. 2011;6(8):e23657. Epub 2011/08/31.
119. Willems E, Spiering S, Davidovics H, Lanier M, Xia Z, Dawson M, et al. Small-molecule inhibitors of the Wnt pathway potently promote cardiomyocytes from human embryonic stem cell-derived mesoderm. Circ Res. 2011;109(4):360–4. Epub 2011/07/09.
120. Zhang Q, Jiang J, Han P, Yuan Q, Zhang J, Zhang X, et al. Direct differentiation of atrial and ventricular myocytes from human embryonic stem cells by alternating retinoid signals. Cell Res. 2011;21(4):579–87. Epub 2010/11/26.
121. Osafune K, Caron L, Borowiak M, Martinez RJ, Fitz-Gerald CS, Sato Y, et al. Marked differences in differentiation propensity among human embryonic stem cell lines. Nat Biotechnol. 2008;26(3):313–5.
122. Karakikes I, Senyei GD, Hansen J, Kong CW, Azeloglu EU, Stillitano F, et al. Small molecule-mediated directed differentiation of human embryonic stem cells toward ventricular cardiomyocytes. Stem Cells Transl Med. 2014;3(1):18–31. Epub 2013/12/11.
123. Bu L, Jiang X, Martin-Puig S, Caron L, Zhu S, Shao Y, et al. Human ISL1 heart progenitors generate diverse multipotent cardiovascular cell lineages. Nature. 2009;460(7251):113–7. Epub 2009/07/03.
124. Hattori F, Chen H, Yamashita H, Tohyama S, Satoh YS, Yuasa S, et al. Nongenetic method for purifying stem cell-derived cardiomyocytes. Nat Methods. 2010;7(1):61–6. Epub 2009/12/01.
125. Tohyama S, Hattori F, Sano M, Hishiki T, Nagahata Y, Matsuura T, et al. Distinct metabolic flow enables large-scale purification of mouse and human pluripotent stem cell-derived cardiomyocytes. Cell Stem Cell. 2013;12(1):127–37. Epub 2012/11/22.
126. Kropp EM, Oleson BJ, Broniowska KA, Bhattacharya S, Chadwick AC, Diers AR, et al. Inhibition of an NAD+ salvage pathway provides efficient and selective toxicity to human pluripotent stem cells. Stem Cells Transl Med. 2015;4(5):483–93. Epub 2015/04/03.
127. Burridge PW, Holmström A, Wu JC. Chemically defined culture and cardiomyocyte differentiation of – human pluripotent stem cells. Curr Protoc Hum Genet. 2015;87:21.3.1–15.
128. Lee AS, Tang C, Rao MS, Weissman IL, Wu JC. Tumorigenicity as a clinical hurdle for pluripotent stem cell therapies. Nat Med. 2013;19(8):998–1004. Epub 2013/08/08.
129. Lindsey ML, Mayr M, Gomes AV, Delles C, Arrell DK, Murphy AM, et al. Transformative impact of proteomics on cardiovascular health and disease: a scientific statement from the American Heart Association. Circulation. 2015;132(9):852–72. Epub 2015/07/22.
130. Bryder D, Rossi DJ, Weissman IL. Hematopoietic stem cells: the paradigmatic tissue-specific stem cell. Am J Pathol. 2006;169(2):338–46. Epub 2006/08/01.
131. Van Hoof D, Dormeyer W, Braam SR, Passier R, Monshouwer-Kloots J, Ward-van Oostwaard D, et al. Identification of cell surface proteins for antibody-based selection of human embryonic stem cell-derived cardiomyocytes. J Proteome Res. 2010;9(3):1610–8.
132. Dubois NC, Craft AM, Sharma P, Elliott DA, Stanley EG, Elefanty AG, et al. SIRPA is a specific cell-surface marker for isolating cardiomyocytes derived from human pluripotent stem cells. Nat Biotechnol. 2011;29(11):1011–8. Epub 2011/10/25.

133. Kropp EM, Bhattacharya S, Waas M, Chuppa SL, Hadjantonakis AK, Boheler KR, et al. N-glycoprotein surfaceomes of four developmentally distinct mouse cell types. Proteomics Clin Appl. 2014;8(7–8):603–9. Epub 2014/06/13.
134. Boheler KR, Bhattacharya S, Kropp EM, Chuppa S, Riordon DR, Bausch-Fluck D, et al. A human pluripotent stem cell surface N-glycoproteome resource reveals markers, extracellular epitopes, and drug targets. Stem Cell Rep. 2014;3(1):185–203. Epub 2014/07/30.
135. Overington JP, Al-Lazikani B, Hopkins AL. How many drug targets are there? Nat Rev Drug Discov. 2006;5(12):993–6. Epub 2006/12/02.
136. Elschenbroich S, Kim Y, Medin JA, Kislinger T. Isolation of cell surface proteins for mass spectrometry-based proteomics. Expert Rev Proteomics. 2010;7(1):141–54. Epub 2010/02/04.
137. Cordwell SJ, Thingholm TE. Technologies for plasma membrane proteomics. Proteomics. 2010;10(4):611–27. Epub 2009/10/17.
138. Sun B, Hood L. Protein-centric N-glycoproteomics analysis of membrane and plasma membrane proteins. J Proteome Res. 2014;13(6):2705–14. Epub 2014/04/24.
139. Wollscheid B, Bausch-Fluck D, Henderson C, O'Brien R, Bibel M, Schiess R, et al. Mass-spectrometric identification and relative quantification of N-linked cell surface glycoproteins. Nat Biotechnol. 2009;27(4):378–86. Epub 2009/04/08.
140. Gundry RL, Riordon DR, Tarasova Y, Chuppa S, Bhattacharya S, Juhasz O, et al. A cell surfaceome map for immunophenotyping and sorting pluripotent stem cells. Mol Cell Proteomics. 2012;11(8):303–16. Epub 2012/04/12.
141. Zeng Y, Ramya TN, Dirksen A, Dawson PE, Paulson JC. High-efficiency labeling of sialylated glycoproteins on living cells. Nat Methods. 2009;6(3):207–9. Epub 2009/02/24.
142. Ramya TN, Weerapana E, Cravatt BF, Paulson JC. Glycoproteomics enabled by tagging sialic acid- or galactose-terminated glycans. Glycobiology. 2013;23(2):211–21. Epub 2012/10/17.
143. Lee SJ, Kim KH, Park JS, Jung JW, Kim YH, Kim SK, et al. Comparative analysis of cell surface proteins in chronic and acute leukemia cell lines. Biochem Biophys Res Commun. 2007;357(3):620–6. Epub 2007/04/24.
144. Weekes MP, Antrobus R, Lill JR, Duncan LM, Hor S, Lehner PJ. Comparative analysis of techniques to purify plasma membrane proteins. J Biomol Techn. 2010;21(3):108–15. Epub 2010/09/03.
145. Bausch-Fluck D, Hofmann A, Wollscheid B. Cell surface capturing technologies for the surfaceome discovery of hepatocytes. Methods Mol Biol. 2012;909:1–16. Epub 2012/08/21.
146. Springer DL, Auberry DL, Ahram M, Adkins JN, Feldhaus JM, Wahl JH, et al. Characterization of plasma membrane proteins from ovarian cancer cells using mass spectrometry. Dis Markers. 2003;19(4–5):219–28. Epub 2004/07/20.
147. Ghosh D, Krokhin O, Antonovici M, Ens W, Standing KG, Beavis RC, et al. Lectin affinity as an approach to the proteomic analysis of membrane glycoproteins. J Proteome Res. 2004;3(4):841–50. Epub 2004/09/14.
148. Wang Y, Ao X, Vuong H, Konanur M, Miller FR, Goodison S, et al. Membrane glycoproteins associated with breast tumor cell progression identified by a lectin affinity approach. J Proteome Res. 2008;7(10):4313–25. Epub 2008/08/30.
149. McDonald CA, Yang JY, Marathe V, Yen TY, Macher BA. Combining results from lectin affinity chromatography and glycocapture approaches substantially improves the coverage of the glycoproteome. Mol Cell Proteomics. 2009;8(2):287–301. Epub 2008/10/17.
150. Kim Y, Elschenbroich S, Sharma P, Sepiashvili L, Gramolini AO, Kislinger T. Use of colloidal silica-beads for the isolation of cell-surface proteins for mass spectrometry-based proteomics. Methods Mol Biol. 2011;748:227–41. Epub 2011/06/28.
151. Rahbar AM, Fenselau C. Integration of Jacobson's pellicle method into proteomic strategies for plasma membrane proteins. J Proteome Res. 2004;3(6):1267–77. Epub 2004/12/15.
152. Cortes LK, Vainauskas S, Dai N, McClung CM, Shah M, Benner JS, et al. Proteomic identification of mammalian cell surface derived glycosylphosphatidylinositol-anchored proteins through selective glycan enrichment. Proteomics. 2014;14(21–22):2471–84. Epub 2014/09/30.

153. Elortza F, Mohammed S, Bunkenborg J, Foster LJ, Nuhse TS, Brodbeck U, et al. Modification-specific proteomics of plasma membrane proteins: identification and characterization of glycosylphosphatidylinositol-anchored proteins released upon phospholipase D treatment. J Proteome Res. 2006;5(4):935–43. Epub 2006/04/11.
154. Tian Y, Koganti T, Yao Z, Cannon P, Shah P, Pietrovito L, et al. Cardiac extracellular proteome profiling and membrane topology analysis using glycoproteomics. Proteomics Clin Appl. 2014;8(7–8):595–602. Epub 2014/06/13.
155. Naba A, Clauser KR, Hoersch S, Liu H, Carr SA, Hynes RO. The matrisome: in silico definition and in vivo characterization by proteomics of normal and tumor extracellular matrices. Mol Cell Proteomics. 2012;11(4):M111.014647. Epub 2011/12/14.
156. Bendz M, Skwark M, Nilsson D, Granholm V, Cristobal S, Kall L, et al. Membrane protein shaving with thermolysin can be used to evaluate topology predictors. Proteomics. 2013; 13(9):1467–80. Epub 2013/03/21.
157. Speers AE, Blackler AR, Wu CC. Shotgun analysis of integral membrane proteins facilitated by elevated temperature. Anal Chem. 2007;79(12):4613–20. Epub 2007/05/16.
158. Frei AP, Jeon OY, Kilcher S, Moest H, Henning LM, Jost C, et al. Direct identification of ligand-receptor interactions on living cells and tissues. Nat Biotechnol. 2012;30(10):997–1001. Epub 2012/09/18.
159. Frei AP, Moest H, Novy K, Wollscheid B. Ligand-based receptor identification on living cells and tissues using TRICEPS. Nat Protoc. 2013;8(7):1321–36. Epub 2013/06/15.
160. Elia G. Biotinylation reagents for the study of cell surface proteins. Proteomics. 2008; 8(19):4012–24. Epub 2008/09/04.
161. Parker BL, Palmisano G, Edwards AV, White MY, Engholm-Keller K, Lee A, et al. Quantitative N-linked glycoproteomics of myocardial ischemia and reperfusion injury reveals early remodeling in the extracellular environment. Mol Cell Proteomics. 2011;10(8):M110 006833. Epub 2011/03/29.
162. Didangelos A, Yin X, Mandal K, Saje A, Smith A, Xu Q, et al. Extracellular matrix composition and remodeling in human abdominal aortic aneurysms: a proteomics approach. Mol Cell Proteomics. 2011;10(8):M111.008128. Epub 2011/05/20.
163. Sun B, Ma L, Yan X, Lee D, Alexander V, Hohmann LJ, et al. N-glycoproteome of E14.Tg2a mouse embryonic stem cells. PLoS One. 2013;8(2):e55722. Epub 2013/02/14.
164. Waas M, Bhattacharya S, Chuppa S, Wu X, Jensen DR, Omasits U, et al. Combine and conquer: surfactants, solvents, and chaotropes for robust mass spectrometry based analyses of membrane proteins. Anal Chem. 2014;86(3):1551–9. Epub 2014/01/08.
165. Chang YH, Gregorich ZR, Chen AJ, Hwang L, Guner H, Yu D, et al. New mass-spectrometry-compatible degradable surfactant for tissue proteomics. J Proteome Res. 2015;14(3):1587–99. Epub 2015/01/16.
166. Zhang X, Miller KW. Dodecyl maltopyranoside enabled purification of active human GABA type A receptors for deep and direct proteomic sequencing. Mol Cell Proteomics. 2015; 14(3):724–38. Epub 2014/12/05.
167. Zhang X. Less is more: membrane protein digestion beyond urea-trypsin solution for next-level proteomics. Mol Cell Proteomics. 2015;14(9):2441–53. Epub 2015/06/18.
168. Tan M, Luo H, Lee S, Jin F, Yang JS, Montellier E, et al. Identification of 67 histone marks and histone lysine crotonylation as a new type of histone modification. Cell. 2011;146(6):1016–28. Epub 2011/09/20.
169. Mir AR, Moinuddin. Glycoxidation of histone proteins in autoimmune disorders. Clin Chim Acta. 2015;450:25–30. Epub 2015/08/04.
170. Leschik J, Caron L, Yang H, Cowan C, Puceat M. A view of bivalent epigenetic marks in two human embryonic stem cell lines reveals a different cardiogenic potential. Stem Cells Dev. 2015;24(3):384–92. Epub 2014/09/10.
171. Oyama K, El-Nachef D, Zhang Y, Sdek P, MacLellan WR. Epigenetic regulation of cardiac myocyte differentiation. Front Genet. 2014;5:375. Epub 2014/11/20.
172. Yang JY, Wang Q, Wang W, Zeng LF. Histone deacetylases and cardiovascular cell lineage commitment. World J Stem Cells. 2015;7(5):852–8. Epub 2015/07/02.

173. Wang Y, Miao X, Liu Y, Li F, Liu Q, Sun J, et al. Dysregulation of histone acetyltransferases and deacetylases in cardiovascular diseases. Oxid Med Cell Longev. 2014;2014:641979. Epub 2014/04/03.
174. Handy DE, Castro R, Loscalzo J. Epigenetic modifications: basic mechanisms and role in cardiovascular disease. Circulation. 2011;123(19):2145–56. Epub 2011/05/18.
175. Wu SS, Lin X, Yuan LQ, Liao EY. The role of epigenetics in arterial calcification. BioMed Res Int. 2015;2015:320849. Epub 2015/07/30.
176. Takemura A, Iijima K, Ota H, Son BK, Ito Y, Ogawa S, et al. Sirtuin 1 retards hyperphosphatemia-induced calcification of vascular smooth muscle cells. Arterioscler Thromb Vasc Biol. 2011;31(9):2054–62. Epub 2011/07/02.
177. Nikolov M, Fischle W. Systematic analysis of histone modification readout. Mol Biosyst. 2013;9(2):182–94. Epub 2012/12/06.
178. Arnaudo AM, Garcia BA. Proteomic characterization of novel histone post-translational modifications. Epigenetics Chromatin. 2013;6(1):24. Epub 2013/08/07.
179. Zheng Y, Tipton JD, Thomas PM, Kelleher NL, Sweet SM. Site-specific human histone H3 methylation stability: fast K4me3 turnover. Proteomics. 2014;14(19):2190–9. Epub 2014/05/16.
180. Shah RV, Januzzi Jr JL. ST2: a novel remodeling biomarker in acute and chronic heart failure. Curr Heart Fail Rep. 2010;7(1):9–14. Epub 2010/04/29.
181. Liddy KA, White MY, Cordwell SJ. Functional decorations: post-translational modifications and heart disease delineated by targeted proteomics. Genome Med. 2013;5(2):20. Epub 2013/03/01.
182. Smith LM, Kelleher NL, Consortium for Top Down Proteomics. Proteoform: a single term describing protein complexity. Nat Methods. 2013;10(3):186–7. Epub 2013/02/28.
183. Peng Y, Ayaz-Guner S, Yu D, Ge Y. Top-down mass spectrometry of cardiac myofilament proteins in health and disease. Proteomics Clin Appl. 2014;8:554–68. Epub 2014/06/20.
184. Savaryn JP, Catherman AD, Thomas PM, Abecassis MM, Kelleher NL. The emergence of top-down proteomics in clinical research. Genome Med. 2013;5(6):53. Epub 2013/06/29.
185. Zhang J, Guy MJ, Norman HS, Chen YC, Xu Q, Dong X, et al. Top-down quantitative proteomics identified phosphorylation of cardiac troponin I as a candidate biomarker for chronic heart failure. J Proteome Res. 2011;10(9):4054–65. Epub 2011/07/15.
186. Peng Y, Yu D, Gregorich Z, Chen X, Beyer AM, Gutterman DD, et al. In-depth proteomic analysis of human tropomyosin by top-down mass spectrometry. J Muscle Res Cell Motil. 2013;34(3–4):199–210. Epub 2013/07/25.
187. Mazur MT, Cardasis HL, Spellman DS, Liaw A, Yates NA, Hendrickson RC. Quantitative analysis of intact apolipoproteins in human HDL by top-down differential mass spectrometry. Proc Natl Acad Sci U S A. 2010;107(17):7728–33. Epub 2010/04/15.
188. Gregorich ZR, Peng Y, Lane NM, Wolff JJ, Wang S, Guo W, et al. Comprehensive assessment of chamber-specific and transmural heterogeneity in myofilament protein phosphorylation by top-down mass spectrometry. J Mol Cell Cardiol. 2015;87:102–12. Epub 2015/08/14.
189. Fellers RT, Greer JB, Early BP, Yu X, LeDuc RD, Kelleher NL, et al. ProSight Lite: graphical software to analyze top-down mass spectrometry data. Proteomics. 2015;15(7):1235–8. Epub 2015/04/02.
190. LeDuc RD, Taylor GK, Kim YB, Januszyk TE, Bynum LH, Sola JV, et al. ProSight PTM: an integrated environment for protein identification and characterization by top-down mass spectrometry. Nucleic Acids Res. 2004;32(Web Server issue):W340–5. Epub 2004/06/25.
191. Guner H, Close PL, Cai W, Zhang H, Peng Y, Gregorich ZR, et al. MASH Suite: a user-friendly and versatile software interface for high-resolution mass spectrometry data interpretation and visualization. J Am Soc Mass Spectrom. 2014;25(3):464–70. Epub 2014/01/05.
192. Petrotchenko EV, Borchers CH. HDX match software for the data analysis of top-down ECD-FTMS hydrogen/deuterium exchange experiments. J Am Soc Mass Spectrom. 2015;26:1895–8. Epub 2015/07/15.
193. Kellie JF, Tran JC, Lee JE, Ahlf DR, Thomas HM, Ntai I, et al. The emerging process of top down mass spectrometry for protein analysis: biomarkers, protein-therapeutics, and achieving high throughput. Mol Biosyst. 2010;6(9):1532–9. Epub 2010/08/17.

194. Catherman AD, Skinner OS, Kelleher NL. Top down proteomics: facts and perspectives. Biochem Biophys Res Commun. 2014;445(4):683–93. Epub 2014/02/22.
195. Brunner AM, Lossl P, Liu F, Huguet R, Mullen C, Yamashita M, et al. Benchmarking multiple fragmentation methods on an orbitrap fusion for top-down phospho-proteoform characterization. Anal Chem. 2015;87(8):4152–8. Epub 2015/03/25.
196. Fornelli L, Parra J, Hartmer R, Stoermer C, Lubeck M, Tsybin YO. Top-down analysis of 30-80 kDa proteins by electron transfer dissociation time-of-flight mass spectrometry. Anal Bioanal Chem. 2013;405(26):8505–14. Epub 2013/08/13.
197. Belov ME, Damoc E, Denisov E, Compton PD, Horning S, Makarov AA, et al. From protein complexes to subunit backbone fragments: a multi-stage approach to native mass spectrometry. Anal Chem. 2013;85(23):11163–73. Epub 2013/11/19.
198. Walmsley SJ, Rudnick PA, Liang Y, Dong Q, Stein SE, Nesvizhskii AI. Comprehensive analysis of protein digestion using six trypsins reveals the origin of trypsin as a significant source of variability in proteomics. J Proteome Res. 2013;12(12):5666–80. Epub 2013/10/15.

Chapter 7
Bottom-Up Proteomics

Dario Di Silvestre, Francesca Brambilla, Giulio Agnetti, and Pierluigi Mauri

Abstract In this chapter we provide an overview of bottom-up proteomic approaches. These allow the identification and characterization of proteins and their amino acid sequences, including post-translational modifications, by proteolytic digestion prior to mass spectrometry (MS) analysis. Intact proteins can be separated by gel electrophoresis followed by in-gel protein digestion to generate peptides which are then analyzed by MS. Alternatively, complex protein mixtures can be digested directly (an approach referred to as "shotgun") and the resulting peptides can be separated by liquid chromatography prior to MS. Following MS analysis, the comparison of the peptides' spectra with those predicted from genomics/proteomics sequence databases, or annotated peptide spectral libraries, allows the identification of peptides which are finally assigned to corresponding proteins. After a description of the separation methods and MS acquisition modes, a relevant part of the chapter will be dedicated to data processing pointing to algorithms, computational tools and strategies useful for researchers in the discovery process. In particular, liquid-chromatography (LC) based approaches, including Multidimensinal Protein Identification Technology (MudPIT), will be taken as reference and different aspects, ranging from database search engines to protein-protein interaction (PPI) network analysis, will be addressed. Potential issues will be

D. Di Silvestre, PhD (✉) • F. Brambilla, PhD
Department of Proteomics and Metabolomics, Institute for Biomedical Technologies-National Research Council, Milan, Segrate, Italy
e-mail: dario.disilvestre@itb.cnr.it

G. Agnetti, PhD, FAHA
Laboratory of Cardiovascular Biochemistry, Division of Cardiology, Department of Medicine, The Johns Hopkins University School of Medicine, Baltimore, MD, USA

Department of Biomedical and Neuromotor Sciences (DIBINEM), University of Bologna, Bologna, Italy

P. Mauri, PhD
Department of Proteomics and Metabolomics, Institute for Biomedical Technologies-National Research Council, Milan, Segrate, Italy

Laboratory of Medical Science, Scuola Superiore Sant'Anna, Institute of Life Sciences, via G. Moruzzi, 1, Pisa 56124, Italy

G. Agnetti et al. (eds.), *Manual of Cardiovascular Proteomics*,
DOI 10.1007/978-3-319-31828-8_7

discussed in the context of cardiovascular research, and specifically the last section will focus on the translational applications (clinical proteomics) of cardiovascular proteomics.

Keywords Liquid chromatography • Mass spectrometry • Heart • Proteomics • MudPIT

Introduction

Historically, proteomics started with two-dimensional gel electrophoresis (2DE) although, in the last two decades a wide range of technologies has been developed; separation approaches, such as electrophoresis and chromatography, have been paired with identification systems, such as mass spectrometry (MS). In recent years MS has evolved rapidly, increasing its impact on proteomic investigations and generating numerous approaches along with the need to revise and update current classifications. These methods are broadly divided into top-down and bottom-up depending on whether whole proteins or the peptides originated from their digestion are analyzed in a MS (respectively) [1–2]. The distinction between top-down *vs.* bottom-up is commonly based on the separation of intact proteins *vs.* digested proteins, respectively. However, we think that "top-down" should only be used to define those methods where intact proteins are carried through the identification step. Another classification is based on the protein separation methods: gel-based *vs.* gel-free. The first uses electrophoresis (mono- or two-dimensional) to separate intact proteins; gel-free methods, such as liquid chromatography LC and capillary electrophoresis (CE), separate the peptides obtained by enzymatic digestion of proteins. More recently LC-MS has been increasingly applied also to intact proteins (top-down). Classical 2DE separation of proteins, followed by MS analysis of peptides obtained from in-gel digestion of single spots is considered a bottom-up approach. However, as the information related to the intact protein size is retained, 2DE may be right on the blurred line between top-down and bottom-up [3]. Table 7.1 summarizes the available proteomic technologies and their classification according the different criteria described above.

As mentioned, this chapter provides an overview of bottom-up methods (section "Separation Methods") according to the definition based on identification, while top-down approaches are addressed in detail in Chapter 8. In section "Acquisition Modes" we address in detail the workflow of the typical MS-based, gel-free and bottom-up approaches, commonly referred to as "shotgun". These methods are based on the enzymatic digestion of protein mixtures, followed by peptide separation by means of LC (mono- or two- dimensional, LC and LC/LC, or 2 DLC, respectively), and coupled to tandem mass spectrometry (MS/MS). This approach is broadly applicable and is also referred to as MudPIT (Multidimensional Protein Identification Technology) [4] or mono long reverse-phase gradient [5].

Table 7.1 Technologies for proteomics and their classification

Proteomics steps			Classification		
Protein treatment	Separation	Identification	Gel-based	MS-based	Top-down
Intact	Gel	Position on the map	+	–	+
		Antibodies	+	–	+
In-gel digestion		MS	+/–	+/–	+/–
Intact	Protein array	Antibodies	–	–	+
	Chromatography	MS	–	+	+
Digestion before separation	Chromatography	Mass spectrometry	–	+	–
	Capillary electrophoresis		–	+	–

Separation Methods

Traditionally, the biological role of proteins has been addressed by studying "one protein at a time". In contrast to this traditional approach, high-throughput methods have been developed in the last decades to study large numbers of molecules, including nucleic acids, proteins, metabolites and others, at once. These global "approaches" are usually named by a "-omics" suffix (e.g., transcriptomics, proteomics, metabolomics, etc). In proteomics, the complete detection and reproducible measurement of proteins content in complex samples, such as biological fluids or tissue biopsies, is a necessary goal to address the mechanisms underlying disease, its diagnosis and cure (i.e biomarker discovery and elucidation of pathogenic mechanisms). For instance, several strategies have been implemented to detect low abundant proteins or those characterized by a very wide dynamic range in terms of copy number or concentration. Because proteins of biological relevance are often present in very low amount, protein fractionation [6–8] for the reduction of sample complexity prior to identification, represent an important requirement for proteomics analysis. Beside tailored, biochemical fractionation (addressed in Chap. 4), the separation of proteins or peptides can be achieved by gel-electrophoresis (gel-based) and/or liquid chromatography (gel-free).

Gel-Based Separation

Several different electrophoretic approaches can be used alone (mono-dimensional separation, 1D) or in combination (two-dimensional separation, 2D) to separate of proteins. As mentioned above, higher dimensionality translates to higher fractionation, resulting in higher selectivity and sensitivity. As sodium dodecyl sulphate is arguably one of the best detergents to maintain proteins in solution, which is necessary for them to be separated, one of the oldest electrophoretic method exploits this feature to separate proteins using polyacrilamide gel electrophoresis (SDS-PAGE). In SDS-PAGE, proteins are resolved based on their size (molecular weight/charge), using the acrylamide matrix as a molecular sieve. Albeit this is a robust technique, its resolving power is limited. For this reason, in order to separate complex samples, two dimensional electrophoresis (2DE) was implemented [9]. In 2DE, proteins are classically separated in the first dimension based on their isoelectric point (pI), commonly by embedding ampholytes in the acrylamide matrix, in order to create a stable pH gradient. Proteins bear a charge until they reach their pI, by definition the pH to which a proteins net charge equals zero. The way that is commonly used to create a pH gradient is IPG (immobilized pH gradient) [10]. In the second dimension, proteins are separated based on SDS-PAGE [11]. The combination of two separation methods applied orthogonally generates a 2D map [12]. After separation, protein spots can be visualized by several methods such as visible (Comassie Blue, Silver Stain, etc.) and fluorescent staining (there are several brand names) [13].

When protein complexes are the objects of investigation, non-denaturing gel electrophoresis (such as blue native) can be employed to separate them. Non-denaturing gel electrophoresis can be coupled with denaturing SDS-PAGE in "non-classical" 2DE. This allows to identify the protein components (including post-translational modifications, PTMs), within a given protein complex [14].

Gel spots may be identified by means of MS or more classically by immunodetection (such as in Western blot). Indeed antibody-based detection is very sensitive [15]; however, due to the limited knowledge of many PTMs it is very difficult to predict their impact on antibodies's affinity. For these reasons, over the years MS became the standard technique for protein identification [16]. MS allows the unbiased identification of proteins from gel bands (1DE) or spots (2DE) by digestion, typically with trypsin. The peptides resulting from digestion are free to move out of the gel and into a liquid phase; sodium dodecyl sulphate may be a problematic contaminant for MS, however it can be removed effectively using organic/acid washes (standard protocols for in-gel digestion are well established) [17]. Classically Matrix Assisted Laser Ionization-Time of Flight (MALDI-TOF) MS was initially utilized to quickly identify the molecular weight of peptides originated by in-gel digestion of a band or spot (see Chap. 2).

As MS technologies progressed, other types of MS, such as ion trap or Q-TOF could be also employed and enable sequencing by means of tandem mass spectrometry (MS/MS). In MALDI, extracted peptides are co-crystallized with an organic matrix on a metal target. A pulsed laser is used to excite the matrix, which causes a rapid thermal heating of the molecules and eventually desorption of ions into the gas phase. Singly charged ions of peptides are then detected by time-of-flight (TOF) mass analyzer and the corresponding protein is identified by "peptide mass fingerprint" on the basis of the *m/z* values of its tryptic peptides [18, 19]. The incomplete recovery of the peptides from gel represents the main drawback of this procedure: to enhance peptide recovery the incubation of the gel pieces with different concentrations of acetonitrile can improve yield.

The main advantage and strength of gel based approaches is their widespread use (1DE), and, when it comes to 2DE, the type of information on intact mass and the direct visual confirmation of changes in protein/post-translational modifications (PTMs) abundance. On the other hand, the main limitations with 2DE are that they are largely dependent on the skills of the operator, that can deeply impact reproducibility, as well as being labor intensive. Also, it is difficult to separate hydrophobic proteins (such as membrane proteins), or those with extreme pI, acidic or basic, and high molecular weight; finally, narrow dynamic range and low-throughput are other disadvantages.

Gel-Free Separation

In recent years, a number of gel-free proteomics methods to separate proteins/peptides mixtures have been developed to overcome the difficulties related to gel-based approaches. In this context, methods based on protein arrays or chromatography are now available. Traditionally, an array coupled with mass spectrometry is SELDI (surface enhanced laser desorption ionization), an integrated high-throughput proteomics technique firstly introduced in 1993 [20], which may be considered as a variation of MALDI [21, 22]; another gel-free separation approach are protein arrays that represents a promising high-throughput approach for a wide variety of applications including the study of the biochemical activities of proteins (identification of protein-protein interactions, protein-ligand interactions, protein-DNA interactions), diagnosis and monitoring of disease states [23–25]. When these data are relevant for the investigations, top-down approaches or hybrids may be required depending upon the goal of the study and the biological question to be addressed.

Liquid chromatography (LC) and capillary electrophoresis (CE) are the two most widely used techniques for the liquid separation of proteins and peptides [26]; in particular, in bottom-up proteomics, both technologies are coupled on-line with MS/MS to fully characterize peptide mixtures obtained by digestion of proteins. Capillary electrophoresis can be based on a number of principles, such as capillary gel electrophoresis (CGE), capillary isoelectric focusing (CIEF), capillary isotachophoresis (CIT) and micellar electrokinetic chromatography (MEKC); CE uses very small glass capillaries (50–100 μm internal diameter long 25–75 cm) where high voltages (10–25 kV) are applied to allow separation of analytes according to their ionic mobility [27]. The rate at which the particles moves is directly proportional to the applied electric field: the greater the field strength, the faster the mobility. In CE a fast (10–20 min) and efficient (theoretical plates, $N > 10^6$ plates) separation is obtained using a different mechanism from reverse phase [28, 29]; it is highly sensitive due to low dilution and high concentration of analytes. Usually, visible/UV detector is used for monitoring separation; moreover, a laser induced fluorescence (LIF) detector enable unparalleled sensitivity. Recently, CE separation system was combined to MS analyzer using a sheath buffer for reducing the ion concentration of CE buffer (at millimolar level), but this introduces a significant dilution of the analytes (peptides) [30].

Liquid Chromatography

Liquid chromatography (LC) is recognized as an indispensable tool in proteomics research since it provides high-speed, high-resolution and high-sensitivity separation of macromolecules, such as proteins and peptides [31]. An attractive feature of LC is the broad selection of stationary and mobile phases, that makes LC extremely versatile to resolve analytes on the basis of different mechanisms of interaction between soluble and stationary phases. Each component in the sample interacts differently with the stationary phase, causing different retention for different analytes, leading to their separation and consequently different elution times (the time the analyte spends "dwelling" in the column). The progressive reduction of both the column internal diameter (ID) and solvent flow rate, from micro-LC to nano-LC, radically reduce sample volume and solvent consumption, while it increases efficiency, resolution and sensitivity.

Reverse-phase (RP), ion-exchange (IEC), affinity and size-exclusion liquid chromatography (SEC) are the most used separation methods to separate peptides and proteins [32]. In particular, reverse-phase high-performance liquid chromatography (RP-HPLC) is the most common gel-free separation method in shotgun proteomics, mainly due to its compatibility with mass spectrometry (MS). Reversed-phase chromatography uses a hydrophobic stationary phase, constituted by alkyl chains covalently bonded to the solid support, and a polar (aqueous) mobile phase. Hydrophobic molecules in the polar mobile phase adsorb to the hydrophobic stationary phase, and hydrophilic molecules in the mobile phase will pass through the column and are eluted first. Hydrophobic molecules can be eluted from the column by decreasing the polarity of the mobile phase using an organic (non-polar) solvent, commonly acetonitrile, which reduces hydrophobic interactions. The more hydrophobic the molecule, the stronger it will bind to the stationary phase, and the higher the concentration of organic solvent that will be required to elute the molecule. Peptide/protein elution by organic solvent in RP allows a simple coupling to mass spectrometry by means of Electrospray Ionization source.

Electrospray Ionization (ESI) is a soft ionization technique capable of imparting a charge to peptides with low internal energy and thus in-source fragmentation [33] (also see Chapter 1 for an historical perspective). Electrospray ionization source works by generating a fine mist of charged analyte droplets at the inlet of the MS at atmospheric pressure. Also, ESI sources, assisted by a stream of external gas (usually nitrogen), determine the transition of peptides from liquid to gas phase. As the liquid evaporates and charge density increases on the surface of the droplets, electrostatic interactions cause the droplet to "burst" generating multiply charged peptide ions in gas phase (for more details see Chap. 2). Then, the mass spectrometer measures the mass-to-charge ratio (*m/z*) of analytes (precursor ion), fragments precursor ions into product ions acquiring MS/MS spectra, and finally the detector records the number of ions (precursors and fragments) at each *m/z* value.

Nano-ESI represents an improvement of ESI for spraying low chromatographic flow (rate at nanoliter/minute) due to the reduction of the inner diameter (lower than 0.1 mm) of the RP column. In this case droplet formation occurs more readily and it is possible to reduce the electrical field and to improve the spray stability. As mentioned above, the nanoflow rate determines reduction of analytes dilution and increases sensitivity.

As a recurring theme in proteomics, the combination of two separating procedures increases resolution and decreases sample complexity, resulting in higher sensitivities and higher numbers of identified proteins. In liquid chromatography, this is obtained by Multidimensional Protein Identification Technology (MudPIT) described in the next section.

A wide variety of samples, ranging from tissue or cells protein extracts and biological fluids, can be analyzed by LC-MS to detect protein changes related to cardiovascular diseases [34–38]. As mentioned, the standard preparation for LC-MS in shotgun proteomics requires the enzymatic digestion, usually by trypsin, of the initial protein mixture. It is important that the extracted proteins stay in solution so that can be digested efficiently by trypsin. However, as many detergents can deeply affect both LC separation and MS analysis these have to be chosen carefully. The most performing belong to a new generation of surfactants: they efficiently keep proteins in solution, and after digestion they are readily decomposed under acidic conditions into breakdown products, that do not interfere with LC-MS analysis. Among them, RapiGest™ was found to be one of the best detergents to obtain the maximum number of protein identifications [39].

Multidimensional Chromatography

Two-dimensional chromatography (LC/LC), coupled to tandem mass spectrometry (MS/MS), also named MudPIT (Multidimensional Protein Identification Technology) [3], was developed to separate and identify a high number of peptides resulting from digestion of very complex protein mixtures. This approach combines strong cation exchange chromatography (SCX) to reversed-phase chromatography (RP). Peptides deriving from the protein sample are loaded on an SCX column and separated by increasing salt (NH_4Cl) concentration steps. Each eluted salt fraction is then separated in the second dimension by RP chromatography using a gradient of organic solvent (typically acetonitrile at acidic pH). Finally, peptides are directly eluted in the MS, ionized by ESI ion source and detected by collection of MS spectra (Fig. 7.1). The SCX and RP phases could be packed in a single biphasic column or in two separate columns [40]. When using separate columns, it is possible to: (1) increase the inner diameter for the first column (the SCX) and therefore loading capacity, (2) employ a desalting pre-column before the RP separation column to decrease the concentration of salt which interferes with MS analysis. Multidimensional separation can be performed in a discontinuous fashion by collecting SCX fractions and subsequently subjecting them to RP at a later time; however this off-line procedure decreases recovery, efficiency and sensitivity. For these reasons, the on-line coupling of SCX to RP, by means of a simple 10 port valve, is to be preferred. Of note, using single biphasic column it is mandatory to employ ammonium acetate/formiate for eluting peptides from SCX phase, because this buffer is eluted in the MS and it requires to be volatile; on the contrary, by using two separated columns it is possible to employ ammonium chloride, that is more selective during the SCX separation, but it is very aggressive for MS instrument and less volatile than ammonium acetate.

MudPIT analysis requires more time (5–12 h/sample) than mono-dimensional chromatography; however, it increases resolution and sensitivity, and it is also automatized. On the other hand, recently, some authors use long acetonitrile gradient (4–8 h) for increasing resolution of monodimensional LC-MS [41].

Acquisition Modes

The on-line coupling between LC and MS provides several advantages; however, the harmonization of the two instruments is not free from limitations either. It takes time for an MS to fragment and analyze peptides (known as duty cycle), while peptides are eluted at a constant rate. Due to the stochastic nature of the ionization process and the competitive nature of their detection, peptides could go undetected due to the duty cycle of the MS; it may cause for peptide to go undetected while the instrument is "busy" fragmenting/analyzing peptides that are in the cue, causing "random" coverage of the proteome. This aspect is highly dependent on the reproducibility of LC runs as different experiments can generate different result due to a drift in sampling. The optimization of acquisition modes we will discuss in the following paragraph.

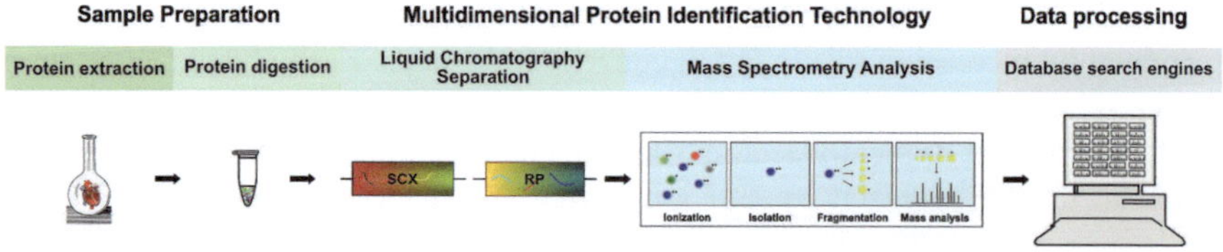

Fig. 7.1 Multidimensional Protein Identification Technology. Main steps involved in the MudPIT approach. *SCX* strong cation exchange chromatography, *RP* reverse phase chromatography

Data Dependent Analysis

Traditionally, bottom-up proteomics are based on a strategy known as Data Dependent Acquisition (DDA) (Fig. 7.2). According to this approach, the selection of peptides to be fragmented is controlled by data themselves during the analysis, and automatically the most abundant ions are selected for fragmentation [42]. In other words, the instrument selects the most abundant precursor ions from the first MS scan acquiring their tandem mass spectra in the following scans. The constant technological advances in MS have generated quicker instruments, with a shorter duty cycle. Therefore, the increased speed of MS analysis translates to a higher number of ions that are isolated and fragmented (e.g. from 3 to 20 parent ions) for each full and MS/MS cycle. The definition "data dependent acquisition" reflects the dependence of MS/MS spectra acquisition on the intensity of the ions detected in the full scan. This means that in DDA approach high-abundant precursor ions are preferentially fragmented, while low abundant could be never sampled; however, by performing technical replicate analyses this phenomenon may decrease. Moreover, to limit the redundant fragmentation of the same ions, these can be excluded (dynamic exclusion) so that ions selected for fragmentation are temporarily excluded in the following scans,

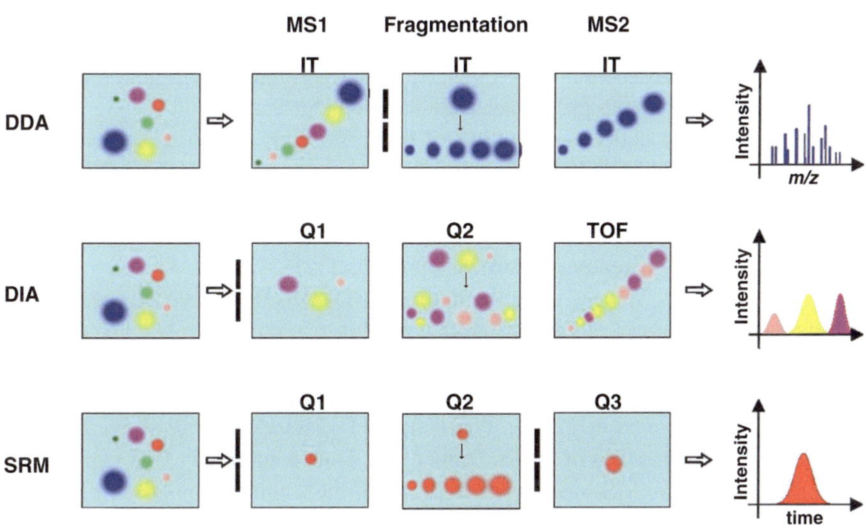

Fig. 7.2 Mass spectrometry acquisition methods. By DDA, ions are selected and fragmented by applying signal intensity and charge state filters. Data acquisition is performed in iterative cycles and fragment ion scans. By DIA, such as SWATH MS, a group of precursor ions is simultaneously selected and fragmented. Precursor ions are sequentially isolated in smaller mass windows (typically 25 m/z), and the composite fragment ion spectra are recorded and then deconvoluted. By SRM/MRM, precursor ions of a predetermined set of peptides are selected, fragmented and the signal intensities recorded over time. *DDA* data dependent acquisition, *DIA* data independent acquisition, *SRM* selected reaction monitoring, *IT* ion trap, *Q* quandrupole, *TOF* time of flight

over a determined time. This approach improves the number of distinct MS/MS spectra and, consequently, the number of identified peptides and proteome coverage [43]. Even though this approach has notably improved DDA methodology, low intensity ions, co-eluting with high-intensity peptides, are usually excluded from the selection and they will not be identified. This is one of the main limitations of DDA, especially in discovery proteomics, which is the initial step in the characterization of unknown samples.

Data Independent Analysis (DIA) and SWHAT Analysis

Alternative methods of acquisition able to generate product ion data (MS/MS spectra) for all co-eluting precursors, regardless of their abundance, have been recently developed. For this purpose, data independent analysis (DIA) has emerged to address some of the limitations with data-dependent analysis (DDA) and selected or multiple reaction monitoring (SRM or MRM, respectively) (Fig. 7.2). This last approach (SRM and MRM are used interchangeably) is an example of targeted proteomics, where the *a priori* knowledge of a peptide mass allows to focus on that mass alone, greatly increasing sensitivity and enabling relative and absolute quantitation (for more details see Chapter 9). Specifically, shotgun proteomics have limited capabilities on very complex samples mainly due to under-sampling [44]. In contrast, targeted proteomics are limited by the number of measurements (up to a few hundreds transitions) per LC-MS/MS run, which is hardly sufficient to provide extensive coverage of the proteome. Data independent acquisition is based on the sequential isolation and fragmentation of precursor ions within a defined "mass window" also named SWATH (sequential window acquisition of all theoretical fragment ion spectra) [45]. In this mode, MS system systematically scans samples by acquiring fragmentation spectra of all precursor ions within sequential isolation windows, cyclically repeating one full spectrum and selected MS/MS windows over a desired *m/z* range. Various implementations of DIA method have been described using isolation windows of various widths, ranging from the complete *m/z* range to few Daltons (Da), in relation to the acquisition of MS experiments. The setup of the MS instrumentation plays an important role as the MS required for DIA are fast; the main methods described in the literature consist of 32 windows (called swaths, also) of 25 Da [46] or 20 swaths of 10–20 Da [47, 48]. Data acquired by DIA can thus be described as iterative MS/MS maps consisting of the fragment ion spectra from all the precursors fragmented in each window. DIA data processing is mainly based on a targeted data extraction from MS/MS maps. A more recent multiplex strategy, the MSX-DIA has been developed for very fast and high-resolution instruments [49]. It is based on the random selection of five separate 4 Da *m/z* large windows per MS/MS acquisition; then a de-multiplexing process allows isolating tandem spectra from the five *m/z* windows to obtain more clean spectra and to improve the number of identified peptides. The identification of peptides has been achieved by using two strategies. In the first strategy, called SWATH-MS [46], data analysis consists of targeted data extraction procedure to query the acquired fragment ion maps of specific peptides of interest, using *a priori* information contained in spectral libraries acquired by sequential DDA experiments. The extraction of fragment ion traces, from data-independently acquired sample sets (libraries), are used for the quantification of formerly identified peptides. The appeal of the approach is that any peptide precursor and product ion data within the limit of detection of the instrument can be *a posteriori* extracted from acquired data. The second strategy is based on the reconstruction of "deduced MS/MS spectra" by the alignment of the retention time of product ions, using their extracted ion chromatograms (EIC); co-eluting product ions are correlated to one precursor ion. Database searching, as in DDA method, completes the procedure [47, 49] together with MSX-DIA and skyline software for data processing.

DIA combines the robustness of shotgun proteomics with the quantitative power of SRM and potentially increases coverage with newer instrumentation. Theoretically, this may allow to monitor all peptides present in a complex biological sample.

Data Handling: From Raw Data to Network Analysis

Mass spectrometry-based proteomics, like MudPIT, allow the high-throughput analysis of complex samples generating big amount of data per experiments, including thousands of spectra, peptides and proteins. These experiments provide a good snapshot of the proteome, and the breadth of information obtained poses challenges in terms of computational power and interpretation of biologically and clinically meaningful information (Fig. 7.3). For example, the processing of raw spectra involves computational procedures, database searching, protein identification and their quantitation, as well as models of sample classification. In addition, peptides and proteins (post-search analysis) are useful for the characterization of proteotypic peptides, while the integration of protein profiles with protein-protein interaction (PPI) databases is used for reconstructing PPI networks and to evaluate the results at the biological and topological levels.

Database Searching Methods

A hallmark of bottom-up proteomics is the identification of proteins by database searching methods that are continuously implemented by several algorithms and specific software products [50]. The experimental masses of the parent peptides and their fragments (MS/MS) are compared with those generated *in-silico,* based on the available genomic/proteomic sequences or annotated in spectral libraries; in this way, peptide sequences are characterized and assembled back into the corresponding proteins (Fig. 7.4). Alternatively, the identification of peptides may be carried out without the use of a database through "de novo sequencing" [51]. However, these two methods may be used in combination to improve the confidence in both database searches and *de novo* identification [52].

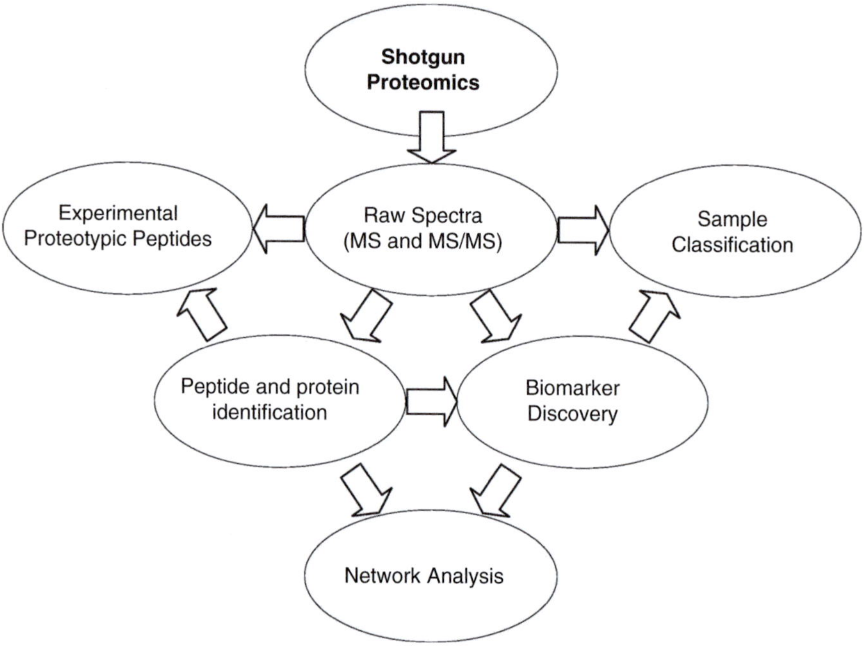

Fig. 7.3 Steps of the discovery process using shotgun proteomic data

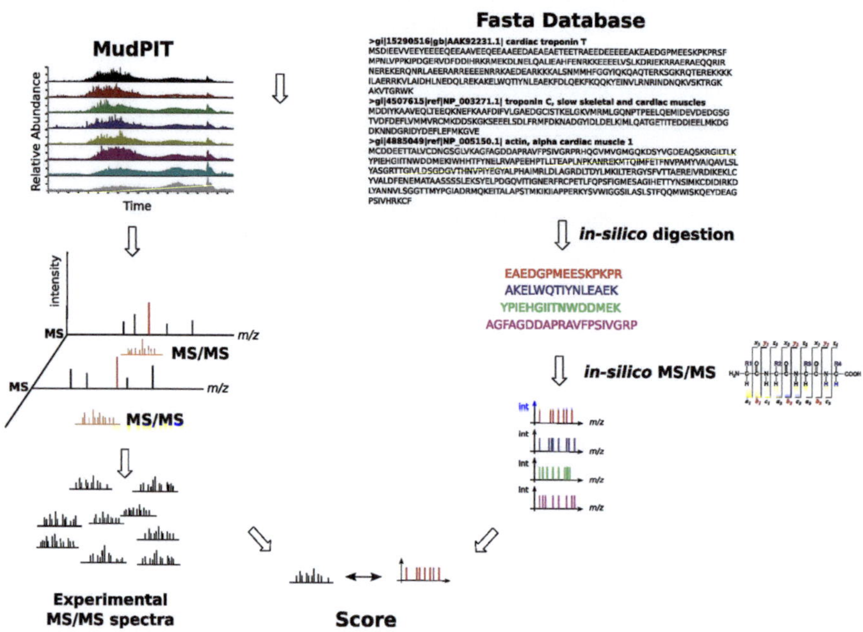

Fig. 7.4 Database search engines method workflow

Among the algorithms for database searching, MASCOT [53] represents one of the most used tools for mass spectra interpretation, while SEQUEST [54] is one of the oldest and best-recognized algorithms: the former became very popular due to its free availability on Matrix Science website for the identification of gel-electrophoresis spots; the second was originally developed in 1993 in the Yates lab at the University of Washington (SEQUEST® is a registered trademark of the University of Washington) and licensed to Thermo Scientific, that inserted the algorithm in its platforms, such as Bioworks. A new version, SEQUEST HT, is now available with Proteome Discoverer 2.0 [55]. Other tools provided by MS instrument manufacturers are available and among the most popular belong Protein Pilot from AB Sciex [56] and Spectrum Mill by Agilent [57]. In addition, a wide range of both proprietary and open-source tools have been developed in the last decade. Examples of open source tools are X!Tandem [58], Myrimatch [59], OMSSA [60], MS-GFDB [61], pFind [62] or Protein Prospector [63]; while among commercial software Peaks DB [64] represents a valid alternative, in particular for the identification of PTMs and for *de novo* sequencing. For more details about search engine tools we re-direct the interested reader to Chaps. 11 and 12 of this book.

The development of several tools to perform database searching was also fuelled by the availability of standard MS file types. In fact, MS manufacturers initially produced raw mass spectra in proprietary file formats, making data exchange difficult. However, over the years, standard formats like mzData, mzXML or mzML have been developed [65], improving exchangeability and promoting the development of open-source platforms. The introduction of these file formats also enabled proteomic scientists to process raw spectra using different search engines and therefore improving the confidence in the identification of peptide spectrum matches (PSMs) and overall proteome coverage. As a consequence, many computational platforms, including Discoverer 2.0 or Peaks DB, now allow the export the raw spectra in these formats and their simultaneous processing with different algorithms. Most of them validate the vast lists of identifications by using a Target-Decoy Approach (TDA) to impose an empirical False Discovery Rate (FDR) at a pre-determined threshold [66], usually 1–3 %, while different platforms use different scoring systems to rank PSMs.

Differential Analysis

An important goal in basic research and clinical applications is the characterization of differentially expressed proteins [67]; the comparison of samples from healthy controls and patients allows to highlight changes in protein levels, useful to decipher disease mechanisms and generate new hypotheses, while new potential protein biomarkers could be discovered and validated for early diagnosis and to monitor the response to a certain treatment.

In 2DE the differential analysis is performed by comparison of the spot abundance across groups of gels using tools dedicated to gel image analysis. An improvement of

2DE is represented by difference gel electrophoresis (DIGE) [68], where the use of labels carrying different fluorophores (Cy2, Cy3 and Cy5) for the different samples (e.g. control and treated or diseased) can be mixed and separated together within the same gel. A laser scanner is then used to record the different signals belonging to different samples and the gel can be post-stained with visible staining or robotic spot pickers can be used to carve out protein spots. Protein changes can be quickly visualized and related to the specific sample or condition by its color component (Cy3, green, or Cy5, red) that will be over represented. Under these conditions, a given protein spot will appear red or green depending on sample to dye assignment. DIGE experimental variability.

As for gel-free approaches, a number of strategies has been developed to allow simultaneous protein identification and quantification. These include stable isotope-labelling and label-free approaches [69, 70]. Labelling approaches are based on the introduction of isotopic or isobaric tags into peptides to create a specific mass signature and measure the ratio of the signal intensities between the differently labelled peptides (Fig. 7.5a). Two main sub-categories distinguish labelling methodologies depending on whether a isotopic or isobaric tag is used to derivatize the proteins before digestion, such as ICAT [71], iTRAQ [72] or TMT [73], or whether isotopes are introduced directly in the protein sequence either during their synthesis, such as SILAC [74], or during digestion, such as the enzymatic incorporation of ^{18}O [75]. All of these are described in greater detail in Chapter 11. The use of both labelling approaches provides highly accurate and reproducible results along with the possibility of multiplexing, but its cost could represent a limitation [76].

Label-free approaches represent a simpler and low-cost alternative; a first group of them is based on the evaluation of the chromatographic peak area of the identified peptides, while other procedures exploit the direct relationship between protein abundance and sampling parameters, such as the spectral count (SpC) [77] (Fig. 7.5b). Although these approaches are considered less accurate due to technical variability, several studies have demonstrated their effectiveness in identifying differentially expressed proteins [45, 78, 79]. A variety of statistical approaches to use SpC values for quantitation across different samples have been published. They mainly rely on the relationship between the amount of a certain protein and the number of times that its peptides are observed [80–83]. Several groups implemented statistical methods to infer protein quantity based on the empirical observation that more peptides correspond to more protein. Zhang and colleagues compared SpC values by means of the statistical G-test [84]. Washburn and co-authors generated a normalized spectral abundance factor (NSAF) [85], while the protein abundance index (PAI or emPAI) [86], calculated by dividing (for each protein) the number of observed peptides for each protein, by the number of all possible detectable tryptic peptides, has been proposed. Finally SEQUEST Score and SpC values are processed by DAve and DCI algorithms [32, 37, 87].

In parallel to the development of biostatistical tools and indices to compare protein levels from shotgun experiments, the need to automate the identification step has driven the development of even more tools. Various software that were initially designed for the interpretation of mass spectra have now evolved in complete

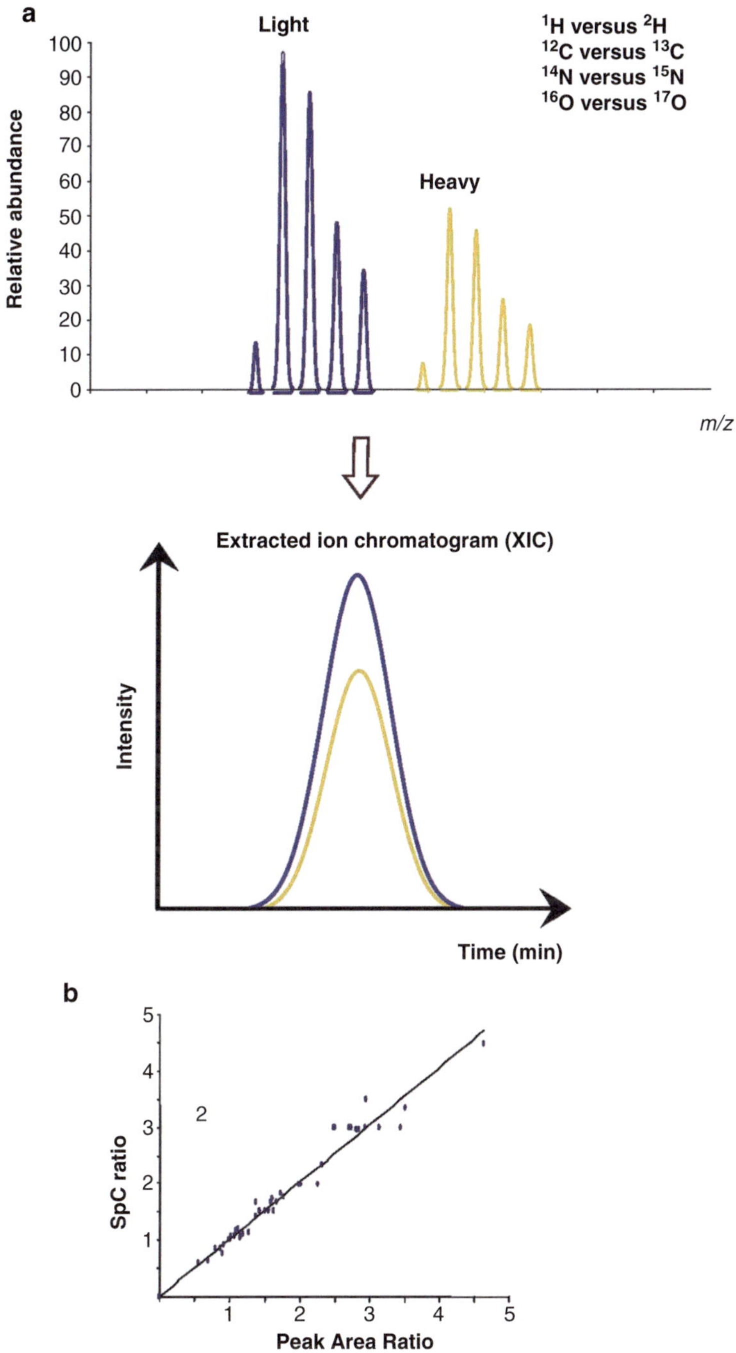

Fig. 7.5 Label and label-free quantitative analysis. (**a**) Differently labelled peptides are identified and their intensity ratio is calculated by extracted ion chromatogram (*XIC*) of light and heavy isotope labelled peptides. (**b**) Relation between spectral count and peak area ratio

platforms, that allow the quantitative and functional evaluation of the identified proteins. As a consequence, the list of products dedicated to label-free quantitative analysis is very long. Among the commercial software, Nonlinear Dynamics Progenesis LC-MS [88], Elucidator [89] and Scaffold [90] allow a protein quantification by both peak intensities and SpC evaluation; while SIEVE [55], ProteinLynx [91] or Peaks DB [52] rely mainly on the former. On the other hand, a considerable number of software are open-source, which include MsInspect [92], OpenMs [93], MSQuant [94], MaxQuant [95], ProteinQuant Suite [96], *MAProMa* [97], Census [98], PatternLab [99] and many others. Among them, *MAProMa*, Census and PatternLab are of interest for shotgun proteomics experiments. *MAProMa* compares SEQUEST Score or SpC values in pairwise fashion to up to 125 protein lists. Census performs protein quantitation by both peak area and SpC, accepts high-resolution MS data and it is designed to analyse both label and label-free data; while PatternLab allows different data normalization strategies, including Total Signal, log preprocessing (by ln) or Z normalization, and the identification of protein differences by implementing ACFold and nSVM methods. Of note, the systems reported above are based on comparison across different samples, representing different conditions, and therefore are referred to relative quantitation.

Feature Selection and Sample Classification

The breadth of information contained in the amount of data generated by bottom-up proteomics experiments is remarkable; these data are used to develop methods for classifying samples according to their phenotype (e.g. treated *vs.* untreated, healthy *vs.* diseased, early *vs.* late stage, etc) with the purpose to cure and prevent disease, improve diagnosis or monitoring disease progression [100–103]. Specifically, the procedure for sample classification consists of 4 different steps including data pre-processing, feature selection, classification and validation. As for MS data, pre-processing is necessary to correct biological and technical variability; matrix effects, issues with chromatographic alignment or differences in signal intensities [104, 105] can be addressed by using dedicated tools, including MZmine [106], MsInspect [92] or a number of R packages [107] as stand alone or inserted in other software packages, such as Bioconductor [108]. In a similar way, issues that could potentially affect the spectral counting, can be usually corrected by data normalization [99].

Since results derived from MudPIT are multi-dimensional and may be formatted in a $m \times n$ matrix (Fig. 7.6), similarly to microarray experiments, many tools, algorithms and strategies for analyzing genomics data are used for proteomics ones as well [109, 110]. For example, when dealing with these -omics data, where the number of conditions (n; samples) is far smaller than the number of observations (p; thousands genes, proteins, peptides or spectra), a problem called "*curse of dimensionality*" need to be addressed to obtain a lower-dimensional space and extracting an informative set of features [111]. For this purpose, a number of methods, including support vector machines (SVM), Principal Components Analysis

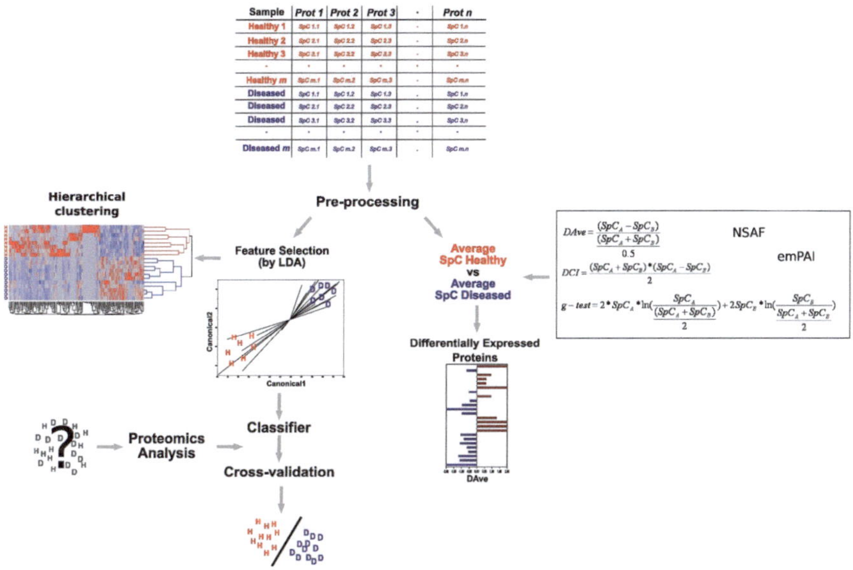

Fig. 7.6 Biomarker discovery and samples classification workflow using MudPIT data. By processing high-dimensional MudPIT data, *linear discriminant analysis* (*LDA*) finds a linear combination of features that best differentiates two or more classes of samples. This method selects proteins whose variations are reproducibly repeated in all considered samples. Otherwise, single samples or pools may be considered by using other statistical indices, including G-test, NSAF, emPAI or DAve and DCI; outlier protein levels might indicate subjective protein levels, and therefore personalized mechanisms of disease development and therapeutic response. *LDA* linear discriminant analysis, *DAve* differential average, *DCI* differential confidence index, *emPAI* exponentially modified protein abundance index, *NSAF* normalized spectral abundance factor

(PCA) artificial neural networks (ANN), random forests (RF), Partial Least Squares (PLS) and Linear Discriminant Analysis (LDA) have been exploited to process proteomic data [112]. Finally, in order to test the adequacy/inadequacy of a classifier, its performance is evaluated through a validation set, previously unseen. The most common measures to evaluate the performance of a classifier are based on standard indices, such as sensitivity, specificity, positive predicted values (PPV), negative predicted values (NPV) and accuracy values [113]; alternatively, a receiver operating characteristic (ROC) can be also applied [109].

Proteotypic Peptides

Experimental data produced by shotgun proteomics represent a source of information for calculating or predicting proteotypic peptides. These are peptides that can be observed in MS, in fact not all peptides possess the biophysical properties to become ionized, etc. Their characterization is useful for targeted proteomics analyses (such as MRM/SRM) as well as for reducing the time required for the interpretation of MS/MS spectra [114]; the advantage derives from the reduction of the actual database size, which contains a reduced number of sequences/spectra. In addition, the simultaneous quantification of multiple proteins in a spectrum, such as those generated by data independent analysis (DIA), via proteotypic peptides, represents an important field of research that needs to be implemented [115].

Proteotypic peptides were initially defined as the most observed peptides by established MS-based proteomics approaches [116], while other authors added the condition of uniqueness for a given protein [117] and a frequency of identification of at least of 50 % of all identifications of the corresponding parent protein [118]. In other words, a proteotypic peptide should be unambiguously related to one specific protein, should have been previously identified and its MS/MS fragmentation pattern should be known. Based on these definitions, proteotyipic peptides could be identified by using experimental data and dedicated software, like EPPI [119]. On the other hand, an increasing number of proteomic studies have driven the creation of specialized repositories. The Global Proteome Machine Database [120] is a representative example which allows to compare experimental data against those previously obtained in other studies. Stored data may be queried by different keys with the possibility to restrict search to a specific data source. Similarly, PeptideAtlas [121] is a publicly accessible source of peptides experimentally identified by tandem mass spectrometry. Users can browse the data and select different sources (i.e. different MS platforms) and few of these need permission to be accessed. In addition, for each protein, a graphical panel indicates unique peptides found and their occurrence, and for each one it is possible to retrieve information like spectra, modifications and genome mapping. The PRIDE database [122] stores experiments, identified proteins and peptides, unique peptides and spectra; in addition to protein name and various identifiers, it is possible to browse PRIDE by species, tissues, cell types, gene ontology (GO) terms and diseases. In the same way, ProteomeCommons [123] is a powerful open source web application designed for storing and exchanging proteomics data. With the same purpose, Proteomexchange [124] is a web-based environment that encourages data exchange and dissemination; its consortium has been set up to provide a single point of submission for MS data shared with the main existing proteomic repositories (at the moment PRIDE, PeptideAtlas and Tranche).

The increasing amount of data collected by proteomics experiments has also allowed the development of tools aiming at predicting and scoring putative proteotypic peptides. For example, Peptide Sieve was developed by studying the physicochemical properties of more than 600,000 peptides [118]. Similarly, the STEPP software, based on Support Vector Machine (SVM), uses a descriptor space based on 35 amino acids properties [125]. Neural networks were used by Tang and colleagues to develop the Detectability Predictor software that uses 175 amino acid properties [126]; these were used to predict proteotypic peptides generated by MudPIT experiments, for a given set of experimental, instrumental and analytical conditions [127].

From Proteomics to Systems Biology

The identification of thousands of peptides and proteins provides an high-definition snapshot of a sample's proteome and allows to investigate it in a global fashion by using accurate models to generate hypotheses for testing [128]. These approaches, called "data-derived", rely on "-omics" datasets and facilitate the process of linking different data to functional relationships among proteins and other biological macromolecules [129–132]. Recent studies have applied proteomics and other systems biology approaches to the interrogation of mechanisms related to cardiovascular diseases [133–136]. The possibility to link protein expression to pathways and protein-protein interaction networks could highlight functional modules that are differently regulated under various conditions (Fig. 7.7). On the other hand, the topological analysis of networks reconstructed from experimental data represent a complementary approach to identify proteins with relevant biological significance, in the context of a particular disease [137]. Other interesting applications combined experimental proteomic data with weighted gene co-expression network analysis [138, 139]. Reconstructed networks provided informative graph properties to determine modules and molecules that correlate with the investigated biological phenotypes [140–142]. Network inference procedures and topological analysis are two promising approaches to extract new potential biomarkers and a sustainable alternative to look into the pathogenetic mechanisms that could improve clinical outcomes through early diagnosis and risk stratification.

An excellent collection of resources that can be useful to investigate -omics data at the systems biology level is listed on the Pathguide website [143], which

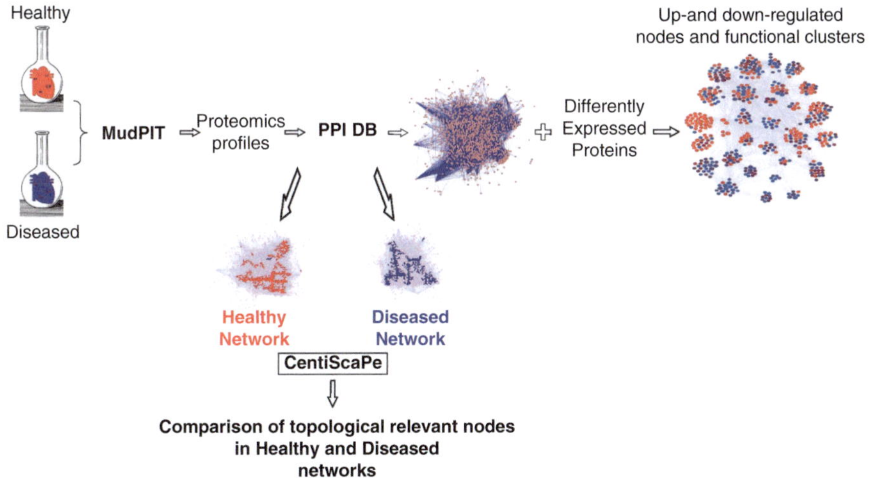

Fig. 7.7 MudPIT data-derived network analysis. The integration of experimental proteomics data and PPI network models is used to identify cluster of functionally related proteins, up- or down-regulated between different phenotypes. In addition, topological analysis of reconstructed networks represents a complementary approach to identify relevant biological nodes

contains links and information related to more than five hundreds resources. Thanks to methodologies like affinity purification-MS and two-yeast hybrid assays, protein-protein interaction (PPI) datasets are continuously improving and reliable data are increasingly available for different organisms, in addition to the well-studied *Homo sapiens*, *Mus musculus*, *E. coli*, *Saccharomyces cerevisiae*. Specialized repositories include String [144], HPRD [145], Reactome [146], IntAct [147] and GeneMania [148]. A number of computational tools to process -omics data, including proteomics ones, are also available. Cytoscape [149], VisANT [150] and Ingenuity Pathway Analysis [151] are some of those worth mentioning. Cytoscape is probably the most used thanks to a growing community of developers which apply the Java technology-based Cytoscape [152]. Additional plugins are freely available to address different questions including network import, visualization, topological evaluation or retrieval of GO annotations. For example, this last feature is available through plugins like Bingo2.44 [153], Mosaic [154] or NOA [155]; others plugins, like MCODE [156] and ClusterMaker [157], allow to rank nodes by local neighbourhood density and graphically display the extracted, significant clusters. As for topological analysis, the Cytoscape plugin CentiScaPe [137], computes several centrality parameters (average distance, diameter, degree, stress, "betweenness", radiality, closeness, centroid value and eccentricity) to identify nodes that display a prominent position in the network architecture.

Clinical Proteomics for Cardiovascular Diseases

Proteomics started as a set of technologies that, when combined, can provide dynamic information about the cellular phenotype as a whole. Its applications to the study of the human proteomes, in terms of the sequences that are actually translated from the human genome, soon became one of the main focuses. It was calculated that the number of human proteins are higher than the estimated 20,000–25,000 human protein-coding genes [158], mainly due to alternative splicing and post-translational modifications (PTMs). The importance of addressing biological macromolecules that relate more closely than genes and transcripts to the phenotype, became evident early on to proteomic scientists. Overall, the proteomic community could first appreciate the unpredictable complexity and heterogeneity of PTMs and therefore the importance of determining the actual level of proteins and their modifications, and functional complexes in biological samples.

In this context, the detailed characterization of the cardiovascular proteomes (such as cardiac and vascular tissues and cells) has greatly improved our knowledge of cardiovascular physiology, and has already contributed to the identification of key features underlying the onset of many diseases, such as cardiomyopathy. For example, Comunian and colleagues [34] characterized and compared the proteome from murine ventricles and atria, and identified and quantified the levels of thousands of proteins, that are differentially expressed among different cardiac chambers using a label-free MudPIT. Proteins related to atrial natriuretic peptide (ANP) were found to be more abundant in the atria than ventricles, whereas titin isoform N2B resulted more abundant in the left ventricle and voltage-dependent anion channel (VDAC) in the right ventricle.

Numerous advancements in bottom-up technologies have expanded the objectives of proteomic investigations from mere protein profiling, to quantitation, identification of PTMs, characterization of new protein functions and protein-protein interactions (Fig. 7.8). One of the goals of proteomics is also the identification of biomarkers for the early diagnosis and risk stratification in diseased populations, along with the improvement of reproducible and quantitative methodologies to trace them in clinical specimen with high sensitivity and specificity. Usually, biomarker discovery involved the analysis of biological fluids that can be collected non-invasively, such as plasma, urine and saliva. Plasma is surely the most popular sample for proteomic investigations, including cardiovascular disease (CVD). However, it should be underscored that albumin and immunoglobulins represent the vast majority of circulating proteins in terms of abundance. As such, the dynamic range of proteins is very wide (the concentrations of plasma proteins span 10 orders of

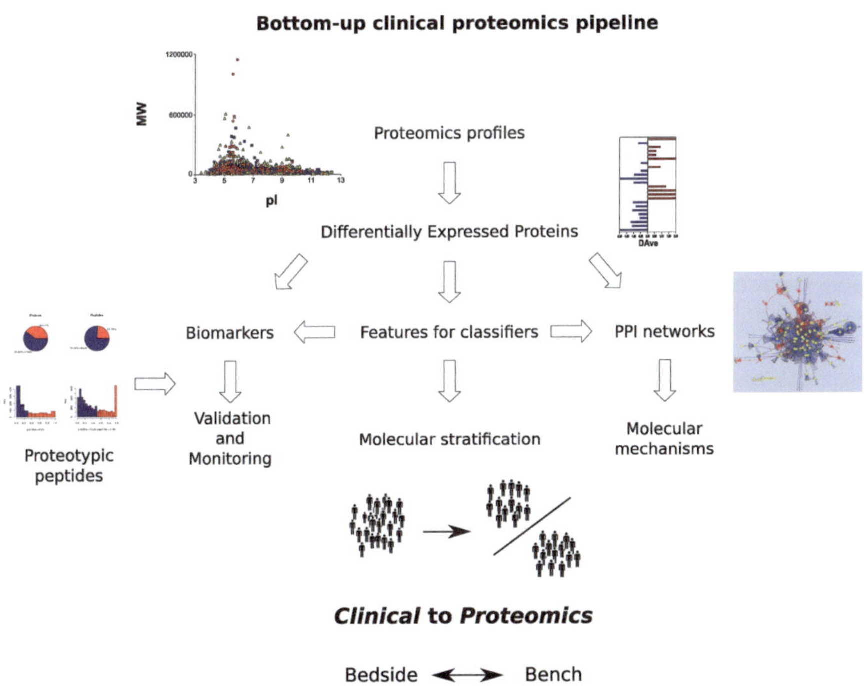

Fig. 7.8 Bottom-up clinical proteomics pipeline. Here are summarized the main steps and information obtained by bottom-up proteomics: 1) profiling and differential analysis for the simultaneous identification of biomarkers, useful for diagnosis; 2) cluster analysis to evaluate patients stratification and sub-typing diseases; 3) pathway and network analysis to elucidate the molecular mechanisms related to disease and/or therapy

magnitude), and potential biomarkers could be present in very low amount; these are only two of the reasons because the identification of new biomarkers from plasma samples can be challenging.

Over a hundred plasma proteins have already been proposed as candidate biomarkers for CVD [159]. Moreover, due to the increased sensitivity and specificity of mass spectrometers, and the creation of new methodologies, the relative and absolute quantitation of very low abundant proteins from plasma is now within reach using SRM/MRM or SWATH. Sample preparation and experimental design are crucial aspects of a proteomic investigation; for example, as many plasma proteins are glycosylated, a remarkable increase in sensitivity can be achieved by simply enriching for glycoproteins because albumin, the most abundant plasma protein, is not glycosylated.

Alternative strategies for biomarker discovery involve the analysis of biopsies, or *in vitro* studies by means of cellular lines, or primary cultures such as cardiomyocytes. The idea being that biomarkers which are progressively diluted down in the bloodstream show higher concentration at the site of production (e.g. cardiac myocytes), and that with *a priori* knowledge of the proteins to be searched, sensitivity can be greatly enhanced by using targeted approaches. Tissue biopsies can therefore be most informative as the changes in protein levels associated with disease is higher; unfortunately, heart biopsies are not always available and this impacts on the possibility of designing large-scale studies. As a consequence, researchers often use animal models, mainly mouse, though the data obtained in these animal models may not always be relevant for humans, due to their heterogeneity and the fact that most CVD patients present with comorbidities.

As for primary cultures, such as cultured cardiomyocytes and fibroblasts, they are necessary to investigate molecular mechanisms and characterizing secreted proteins. The latter could be very important as potential circulating biomarkers detectable in plasma or urine. For example, using the MudPIT methodology, Varrone et al. performed a proteomic analysis of the secretome from cultured cardiomyocytes overexpressing miR-1 [36]; in this way an inverse relationship between myocardial expression of miR-1 and circulating levels of FABP3 was found both *in vitro* and *in vivo* under various pathological conditions, and assessment of FABP3 plasma levels in human patients may be used to indirectly measure cardiac miR-1 activity.

The discovery of biomarkers for early diagnosis is an important goal of cardiovascular proteomics, for instance for the differential diagnosis of myocardial infarction. panel of several proteins may be much more sensitive and specific than a single biomarker. The identification of disease mechanisms, via the characterizion of protein functions or regulatory pathways, is also compelling as these could provide new therapeutic targets. Simioniuc and colleagues combined *in* vivo magnetic resonance imaging (MRI) with proteomic and histological analysis, and demonstrated the regenerative potential of pre-treating placenta-derived human mesenchymal stem cells (FMhMSCs) with a hyaluronan mixed ester of butyric and retinoic acids (HBR) in a clinically relevant animal model of myocardial infarction [37]. In particular, the proteomic data showed that HBR pre-treatment of FMhMSCs allows the reduction of extracellular matrix proteins such as lumican, a member of the small leucine-rich proteoglycan family which contribute to the assembly of collagen fibers and regulates fibrosis. Moreover, HBR-FMhMSC treatment induce an increase of mitochondrial proteins, such as NADH dehydrogenase and cytochrome complexes, and a group of proteins that facilitates transport of fatty acid into cardiomyocytes. On the other hand, recent studies evidenced that the predominant mechanism by which stem cells participate to heart tissue repair is through paracrine activity [160, 161]. As reported by Chimenti and colleagues [162], the therapeutic effects of stem cells may be related to secreted molecules (secretome, including proteins). Li and colleagues analyzed the secretome of rat cardiomyoblast cells, subjected to hypoxia and re-oxygenation, identifying about two thousands proteins by means of a label-free proteomic approach; by analyzing the same samples by means of iTRAQ, about 900 proteins were identified. The authors suggested a link between hypoxia, protein related to angiogenesis, inflammation and remodeling of the extracellular matrix; while re-oxygenation was associated with secreted proteins involved in the suppression of inflammation and reduction of anti-apoptosis proteins. These findings suggest that hypoxia and re-oxygenation act through unique cardiomyocytes secretomes in order to reduce cellular injury and promote healing [163]. Other authors, used MRM to monitor the abundance of peptides containing oxidized methionine148 (M148) in plasmatic Apo A-I, and reported an increase of oxidized M148 peptides in the HDL of CVD patients [164]. These findings suggest a potential relationship between oxidation of M148 and CVD.

In conclusion, the identification of disease mechanisms along with the characterization of new protein functions and regulatory pathways through new hubs or nodes, will greatly enhance our understanding of CVD [165]. Network analysis is allowing to combine proteomic data with those obtained by transcriptomics, metabolomics, and other platform, into single models [166]. Finally, the increased specificity, selectivity and sensitivity of proteomics coupled to clinical phenotyping will improve clinical practice, once the assays are properly validated by means of targeted functional experiments and using large patients cohorts.

References

1. Reid GE, McLuckey SA. 'Top down' protein characterization via tandem mass spectrometry. J Mass Spectrom. 2002;37:663–75.
2. Lin D, Tabb DL, Yates 3rd JR. Large-scale protein identification using mass spectrometry. Biochim Biophys Acta. 2003;1646(1–2):1–10.
3. Oliveira BM, Coorssen JR, Martins-de-Souza D. 2DE: the phoenix of proteomics. J Proteomics. 2014;104:140–50. doi:10.1016/j.jprot.2014.03.035.
4. Washburn MP, Wolters D, Yates JR. Large-scale analysis of the yeast proteome by multidimensional protein identification technology. Nat Biotechnol. 2001;19(3):242–7.

5. Wang H, Yang Y, Li Y, Bai B, Wang X, Tan H, Liu T, Beach TG, Peng J, Wu Z. Systematic optimization of long gradient chromatography mass spectrometry for deep analysis of brain proteome. J Proteome Res. 2015;14(2):829–38.
6. Warren CM, Geenen DL, Helseth Jr DL, Xu H, Solaro RJ. Sub-proteomic fractionation, iTRAQ, and OFFGEL-LC-MS/MS approaches to cardiac proteomics. J Proteomics. 2010;73(8):1551–61. doi:10.1016/j.jprot.2010.03.016.
7. Didangelos A, Yin X, Mayr M. Method for protein subfractionation of cardiovascular tissues before DIGE analysis. Methods Mol Biol. 2012;854:287–97. doi:10.1007/978-1-61779-573-2_20.
8. Atanassov I, Urlaub H. Increased proteome coverage by combining PAGE and peptide isoelectric focusing: comparative study of gel-based separation approaches. Proteomics. 2013;13(20):2947–55. doi:10.1002/pmic.201300035.
9. O'Farrell PH. High resolution two-dimensional electrophoresis of proteins. J Biol Chem. 1975;250:4007–21.
10. Gianazza E, Righetti PG. Immobilized pH gradients. Electrophoresis. 2009;30 Suppl 1:S112–21.
11. Schägger H. Tricine–SDS-PAGE. Nat Protoc. 2006;1:16–22. doi:10.1038/nprot.2006.4.
12. Magdeldin S, Enany S, Yoshida Y, Xu B, Zhang Y, Zureena Z, Lokamani I, Yaoita E, Yamamoto T. Basics and recent advances of two dimensional- polyacrylamide gel electrophoresis. Clin Proteomics. 2014;11(1):16. doi:10.1186/1559-0275-11-16.
13. Patton WF. A thousand points of light: the application of fluorescence detection technologies to two-dimensional gel electrophoresis and proteomics. Electrophoresis. 2000;21(6):1123–44.
14. Maiorino M, Roveri A, Benazzi L, Bosello V, Mauri P, Toppo S, Tosatto SC, Ursini F. Functional interaction of phospholid hydroperoxide glutathione peroxidase with sperm mitochondrion-associated cysteine-rich protein discloses the adjacent cysteine motif as a new substrate of the SE-peroxidase. J Biol Chem. 2005;280:38395–402.
15. Ford PJ. Immunological techniques: ELISA, flow cytometry, and immunohistochemistry. Methods Mol Biol. 2010;666:327–43. doi:10.1007/978-1-60761-820-1_20.
16. Scherl A. Clinical protein mass spectrometry. Methods. 2015;81:3–14. doi:10.1016/j.ymeth.2015.02.015.
17. Shevchenko A, Tomas H, Havlis J, Olsen JV, Mann M. In-gel digestion formass spectrometric characterization of proteins and proteomes. Nat Protoc. 2006;1:2856–60.
18. Kaufmann R. Matrix-assisted laser desorption ionization (MALDI) mass spectrometry: a novel analytical tool in molecular biology and biotechnology. J Biotechnol. 1995;41(2–3):155–75.
19. Haas S, Jahnke HG, Moerbt N, von Bergen M, Aharinejad S, Andrukhova O, Robitzki AA. DIGE proteome analysis reveals suitability of ischemic cardiac in vitro model for studying cellular response to acute ischemia and regeneration. PLoS One. 2012;7(2):e31669. doi:10.1371/journal.pone.0031669.
20. Hutchens TW, Yip TT. New desorption strategies for the mass spectrometric analysis of macromolecules. Rapid Commun Mass Spectrom. 1993;7:576–80.
21. Scott PA, Zeidan B, Ng LL, Zeb M, Rosengarten JA, Garbis S, Curzen NP, Morgan JM, Townsend PA. Proteomic profiling to identify prognostic biomarkers in heart failure. In Vivo. 2012;26(6):875–82.
22. Pinet F, Beseme O, Cieniewski-Bernard C, Drobecq H, Jourdain S, Lamblin N, Amouyel P, Bauters C. Predicting left ventricular remodeling after a first myocardial infarction by plasma proteome analysis. Proteomics. 2008;8(9):1798–808. doi:10.1002/pmic.200700781.
23. Horn S, Lueking A, Murphy D, Staudt A, Gutjahr C, Schulte K, König A, Landsberger M, Lehrach H, Felix SB, Cahill DJ. Profiling humoral autoimmune repertoire of dilated cardiomyopathy (DCM) patients and development of a disease-associated protein chip. Proteomics. 2006;6(2):605–13.
24. Abel L, Kutschki S, Turewicz M, Eisenacher M, Stoutjesdijk J, Meyer HE, Woitalla D, May C. Autoimmune profiling with protein microarrays in clinical applications. Biochim Biophys Acta. 2014;1844(5):977–87. doi:10.1016/j.bbapap.2014.02.023.

25. Amartely H, Iosub-Amir A, Friedler A. Identifying protein-protein interaction sites using peptide arrays. J Vis Exp. 2014;18(93):e52097. doi:10.3791/52097.
26. Gamagedara S, Ma Y. Biomarker analysis for prostate cancer diagnosis using LC-MS and CE-MS. Bioanalysis. 2011;3(18):2129–42. doi:10.4155/bio.11.203. Review.
27. Albalat A, Husi H, Siwy J, Nally JE, McLauglin M, Eckersall PD, Mullen W. Capillary electrophoresis interfaced with a mass spectrometer (CE-MS): technical considerations and applicability for biomarker studies in animals. Curr Protein Pept Sci. 2014;15(1):23–35.
28. Ibáñez C, Simó C, García-Cañas V, Cifuentes A, Castro-Puyana M. Metabolomics, peptidomics and proteomics applications of capillary electrophoresis-mass spectrometry in Foodomics: a review. Anal Chim Acta. 2013;802:1–13. doi:10.1016/j.aca.2013.07.042.
29. Zhang Z, Staessen JA, Thijs L, Gu Y, Liu Y, Jacobs L, Koeck T, Zürbig P, Mischak H, Kuznetsova T. Left ventricular diastolic function in relation to the urinary proteome: a proof-of-concept study in a general population. Int J Cardiol. 2014;176(1):158–65. doi:10.1016/j.ijcard.2014.07.014.
30. Liangliang S, Guijie Z, Xiaojing Y, Champion MM, Dovichi NJ. Capillary zone electrophoresis for analysis of complex proteomes using an electrokinetically pumped sheath flow nanospray interface. Proteomics. 2014;14:622–8.
31. Šesták J, Moravcová D, Kahle V. Instrument platforms for nano liquid chromatography. J Chromatogr A. 2015. pii:S0021–9673(15)01078-X. doi:10.1016/j.chroma.2015.07.090.
32. Liu H, Lin D, Yates JR. Multidimensional separations for protein/peptide analysis in the post-genomic era. Biotechniques. 2002;32(4):898, 900, 902 passim.
33. Ho CS, Lam CW, Chan MH, Cheung RC, Law LK, Lit LC, Ng KF, Suen MW, Tai HL. Electrospray ionisation mass spectrometry: principles and clinical applications. Clin Biochem Rev. 2003;24(1):3–12.
34. Comunian C, Rusconi F, De Palma A, Brunetti P, Catalucci D, Mauri PL. A comparative MudPIT analysis identifies different expression profiles in heart compartments. Proteomics. 2011;11:2320–8.
35. Di Silvestre D, Brambilla F, Mauri PL. Multidimensional protein identification technology for direct-tissue proteomics of heart. Methods Mol Biol. 2013;38(1005):1025.
36. Varrone F, Gargano B, Carullo P, Di Silvestre D, De Palma A, Grasso L, Di Somma C, Mauri P, Benazzi L, Franzone A, Jotti GS, Bang ML, Esposito G, Colao A, Condorelli G, Catalucci D. The circulating level of FABP3 is an indirect biomarker of microRNA-1. J Am Coll Cardiol. 2013;61(1):88–95. doi:10.1016/j.jacc.2012.08.1003.
37. Simioniuc A, Campan M, Lionetti V, Marinelli M, Aquaro GD, Cavallini C, Valente S, Di Silvestre D, Cantoni S, Bernini F, Simi C, Pardini S, Mauri P, Neglia D, Ventura C, Pasquinelli G, Recchia FA. Placental stem cells pre-treated with a hyaluronan mixed ester of butyric and retinoic acid to cure infarcted pig hearts: a multimodal study. Cardiovasc Res. 2011;90:546–56.
38. Lionetti V, Matteucci M, Ribezzo M, Di Silvestre D, Brambilla F, Agostini S, Mauri P, Padeletti L, Pingitore A, Delsedime L, Rinaldi M, Recchia FA, Pucci A. Regional mapping of myocardial hibernation phenotype in idiopathic end-stage dilated cardiomyopathy. J Cell Mol Med. 2014;18(3):396–414. doi:10.1111/jcmm.12198.
39. Winter D, Steen H. Optimization of cell lysis and protein digestion protocols for the analysis of HeLa S3 cells by LC-MS/MS. Proteomics. 2011;11:4726–30.
40. Motoyama A, Yates JR. Multidimensional LC separations in shotgun proteomics. Anal Chem. 2008;80:7187–93.
41. Thakur SS, Geiger T, Chatterjee B, Bandilla P, Fröhlich F, Cox J, Mann M. Deep and highly sensitive proteome coverage by LC-MS/MS without prefractionation. Mol Cell Proteomics. 2011;10(8):1–9.
42. Mann M, Hendrickson RC, Pandey A. Analysis of proteins and proteomes by mass spectrometry. Annu Rev Biochem. 2001;70:437–73.
43. Pirmoradian M, Budamgunta H, Chingin K, Zhang B, Astorga-Wells J, Zubarev RA. Rapid and deep human proteome analysis by single-dimension shotgun proteomics. Mol Cell Proteomics. 2013;12(11):3330–8.

44. Michalski A, Cox J, Mann M. More than 100,000 detectable peptide species elute in single shotgun proteomics runs but the majority is inaccessible to data-dependent LC-MS/MS. J Proteome Res. 2011;10(4):1785–93.
45. Venable JD, Dong MQ, Wohlschlegel J, Dillin A, Yates JR. Automated approach for quantitative analysis of complex peptide mixtures from tandem mass spectra. Nat Methods. 2004;1:39–45.
46. Gillet LC, Navarro P, Tate S, Röst H, Selevsek N, Reiter L, Bonner R, Aebersold R. Targeted data extraction of the MS/MS spectra generated by data-independent acquisition: a new concept for consistent and accurate proteomeanalysis. Mol Cell Proteomics. 2012;11:1–17.
47. Bern M, Finney G, Hoopmann MR, Merrihew G, Toth MJ, MacCoss MJ. Deconvolution of mixture spectra from ion-trap data-independent-acquisition tandem mass spectrometry. Anal Chem. 2010;82(3):833–41.
48. Carvalho PC, Han X, Xu T, Cociorva D, Carvalho Mda G, Barbosa VC, Yates JR. XDIA: improving on the label-free data-independent analysis. Bioinformatics. 2010;26(6):847–8.
49. Egertson JD, Kuehn A, Merrihew GE, Bateman NW, MacLean BX, Ting YS, Canterbury JD, Marsh DM, Kellmann M, Zabrouskov V, Wu CC, Mac-Coss MJ. Multiplexed MS/MS for improved data-independent acquisition. Nat Methods. 2013;8:744–6.
50. Eng JK, Searle BC, Clauser KR, Tabb DL. A face in the crowd: recognizing peptides through database search. Mol Cell Proteomics. 2011;10(11):R111.009522. doi:10.1074/mcp.R111.009522.
51. Allmer J. Algorithms for the de novo sequencing of peptides from tandem mass spectra. Expert Rev Proteomics. 2011;8(5):645–57.
52. Zhang J, Xin L, Shan B, Chen W, Xie M, Yuen D, Zhang W, Zhang Z, Lajoie GA, Ma B. PEAKS DB: de novo sequencing assisted database search for sensitive and accurate peptide identification. Mol Cell Proteomics. 2012;11(4):M111.010587.
53. Perkins DN, Pappin DJ, Creasy DM, Cottrell JS. Probability-based protein identification by searching sequence databases using mass spectrometry data. Electrophoresis. 1999;20(18):3551–67.
54. Eng JK, McCormack AL, Yates JR. An approach to correlate tandem mass spectral data of peptides with amino acid sequences in a protein database. J Am Soc Mass Spectrom. 1994;5(11):976–89. doi:10.1016/1044-0305(94)80016-2.
55. http://portal.thermo-brims.com/.
56. http://sciex.com/products/software/proteinpilot-software.
57. http://www.agilent.com/en-us/products/software-informatics/masshunter-workstation-software/spectrum-mill.
58. Muth T, Vaudel M, Barsnes H, Martens L, Sickmann A. X!Tandem Parser: an open-source library to parse and analyse X!Tandem MS/MS search results. Proteomics. 2010;10(7):1522–4.
59. Tabb DL, Fernando CG, Chambers MC. MyriMatch: highly accurate tandem mass spectral peptide identification by multivariate hypergeometric analysis. J Proteome Res. 2007;6(2):654–61.
60. Geer LY, Markey SP, Kowalak JA, Wagner L, Xu M, Maynard DM, Yang X, Shi W, Bryant SH. Open mass spectrometry search algorithm. J Proteome Res. 2004;3(5):958–64.
61. Kim S, Mischerikow N, Bandeira N, Navarro JD, Wich L, Mohammed S, Heck AJ, Pevzner PA. The generating function of CID, ETD, and CID/ETD pairs of tandem mass spectra: applications to database search. Mol Cell Proteomics. 2010;9(12):2840–52. doi:10.1074/mcp.M110.003731.
62. Wang LH, Li DQ, Fu Y, Wang HP, Zhang JF, Yuan ZF, Sun RX, Zeng R, He SM, Gao W. pFind 2.0: a software package for peptide and protein identification via tandem mass spectrometry. Rapid Commun Mass Spectrom. 2007;21(18):2985–91.
63. Chalkley RJ, Baker PR, Huang L, Hansen KC, Allen NP, Rexach M, Burlingame AL. Comprehensive analysis of a multidimensional liquid chromatography mass spectrometry dataset acquired on a quadrupole selecting, quadrupole collision cell, time-of-flight mass spectrometer: II. New developments in Protein Prospector allow for reliable and comprehensive automatic analysis of large datasets. Mol Cell Proteomics. 2005;4(8):1194–204. Epub 2005 Jun 3.

64. Ma B, Zhang K, Hendrie C, Liang C, Li M, Doherty-Kirby A, Lajoie G. PEAKS: powerful software for peptide de novo sequencing by tandem mass spectrometry. Rapid Commun Mass Spectrom. 2003;17(20):2337–42.
65. Orchard S, Albar J-P, Deutsch EW, Eisenacher M, Binz P-A, Hermjakob H. Implementing data standards: a report on the HUPOPSI workshop September 2009, Toronto, Canada. Proteomics. 2010;10:1895–8.
66. Jeong K, Kim S, Bandeira N. False discovery rates in spectral identification. BMC Bioinform. 2012;13(Suppl 16):S2. doi:10.1186/1471-2105-13-S16-S2.
67. Parker CE, Borchers CH. Mass spectrometry based biomarker discovery, verification, and validation – Quality assurance and control of protein biomarker assays. Mol Oncol. 2014. pii:S1574-7891(14)00054-4.
68. Alban A, David SO, Bjorkesten L, Andersson C, Sloge E, Lewis S, Currie I. A novel experimental design for comparative two-dimensional gel analysis: two-dimensional difference gel electrophoresis incorporating a pooled internal standard. Proteomics. 2003;3(1):36–44.
69. Patel VJ, Thalassinos K, Slade SE, Connolly JB, Crombie A, Murrell JC, Scrivens JH. A comparison of labeling and label-free mass spectrometry-based proteomics approaches. J Proteome Res. 2009;8(7):3752–9. doi:10.1021/pr900080y.
70. Mauri P, Scigelova M. Multidimensional protein identification technology for clinical proteomic analysis. Clin Chem Lab Med. 2009;47(6):636–46. doi:10.1515/CCLM.2009.165.
71. Gygi SP, Rist B, Gerber SA, Turecek F, Gelb MH, Aebersold R. Quantitative analysis of complex protein mixtures using isotope-coded affinity tags. Nat Biotechnol. 1999;17:994–9.
72. Aggarwal K, Choe LH, Lee KH. Shotgun proteomics using the iTRAQ isobaric tags. Brief Funct Genomic Proteomic. 2006;5:112–20.
73. Werner T, Sweetman G, Savitski MF, Mathieson T, Bantscheff M, Savitski MM. Ion coalescence of neutron encoded TMT 10-plex reporter ions. Anal Chem. 2014;86(7):3594–601. doi:10.1021/ac500140s.
74. Mann M. Functional and quantitative proteomics using SILAC. Nat Rev. 2006;7:952–8.
75. Niles R, Witkowska HE, Allen S, Hall SC, Fisher SJ, Hardt M. Acid-catalyzed oxygen-18 labeling of peptides. Anal Chem. 2009;81(7):2804–9.
76. Evans C, Noirel J, Ow SY, Salim M, Pereira-Medrano AG, Couto N, Pandhal J, Smith D, Pham TK, Karunakaran E, Zou X, Biggs CA, Wright PC. An insight into iTRAQ: where do we stand now? Anal Bioanal Chem. 2012;404(4):1011–27.
77. Ahrné E, Molzahn L, Glatter T, Schmidt A. Critical assessment of proteome-wide label-free absolute abundance estimation strategies. Proteomics. 2013;13(17):2567–78.
78. Mauri P, Scarpa A, Nascimbeni AC, Benazzi L, Parmagnani E, Mafficini A, Della Peruta M, Bassi C, Miyazaki K, Sorio C. Identification of proteins released by pancreatic cancer cells by multidimensional protein identification technology: a strategy for identification of novel cancer markers. FASEB J. 2005;19(9):1125–7.
79. Brambilla F, Lavatelli F, Di Silvestre D, Valentini V, Palladini G, Merlini G, Mauri PL. Shotgun protein profile of human adipose tissue and its changes in relation to systemic amyloidoses. J Proteome Res. 2013;12(12):5642–55. doi:10.1021/pr400583h.21.
80. Lundgren DH, Hwang SI, Wu L, Han DK. Role of spectral counting in quantitative proteomics. Expert Rev Proteomics. 2010;7(1):39–53. doi:10.1586/epr.09.69.
81. Gao J, Opiteck GJ, Friedrichs MS, Dongre AR, Hefta SA. Changes in the protein expression of yeast as a function of carbon source. J Proteome Res. 2003;2:643–9.
82. Liu H, Sadygov RG, Yates 3rd JR. A model for random sampling and estimation of relative protein abundance in shotgun proteomics. Anal Chem. 2004;76:4193–201.
83. Regonesi ME, Del Favero M, Basilico F, Briani F, Benazzi L, Tortora P, Mauri P, Dehò G. Analysis of the Escherichia coli RNA degradosome composition by a proteomic approach. Biochimie. 2006;88(2):151–61.
84. Zhang B, VerBerkmoes NC, Langston MA, Uberbacher E, Hettich RL, Samatova NF. Detecting differential and correlated protein expression in label-free shotgun proteomics. J Proteome Res. 2006;5:2909–18.

85. Zybailov B, Mosley AL, Sardiu ME, Coleman MK, Florens L, Washburn MP. Statistical analysis of membrane proteome expression changes in Saccharomyces cerevisiae. J Proteome Res. 2006;5:2339–47.
86. Ishihama Y, Oda Y, Tabata T, Sato T, Nagasu T, Rappsilber J, Mann M. Exponentially modified protein abundance index (emPAI) for estimation of absolute protein amount in proteomics by the number of sequenced peptides per protein. Mol Cell Proteomics. 2005;4:1265–72.
87. Bergamini G, Di Silvestre D, Mauri P, Cigana C, Bragonzi A, De Palma A, Benazzi L, Döring G, Assael BM, Melotti P, Sorio C. MudPIT analysis of released proteins in Pseudomonas aeruginosa laboratory and clinical strains in relation to pro-inflammatory effects. Integr Biol (Camb). 2012;4(3):270–9.
88. http://www.nonlinear.com/.
89. http://www.rosettabio.com/.
90. http://www.proteomesoftware.com/products/scaffold/.
91. http://www.waters.com/.
92. Fitzgibbon M, Law W, May D, Detter A, McIntosh M. Open-source platform for the analysis of liquid chromatography-mass spectrometry (LC-MS) data. Methods Mol Biol. 2008;428:369–82.
93. Sturm M, Bertsch A, Gröpl C, Hildebrandt A, Hussong R, Lange E, Pfeifer N, Schulz-Trieglaff O, Zerck A, Reinert K, Kohlbacher O. OpenMS – an open-source software framework for mass spectrometry. BMC Bioinformatics. 2008;9:163.
94. Mortensen P, Gouw JW, Olsen JV, Ong SE, Rigbolt KT, Bunkenborg J, Cox J, Foster LJ, Heck AJ, Blagoev B, Andersen JS, Mann M. MSQuant, an open source platform for mass spectrometry-based quantitative proteomics. J Proteome Res. 2010;9(1):393–403.
95. Cox J, Mann M. MaxQuant enables high peptide identification rates, individualized p.p.b.-range mass accuracies and proteome-wide protein quantification. Nat Biotechnol. 2008;26(12):1367–72.
96. Mann B, Madera M, Sheng Q, Tang H, Mechref Y, Novotny MV. ProteinQuant Suite: a bundle of automated software tools for label-free quantitative proteomics. Rapid Commun Mass Spectrom. 2008;22(23):3823–34.
97. Mauri PL, Dehò G. A proteomic approach to the analysis of RNA degradosome composition in Escherichia coli. Methods Enzymol. 2008;447:99–117.
98. Park SK, Venable JD, Xu T, Yates JR. A quantitative analysis software tool for mass spectrometry-based proteomics. Nat Methods. 2008;5(4):319–22.
99. Carvalho PC, Fischer JS, Chen EI, Yates 3rd JR, Barbosa VC. PatternLab for proteomics: a tool for differential shotgun proteomics. BMC Bioinformatics. 2008;9:316.
100. Palmblad M, Tiss A, Cramer R. Mass spectrometry in clinical proteomics – from the present to the future. Proteomics Clin Appl. 2009;3:6–17.
101. Simpson KL, Whetton AD, Dive C. Quantitative mass spectrometry-based techniques for clinical use: biomarker identification and quantification. J Chromatogr B Analyt Technol Biomed Life Sci. 2009;877:1240–9.
102. Brambilla F, Lavatelli F, Di Silvestre D, Valentini V, Rossi R, Palladini G, Obici L, Verga L, Mauri P, Merlini G. Reliable typing of systemic amyloidoses through proteomic analysis of subcutaneous adipose tissue. Blood. 2012;119:1844–7.
103. Di Silvestre D, Zoppis I, Brambilla F, Bellettato V, Mauri G, Mauri P. Availability of MudPIT data for classification of biological samples. J Clin Bioinforma. 2013;3(1):1.66.
104. Yu W, Li X, Liu J, Wu B, Williams KR, Zhao H. Multiple peak alignment in sequential data analysis: a scale-space-based approach. IEEE/ACM Trans Comput Biol Bioinform. 2006;3:208–19.
105. Arneberg R, Rajalahti T, Flikka K, Berven FS, Kroksveen AC, Berle M, Myhr K-M, Vedeler CA, Ulvik RJ, Kvalheim OM. Pretreatment of mass spectral profiles: application to proteomic data. Anal Chem. 2007;79:7014–26.
106. Pluskal T, Castillo S, Villar-Briones A, Oresic M. MZmine 2: modular framework for processing, visualizing, and analyzing mass spectrometry-based molecular profile data. BMC Bioinformatics. 2010;11:395.

107. https://www.r-project.org/.
108. http://www.bioconductor.org/.
109. Ressom HW, Varghese RS, Zhang Z, Xuan J, Clarke R. Classification algorithms for phenotype prediction in genomics and proteomics. Front Biosci. 2008;13:691–708.
110. Dakna M, He Z, Yu WC, Mischak H, Kolch W. Technical, bioinformatical and statistical aspects of liquid chromatography-mass spectrometry (LC-MS) and capillary electrophoresis-mass spectrometry (CE-MS) based clinical proteomics: a critical assessment. J Chromatogr B Analyt Technol Biomed Life Sci. 2009;877:1250–8.
111. Guyon I, Gunn S, Nikravesh M, Zadeh LA. Feature extraction: foundations and applications. Berlin Heidelberg: Springer-Verlag; 2006.
112. Sampson DL, Parker TJ, Upton Z, Hurst CP. A comparison of methods for classifying clinical samples based on proteomics data: a case study for statistical and machine learning approaches. PLoS One. 2011;6(9), e24973.
113. Di Silvestre D, Zoppis I, Mauri G, Mauri PL. Stratification of biological samples based on proteomics data. OA Bioinformatics. 2013;1(1):1.
114. Alves G, Ogurtsov AY, Yu YK. Assigning statistical significance to proteotypic peptides via database searches. Proteomics. 2011;74(2):199–211. doi:10.1016/j.jprot.2010.10.005.
115. Dittrich J, Becker S, Hecht M, Ceglarek U. Sample preparation strategies for targeted proteomics via proteotypic peptides in human blood using liquid chromatography tandem mass spectrometry. Proteomics Clin Appl. 2015;9(1-2):5–16. doi:10.1002/prca.201400121.
116. Craig R, Cortens JP, Beavis RC. The use of proteotypic peptide libraries for protein identification. Rapid Commun Mass Spectrom. 2005;19:1844–50.
117. Kuster B, Schirle M, Mallick P, Aebersold R. Scoring proteomes with proteotypic peptide probes. Nat Rev Mol Cell Biol. 2005;6:577–83.
118. Mallick P, Schirle M, Chen SS, Flory MR, Lee H, Martin D, Ranish J, Raught B, Schmitt R, Werner T, Kuster B, Aebersold R. Computational prediction of proteotypic peptides for quantitative proteomics. Nat Biotechnol. 2007;25:125–31.
119. Di Silvestre D, Brunetti P, Vella D, Brambilla F, De Palma F, Mauri PL. Automated extraction of proteotypic peptides by shotgun proteomic experiments: a new computational tool and two actual cases. Current Biotechnology. 2015;4:39–45.
120. Craig R, Cortens JP, Beavis RC. Open source system for analyzing, validating, and storing protein identification data. J Proteome Res. 2004;3(6):1234–42.
121. Desiere F. The peptide atlas project. Nucleic Acids Res. 2006;34:D655–8.
122. Martens L, Hermjakob H, Jones P, Adamski M, Taylor C, States D, Gevaert K, Vandekerckhove J, Apweiler R. PRIDE: the proteomics identifications database. Proteomics. 2005;5:3537–45.
123. Hill JA, Smith BE, Papoulias PG, Andrews PC. ProteomeCommons.org collaborative annotation and project management resource integrated with the Tranche repository. J Proteome Res. 2010;9:2809–11.
124. Hermjakob H, Apweiler R. The Proteomics Identifications Database (PRIDE) and the ProteomExchange Consortium: making proteomics data accessible. Expert Rev Proteomics. 2006;3:1–3.
125. Webb-Robertson B-JM. Support vector machines for improved peptide identification from tandem mass spectrometry database search. Methods Mol Biol. 2009;492:453–60.
126. Tang H, Arnold RJ, Alves P, Xun Z, Clemmer DE, Novotny MV, Reilly JP, Radivojac P. A computational approach toward label-free protein quantification using predicted peptide detectability. Bioinformatics. 2006;22:e481–8.
127. Sanders WS, Bridges SM, McCarthy FM, Nanduri B, Burgess SC. Prediction of peptides observable by mass spectrometry applied at the experimental set level. BMC Bioinform. 2007;8(Suppl 7):S23.
128. Barabási A-L, Oltvai ZN. Network biology: understanding the cell's functional organization. Nat Rev Genet. 2004;5:101–13.

129. Rho S, You S, Kim Y, Hwang D. From proteomics toward systems biology: integration of different types of proteomics data into network models. BMB Rep. 2008;41:184–93.
130. Li J, Zimmerman LJ, Park B-H, Tabb DL, Liebler DC, Zhang B. Network-assisted protein identification and data interpretation in shotgun proteomics. Mol Syst Biol. 2009;5:303.
131. Jianu R, Yu K, Cao L, Nguyen V, Salomon AR, Laidlaw DH. Visual integration of quantitative proteomic data, pathways, and protein interactions. IEEE Trans Vis Comput Graph. 2010;16:609–20.
132. Pflieger D, Gonnet F, de la Fuente van Bentem S, Hirt H, de la Fuente A. Linking the proteins—elucidation of proteome-scale networks using mass spectrometry. Mass Spectrom Rev. 2011;30:268–97.
133. Wheelock CE, Wheelock AM, Kawashima S, Diez D, Kanehisa M, van Erk M, Kleemann R, Haeggström JZ, Goto S. Systems biology approaches and pathway tools for investigating cardiovascular disease. Mol Biosyst. 2009;5:588–602.
134. Isserlin R, Merico D, Alikhani-Koupaei R, Gramolini A, Bader GD, Emili A. Pathway analysis of dilated cardiomyopathy using global proteomic profiling and enrichment maps. Proteomics. 2010;10:1316–27.
135. Drozdov I, Didangelos A, Yin X, Zampetaki A, Abonnenc M, Murdoch C, Zhang M, Ouzounis CA, Mayr M, Tsoka S, Shah AM. Gene network and proteomic analyses of cardiac responses to pathological and physiological stress. Circ Cardiovasc Genet. 2013;6(6):588–97. doi:10.1161/CIRCGENETICS.113.000063.
136. Nguyen NT, Zhang X, Wu C, Lange RA, Chilton RJ, Lindsey ML, Jin YF. Integrative computational and experimental approaches to establish a post-myocardial infarction knowledge map. PLoS Comput Biol. 2014;10(3), e1003472. doi:10.1371/journal.pcbi.1003472.
137. Scardoni G, Petterlini M, Laudanna C. Analyzing biological network parameters with CentiScaPe. Bioinformatics. 2009;25:2857–9.
138. Poultney CS, Greenfield A, Bonneau R. Integrated inference and analysis of regulatory networks from multi-level measurements. Methods Cell Biol. 2012;110:19–56. doi:10.1016/B978-0-12-388403-9.00002-3.
139. Gibbs DL, Baratt A, Baric RS, Kawaoka Y, Smith RD, Orwoll ES, Katze MG, McWeeney SK. Protein co-expression network analysis (procona). J Clin Bioinforma. 2013;3(1):11.
140. Vinayagam A, Zirin J, Roesel C, Hu Y, Yilmazel Y, Samsonova AA, Neumuller RA, Mohr SE, Perrimon P. Integrating protein-protein interaction networks with phenotypes reveals signs of interactions. Nat Methods. 2014;11(1):94–9.
141. Currie HN, Vrana JA, Han AA, Scardoni G, Boggs N, Boyd JW. An approach to investigate intracellular protein network responses. Chem Res Toxicol. 2014;27(1):17–26.
142. Kim Y, Han S, Choi S, Hwang D. Inference of dynamic networks using time-course data. Brief Bioinform. 2014;15(2):212–28. doi:10.1093/bib/bbt028.
143. http://www.pathguide.org/.
144. Franceschini A, Szklarczyk D, Frankild S, Kuhn M, Simonovic M, Roth A, Lin J, Minguez P, Bork P, von Mering C, Jensen LJ. STRING v9.1: protein-protein interaction networks, with increased coverage and integration. Nucleic Acids Res. 2013;41(Database issue):D808–15.
145. Keshava Prasad TS, Goel R, Kandasamy K, Keerthikumar S, Kumar S, Mathivanan S, Telikicherla D, Raju R, Shafreen B, Venugopal A, Balakrishnan L, Marimuthu A, Banerjee S, Somanathan DS, Sebastian A, Rani S, Ray S, Harrys Kishore CJ, Kanth S, Ahmed M, Kashyap MK, Mohmood R, Ramachandra YL, Krishna V, Rahiman BA, Mohan S, Ranganathan P, Ramabadran S, Chaerkady R, Pandey A. Human Protein Reference Database—2009 update. Nucleic Acids Res. 2009;37(Database issue):D767–72.
146. Croft D, Mundo AF, Haw R, Milacic M, Weiser J, Wu G, Caudy M, Garapati P, Gillespie M, Kamdar MR, Jassal B, Jupe S, Matthews L, May B, Palatnik S, Rothfels K, Shamovsky V, Song H, Williams M, Birney E, Hermjakob H, Stein L, D'Eustachio P. The Reactome pathway knowledgebase. Nucleic Acids Res. 2014;42(Database issue):D472–7.
147. Kerrien S, Alam-Faruque Y, Aranda B, Bancarz I, Bridge A, Derow C, Dimmer E, Feuermann M, Friedrichsen A, Huntley R, Kohler C, Khadake J, Leroy C, Liban A, Lieftink C, Montecchi-Palazzi L, Orchard S, Risse J, Robbe K, Roechert B, Thorneycroft D, Zhang Y,

Apweiler R, Hermjakob H. IntAct-open source resource for molecular interaction data. Nucleic Acids Res. 2007;35:D561–5.
148. Zuberi K, Franz M, Rodriguez H, Montojo J, Lopes CT, Bader GD, Morris Q. GeneMANIA prediction server 2013 update. Nucleic Acids Res. 2013;41(Web Server issue):W115–22. doi:10.1093/nar/gkt533
149. Shannon P, Markiel A, Ozier O, Baliga NS, Wang JT, Ramage D, Amin N, Schwikowski B, Ideker T. Cytoscape: a software environment for integrated models of biomolecular interaction networks. Genome Res. 2003;13(11):2498–504.
150. Hu Z, Chang YC, Wang Y, Huang CL, Liu Y, Tian F, Granger B, Delisi C. VisANT 4.0: Integrative network platform to connect genes, drugs, diseases and therapies. Nucleic Acids Res. 2013;41(Web Server issue):W225–31.
151. http://www.ingenuity.com.
152. Saito R, Smoot ME, Ono K, Ruscheinski J, Wang PL, Lotia S, Pico AR, Bader GD, Ideker T. A travel guide to Cytoscape plugins. Nat Methods. 2012;9(11):1069–76.
153. Maere S, Heymans K, Kuiper M. BiNGO: a Cytoscape plugin to assess overrepresentation of gene ontology categories in biological networks. Bioinformatics. 2005;21(16):3448–9.
154. Zhang C, Hanspers K, Kuchinsky A, Salomonis N, Xu D, Pico AR. Mosaic: making biological sense of complex networks. Bioinformatics. 2012;28(14):1943–4.
155. Zhang C, Wang J, Hanspers K, Xu D, Chen L, Pico AR. NOA: a cytoscape plugin for network ontology analysis. Bioinformatics. 2013;29(16):2066–7.
156. Bader GD, Hogue CWV. An automated method for finding molecular complexes in large protein interaction networks. BMC Bioinformatics. 2003;4:2.
157. Morris JH, Apeltsin L, Newman AM, Baumbach J, Wittkop T, Su G, Bader GD, Ferrin TE. ClusterMaker: a multi-algorithm clustering plugin for Cytoscape. BMC Bioinformatics. 2011;12:436.
158. Lander ES, Linton LM, Birren B, Nusbaum C, Zody MC, Baldwin J, et al. International Human Genome Sequencing Consortium. Initial sequencing and analysis of the human genome. Nature. 2001;409(6822):860–921.
159. Anderson NL. The clinical plasma proteome: a survey of clinical assays for proteins in plasma and serum. Clin Chem. 2010;56:177–85.
160. Gallina C, Turinetto V, Giachino C. A new paradigm in cardiac regeneration: the mesenchymal stem cell secretome. Stem Cells Int. 2015;2015:765846. doi:10.1155/2015/765846.
161. Maguire G. Stem cell therapy without the cells. Commun Integr Biol. 2013;6(6):e26631. doi:10.4161/cib.26631.
162. Chimenti I, Smith RR, Li TS, Gerstenblith G, Messina E, Giacomello A, Marbán E. Relative roles of direct regeneration versus paracrine effects of human cardiosphere-derived cells transplanted into infarcted mice. Circ Res. 2010;106(5):971–80. doi:10.1161/CIRCRESAHA.109.210682.
163. Li X, Ren Y, Sorokin V, Poh KK, Ho HH, Lee CN, de Kleijn D, Lim SK, Tam JP, Sze SK. Quantitative profiling of the rat heart myoblast secretome reveals differential responses to hypoxia and re-oxygenation stress. J Proteomics. 2014;98:138–49. doi:10.1016/j.jprot.2013.12.025.
164. Yassine HN, Jackson AM, Reaven PD, Nedelkov D, Nelson RW, Lau SS, Borchers CH. The application of multiple reaction monitoring to assess Apo A-I methionine oxidations in diabetes and cardiovascular disease. Transl Proteom. 2014;4–5:18–24.
165. Lindsey ML, Mayr M, Gomes AV, Delles C, Arrell DK, Murphy AM, Lange RA, Costello CE, Jin YF, Laskowitz DT, Sam F, Terzic A, Van Eyk J, Srinivas PR. Transformative impact of proteomics on cardiovascular health and disease: a scientific statement from the American Heart Association. Circulation. 2015;132(9):852–72. doi:10.1161/CIR.0000000000000226.
166. Hood L, Flores M. A personal view on systems medicine and the emergence of proactive P4 medicine: predictive, preventive, personalized and participatory. N Biotechnol. 2012;29(6):613–24. doi:10.1016/j.nbt.2012.03.004.

Chapter 8
Top-Down Proteomics

Nicole M. Lane, Zachery R. Gregorich, and Ying Ge

Abstract The field of proteomics, particularly top-down mass spectrometry (MS), holds great promise for cardiovascular research and diagnosis. Top-down MS provides a "bird's eye view" of the entire protein. This is in contrast to bottom-up MS, which analyzes peptides. By studying intact proteins, more information can be gleaned. In particular, top-down proteomics provides valuable insights into what modifications are present on a protein of interest, including PTMs and sequence variations, even when *a priori* knowledge is lacking. PTMs, such as phosphorylation, have increasingly been linked to numerous cardiovascular diseases. Furthermore, changes in the expression levels of certain proteins have also been linked to disease. Top-down MS is able to quantify these changes, even when doing so necessitates distinguishing between various biologically relevant isoforms and proteoforms, which have proven difficult to differentiate using other methods. This chapter will explore how to prepare samples for top-down MS, the

N.M. Lane, BS, BA
Cellular and Molecular Pathology Training Program, University of Wisconsin-Madison, Madison, WI 53706, USA

Department of Cell and Regenerative Biology, University of Wisconsin-Madison, Madison, WI 53706, USA

Z.R. Gregorich, BS
Department of Cell and Regenerative Biology, University of Wisconsin-Madison, Madison, WI 53706, USA

Molecular Pharmacology Training Program, University of Wisconsin-Madison, Madison, WI 53706, USA

Y. Ge, PhD
Cellular and Molecular Pathology Training Program, University of Wisconsin-Madison, Madison, WI 53706, USA

Human Proteomics Program, University of Wisconsin-Madison, 1300 University Ave., SMI 130, Madison, WI 53706, USA
e-mail: ge2@wisc.edu

G. Agnetti et al. (eds.), *Manual of Cardiovascular Proteomics*,
DOI 10.1007/978-3-319-31828-8_8

instrumentation required, data analysis, current applications, and future directions of this valuable technique.

Keywords Top-down proteomics • Mass spectrometry • Chromatography methods • Fragmentation methods • Cardiovascular disease • Post-translational modifications • Phosphorylation • Myofilament proteins • Cardiac troponin • Tropomyosin

What Is Top-Down Proteomics?

There are two primary mass spectrometry (MS)-based approaches in modern proteomics [1–3]. The conventional peptide-based bottom-up approach, which involves protein digestion and MS analysis of the resulting peptides, has been well-developed and is now routinely used for protein identification and quantification with high throughput and automation [4]. However, the bottom-up approach has several shortcomings. In particular, the digestion of proteins into peptides increases sample complexity and can result in the loss of information pertaining to the sequence of the protein and/or any post-translational modifications (PTMs) [5]. Loss of information regarding the protein sequence (i.e., when a single or small number of peptides are detected for a specific protein) can make it difficult to distinguish closely related proteins or protein isoforms that have high sequence similarity. Furthermore, PTMs are often missed if the sequence coverage is low. Consequently, bottom-up proteomics is sub-optimal for the analysis of proteoforms—a term encompassing the myriad protein forms arising from a single gene, including those containing sequence variations as a consequence of mutations, polymorphisms, alternative splicing, or truncations, as well as those harboring PTMs [6]. Such modifications have increasingly been shown to play an important role in many diseases, including cardiovascular diseases [7].

In the protein-based top-down proteomics, intact proteins are analyzed, rather than peptides, which provides a "bird's eye view" of all protein modifications [5, 8]. Subsequently, intact proteoforms of interest can be isolated and fragmented in the gas phase to glean sequence information and localize amino acid changes and PTMs. Therefore, top-down proteomics is particularly well-suited to distinguishing between proteoforms (even those differing by a single amino acid) and localizing key PTMs [9]. This ability has already been employed in cardiovascular research and has provided important insights into the mechanisms of cardiovascular diseases, including ischemia, hypertension, and heart failure [10–12]. Nevertheless, due to the relatively recent innovation of top-down proteomics, this approach still faces many challenges, including difficulty in solubilizing intact proteins (particularly large proteins and very hydrophobic proteins such as membrane proteins), under-developed protein separation methods, and a lack of comprehensive data analysis tools [1]. This chapter will expand on top-down proteomics, specifically the workflow and application of this approach to gain insights into the molecular mechanisms that underlie appropriate cardiovascular function and disease, as well as the challenges currently facing top-down proteomics.

How Do You Prepare Samples for Top-Down Proteomics?

With the exception of protein digestion, the general workflow used for top-down proteomics is similar in nature to that used in bottom-up proteomics (Fig. 8.1) [1]. The top-down workflow entails protein extraction and separation, followed by MS and MS/MS analyses. The resulting MS and MS/MS data are then examined and bioinformatics analysis can be performed for protein identification and quantification. However, the analysis of intact proteins, rather than peptides, in top-down proteomics necessitates the use of unique approaches for protein solubilization and

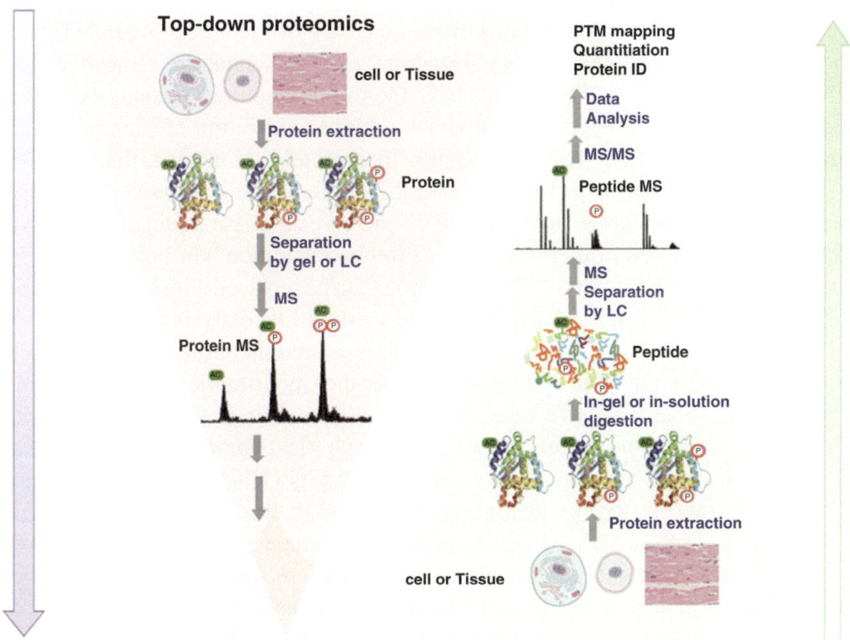

Fig. 8.1 Schematic depicting the basic workflow for top-down proteomics as compared with that of bottom-up proteomics (With kind permission from Springer Science+Business Media: Gregorich et al. [1])

separation, as well as high-resolution mass spectrometers and novel data analysis tools. In the following sections this workflow will be discussed in greater detail.

Protein Extraction

Proteins to be analyzed by top-down proteomics can be extracted from a variety of biological sources, including tissues, cultured cells, and biofluids [5]. However, as is the case in bottom-up proteomics, many of the traditional buffers (those employing salt concentrations in excess of 100 mM) [13] and buffer additives (detergents such as SDS) used to extract and solubilize proteins for biological or biochemical experiments interfere with the analysis of proteins by MS in several ways. This is a less pressing issue for bottom-up MS because protein digestion usually results in at least some peptides that are soluble without the addition of surfactant. However, for top-down proteomics, if an intact protein is not soluble without surfactant, it is particularly challenging to analyze. For this reason, buffers employing high concentrations of salt (>100 mM) and/or surfactants are frequently used during sample preparation in MS-based proteomics due to the fact that protein solubilization is greatly aided [1]. This is particularly true in the case of membrane-bound proteins due to their hydrophobicity [14]. However, it is essential to remove excess MS-incompatible salts and detergents from the solution before MS analysis. This is because salts and detergents typically have a higher ionization efficiency and are present in solution at a higher concentration than proteins of interest, thus allowing them to outcompete proteins for ionization [5]. Buffers with a high salt content can be removed via a desalting step utilizing offline reverse-phase liquid chromatography (RPLC: discussed further in Section "Protein Separation") or buffer exchange. Similarly, the biological salts used in extractions can be replaced with MS-compatible volatile salt buffers using a buffer exchange method like dialysis or ultrafiltration [5]. However, these practices consist of extra steps, and therefore extra time, spent in sample preparation. They can also increase the opportunities for sample loss. Alternatively, some labs have been focusing on developing MS-compatible salts. Xiu et al. identified ammonium tartrate as one such MS-compatible salt, which can be used with hydrophobic interaction chromatography (HIC; described in Section "Protein Separation") to provide high-resolution separation of intact proteins [15].

Detergents, such as SDS, can form adducts that reduce the protein signal, as well as outcompete the protein for ionization [5]. Therefore, they need to be removed prior to MS analysis. In bottom-up MS, gel-based methods can also be used to remove the surfactant. However, intact protein extraction from gel matrices is technically difficult and, therefore, this method is rarely used in top-down applications [5]. Instead, top-down MS methods precipitate the proteins in a sample using organic solvents (e.g., acetone), remove the organic solvent (as well as the surfactant, which is soluble in the organic solvent), and resolubilize the proteins in a detergent-free buffer [16]. However, both of these approaches can result in sample loss (especially as the sample may precipitate out in detergent-free buffer), and require additional time and steps [5]. Currently, efforts are underway to develop MS-compatible detergents to replace those currently in use and eradicate this problem [17, 18]. Such surfactants currently exist for use in bottom-up MS, including RapiGest, ProteaseMAX, and MaSDeS [16]. Various labs have found other methods for avoiding this problem. Whitelegge et al. made use of high concentrations of formic acid to maintain protein solubility of membrane proteins [19]. However, prolonged storage in formic acid can introduce artificial modifications into proteins. Carroll et al. employed a similar principle, using a high percentage of organic solvent with chaotropes, which disrupt hydrogen bonds to weaken the hydrophobic effect, to solubilize membrane proteins. They then fractionated in the same solvents using hydrophilic interaction chromatography [20]. The Robinson lab has used nonionic detergents to analyze membrane protein complexes via MS [21]. Unfortunately, these surfactants are often weak (they are not able to solubilize proteins to the same degree as SDS, the gold standard surfactant), so it is essential to develop stronger top-down MS-compatible detergents.

Protein Separation

The extreme complexity of the proteome necessitates its separation or fractionation into smaller subproteomes containing fewer proteins prior to MS analysis [5, 22]. Following extraction, proteins can be separated based on their physiochemical properties, including biospecificity, charge, hydrophilicity/hydrophobicity, and size. Traditionally, SDS-PAGE has been an important protein separation tool. However, as previously mentioned, it is technically difficult to extract intact proteins from SDS-PAGE gels, often resulting in a low recovery rate and making it a poor choice for protein separation in top-down proteomics [23]. Instead, numerous separation methods for intact proteins have been developed to separate proteins based on the above properties, including multiple liquid chromatography (LC) methods and electrophoretic methods, specifically isoelectric focusing (IEF) and the new gel-eluted liquid fraction entrapment electrophoresis (GELFrEE) method [3]. These approaches will be discussed in greater detail below.

LC covers a broad range of chromatographic methods that are based on dissolving the sample in a solvent and then passing that solvent through a second solid or liquid phase [24]. Top-down proteomics has employed numerous LC methods to achieve protein separation, including affinity chromatography, reverse phase liquid chromatography (RPLC), and size-exclusion chromatography (SEC) [5]. Affinity chromatography is a very effective method to separate proteins because it is based on specific biological interactions (i.e., antibody-antigen) [25]. This allows for specific proteins or complexes to be separated from a complex mixture [26]. Furthermore, affinity purification is robust, as the antibodies can be used many times with proper column maintenance. Unfortunately, the high specificity has the disadvantage of being low throughput, as only a single or small number of proteins can be analyzed at a time. Furthermore, affinity chromatography is difficult to perform online due to the necessity of a desalting step. This is often accomplished by coupling offline affinity chromatography with RPLC, which is based on separation by hydrophobicity, to provide an additional dimension of separation and to desalt the sample prior to MS analysis. RPLC is usually used as the final separation step as it has the added benefit of being able to connect online to the mass spectrometer [5]. This is because RPLC employs "MS friendly" buffers, which are composed of volatile solvents such as methanol or acetonitrile. Of note, SEC is also capable of being connected directly to MS [27]. This is because separation using SEC is based solely on molecular weight, so there is no interaction between the proteins and the stationary phase. This allows for a wide variety of solvents to be used in the mobile phase, including ones that are compatible with MS. Furthermore, use of ultra-high pressure SEC results in a relatively high-throughput method of separation for intact proteins [28]. Nevertheless, the resolution of this method is still relatively low in comparison to other separation methods, such as RPLC; this, in addition to sample dilution, has limited the use of SEC in top-down proteomics analyses. An encouraging new chromatography method is chromatofocusing, which is a promising extension of salt-gradient ion-exchange chromatography (IEC) [29]. IEC separates polar molecules by their affinity to an ion exchanger while chromatofocusing separates proteins based on differences in their isoelectric points [30]. Additionally, MS-compatible salts have opened up many new possibilities for protein separation in top-down proteomics. Notably, the use of HIC for protein separation for top-down proteomics has recently benefited from MS-compatible salts. HIC separates proteins based on hydrophobicity and has the capability to provide high-resolution separation of intact proteins; however, the salts typically used are incompatible with MS as they result in adductions and other issues. Valeja et al. used one such MS-compatible salt as part of a 3D LC scheme involving HIC (IEC-HIC-RPLC) that enabled the detection of 640 proteins from one of 35 IEC fractions from whole cell lysate, as compared to only 47 proteins detected using a conventional 2D approach (IEC-RPLC) [22].

Electrophoretic separation methods, in contrast to LC methods, are based on the movement of proteins in an electric field, rather than through a stationary phase. IEF separates proteins based on their isoelectric point, pI (the pH at which a protein has no net charge) [31]. Zhang et al. coupled IEF with superficially porous silica LC to separate cardiac myofibrils, thereby gaining about six-fold increase in the unique monoisotopic masses observed below 30 kDa. Additionally, a four-fold increase in the mass range allowed for the observation of proteins with masses greater than 200 kDa [32]. Recently, IEF has been coupled with elution tube gel electrophoresis to produce the promising GELFrEE method developed by Tran et al. [33]. GELFrEE uses solution-based IEF coupled with a multiplex tube gel electrophoresis separation device. This allows proteins to be separated based on molecular mass (10–250 kDa), while providing high-resolution protein separation and high protein recovery rates [29, 34]. Unfortunately, this method utilizes the detergent SDS and, thus, detergent removal prior to MS analysis is necessary. Despite this limitation, GELFrEE has been employed effectively in conjunction with other separation methods, as described below

As has been touched on in the above paragraphs, it is often useful to couple two or more methods of separation, producing a multidimensional separation strategy. This has several advantages: extra dimensions of separation can serve to reduce sample complexity, increase the number of unique proteins observed, and expand the mass range covered [32]. Notably, the Kelleher lab used a 4D method (IEF-GELFrEE-nanocapillary LC-MS) to identify 1043 proteins, and 3093 proteoforms, including proteins up to 105 kDa proteins and transmembrane proteins, both of which are notoriously difficult to detect in top-down proteomic analyses [35]. Two years later, they used this same strategy to conduct the largest top-down proteomics study to date, identifying 1220 proteins and 5000 proteoforms [36]. These studies clearly demonstrate the power of multidimensional separation methods. Unfortunately, separating the sample in this way results in a large number of proteome fractions. To analyze each of these would be incredibly time-consuming. Therefore, many studies focus on only one or a few of these fractions [1]. However, this leads to selection bias as to which fractions are analyzed and which are ignored. Fortunately, new mass spectrometers provide higher resolution and are able to separate more complex mixtures of proteins, combating this issue by requiring fewer fractions. For instance, Ge et al. used a Fourier transform ion cyclotron resonance (FT-ICR) mass spectrometer to resolve more than one hundred protein components in a single spectrum [37].

What Instrumentation Is Needed for Top-Down MS Analysis?

For top-down MS, the need to be able to analyze whole proteins dictates the type of fragmentation methods and instrumentation that are commonly used. The ability of a given mass spectrometer to accommodate a preferred fragmentation method, as well as its resolution, dictate the usefulness of that spectrometer for top-down applications. For instance, mass spectrometers with electrospray ionization (ESI) are preferred over matrix-assisted laser desorption/ionization (MALDI) due to their ability to produce multiply charged precursor ions [38]. This is important for electron-based dissociation methods, such as electron-capture dissociation (ECD) [39] and electron transfer dissociation (ETD) [40], which are able to preserve labile PTMs (e.g. phosphorylation, glycosylation, etc.). The fragmentation efficiency of the electron-based dissociation methods increases with increasing charge; thus, multiply charged precursors will fragment more efficiently and yield more fragment ions than singly (or low) charged precursors [41]. If the protein is known, the experimentally determined molecular mass can be compared to the predicted molecular mass of the protein (based on the DNA sequence) to determine what, if any, modifications are present [5]. Following MS analysis, proteoforms of interest can be isolated and fragmented in a similar way using MS/MS to determine the amino acid sequence and localize PTMs or sequence variations. This section will discuss various mass spectrometers and fragmentation methods commonly employed in top-down proteomics.

Commonly Used Mass Spectrometers

Traditionally, top-down MS has relied on FT-ICR mass spectrometers [42]. These instruments provide the high-resolution and, therefore, high mass accuracy necessary for intact protein analysis and confident protein identification and PTM characterization. This is important because it allows co-eluting proteins with similar mass-to-charge (*m/z*) ratios to be distinguished and aids in accurate mass measurements, charge state determination, and quantitation [43]. Accurate mass measurements of the fragment ions also add valuable information in the identification of proteins and PTMs, and the localization of said PTMs and sequence variations. Unfortunately, FT-ICR instruments rely on superconducting magnets, which have high maintenance requirements. This has limited the use of these instruments to specialized labs. However, although FT-ICR remains the gold-standard in terms of resolving power, advances in instrumentation, particularly the development of Orbitrap mass spectrometers [44] and high-end time-of-flight (TOF) instruments [45], have enabled other mass spectrometers to provide the high resolution and high mass accuracy necessary for top-down MS studies [8].

In addition to the above instruments with a single detector, hybrid mass spectrometers, such as the LTQ/FT, the LTQ-Orbitrap, and the QTOF have also been used in top-down proteomics [5]. These advances in instrumentation have opened the field of top-down proteomics by providing instruments with resolution that is high enough for top-down proteomics analyses, but without the superconducting magnet employed in FT-ICR MS, and the associated maintenance and costs [43]. Of note, depending on the research question, high-resolution mass spectrometers may not be required. For instance, Coelho Graça et al. utilized a low-resolution ion trap mass spectrometer to distinguish between hemoglobin variants [46]. One benefit of low-resolution mass spectrometers is the fast scan time, which allows rapid results, an important feature for clinical applications, such as Coelho Graça explored.

Fragmentation Methods

Following MS analysis, specific proteoforms of interest can be isolated in the gas phase and fragmented using a variety of fragmentation methods. The resulting fragment ions can be detected to obtain sequence information and localize PTMs, even when *a priori* knowledge is lacking. There are two main categories of fragmentation techniques, as illustrated in Fig. 8.2. The first is energetic dissociation. This category includes collision-induced dissociation (CID) (also known as collisionally-activated dissociation, CAD) [47], high-energy collision dissociation (HCD) [48], infrared multiphoton dissociation (IRMPD) [49], and post-source decay (PSD) [50]. All energetic dissociation methods result in the cleavage of bonds having the lowest activation energy. This normally results in the cleavage of the CO-NH bonds in the protein backbone (producing *b* and *y* ions). However, if labile PTMs (such as

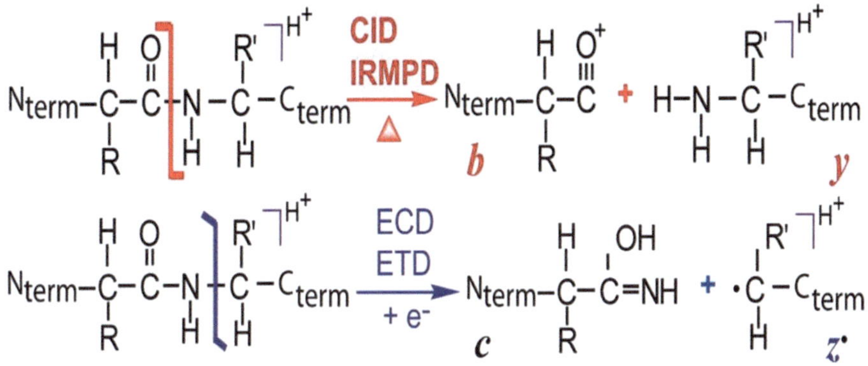

Fig. 8.2 Cartoon of the peptides produced during energetic dissociation (*above*) vs. electron-based dissociation (*below*). This results in fragments containing the N-terminus (fragments *b* or *c*) and fragments containing the C-terminus (fragments *y* or *z*) (Reprinted with permission from Zhang and Ge [5])

phosphorylation) are present, these modifications will be lost, thus precluding their localization [1]. Recently, the Brodbelt lab modified an ultra-violet photodissociation (UVPD) method for use with top-down MS [51]. Although UVPD is a variant of IRMPD, it produces many fragments, including *a*˙/*x*, *b*/*y*, and *c*/*z*˙ ions [51]. This will likely prove useful for the fragmentation of larger intact proteins.

The second category is electron-based dissociation, which produces *c* and *z*˙ ions (Fig. 8.2). This includes ECD [39] and ETD [40]. Fragmentation using these methods results almost exclusively in the cleavage of NH-CHR bonds in the protein backbone, through a mechanism that remains in debate. Thus, the primary benefit of the electron-based fragmentation methods is that labile PTMs are preserved, which allows their location(s) to be determined [5]. Moreover, ECD has been shown to produce more cleavages than CID, which enhances top-down MS's ability to localize PTMs and sequence variants [52]. Collectively, these benefits make electron-based dissociation methods the preferred dissociation methods for top-down proteomic analyses. Furthermore, ECD has proved very useful in the fragmentation of large proteins. Ge et al. utilized top-down ECD MS/MS to isotopically resolve and fragment a 115 kDa fragment of cardiac myosin binding protein C (cMyBP-C) [53]. To date, the largest protein to be isotopically resolved is a ~148 kDa monoclonal antibody (IgG1κ), fragmented using ECD on all charge states simultaneously [54]. Unfortunately, ECD is only available on FT-ICR mass spectrometers. However, ETD is available on ion trap mass spectrometers; thus, the development of hybrid instruments integrating ion trap mass spectrometers with other MS detectors has greatly aided access to these powerful fragmentation methods [55].

Despite these improvements, the fragmentation of large proteins for top-down analysis remains a challenge. The gas phase structures of large proteins can form compact "balls of spaghetti", which prove difficult to fragment [56]. To combat this, Han et al. developed prefolding dissociation, which uses variable thermal and collisional activation directly following ESI to delay the refolding of proteins once in the gas phase [57]. This allowed the authors to fragment proteins greater than 200 kDa in mass (although fragmentation was primarily restricted to the N- and C-termini). In the same vein, fragmentation efficiency drops with increasing molecular weight due to an increase in intramolecular interactions in the tertiary structure of the proteins which make them more stable as MW increases. This contributes to the difficulty in fragmenting large proteins during tandem MS [5]. Even after the protein backbone is cleaved, intramolecular forces may hold large portions of the protein together. Peng et al. used a combination of ECD and CAD fragmentation methods to enhance fragmentation [58].

Alternatively, strategies that deplete the most highly abundant isotopes or increase the charge of electrosprayed proteins (so called supercharging reagents) have been explored as potential options for enhancing the detection and fragmentation of high molecular mass proteins. Depletion of highly abundant isotopes is effective, but limited because at high molecular mass, isotopes have a decreased role in determining signal-to-noise (S/N), whereas multiple charge states play an increasing role [59]. Therefore, supercharging, which preferentially forms higher charge states, seems more promising [59]. Unfortunately, many supercharging reagents interfere with LC separation [14]. To address this, Valeja et al. have identified several supercharging reagents that are effective for proteins up to 78 kDa and that do not affect LC [60]. Miladinovic et al. addressed the problem from a different angle by developing a technique, called "in-spray supercharging", which utilizes dual-sprayer ESI to inject the supercharging reagents after LC separation [61]. Despite these advances, most top-down MS research is still limited to focusing on proteins below 70 kDa due to difficulties with fragmentation [59]. This is, of course, an important problem as many proteins in the proteome are larger than this.

How Do You Interpret Top-Down Proteomics Data?

Large-scale MS-based proteomics experiments generate a large amount of very complex data. Therefore, several software tools exist for the analysis of this data to aid in the interpretation of the mass and tandem mass spectra, and protein identification, quantification, and characterization. These fall under one of two categories--tools for spectral deconvolution and tools for protein identification and characterization. However, although multiple programs exist to do various pieces of top-down data analysis, there is a dearth of software that is capable of handling deconvolution, identification, quantification, and visual validation within one program. This hinders analysis of the complex tandem mass spectra produced by top-down proteomics [1]. Therefore, many labs are working to expand and improve the existing top-down MS software to address these concerns. The current software, as well as looked-for improvements, will be discussed below.

Current Tools and Algorithms for Spectral Deconvolution

Software for spectral deconvolution is particularly useful for the analysis of complicated high-resolution ESI mass and tandem mass spectra. ESI produces multiple charge states, while MS/MS produces many fragment ions, both of which result in very complex spectra. With sufficiently high resolution, the isotopomer envelopes—a set of peaks all corresponding to the same protein, but with different numbers of chemical isotopes—can be resolved, allowing overlapping fragment ions to be distinguished [62]. Spectral deconvolution aims to group MS peaks into isotopomer envelopes so that charge state and mass can be determined, simplifying the spectra and aiding analysis. Horn et al. developed the first fully automated computer algorithm for the analysis of high-resolution mass spectra, known as THRASH (thorough high resolution analysis of spectra by Horn) [63]. THRASH groups peaks into isotopomer envelopes and determines the charge of each protein species [63]. Unfortunately, THRASH is not compatible with many common operating systems. To combat this, the Smith lab developed Decon2LS [64] and DeconMSn [65], which are based on THRASH and are Mac and PC compatible. Furthermore, this is open-source software, allowing other labs to freely download it and adapt it to their own needs [64]. Guner et al. used this open-source code to develop the more user-friendly MASH Suite, which is able to determine charge state, calculate monoisotopic and most abundant masses, and determine how well the experimental isotopic distribution fits a theoretical model [66]. Thermo has developed a similar tool called ManualXtract [5]. The Pevzner lab created a new algorithm for their program MS-Deconv, which outperformed the THRASH algorithm in terms of true positives identified and time [67].

Informatics and Software Tools for the Identification of Proteoforms

Bioinformatics tools are routinely used in MS-based proteomics experiments to not only identify proteins, but also to identify and locate any PTMs that may be present. ProSight was developed by the Kelleher lab using the THRASH algorithm and shotgun annotated databases [68]. It was the first major top-down proteomics search engine and remains the most common. With it, the user queries the database using both precursor ion MS and product ion MS/MS data, allowing for accurate identifications. Furthermore, it allows for the mapping of PTMs [68]. However, ProSight's database setup is not ideal for identification of unknown PTMs and the size of the database greatly reduces search speed. The Kelleher lab recently released ProSight Lite to address these concerns [69]. PIITA, developed by Tsai et al. makes use of a "precursor ion-independent top-down algorithm" to compare tandem mass spectra to all possible theoretical spectra based on a genomic sequence database [70]. After the protein is identified, differences between the measured and theoretical masses are used to identify PTMs and sequence variations, allowing identification even without *a priori* knowledge [70]. MascotTD, or "BIG Mascot", is an extended version of Mascot, software initially designed to aid bottom-up MS, which has been optimized for top-down MS [71]. The new database includes proteins up to 110 kDa in size [71]. By inputting both precursor and fragment ion data generated in top-down MS, BIG Mascot is able to locate both PTMs and sequence variants, although these potential modifications must be specified before the search [71]. MS-TopDown utilizes spectral alignment to identify the proteins present in a given spectrum, allowing for the identification of unexpected PTMs [72]. However, it lacks statistical tools and has a slow search time. MS-Align+ makes use of spectral alignment, rather than shotgun annotated databases, and compares favorably to software using such methods for protein and PTM identification (particularly for unknown PTMs) [73]. However, MS-Align+, along with the other existing software for top-down spectral analysis, cannot integrate identification with quantification and PTM characterization and lacks visual output for manual validation (which can minimize false-positives and mis-assigned peaks). The Ge lab is working to expand their MASH Suite software to MASH Suite Pro, in an attempt fill these voids. MASH Suite Pro will support multiple deconvolution methods for mass and tandem mass spectra, characterization of PTMs and sequence variants, relative quantitation of proteoforms, visual validation, and graphical output [74].

What Can Top-Down Proteomics Be Used for in Cardiovascular Research?

Due to its ability to provide comprehensive information regarding sequence variations and PTMs, top-down MS is quickly becoming the method of choice for unraveling the inherent complexities of the proteome. In particular, mutations and altered protein activity/function as a consequence of PTMs have been implicated in diseases ranging from cancer, to neurodegenerative disorders, to diabetes, to infectious diseases [14]. This section will focus specifically on the use of top-down MS in cardiovascular research. It is worth noting that, although many of the applications discussed herein are in research settings, top-down MS also holds great promise for clinical use. For instance, a number of labs have developed top-down MS-based methodologies for the detection of single amino acid mutations in blood proteins that are responsible for sickle cell disease and amyloidosis, which have great diagnostic potential [46, 75, 76].

Computing Accurate Mass Measurements to Reveal Cardiac Protein Complexity

Mass accuracy is vital in top-down proteomics—high mass accuracy results in higher confidence in protein and PTM identifications. Today's high-resolution mass spectrometers are able to achieve a mass accuracy on the order of a couple parts per million (ppm) [43]. This level of accuracy, at both the intact and fragment ion stages, greatly narrows the list of potential proteins an observed mass could correspond to. This allows for the precise determination of the identity of a peak of interest with high confidence, including what modifications might be present through comparison of the theoretical (when available) and experimentally determined masses. However, to achieve these levels of certainty, it is necessary to have both a high-resolution mass spectrometer and accurate deconvolution software. Accurate deconvolution software enables isotopomer envelopes to be precisely identified and, therefore, the monoisotopic mass to be reliably and quickly calculated. Although this can be done by hand, it is very low throughput, prohibitively so for large-scale proteomics analyses. Accurate mass measurements are also indispensable for proper protein identification, as well as for the localization of PTMs and sequence variations. Using this ability, Zabrouskov et al. were able to identify 36 modified molecular ions in an FT-ICR mass spectrum of commercial cardiac troponin I (cTnI) from healthy human hearts [77]. Although this sample runs as a single band on SDS-PAGE, it is clear upon MS analysis that the sample is actually quite complex—containing various truncations, oxidations, and other PTMs. The authors were able to identify each of these proteoforms, despite their similar sizes. Figure 8.3 details these results and highlights the ability of top-down MS to distinguish between different proteoforms [5].

Identification of Cardiovascular Proteins from Complex Mixtures

In analyzing complex mixtures, top-down MS has the useful advantage of analyzing whole proteins, unlike bottom-up MS. This allows top-down MS to maintain connections between disparate portions of each protein (rather than breaking all proteins present into multiple peptides, as in bottom-up MS), thereby minimizing sample complexity and preserving valuable sequence and modification information for each protein present in the mixture. This advantage, along with the high mass accuracy afforded by FT-ICR MS was put to good use by Peng et al. to investigate the presence of Tm isoforms in human heart [78]. From cardiac tissue samples, they were able to identify α-Tm, β-Tm, and κ-Tm, and were able to determine that the relative expression levels and PTMs of each varied in relation to the region of the heart the sample was taken from.

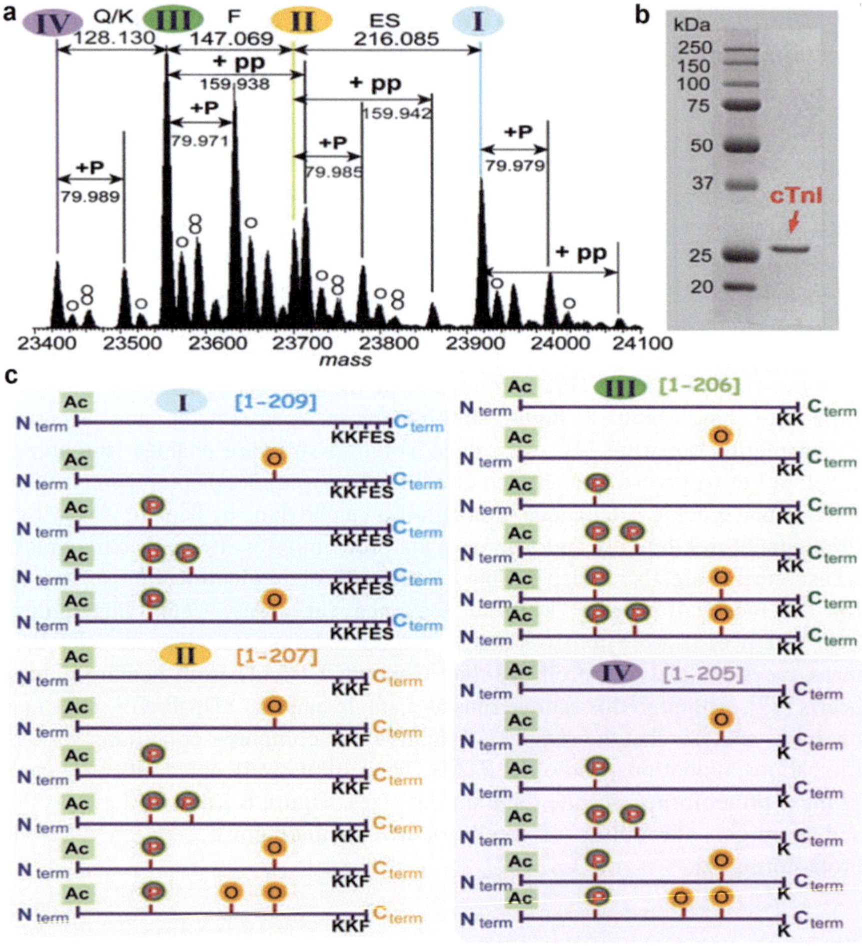

Fig. 8.3 Top-down MS data from human cTnI. (**a**) Deconvoluted data revealed 36 modified ions of human cTnI. These were formed from acetylations, phosphorylations, and truncations. I indicates the full-length protein. II-IV are different truncations. (**b**) SDS-PAGE gel depicting how the commercial cTnI ran as a single band. (**c**) Cartoon of the modified forms identified in A to more clearly demonstrate the various combinations of modifications that were able to be identified using top-down MS (Reprinted with permission from Zhang and Ge [5])

Despite these encouraging results, top-down MS continues to have difficulty identifying proteins of low abundance in mixtures. This problem is being addressed through various enrichment methods. Hwang et al. developed an enrichment method targeting intact phosphoproteins, which uses superparamagnetic nanoparticles covered in functionalized multivalent ligand molecules that bind phosphate groups [79]. The workflow for this method involves binding the phosphoproteins to the nanoparticles, removing the unbound proteins, and eluting the phosphoproteins. The authors showed that this strategy was compatible with top-down MS and was able to significantly increase the detection of phosphoproteins in subsequent MS analysis. Another method, termed ion parking, was developed by McLuckey et al. Ion parking allows an ion's signal to be concentrated into a single, or a few, charge state(s). This increases the S/N ratio, allowing detection of low abundance proteins [80, 81]. However, these new methods have not yet been applied to cardiovascular samples.

Identification and Characterization of Biologically Relevant Sequence Variations

In addition to simple protein identification, top-down MS may be used to distinguish between biologically relevant isoforms and proteoforms, which can otherwise be difficult to separate. For instance, changes in the relative expression levels of α-cardiac actin and α-skeletal actin in the heart have been associated with cardiac dysfunction [82]. However, these two isoforms vary by only 32 Da as a result of differences in two juxtaposed amino acids [82]. Despite this, Chen et al. were able to use top-down MS to distinguish between these isoforms in both human and swine heart tissue and thereby accurately measure changes in the relative expression levels in healthy hearts and failing hearts with dilated cardiomyopathy (Fig. 8.4) [83]. This method, which requires only a small section of tissue, provides rapid, easy, and reliable quantification of the relative expression levels of α-cardiac actin and α-skeletal actin within human hearts, allowing this to be used as a potential biomarker for cardiac dysfunction.

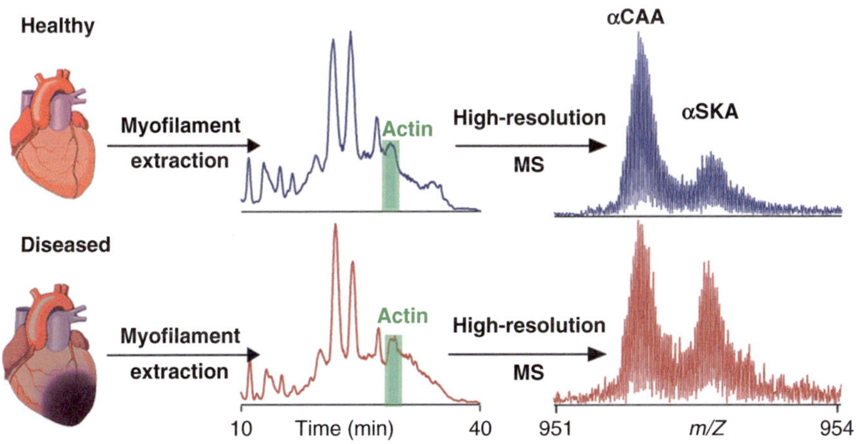

Fig. 8.4 LC/MS analysis of actin from human heart samples showing the levels of both αCAA and αSKA. Note that the relative level of αCAA is higher under normal conditions (Reprinted with permission from Chen et al. [83])

Additionally, one of the great benefits of top-down MS is that it does not require *a priori* knowledge of the modifications (PTMs or sequence variants) present. For instance, Zhang et al. discovered a previously unknown single amino acid polymorphism, V116A, in swine cTnI [9]. MS analysis of their protein revealed a −28 Da difference from the calculated mass based on the cTnI sequence from the database. Using MS/MS, Zhang et al. were able to determine that this discrepancy was located between V116-I122. The only modification that could account for this difference in molecular mass was the substitution V116A [9]. Top-down MS is capable of finding such modifications even if the molecular mass is as predicted (due to events that result in a net mass difference of 0). This is demonstrated by Sze et al., who were able to distinguish modifications to within one residue, allowing them to distinguish two compensatory amino acid substitutions [84]. In another example, Peng et al. identified a "novel" swine tropomyosin (Tm) isoform [58]. Utilizing top-down MS, the isoform was determined to exactly match mouse β-Tm. Therefore, the authors were able to identify this protein as swine β-Tm.

Identification and Characterization of Cardiovascular Disease-Related PTMs

As mentioned above, top-down MS is also ideal for identifying and localizing the full complement of protein PTMs. Figure 8.3, in addition to showing the sequence variations identified by Zabrouskov, also illustrates the various phosphorylation and acetylation sites of human cTnI [77]. Figure 8.5, below, further explains how

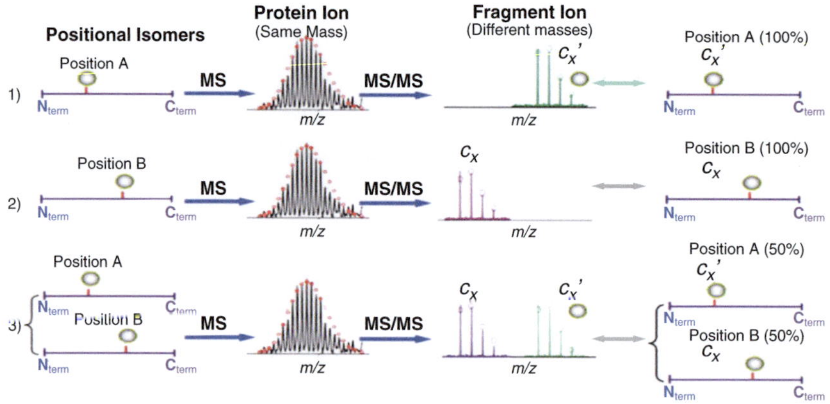

Fig. 8.5 Schematic depicting how top-down MS is able to determine the location of PTMs on a hypothetical protein. Three positional isomers with PTMs at either position A, position B, or a mix of the two are shown. As can be seen, the MS data for these is the same, as they are all the same protein with the same modification, just at different locations. This allows for identification of the protein and the modification, but not its location. Subsequent tandem MS data is able to distinguish between the three possible locations (Reprinted with permission from Zhang and Ge [5])

top-down MS is able to determine the locations of modifications, even for positional isomers, which vary only in the position of the modification. MS allows identification of PTMs present, while MS/MS allows for the identification of the protein, as well as the localization of PTMs. Using this ability, researchers have located many modifications implicated in disease. For instance, Ser43/45 phosphorylation has been identified in *in vitro* phosphorylation assays of cTnI [85]. These are believed to be phosphorylation sites of protein kinase C (PKC) and have been linked to cardiac dysfunction through the use of transgenic mice that overexpress PKC [86, 87]. Dong et al. confirmed phosphorylation sites at Ser43/45 of cTnI *in vivo* from spontaneously hypertensive heart failure rats using top-down MS [12]. Furthermore, Zhang et al. used a top-down proteomics approach to identify altered cTnI phosphorylation as a potential tissue biomarker for chronic heart failure [10]. Through top-down MS analysis of human heart tissue samples, they showed that there is a decline in phosphorylation of cTnI correlating with a progression towards heart failure. This is an especially promising discovery as cTnI is released into the blood following cardiac injury and is thus an easily accessible biomarker for chronic heart disease [10].

Importantly, top-down MS is able to identify modifications that are unexpected. As Peng et al. demonstrated, top-down MS can yield unforeseen results [11]. In this study, the researchers were investigating the phosphorylation of cardiac myofilament proteins from the hearts of swine with acute myocardial infarction (AMI). In addition to noting reduced phosphorylation of cTnI and MLC2, they saw that phosphorylation was also significantly reduced in enigma homolog isoform 2 (ENH2), a Z-disc protein [11]. ENH2 was not previously known to be a phosphoprotein. This was the first study to implicate the phosphorylation of a Z-disc protein in cardiac dysfunction following myocardial infarction.

In addition to mapping PTMs, top-down MS allows researchers to determine in which order those modifications occurred, something that bottom-up MS is incapable of doing. This is done by looking at each modified form and determining the site(s) of modification. For instance, if a protein exists in un-, mono-, and bis-phosphorylated forms, top-down MS can isolate each form and determine the site(s) of phosphorylation. If the mono-phosphorylated form is always phosphorylated at position A, and the bis-phosphorylated form is phosphorylated at both positions A and B, it may be concluded that position A must be phosphorylated before position B [5]. Using this logic, Zhang et al. were able to determine that phosphorylation of cTnI isolated from the myocardium of patients with chronic heart failure is first phosphorylated at Ser22 followed by phosphorylation at Ser23 [10]. Similarly, Ge et al. determined the order of phosphorylation in recombinant mouse cMyBP-C [53]. cMyBP-C is located in the sarcomere's thick filament and is involved in the regulation of cardiac contractility [88]. Its phosphorylation is known to be altered in heart failure [89]. Ge et al. identified the phosphorylation sites in cMyBP-C as Ser283, Ser292, and Ser312, with phosphorylation of Ser292 occurring prior to phosphorylation of Ser283 and Ser312 [53]. This is likely due to phosphorylation of Ser292 causing a conformational change that allows the two other sites to be accessible [90]. Interestingly, Ge et al. discovered that truncation influenced the sites of phosphorylation in cMyBP-C. The phosphorylation sites in truncated cMyBC-C were determined to be Ser292, Ser312, and Ser484 [53]. Furthermore, for these phosphorylation sites, it appears as though order does not matter. This finding suggests that truncations can alter the PTM state of a protein.

Of interest, Borges et al. applied top-down MS to the pressing issue of type II diabetes with concomitant cardiovascular disease [91]. Certain anti-diabetes drugs have been linked to cardiovascular complications in susceptible individuals. Therefore, Borges et al. endeavored to use top-down proteomics to identify biomarkers which could distinguish the spectrum of cardiovascular disease and type II diabetes comorbidities. They identified protein oxidation, typically methionine sulfoxidation on apoAI and apoCI, as indicative of cardiovascular disease. Conversely, increases in RANTES and apoCI protein truncations were indicative of type II diabetes [91].

Quantification of Proteoforms to Elucidate Biologically Relevant Changes

Top-down MS is considered semi-quantitative due to the ability to compare relative amounts of modified and un-modified (or multiply modified) proteoforms present within the same spectrum. This is done by comparing relative signal intensities and is possible because the addition of PTMs or minor sequence variations have little impact on the physiochemical properties of the whole protein, meaning that its ionization efficiency is not greatly affected [5]. This is in contrast to bottom-up MS, where the addition of modifications alters the ionization efficiency of peptides [92]. Zhang et al. used this ability to quantify changes between levels of un-, mono-, and bis-phosphorylated cTnI in the progression of chronic heart failure [10]. This revealed that as the disease progressed, the level of cTnI phosphorylation decreased. Peng et al. took a similar approach to investigate the phosphorylation levels in several myofilament proteins under conditions of AMI [11]. Figure 8.6, below, details their results. As can be seen in Fig. 8.6e, the level of phosphorylation decreased in the AMI model for multiple vital myofilament proteins, notably cTnI, MLC2, and ENH2. Dong et al. used the quantification ability of top-down MS to show that phosphorylation of cTnI, at Ser(22/23) and Ser(42/44), occurs at a higher level in a spontaneously hypertensive rat model as compared with normotensive age-matched rats [12].

This same method was applied to a slightly different question by Gregorich et al. [93]. The authors investigated whether or not chamber-specific or transmural variations exist in the phosphorylation of myofilament proteins in the heart. They were able to show the basal phosphorylation of certain proteins (cTnI and α-Tm, specifically) varies in a chamber-specific manner. However, they did not see any differences based on transmural location in the proteins they investigated. It has been well shown that PTMs, especially phosphorylation, play major roles in cardiovascular disease [93]. This study highlights how consistent sampling of cardiac tissue is essential to elucidate PTM-associated disease mechanisms and identify disease biomarkers in order to minimize unrelated variability which could obscure results (Fig. 8.7).

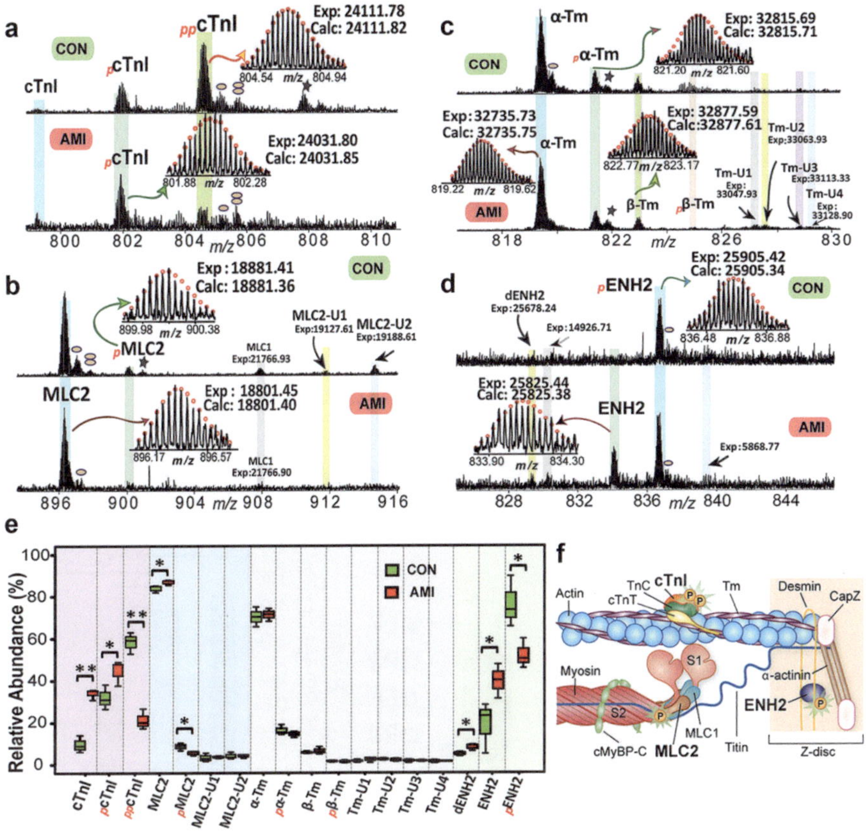

Fig. 8.6 Quantitative analysis of myofilament proteins using high-resolution MS. (**a–d**) Representative high-resolution spectra of cTnI (**a**), MLC2 (**b**), Tm (**c**), and ENH2 (**d**) under control (CON) and acute myocardial infarction (AMI) conditions. Circles: theoretical isotopic abundances. Ellipses: oxidation. Stars: non-covalent phosphoric acid adducts. (**e**) Relative quantification of the proteoforms under control and disease conditions. (**f**) Schematic of relevant myofilament and Z-disc protein interactions in the sarcomere (This research was originally published in Molecular and Cellular Proteomics. Peng et al. [11]. © the American Society for Biochemistry and Molecular Biology)

Conclusions and Future Directions

Top-down MS is an emerging field that holds great promise for proteomics research due to its ability to view whole proteins. It can provide information on the PTMs present on a protein and their locations, protein sequence variations and truncations, and the relative quantity of various proteoforms present, all without *a priori* knowledge. However, it is still a new technology and, therefore, still faces challenges before it can reach its full potential. Issues with protein solubility and separation, fragmentation and detection of certain proteins (particularly those that are large or

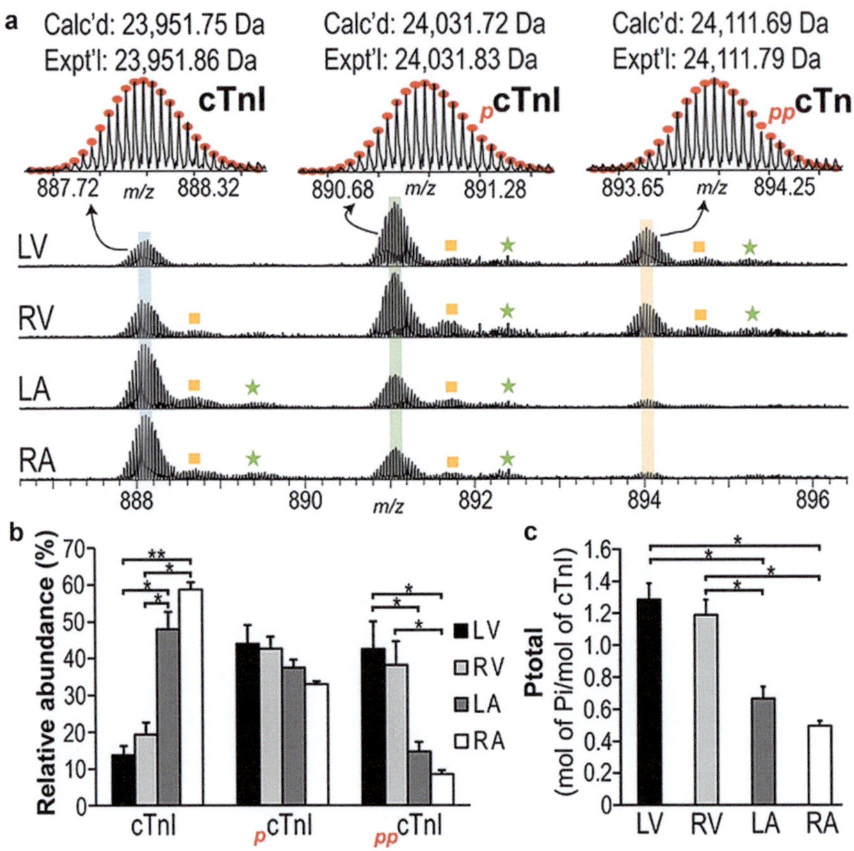

Fig. 8.7 Basal cTnI phosphorylation varies regionally but not transmurally in the heart. (**a**) Representative MS spectra for cTnI from the four chambers of the heart. *Circles*: theoretical isotopic abundance distribution of isotopomer peaks corresponding to assigned mass. *Squares*: oxidation. *Stars*: potassium adducts. (**b**) Graph showing relative abundances of cTnI, $_{p}$cTnI, and $_{pp}$cTnI in the four chambers of the heart. (**c**) Graph showing total cTnI phosphorylation in the left ventricle (*LV*), right ventricle (*RV*), left atrium (*LA*), and right atrium (*RA*). (**d**) Representative MS spectra for cTnI from the three layers of the LV free wall; epicardium (*epi*), myocardium (*myo*), endocardium (*endo*). Stars: potassium adducts. (**e**) Graph showing relative abundances of cTnI, $_{p}$cTnI, and $_{pp}$cTnI in the epicardium, myocardium, and endocardium of the LV free wall. (**f**) Graph showing total cTnI phosphorylation in the three layers of the LV free wall. Data are from three swine hearts with values reported as mean ± SEM. *$p<0.05$, **$p<0.001$ (Reprinted with permission from [93])

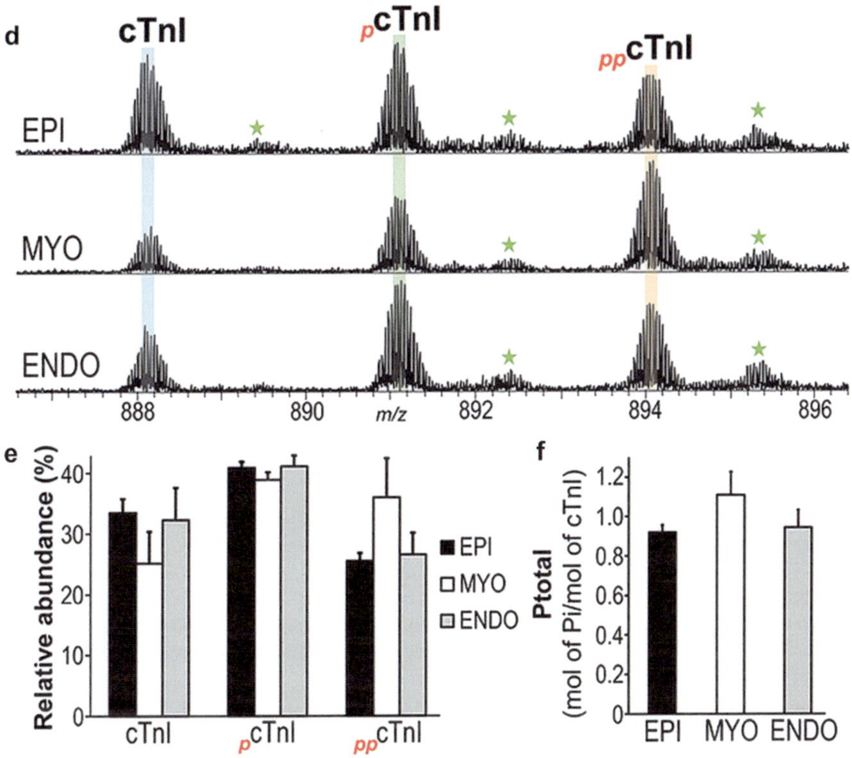

Fig. 8.7 Continued

hydrophobic), low throughput, and a lack of comprehensive data analysis tools hamper research using top-down proteomics. However, important advances are rapidly being made in these areas, allowing top-down MS to provide a wealth of information. Ongoing research combining both top-down MS and bottom-up MS data, as well as data provided by various other fields, such as biomedical research, functional analysis, metabolomics, and transcriptomics, will provide invaluable insights into innumerable biological systems and disease processes. Hopefully, this will lead to the development of novel therapeutic targets and diagnostic biomarkers, not just in cardiology, but in all areas of health. Promisingly, top-down MS has already aided in the identification of biomarkers for cardiovascular diseases and has provided insights into PTM-associated disease mechanisms. Future research and advances will undoubtedly lead to even more discoveries.

References

1. Gregorich ZR, Chang YH, Ge Y. Proteomics in heart failure: top-down or bottom-up? Eur J Physiol. 2014;466(6):1199–209. doi:10.1007/s00424-014-1471-9.
2. Aebersold R, Mann M. Mass spectrometry-based proteomics. Nature. 2003;422(6928):198–207. doi:10.1038/nature01511.
3. Arrell DK, Neverova I, Van Eyk JE. Cardiovascular proteomics: evolution and potential. Circ Res. 2001;88(8):763–73.
4. Steen H, Mann M. The ABC's (and XYZ's) of peptide sequencing. Nat Rev Mol Cell Biol. 2004;5(9):699–711. doi:10.1038/nrm1468.
5. Zhang H, Ge Y. Comprehensive analysis of protein modifications by top-down mass spectrometry. Circ Cardiovasc Genet. 2011;4(6):711. doi:10.1161/CIRCGENETICS.110.957829.
6. Smith LM, Kelleher NL, Consortium for Top Down P. Proteoform: a single term describing protein complexity. Nat Methods. 2013;10(3):186–7. doi:10.1038/nmeth.2369.
7. Jin W, Brown AT, Murphy AM. Cardiac myofilaments: from proteome to pathophysiology. Proteomics Clin Appl. 2008;2(6):800–10. doi:10.1002/prca.200780075.
8. Siuti N, Kelleher NL. Decoding protein modifications using top-down mass spectrometry. Nat Methods. 2007;4(10):817–21. doi:10.1038/nmeth1097.
9. Zhang J, Dong X, Hacker TA, Ge Y. Deciphering modifications in swine cardiac troponin I by top-down high-resolution tandem mass spectrometry. J Am Soc Mass Spectrom. 2010;21(6):940–8. doi:10.1016/j.jasms.2010.02.005.
10. Zhang J, Guy MJ, Norman HS, Chen YC, Xu Q, Dong X, Guner H, Wang S, Kohmoto T, Young KH, Moss RL, Ge Y. Top-down quantitative proteomics identified phosphorylation of cardiac troponin I as a candidate biomarker for chronic heart failure. J Proteome Res. 2011;10(9):4054–65. doi:10.1021/pr200258m.
11. Peng Y, Gregorich ZR, Valeja SG, Zhang H, Cai W, Chen YC, Guner H, Chen AJ, Schwahn DJ, Hacker TA, Liu X, Ge Y. Top-down proteomics reveals concerted reductions in myofilament and Z-disc protein phosphorylation after acute myocardial infarction. Mol Cell Proteom. 2014;13(10):2752–64. doi:10.1074/mcp.M114.040675.
12. Dong X, Sumandea CA, Chen YC, Garcia-Cazarin ML, Zhang J, Balke CW, Sumandea MP, Ge Y. Augmented phosphorylation of cardiac troponin I in hypertensive heart failure. J Biol Chem. 2012;287(2):848–57. doi:10.1074/jbc.M111.293258.
13. Wilm M, Mann M. Analytical properties of the nanoelectrospray ion source. Anal Chem. 1996;68(1):1–8.
14. Gregorich ZR, Ge Y. Top-down proteomics in health and disease: challenges and opportunities. Proteomics. 2014;14(10):1195–210. doi:10.1002/pmic.201300432.
15. Xiu L, Valeja SG, Alpert AJ, Jin S, Ge Y. Effective protein separation by coupling hydrophobic interaction and reverse phase chromatography for top-down proteomics. Anal Chem. 2014;86(15):7899–906. doi:10.1021/ac501836k.
16. Chang YH, Gregorich ZR, Chen AJ, Hwang L, Guner H, Yu D, Zhang J, Ge Y. New mass-spectrometry-compatible degradable surfactant for tissue proteomics. J Proteome Res. 2015;14(3):1587–99. doi:10.1021/pr5012679.
17. Chen EI, McClatchy D, Park SK, Yates 3rd JR. Comparisons of mass spectrometry compatible surfactants for global analysis of the mammalian brain proteome. Anal Chem. 2008;80(22):8694–701. doi:10.1021/ac800606w.
18. Meng F, Cargile BJ, Patrie SM, Johnson JR, McLoughlin SM, Kelleher NL. Processing complex mixtures of intact proteins for direct analysis by mass spectrometry. Anal Chem. 2002;74(13):2923–9.
19. Whitelegge J, Halgand F, Souda P, Zabrouskov V. Top-down mass spectrometry of integral membrane proteins. Expert Rev Proteomics. 2006;3(6):585–96. doi:10.1586/14789450.3.6.585.
20. Carroll J, Fearnley IM, Walker JE. Definition of the mitochondrial proteome by measurement of molecular masses of membrane proteins. Proc Natl Acad Sci U S A. 2006;103(44):16170–5. doi:10.1073/pnas.0607719103.

21. Laganowsky A, Reading E, Hopper JT, Robinson CV. Mass spectrometry of intact membrane protein complexes. Nat Protoc. 2013;8(4):639–51. doi:10.1038/nprot.2013.024.
22. Valeja SG, Xiu L, Gregorich ZR, Guner H, Jin S, Ge Y. Three dimensional liquid chromatography coupling ion exchange chromatography/hydrophobic interaction chromatography/reverse phase chromatography for effective protein separation in top-down proteomics. Anal Chem. 2015;87(10):5363–71. doi:10.1021/acs.analchem.5b00657.
23. Fridriksson EK, Baird B, McLafferty FW. Electrospray mass spectra from protein electroeluted from sodium dodecylsulfate polyacrylamide gel electrophoresis gels. J Am Soc Mass Spectrom. 1999;10(5):453–5. doi:10.1016/S1044-0305(99)00012-4.
24. Neverova I, Van Eyk JE. Role of chromatographic techniques in proteomic analysis. J Chromatogr B Analyt Technol Biomed Life Sci. 2005;815(1–2):51–63. doi:10.1016/j.jchromb.2004.11.009.
25. Hage DS. Affinity chromatography: a review of clinical applications. Clin Chem. 1999;45(5):593–615.
26. Ayaz-Guner S, Zhang J, Li L, Walker JW, Ge Y. In vivo phosphorylation site mapping in mouse cardiac troponin I by high resolution top-down electron capture dissociation mass spectrometry: Ser22/23 are the only sites basally phosphorylated. Biochemistry. 2009;48(34):8161–70. doi:10.1021/bi900739f.
27. Doucette AA, Tran JC, Wall MJ, Fitzsimmons S. Intact proteome fractionation strategies compatible with mass spectrometry. Expert Rev Proteomics. 2011;8(6):787–800. doi:10.1586/epr.11.67.
28. Chen X, Ge Y. Ultrahigh pressure fast size exclusion chromatography for top-down proteomics. Proteomics. 2013;13(17):2563–6. doi:10.1002/pmic.201200594.
29. Kellie JF, Tran JC, Lee JE, Ahlf DR, Thomas HM, Ntai I, Catherman AD, Durbin KR, Zamdborg L, Vellaichamy A, Thomas PM, Kelleher NL. The emerging process of Top down mass spectrometry for protein analysis: biomarkers, protein-therapeutics, and achieving high throughput. Mol Biosyst. 2010;6(9):1532–9. doi:10.1039/c000896f.
30. Sluyterman LAE. Chromatofocusing – a preparative protein separation method. Trends Biochem Sci. 1982;7(5):168–70. doi:10.1016/0968-0004(82)90129-3.
31. Righetti PG. Isoelectric focusing: theory, methodology, and applications. Amsterdam: Elsevier; 1983.
32. Zhang J, Roth MJ, Chang AN, Plymire DA, Corbett JR, Greenberg BM, Patrie SM. Top-down mass spectrometry on tissue extracts and biofluids with isoelectric focusing and superficially porous silica liquid chromatography. Anal Chem. 2013;85(21):10377–84. doi:10.1021/ac402394w.
33. Tran JC, Doucette AA. Gel-eluted liquid fraction entrapment electrophoresis: an electrophoretic method for broad molecular weight range proteome separation. Anal Chem. 2008;80(5):1568–73. doi:10.1021/ac702197w.
34. Vellaichamy A, Tran JC, Catherman AD, Lee JE, Kellie JF, Sweet SM, Zamdborg L, Thomas PM, Ahlf DR, Durbin KR, Valaskovic GA, Kelleher NL. Size-sorting combined with improved nanocapillary liquid chromatography-mass spectrometry for identification of intact proteins up to 80 kDa. Anal Chem. 2010;82(4):1234–44. doi:10.1021/ac9021083.
35. Tran JC, Zamdborg L, Ahlf DR, Lee JE, Catherman AD, Durbin KR, Tipton JD, Vellaichamy A, Kellie JF, Li M, Wu C, Sweet SM, Early BP, Siuti N, LeDuc RD, Compton PD, Thomas PM, Kelleher NL. Mapping intact protein isoforms in discovery mode using top-down proteomics. Nature. 2011;480(7376):254–8. doi:10.1038/nature10575.
36. Catherman AD, Durbin KR, Ahlf DR, Early BP, Fellers RT, Tran JC, Thomas PM, Kelleher NL. Large-scale top-down proteomics of the human proteome: membrane proteins, mitochondria, and senescence. Mol Cell Proteom. 2013;12(12):3465–73. doi:10.1074/mcp.M113.030114.
37. Ge Y, Lawhorn BG, ElNaggar M, Strauss E, Park JH, Begley TP, McLafferty FW. Top down characterization of larger proteins (45 kDa) by electron capture dissociation mass spectrometry. J Am Chem Soc. 2002;124(4):672–8.
38. Scherperel G, Reid GE. Emerging methods in proteomics: top-down protein characterization by multistage tandem mass spectrometry. Analyst. 2007;132(6):500–6. doi:10.1039/b618499p.
39. Zubarev RA, Kelleher NL, McLafferty FW. Electron capture dissociation of multiply charged protein cations. A nonergodic process. J Am Chem Soc. 1998;120(13):3265–6.

40. Syka JE, Coon JJ, Schroeder MJ, Shabanowitz J, Hunt DF. Peptide and protein sequence analysis by electron transfer dissociation mass spectrometry. Proc Natl Acad Sci U S A. 2004;101(26):9528–33.
41. Iavarone AT, Paech K, Williams ER. Effects of charge state and cationizing agent on the electron capture dissociation of a peptide. Anal Chem. 2004;76(8):2231–8. doi:10.1021/ac035431p.
42. Marshall AG, Hendrickson CL, Jackson GS. Fourier transform ion cyclotron resonance mass spectrometry: a primer. Mass Spectrom Rev. 1998;17(1):1–35. doi:10.1002/(SICI)1098-2787(1998)17:1<1::AID-MAS1>3.0.CO;2-K.
43. Mann M, Kelleher NL. Precision proteomics: the case for high resolution and high mass accuracy. Proc Natl Acad Sci U S A. 2008;105(47):18132–8. doi:10.1073/pnas.0800788105.
44. Makarov A. Electrostatic axially harmonic orbital trapping: a high-performance technique of mass analysis. Anal Chem. 2000;72(6):1156–62.
45. Cotter RJ. Time-of-flight mass-spectrometry – preface. ACS Symp Ser. 1994;549:R9–R12.
46. Coelho Graca D, Lescuyer P, Clerici L, Tsybin YO, Hartmer R, Meyer M, Samii K, Hochstrasser D, Scherl A. Electron transfer dissociation mass spectrometry of hemoglobin on clinical samples. J Am Soc Mass Spectrom. 2012;23(10):1750–6.
47. Wells JM, McLuckey SA. Collision-induced dissociation (CID) of peptides and proteins. Methods Enzymol. 2005;402:148–85. doi:10.1016/S0076-6879(05)02005-7.
48. Olsen JV, Macek B, Lange O, Makarov A, Horning S, Mann M. Higher-energy C-trap dissociation for peptide modification analysis. Nat Methods. 2007;4(9):709–12. doi:10.1038/nmeth1060.
49. Little DP, Speir JP, Senko MW, O'Connor PB, McLafferty FW. Infrared multiphoton dissociation of large multiply charged ions for biomolecule sequencing. Anal Chem. 1994;66(18):2809–15.
50. Chaurand P, Luetzenkirchen F, Spengler B. Peptide and protein identification by matrix-assisted laser desorption ionization (MALDI) and MALDI-post-source decay time-of-flight mass spectrometry. J Am Soc Mass Spectrom. 1999;10(2):91–103. doi:10.1016/S1044-0305(98)00145-7.
51. Shaw JB, Li W, Holden DD, Zhang Y, Griep-Raming J, Fellers RT, Early BP, Thomas PM, Kelleher NL, Brodbelt JS. Complete protein characterization using top-down mass spectrometry and ultraviolet photodissociation. J Am Chem Soc. 2013;135(34):12646–51. doi:10.1021/ja4029654.
52. Bakhtiar R, Guan Z. Electron capture dissociation mass spectrometry in characterization of post-translational modifications. Biochem Biophys Res Commun. 2005;334(1):1–8. doi:10.1016/j.bbrc.2005.05.138.
53. Ge Y, Rybakova IN, Xu Q, Moss RL. Top-down high-resolution mass spectrometry of cardiac myosin binding protein C revealed that truncation alters protein phosphorylation state. Proc Natl Acad Sci U S A. 2009;106(31):12658–63. doi:10.1073/pnas.0813369106.
54. Mao Y, Valeja SG, Rouse JC, Hendrickson CL, Marshall AG. Top-down structural analysis of an intact monoclonal antibody by electron capture dissociation-Fourier transform ion cyclotron resonance-mass spectrometry. Anal Chem. 2013;85(9):4239–46. doi:10.1021/ac303525n.
55. McAlister GC, Phanstiel D, Good DM, Berggren WT, Coon JJ. Implementation of electron-transfer dissociation on a hybrid linear ion trap-orbitrap mass spectrometer. Anal Chem. 2007;79(10):3525–31. doi:10.1021/ac070020k.
56. McLafferty FW, Breuker K, Jin M, Han X, Infusini G, Jiang H, Kong X, Begley TP. Top-down MS, a powerful complement to the high capabilities of proteolysis proteomics. FEBS J. 2007;274(24):6256–68. doi:10.1111/j.1742-4658.2007.06147.x.
57. Han X, Jin M, Breuker K, McLafferty FW. Extending top-down mass spectrometry to proteins with masses greater than 200 kilodaltons. Science. 2006;314(5796):109–12. doi:10.1126/science.1128868.
58. Peng Y, Chen X, Zhang H, Xu Q, Hacker TA, Ge Y. Top-down targeted proteomics for deep sequencing of tropomyosin isoforms. J Proteome Res. 2013;12(1):187–98. doi:10.1021/pr301054n.
59. Compton PD, Zamdborg L, Thomas PM, Kelleher NL. On the scalability and requirements of whole protein mass spectrometry. Anal Chem. 2011;83(17):6868–74. doi:10.1021/ac2010795.

60. Valeja SG, Tipton JD, Emmett MR, Marshall AG. New reagents for enhanced liquid chromatographic separation and charging of intact protein ions for electrospray ionization mass spectrometry. Anal Chem. 2010;82(17):7515–9. doi:10.1021/ac1016858.
61. Miladinovic SM, Fornelli L, Lu Y, Piech KM, Girault HH, Tsybin YO. In-spray supercharging of peptides and proteins in electrospray ionization mass spectrometry. Anal Chem. 2012;84(11):4647–51. doi:10.1021/ac300845n.
62. Kou Q, Wu S, Liu X. A new scoring function for top-down spectral deconvolution. BMC Genomics. 2014;15:1140. doi:10.1186/1471-2164-15-1140.
63. Horn DM, Zubarev RA, McLafferty FW. Automated reduction and interpretation of high resolution electrospray mass spectra of large molecules. J Am Soc Mass Spectrom. 2000;11(4):320–32.
64. Jaitly N, Mayampurath A, Littlefield K, Adkins JN, Anderson GA, Smith RD. Decon2LS: an open-source software package for automated processing and visualization of high resolution mass spectrometry data. BMC Bioinformatics. 2009;10:87. doi:10.1186/1471-2105-10-87.
65. Mayampurath AM, Jaitly N, Purvine SO, Monroe ME, Auberry KJ, Adkins JN, Smith RD. DeconMSn: a software tool for accurate parent ion monoisotopic mass determination for tandem mass spectra. Bioinformatics. 2008;24(7):1021–3. doi:10.1093/bioinformatics/btn063.
66. Guner H, Close PL, Cai W, Zhang H, Peng Y, Gregorich ZR, Ge Y. MASH Suite: a user-friendly and versatile software interface for high-resolution mass spectrometry data interpretation and visualization. J Am Soc Mass Spectrom. 2014;25(3):464–70. doi:10.1007/s13361-013-0789-4.
67. Liu X, Inbar Y, Dorrestein PC, Wynne C, Edwards N, Souda P, Whitelegge JP, Bafna V, Pevzner PA. Deconvolution and database search of complex tandem mass spectra of intact proteins: a combinatorial approach. Mol Cell Proteom. 2010;9(12):2772–82. doi:10.1074/mcp.M110.002766.
68. LeDuc RD, Taylor GK, Kim YB, Januszyk TE, Bynum LH, Sola JV, Garavelli JS, Kelleher NL. ProSight PTM: an integrated environment for protein identification and characterization by top-down mass spectrometry. Nucleic Acids Res. 2004;32(Web server issue):W340–5. doi:10.1093/nar/gkh447.
69. Fellers RT, Greer JB, Early BP, Yu X, LeDuc RD, Kelleher NL, Thomas PM. ProSight Lite: graphical software to analyze top-down mass spectrometry data. Proteomics. 2015;15(7):1235–8. doi:10.1002/pmic.201570050.
70. Tsai YS, Scherl A, Shaw JL, MacKay CL, Shaffer SA, Langridge-Smith PR, Goodlett DR. Precursor ion independent algorithm for top-down shotgun proteomics. J Am Soc Mass Spectrom. 2009;20(11):2154–66. doi:10.1016/j.jasms.2009.07.024.
71. Karabacak NM, Li L, Tiwari A, Hayward LJ, Hong P, Easterling ML, Agar JN. Sensitive and specific identification of wild type and variant proteins from 8 to 669 kDa using top-down mass spectrometry. Mol Cell Proteom. 2009;8(4):846–56. doi:10.1074/mcp.M800099-MCP200.
72. Frank AM, Pesavento JJ, Mizzen CA, Kelleher NL, Pevzner PA. Interpreting top-down mass spectra using spectral alignment. Anal Chem. 2008;80(7):2499–505. doi:10.1021/ac702324u.
73. Liu X, Sirotkin Y, Shen Y, Anderson G, Tsai YS, Ting YS, Goodlett DR, Smith RD, Bafna V, Pevzner PA. Protein identification using top-down. Mol Cell Proteom. 2012;11(6):M111 008524. doi:10.1074/mcp.M111.008524.
74. Cai W, Guner H, Chen AJ, Ayaz-Guner S, Peng Y, Valeja SG, Liu Z, Gregorich Z, Ge Y. MASH Suite Pro: a comprehensive software tool for top-down proteomics. Mole Cell Proteomics. 2016;15(2):703–14. doi: 10.1074/mcp.O115.054387.
75. Edwards RL, Griffiths P, Bunch J, Cooper HJ. Top-down proteomics and direct surface sampling of neonatal dried blood spots: diagnosis of unknown hemoglobin variants. J Am Soc Mass Spectrom. 2012;23(11):1921–30. doi:10.1007/s13361-012-0477-9.
76. Theberge R, Infusini G, Tong W, McComb ME, Costello CE. Top-down analysis of small plasma proteins using an LTQ-orbitrap. Potential for mass spectrometry-based clinical assays for transthyretin and hemoglobin. Int J Mass Spectrom. 2011;300(2–3):130–42. doi:10.1016/j.ijms.2010.08.012.

77. Zabrouskov V, Ge Y, Schwartz J, Walker JW. Unraveling molecular complexity of phosphorylated human cardiac troponin I by top down electron capture dissociation/electron transfer dissociation mass spectrometry. Mol Cell Proteom. 2008;7(10):1838–49. doi:10.1074/mcp.M700524-MCP200.
78. Peng Y, Yu D, Gregorich Z, Chen X, Beyer AM, Gutterman DD, Ge Y. In-depth proteomic analysis of human tropomyosin by top-down mass spectrometry. J Muscle Res Cell Motil. 2013;34(3–4):199–210. doi:10.1007/s10974-013-9352-y.
79. Hwang L, Ayaz-Guner S, Gregorich ZR, Cai W, Valeja SG, Jin S, Ge Y. Specific enrichment of phosphoproteins using functionalized multivalent nanoparticles. J Am Chem Soc. 2015;137(7):2432–5. doi:10.1021/ja511833y.
80. McLuckey SA, Reid GE, Wells JM. Ion parking during ion/ion reactions in electrodynamic ion traps. Anal Chem. 2002;74(2):336–46.
81. Huang TY, McLuckey SA. Top-down protein characterization facilitated by ion/ion reactions on a quadrupole/time of flight platform. Proteomics. 2010;10(20):3577–88. doi:10.1002/pmic.201000187.
82. Copeland O, Nowak KJ, Laing NG, Ravenscroft G, Messer AE, Bayliss CR, Marston SB. Investigation of changes in skeletal muscle alpha-actin expression in normal and pathological human and mouse hearts. J Muscle Res Cell Motil. 2010;31(3):207–14. doi:10.1007/s10974-010-9224-7.
83. Chen YC, Ayaz-Guner S, Peng Y, Lane NM, Locher MR, Kohmoto T, Larsson L, Moss RL, Ge Y. Effective Top-down LC/MS+ method for assessing actin isoforms as a potential cardiac disease marker. Anal Chem. 2015;87(16):8399–406. doi:10.1021/acs.analchem.5b01745.
84. Sze SK, Ge Y, Oh H, McLafferty FW. Top-down mass spectrometry of a 29-kDa protein for characterization of any posttranslational modification to within one residue. Proc Natl Acad Sci U S A. 2002;99(4):1774–9. doi:10.1073/pnas.251691898.
85. Solaro RJ, van der Velden J. Why does troponin I have so many phosphorylation sites? Fact and fancy. J Mol Cell Cardiol. 2010;48(5):810–6. doi:10.1016/j.yjmcc.2010.02.014.
86. Huang L, Wolska BM, Montgomery DE, Burkart EM, Buttrick PM, Solaro RJ. Increased contractility and altered Ca(2+) transients of mouse heart myocytes conditionally expressing PKCbeta. Am J Physiol Cell Physiol. 2001;280(5):C1114–20.
87. Goldspink PH, Montgomery DE, Walker LA, Urboniene D, McKinney RD, Geenen DL, Solaro RJ, Buttrick PM. Protein kinase Cepsilon overexpression alters myofilament properties and composition during the progression of heart failure. Circ Res. 2004;95(4):424–32. doi:10.1161/01.RES.0000138299.85648.92.
88. de Tombe PP. Myosin binding protein C in the heart. Circ Res. 2006;98(10):1234–6. doi:10.1161/01.RES.0000225873.63162.c4.
89. Barefield D, Sadayappan S. Phosphorylation and function of cardiac myosin binding protein-C in health and disease. J Mol Cell Cardiol. 2010;48(5):866–75. doi:10.1016/j.yjmcc.2009.11.014.
90. Gautel M, Zuffardi O, Freiburg A, Labeit S. Phosphorylation switches specific for the cardiac isoform of myosin binding protein-C: a modulator of cardiac contraction? EMBO J. 1995;14(9):1952–60.
91. Borges CR, Oran PE, Buddi S, Jarvis JW, Schaab MR, Rehder DS, Rogers SP, Taylor T, Nelson RW. Building multidimensional biomarker views of type 2 diabetes on the basis of protein microheterogeneity. Clin Chem. 2011;57(5):719–28. doi:10.1373/clinchem.2010.156976.
92. Steen H, Jebanathirajah JA, Rush J, Morrice N, Kirschner MW. Phosphorylation analysis by mass spectrometry: myths, facts, and the consequences for qualitative and quantitative measurements. Mol Cellular Proteom. 2006;5(1):172–81. doi:10.1074/mcp.M500135-MCP200.
93. Gregorich ZR, Peng Y, Lane NM, Wolff JJ, Wang S, Guo W, Guner H, Doop J, Hacker TA, Ge Y. Comprehensive assessment of chamber-specific and transmural heterogeneity in myofilament protein phosphorylation by top-down mass spectrometry. J Mol Cell Cardiol. 2015;87:102–12. doi:10.1016/j.yjmcc.2015.08.007.

Chapter 9
Targeted Proteomics (MRM) in Cardiovascular Research

Genaro Pimienta, Mingguo Xu, Anne M. Murphy, and Genaro Antonio Ramirez-Correa

Abstract Multiple reaction monitoring (MRM) is a powerful proteomic technique that is hypothesis driven. MRM is gaining increasing attention among cardiovascular researchers because it has begun to provide mechanistic insights into heart failure and cardiomyopathies, and promises to develop into assays for clinical management. In this chapter we provide an overlook at the background of MRM considering basic principles of proteomics workflows, instrumentation and basic MRM assay design and compare it to shotgun proteomics. As case studies we highlight how MRM has been successfully employed to quantify myofilament post-translational modifications, more specifically phosphorylation of cardiac Troponin I, myofilament mutants replacement in cardiomyopathies and as a potential biomarker detection pipeline in heart failure.

Keywords Cardiovascular diseases • Targeted proteomics • Multiple reaction monitoring • Biomarker • Diagnosis • Heart failure • Cardiac myofilaments • Post-translational modifications • Phosphorylation • Educational protocol

G. Pimienta, PhD, MSc
Biomedical Innovation Department, Center for Scientific Research and Higher Education at Ensenada, Ensenada, Baja California, Mexico

M. Xu, MD, PhD
Department of Pediatric Cardiology, Shenzhen Children's Hospital, Shen Zhen, Guangdong, China

Division of Cardiology, Department of Pediatrics, Johns Hopkins University School of Medicine, Baltimore, MD, USA

A.M. Murphy, MD • G.A. Ramirez-Correa, MD, PhD (✉)
Division of Cardiology, Department of Pediatrics, Johns Hopkins University School of Medicine, Baltimore, MD, USA
e-mail: gramirez@jhmi.edu; tony.ramirez75@gmail.com

G. Agnetti et al. (eds.), *Manual of Cardiovascular Proteomics*, .
DOI 10.1007/978-3-319-31828-8_9

Brief Introduction to Mass Spectrometry-Based Proteomics

The sequence of the human genome indirectly gave birth to the field of proteomics, a word that first appeared late mid 1990s and refers – in analogy to genomics – to the collective study of all proteins in a system, either a cellular compartment, a cell type or an organism [1]. Liquid chromatography (LC) coupled in-line to electrospray ionization (ESI) and tandem mass spectrometry (MS/MS) – (LC-ESI-MS/MS) – is currently the work-horse in peptide and protein sequencing (Fig. 9.1b) [2]. For technical reasons most proteomics studies are "bottom-up" experiments (Fig. 9.1a). In a bottom-up experiment one or more proteins are first digested, typically with trypsin – the preference for Trypsin in protein sequencing is discussed in detail in the literature [3–5]. Next, tryptic peptide sequences are determined by MS/MS and bioinformatics analysis. Based on the MS/MS data generated, bioinformatics tools aided by predefined gene model databases are used to identify the protein(s) from which the sequenced peptide derived (Fig. 9.1b) [3, 4].

A bottom-up proteomics experiment can be of either two types: discovery (shotgun)- or hypothesis-based (targeted) [6]. The two proteomics strategies are rather complementary and reflect the technological advances in the field – shotgun proteomics appeared first, prompted by the genomics era, but was followed by the emergence of targeted proteomics, which has made proteomics a hypothesis-based experimental tool [6]. As we will explain in more detail below, current targeted proteomics approaches are highly reproducible and quantitative (Figs. 9.2a, b). Discovery based proteome analysis; such as recently reported studies mapping the human proteome have used Orbitrap instruments [7–9].

Shotgun and targeted proteomics experiments each rely on highly sensitive instruments. In a discovery-based proteomics experiment, a fast-scanning linear ion trap quadrupole (LTQ) is coupled to a high-resolution mass analyzer (Orbitrap) [7]. The experiment entails a so-called "survey scan" in "data-dependent mode" in which the mass spectrometer connected in-line to a nano-LC system scans the peptides eluted off the LC capillary column as they are ionized by ESI and selects the most abundant precursor (peptide) ions for mass-to-charge ratio (m/z) determination and downstream collision-induced dissociation (MS/MS) (Fig. 9.3) [6]. In complex samples (e.g. eukaryotic proteomes), the data generated is therefore dominated by the most abundant proteins and is subject to the stochastic nature of chromatographic elution profiles (peak width and height) that can vary in-between technical replicates (Fig. 9.3) [6]. Currently, derivatives of LTQ-Orbitrap instruments are among the best option in discovery-based proteomics, given their high scan rate and high mass resolution and accuracy [10].

Shotgun proteomics due to its discovery power has expanded the repertoire of proteins that can be studied with hypothesis-based biochemistry, which ranges from the traditional western-blots to genetic gene deletion/inactivation experiments [8]. Experimental improvements in sample preparation protocols, quantitation strategies (either metabolically- and chemically-labeled isotopic reference peptide tags, or label-free), and the development of faster-scanning and high mass accuracy mass

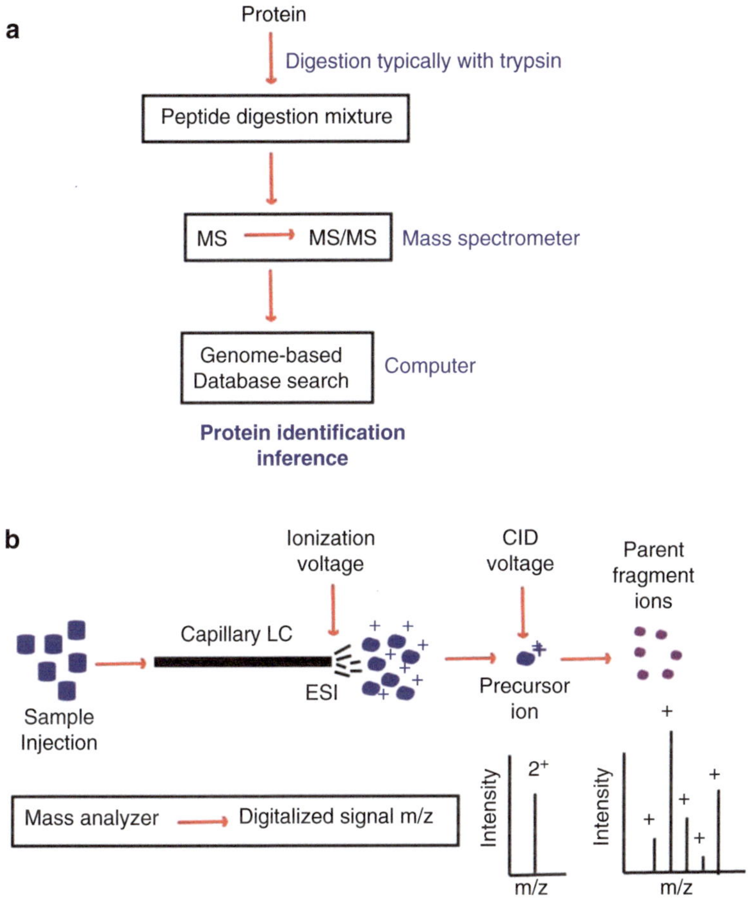

Fig. 9.1 Basic proteomics principles. This figure shows two major features commonly found in a proteomics strategy. Panel (**a**) is a bottom-up proteomics approach summarized as a global workflow, which includes sample processing (trypsinization, desalting), instrument analysis (*MS and MS/MS*) and finally the computer-based protein identification. Emphasis is posed on the need for a specific genome-based database in standard protein identification by mass spectrometry. Panel (**b**) depicts peptide electro spray ionization (*ESI*) technique, where regular peptides acquire charge after the application of current, which is then followed by collision-induced dissociation (*CID*). Here high-energy bombardment is applied to fragment the precursor ions (ionized peptides) into parent fragment ions. This is shown as an instrument standard workflow regardless of the tandem mass spectrometer that would be used to analyze one or more proteins

spectrometers –mainly the ever-advancing series of LTQ-Orbitrap technology – have resulted in the deep proteome analysis of key model systems, especially the small ones like yeast and bacteria; along with human cell lines and tissues [8, 9, 11]. A recent major accomplishment in discovery-based proteomics has been the first draft of the human proteome [8, 9]. Despite these achievements, shotgun proteomics is a global approach that through its discovery power generates hypotheses, but does

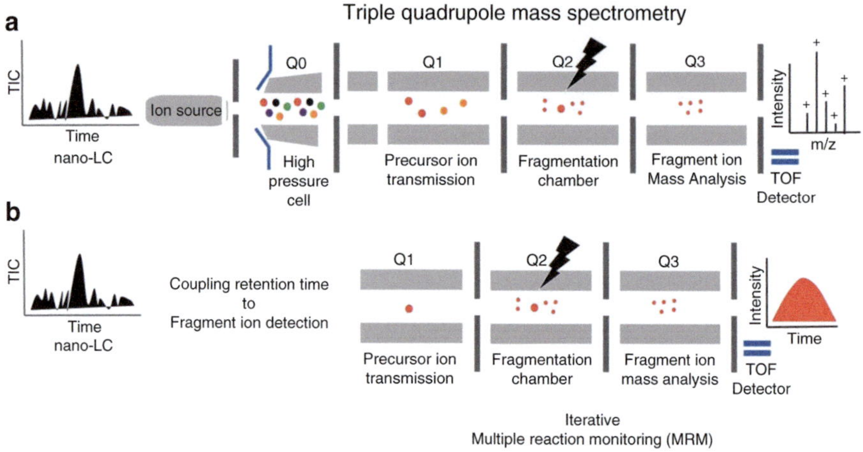

Fig. 9.2 QQQ-TOF mass spectrometer – MRM/SRM. This figure depicts in a simplistic manner the scheme of a typical QQQ-TOF mass spectrometer. Panel (**a**) is a general sketch an instrument in discovery mode, indicating the nano-liquid chromatography that provides the retention or elution time, coupled to an ion source such as electro spray ionization (ESI). Q0 is the high pressure cell from which ions pass a slit on Q1 and is called precursor ion transmission, ions are passed to another slit on Q2 and they are fragmented by high-energy bombardment. Finally, fragmented ions pass the slit on Q3 and the obtained signal is digitalized after (TOF detector) in the form of ion intensities of precursor and fragment (parent) ions as a function of their time of time-of-flight (*TOF*), which is proportional to their mass/charge (m/z) ratio. In Panel (**b**) the same instrument can be used in targeted mode, in this type of analysis there is done by linking precursor ions between their retention time and mass to charge ration (m/z), the first step of selection occurs at the slit of Q1, selected ions undergo collision-induced dissociation (CID) at the fragmentation chamber (Q2) and Q3 further refine the selection to specific ion pairs (transitions). To show how MRM/SRM experiments are performed at the TOF detector the selection analysis is depicted as a function of retention time and fragment ions intensity

not test them [6]. Furthermore, the stochastic nature with which data-dependent mass spectrometry selects the most abundant peptides eluting off a chromatographic LC elution profile makes shotgun proteomics not suitable for quantification. We refer the reader to thorough reviews on the subject [6, 12].

Targeted mass spectrometry has made proteomics a hypothesis-driven high-throughput protein identification and quantitation tool [6]. Originally developed to quantify small molecules, the importance of targeted proteomics in life sciences is underscored by its highlight as the *"Method of the Year"* at the end of 2012 by Nature Methods [6]. The basic experiment in targeted quantitative proteomics is the so-called selected ion-reaction monitoring (SRM) or multiple reaction monitoring (MRM) assay (Fig. 9.4) [12, 13]. We will use the term MRM for the purpose of this chapter. In an MRM experiment, the *"tour de force"* instruments are the derivatives of a triple quadrupole coupled to a time of flight mass analyzer (QQQ-TOF) [12]. In targeted proteomics the instrument operates in a "data-independent mode". This means that instead of stochastically choosing the most intense peptide ions eluting off a capillary LC column, the QQQ-TOF is set to select one or more predetermined

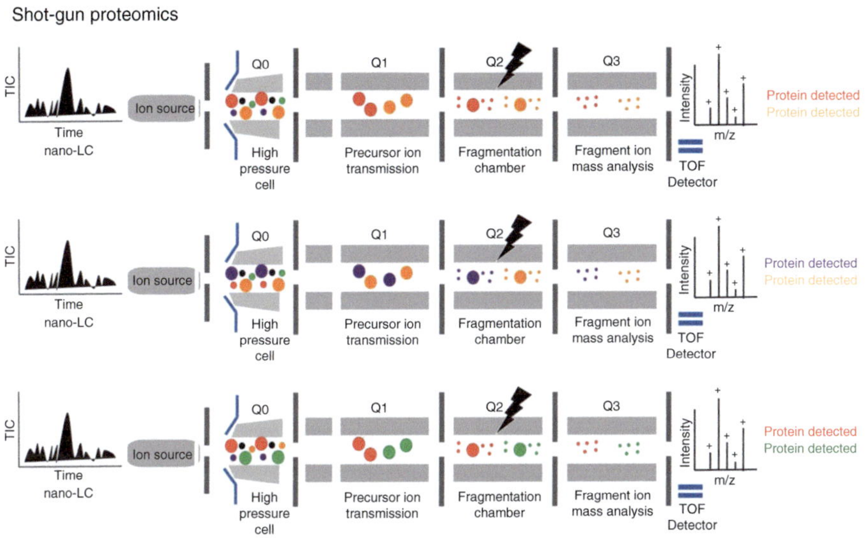

Fig. 9.3 Shot-gun proteomics. QQQ-TOF instruments can also be operated in data-dependent mode or shotgun proteomics; using the same instrument scheme as Fig. 9.2 here we show the stochastic nature of the data obtained. **a**, **b**, and **c** are technical replicates, however due to in-between replicates variability in retention time and peptide ions intensity, which in turn influences the ions that are the most abundant in a data-dependent scan, the proteins identified are not the same. The different proteins are represented in different colors. Protein red is identified in panel **a** and **c**, whereas protein orange is detected in **a** and **b** but not **c**, etcetera

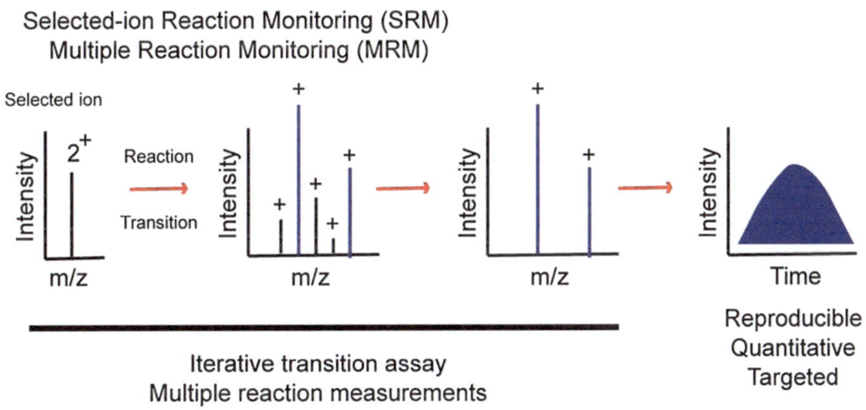

Fig. 9.4 Transition assay. This figure depicts the core principle of targeted proteomics, when using the instrument best suited for this task, a QQQ-TOF in targeted mode or "data-independent mode". A given parent ion (precursor ion) is selected in Q1 from Fig. 9.2b, its fragmentation in Q2 from Fig. 9.2b, will give several product ions that are selected in Q3. A pair of precursor ion and a product ion is called a transition. A repeated measurement of the ion intensities of a particular transition across the chromatographic elution time is a multiple reaction monitoring (*MRM*) or selected-ion reaction monitoring (*SRM*) assay

precursor ions (ionized peptides), based on their carefully calibrated chromatographic elution retention time and expected m/z value. The selected precursor ions are next targeted for downstream fragmentation (MS/MS) and ion intensity quantification [12]. MRM assays have made mass spectrometry-based protein analysis reproducible, truly quantitative (unlike data-dependent discovery mass spectrometry) and highly sensitive. The main applications of targeted proteomics are in systems biology [11] and biomarker validation [13].

Principles of an MRM Experiment

Several reviews and tutorials have been published that provide an overview on use of MRM [14, 15]. During a typical MRM experiment, the first quadrupole (Q1) of a QQQ-TOF is set to monitor one or more pre-defined precursor ions (ionized peptides), each chosen based on its expected chromatographic retention time and m/z values. Each precursor ion is next guided to the second quadrupole (Q2), where the collision-induced dissociation (CID) reaction produces a set of parent ions – the fragment products. Next, more than one representative parent ion produced in Q2 is isolated in the third quadrupole (Q3), where their m/z and ion intensity are accurately measured. The concomitant measurement of m/z as an identity tag and ion intensity of the precursor and parent ions in Q1 and Q3 respectively, is known as the transition.

Measurements of select transitions are the central principle in targeted proteomics that confers a reproducible quantitative dimension (Fig. 9.4). MRM assays involve measurements of transitions that produce a plot composed by the chromatographic retention time of a selected precursor ion and the ion intensities measured in a specific transition assay, given the selected parent ions. An advantage of an MRM assay in data-independent mode over a discovery-based experiment is that the pre-selection of parent ions increases the sensitivity of the experiment and produces a linear response over a wide dynamic range. This allows detection of low abundance proteins in complex mixture samples. In order to provide absolute quantification based on transitions in an MRM experiment, an assays incorporates an isotopically labeled (^{13}C-, ^{15}N-) derivative of each proteotypic precursor ion chosen to be analyzed in a given targeted proteomics experiment. Attention to the proper selection of proteotypic peptide is an important facet of the assay. These peptides must have suitable properties, mainly reproducible ESI ionization, a single prevalent charge state and at least 2 parent ions with good signal [12].

The isobaric properties of each proteotypic peptide and its isotopically labeled reference peptide, implies that if mixed prior to biochemical processing, they will be undistinguishable during the sample preparation process and co-elute in the capillary LC step. Due to their isotopic differences however, they will be sorted out once they ionize by ESI and are analyzed by MS/MS in a QQQ-TOF instrument.

MRM Assay Development

In addition to the excellent methodological papers that address MRM protocols [14, 15], an effort is currently underway to document "good practice" guidelines in the applications of MRM assays to development of clinical assays [16]. Briefly speaking, two critical steps need to be carefully optimized, to reliably discriminate and quantify the protein being analyzed: (a) Selection of two or more so-called proteotypic peptides per protein that reproducibly ionize well by ESI and have an m/z value distinguishable from co-eluting peptides; and (b) Selection of 2–4 parent ions per precursor ion with optimal ion intensity and reproducibility in the transition reaction – typically the most abundant fragment ions [14, 15]. We describe below the steps that should be followed when developing an MRM assay.

Step 1: Proteotypic Peptide Selection

Thorough basic rules have been published for reference peptide selection in targeted proteomics [14, 15, 17]. A proteotypic peptide is the predominant charged state of a precursor ion, that ionizes reproducibly well by ESI, has an average of ten amino acids in length and an amino acid composition not too hydrophilic or hydrophobic [12]. Peptides containing a Pro are preferred due to their typical observed intensity. Missed-trypsin cleavages that contain contiguous Arg/Lys residues are problematic, due to their low abundance and their irregular appearance. The same holds true for peptides that contain amino acids prone to oxidation or deamination reactions, such as Cys, Met and Trp. Following these rules does however not warranty that the peptide selected to work with, will ionize well by ESI. There are three ways of identifying optimal proteotypic peptides that have detectable ion intensity [14, 15]. (a) Experimentally by analyzing the tryptic digest of a pure protein of interest in a QQQ-TOF; (b) Manual inspection of shotgun spectra database, such as SRMAtlas (http://www.srmatlas.org/) [18] and (c) by computational prediction tools, in the case of proteins not documented in shotgun databases [19].

Step 2: Transition Assay Selection

Once two or more different proteotypic peptides per protein have been defined, the next step is to select the transition reactions that behave the best, based primarily on the reproducibility, ion intensity, and m/z value uniqueness of the parental ions per precursor analyzed. As we mentioned above, the combined m/z settings on Q1 (filters a precursor ion) and Q3 (filters 2–4 parent ions per precursor ion) of the QQQ-TOF mass spectrometer are referred to as a transition assay. In other words, a transition assay measures the m/z and ion intensity of the predominant charge of the precursor ion in a narrow mass window, typically 1 Da or narrower. This measurement is coupled to the determination of the m/z and ion intensity of the 2–4 best parent ions derived uniquely from the precursor ion. While tryptic precursor ions commonly have a double or triple charge, CID-derived parent ions are typically single charged. To avoid singly charged background artifacts, typically with a mass smaller than the precursor ion, a good-practice recommendation is to choose parent ions with an m/z larger than the precursor ion. The selection of the transition from precursor ion–to-fragment ion pairs is done experimentally and is usually guided by computational interfaces. A popular one is SpectraST, which is based on a database of data generated by QQQ-TOF instruments [20].

Step 3: Transition Assay Validation

After having identified a potentially well-behaved transition reaction, it is important to verify that the precursor and parent ions analyzed indeed derive from the protein of interest. Verification is important because in a complex biological sample, peptide co-elution is common and may confound the MRM assay, leading to misleading results [14, 15]. The best practice is to incorporate at the beginning or the end of the sample processing protocol an isotopically labeled (^{13}C-, ^{15}N-) derivative of each proteotypic precursor ion analyzed [14, 15]. A complementary option is to acquire a full spectrum of the precursor ions produced in Q2. The information produced by a full spectrum is next used to verify the peptide sequence by means of a computational database search [14].

Step 4: Optimization of the MRM Assay

Once the specificity of the transition assay is determined, peptide ESI ionization (declustering potential/DP) and dissociation (collision energy/CE) voltages should be optimized in order to increase signal sensitivity. It is also important to determine the lower limit of detection (LOD) and quantification (LOQ), both active components of quantitative accuracy in an MRM experiment. LOD and LOQ are highly dependent on the chromatography and instrument used, which have an effect on peak width, ionization and fragmentation quality. To determine the LOD and LOQ parameters per transition assay, a dilution series of isotopically labeled (^{13}C-, ^{15}N-) derivatives of each proteotypic precursor ion are tested. Ideally, this assay is performed in conditions that mimic the experimental matrix the closest – tissue sample, cell lysate, etc. In addition to the protein of interest, approximately housekeeping controls should be included to normalize the values quantified. Skyline [21] or MRMer [22] are freely available software to process data for quantification.

Step 5: Multiplexing MRM Assays

In a multi-protein MRM assay, the number of proteins targeted with enough sensitivity and accuracy is dependent on the MRM cycles and dwell time (reviewed in [14, 15]). A cycle is the amount of time it takes the QQQ-TOF to assay all transitions programed per chromatographic elution peak. The dwell time is the fraction of a second that the instrument spends accumulating data on a given transition every MRM cycle and has to be long enough to achieve sensitivity – enough signal. Longer dwell times increase the signal-to-noise ratio per transition assay, but compromise the number of transitions per protein in each cycle, thus sacrificing specificity. A compromise to optimal cycle and dwell times is the restriction of the retention time peak width analyzed. A good quality transition assay chromatographic trace plot (retention time *vs.* transition assay ion intensity) should have at least 8 quantification events across elution profile [14, 15].

Specialized MRM strategies are optionally used to optimize the experiment, such as parallel transition acquisition and MRM-triggered scanning. These specialized MRM protocols are helpful when analyzing peptides containing post-translational modifications (PTMs). When dealing with PTMs the structure and mass is required in order to trigger full fragmentation spectrum acquisition in an MRM-triggered scanning [14, 15].

MRM in Practice: MRM Assay Design for Use in Human Failing Myocardium

For many applications use of immunoblotting or an established ELISA will be sufficient to determine changes in selected proteins between normal and diseased tissue. However recognizing that the power of MRM based assays could be applied to simultaneously assay for changes in the phosphoproteome of a key myofilament protein our group developed MRM assays to simultaneously assay changes in multiple phosphorylation sites of troponin I (TnI) in heart failure. In this approach, Zhang and colleagues first used an informatics approach to search for every possible phosphorylation site on human cardiac TnI then used LC MS/MS on samples to search for and detect all known TnI sites as well as discovering novel sites based on theoretical phosphorylation motifs [23]. Finally specific MRM assays were developed and tested for each phosphorylation site. Following this, a set of samples from end stage heart failure patients who had hearts explanted at the time of heart transplant were measured using the MRM assays. In order to accurately quantify each site stable isotopically labeled phosphopeptides were utilized to develop standard curves for accurate quantification [23]. Prior physiologic studies provided an understanding of the contribution of many sites to the pathophysiology of altered phosphorylation of the sarcomere in heart failure [24–28], and new studies since publication have elucidated the potential impact of some of some of the novel sites [29, 30].

More recently other groups have begun to use discovery-based proteomics followed by use of MRM assays in plasma related to heart failure. For example, Hollander et al. used a proteomic approach to study a cohort of heart failure patients pre and post heart transplant as well a control healthy cohort and did a comparative analysis of plasma employing iTRAQ mass spectrometry to compare differences. [31]. They then employed MRM assays to further validate protein markers of heart function recovery in a different cohort of heart failure patients who demonstrated recovery after medical therapy compared to a cohort that did not have recovery with treatment. The cohorts were very small and obviously the markers would need further validation in additional larger cohorts. This group also applied a similar approach to examine plasma biomarkers in Andersen-Fabry disease, an X-linked inherited form of cardiomyopathy with differential gender specific presentation [32].

Use of Targeted Proteomic Assays in Hypertrophic Cardiomyopathy

Hypertrophic cardiomyopathy is an autosomal dominant inherited disorder with variable penetrance. In autosomal dominant disorders the mutant allele protein expression may be depressed due to degradation of transcripts with non-sense codons or degradation or diminished sarcomere incorporation of truncated protein. This question has been largely unexplored experimentally, though Helms et al. recently presented data on allele specific protein expression in hypertrophic cardiomyopathy using both immunoblotting and MRM assays to determine relative ratios of normal and mutant protein with synthesized (AQUA) control peptides. This suggested the relative content of the mutant sarcomere protein was variable and in some cases the mutant sarcomere protein was expressed at relatively low levels [33]. Preliminary proof of concept experiments have suggested that MRM assays might permit accurate quantification of the presence of normal and mutant protein in samples from human HCM myocardial tissue with truncating mutations of myosin binding protein C [34].

Application of MRM Assays to Clinical Practice: Workflow

In order to apply MRM assays of proteins or modified sites on proteins several details must be considered in the workflow towards assay development. First, selection of a well-characterized initial cohort of patient samples is crucial. Another crucial factor is establishing a standard operating procedure for sample collection and storage. This may limit use of established banks of samples from previously conducted clinical trials. The next concern relates to the dynamic range of the proteome and relative abundance of potential markers. This is particularly relevant to serum or plasma markers in which low abundance markers may be masked by the high quantity of albumin. Methods have been developed to remove albumin, but a caveat is that some potential biomarkers may be bound to albumin [35]. Discovery cohorts will also need to have an appropriate control group, with consideration for appropriate gender, age and racial/ethnicity matching. Stage of disease of the cohort may also be relevant to design. Prior to expansion to study of a very large population based cohort, it is also necessary to perform assays in secondary validation cohorts to confirm preliminary results.

Regulatory and Practical Issues in Use of MRM Type Assays in Clinical Laboratories

A number of hurdles must be addressed in order to translate targeted proteomic assays from small cohort assays performed for initial proof of principle to biomarker assays suitable for use in a clinical diagnostic assay. Carr et. al reported on a workshop organized by NIH to address this issue [16]. This group distinguished 3 tiers of assays; from the lowest tier 3 being discovery based targeted assays for use in exploratory studies to tier 1 clinical diagnostic assays. Tier 1 assays must have specificity, precision, reproducibility and quantitative accuracy. High levels of quality control are necessary in a diagnostic laboratory setting and internal standards must be employed. Targeted MRM assays have not yet reached this standard and a great deal of product development is necessary to reach this standard.

Future of MRM in Cardiovascular Proteomics

Despite the many challenges outlined in using targeted proteomics in cardiovascular medicine, the application in small pilot clinical or model animal studies is likely to provide insight into cardiovascular disease mechanisms. System based proteomics employing the results of individual analyte assays from proteomics based targeted assays into pathway analysis and transcriptome studies may offer insight into novel disease mechanisms. The power of interfacing genomics with proteomics data was demonstrated by recent maps of the human proteome, which combine genome data, and use of mass spectrometry based tissue profiling [8, 9]. Quantitative data from selected MRM experiments could be integrated into known pathway maps to develop testable hypotheses for ongoing experimentation.

Conclusions and Summary

Targeted proteomics using MRM based assays has begun to provide insight into heart failure and cardiomyopathy. These promising early studies however will need to be further developed in order to be employed in clinical management. This development pathway will involve more robust and larger studies in human cohorts as well as extensive development of assays that are suitable and properly vetted as clinical assays. Despite many gaps that have yet to be addressed, there is great promise for employment of targeted assays to elucidate disease mechanism as well as to serve as clinical/diagnostic tools.

References

1. Wilkins MR, Pasquali C, Appel RD, Ou K, Golaz O, Sanchez JC, Yan JX, Gooley AA, Hughes G, Humphery-Smith I, Williams KL, Hochstrasser DF. From proteins to proteomes: large scale protein identification by two-dimensional electrophoresis and amino acid analysis. Biotechnology (N Y). 1996;14:61–5.
2. Altelaar AF, Munoz J, Heck AJ. Next-generation proteomics: towards an integrative view of proteome dynamics. Nat Rev Genet. 2013;14:35–48.
3. Ahrens CH, Brunner E, Qeli E, Basler K, Aebersold R. Generating and navigating proteome maps using mass spectrometry. Nat Rev Mol Cell Biol. 2010;11:789–801.
4. Johnson RS, Davis MT, Taylor JA, Patterson SD. Informatics for protein identification by mass spectrometry. Methods. 2005;35:223–36.
5. Olsen JV, Ong SE, Mann M. Trypsin cleaves exclusively C-terminal to arginine and lysine residues. Mol Cell Proteomics. 2004;3:608–14.
6. Picotti P, Bodenmiller B, Aebersold R. Proteomics meets the scientific method. Nat Methods. 2013;10:24–7.
7. Eliuk S, Makarov A. Evolution of orbitrap mass spectrometry instrumentation. Annu Rev Anal Chem (Palo Alto Calif). 2015;8:61–80.
8. Kim MS, Pinto SM, Getnet D, Nirujogi RS, Manda SS, Chaerkady R, Madugundu AK, Kelkar DS, Isserlin R, Jain S, Thomas JK, Muthusamy B, Leal-Rojas P, Kumar P, Sahasrabuddhe NA, Balakrishnan L, Advani J, George B, Renuse S, Selvan LD, Patil AH, Nanjappa V, Radhakrishnan A, Prasad S, Subbannayya T, Raju R, Kumar M, Sreenivasamurthy SK, Marimuthu A, Sathe GJ, Chavan S, Datta KK, Subbannayya Y, Sahu A, Yelamanchi SD, Jayaram S, Rajagopalan P, Sharma J, Murthy KR, Syed N, Goel R, Khan AA, Ahmad S, Dey G, Mudgal K, Chatterjee A, Huang TC, Zhong J, Wu X, Shaw PG, Freed D, Zahari MS, Mukherjee KK, Shankar S, Mahadevan A, Lam H, Mitchell CJ, Shankar SK, Satishchandra P, Schroeder JT, Sirdeshmukh R, Maitra A, Leach SD, Drake CG, Halushka MK, Prasad TS, Hruban RH, Kerr CL, Bader GD, Iacobuzio-Donahue CA, Gowda H, Pandey A. A draft map of the human proteome. Nature. 2014;509:575–81.
9. Wilhelm M, Schlegl J, Hahne H, Moghaddas Gholami A, Lieberenz M, Savitski MM, Ziegler E, Butzmann L, Gessulat S, Marx H, Mathieson T, Lemeer S, Schnatbaum K, Reimer U, Wenschuh H, Mollenhauer M, Slotta-Huspenina J, Boese JH, Bantscheff M, Gerstmair A, Faerber F, Kuster B. Mass-spectrometry-based draft of the human proteome. Nature. 2014;509:582–7.
10. Mann M, Kulak NA, Nagaraj N, Cox J. The coming age of complete, accurate, and ubiquitous proteomes. Mol Cell. 2013;49:583–90.
11. Picotti P, Clement-Ziza M, Lam H, Campbell DS, Schmidt A, Deutsch EW, Rost H, Sun Z, Rinner O, Reiter L, Shen Q, Michaelson JJ, Frei A, Alberti S, Kusebauch U, Wollscheid B, Moritz RL, Beyer A, Aebersold R. A complete mass-spectrometric map of the yeast proteome applied to quantitative trait analysis. Nature. 2013;494:266–70.
12. Picotti P, Aebersold R. Selected reaction monitoring-based proteomics: workflows, potential, pitfalls and future directions. Nat Methods. 2012;9:555–66.
13. Chambers AG, Percy AJ, Simon R, Borchers CH. MRM for the verification of cancer biomarker proteins: recent applications to human plasma and serum. Expert Rev Proteomics. 2014;11:137–48.

14. Gallien S, Duriez E, Domon B. Selected reaction monitoring applied to proteomics. J Mass Spectrom. 2011;46:298–312.
15. Lange V, Picotti P, Domon B, Aebersold R. Selected reaction monitoring for quantitative proteomics: a tutorial. Mol Syst Biol. 2008;4:222.
16. Carr SA, Abbatiello SE, Ackermann BL, Borchers C, Domon B, Deutsch EW, Grant RP, Hoofnagle AN, Huttenhain R, Koomen JM, Liebler DC, Liu T, MacLean B, Mani DR, Mansfield E, Neubert H, Paulovich AG, Reiter L, Vitek O, Aebersold R, Anderson L, Bethem R, Blonder J, Boja E, Botelho J, Boyne M, Bradshaw RA, Burlingame AL, Chan D, Keshishian H, Kuhn E, Kinsinger C, Lee JS, Lee SW, Moritz R, Oses-Prieto J, Rifai N, Ritchie J, Rodriguez H, Srinivas PR, Townsend RR, Van Eyk J, Whiteley G, Wiita A, Weintraub S. Targeted peptide measurements in biology and medicine: best practices for mass spectrometry-based assay development using a fit-for-purpose approach. Mol Cell Proteomics. 2014;13:907–17.
17. Bronstrup M. Absolute quantification strategies in proteomics based on mass spectrometry. Expert Rev Proteomics. 2004;1:503–12.
18. Kusebauch U, Deutsch EW, Campbell DS, Sun Z, Farrah T, Moritz RL. Using PeptideAtlas, SRMAtlas, and PASSEL: comprehensive resources for discovery and targeted proteomics. Curr Protoc Bioinformatics. 2014;46:13 25 11–28.
19. Mallick P, Schirle M, Chen SS, Flory MR, Lee H, Martin D, Ranish J, Raught B, Schmitt R, Werner T, Kuster B, Aebersold R. Computational prediction of proteotypic peptides for quantitative proteomics. Nat Biotechnol. 2007;25:125–31.
20. Lam H, Deutsch EW, Eddes JS, Eng JK, Stein SE, Aebersold R. Building consensus spectral libraries for peptide identification in proteomics. Nat Methods. 2008;5:873–5.
21. MacLean B, Tomazela DM, Shulman N, Chambers M, Finney GL, Frewen B, Kern R, Tabb DL, Liebler DC, MacCoss MJ. Skyline: an open source document editor for creating and analyzing targeted proteomics experiments. Bioinformatics. 2010;26:966–8.
22. Martin DB, Holzman T, May D, Peterson A, Eastham A, Eng J, McIntosh M. MRMer, an interactive open source and cross-platform system for data extraction and visualization of multiple reaction monitoring experiments. Mol Cell Proteomics. 2008;7:2270–8.
23. Zhang P, Kirk JA, Ji W, dos Remedios CG, Kass DA, Van Eyk JE, Murphy AM. Multiple reaction monitoring to identify site-specific troponin I phosphorylated residues in the failing human heart. Circulation. 2012;126:1828–37.
24. Bilchick KC, Duncan JG, Ravi R, Takimoto E, Champion HC, Gao WD, Stull LB, Kass DA, Murphy AM. Heart failure-associated alterations in troponin I phosphorylation impair ventricular relaxation-afterload and force-frequency responses and systolic function. Am J Physiol Heart Circ Physiol. 2007;292:H318–25.
25. Ramirez-Correa GA, Cortassa S, Stanley B, Gao WD, Murphy AM. Calcium sensitivity, force frequency relationship and cardiac troponin I: critical role of PKA and PKC phosphorylation sites. J Mol Cell Cardiol. 2010;48:943–53.
26. Takimoto E, Soergel DG, Janssen PM, Stull LB, Kass DA, Murphy AM. Frequency- and afterload-dependent cardiac modulation in vivo by troponin I with constitutively active protein kinase A phosphorylation sites. Circ Res. 2004;94:496–504.
27. Wijnker PJ, Foster DB, Tsao AL, Frazier AH, dos Remedios CG, Murphy AM, Stienen GJ, van der Velden J. Impact of site-specific phosphorylation of protein kinase A sites Ser23 and Ser24 of cardiac troponin I in human cardiomyocytes. Am J Physiol Heart Circ Physiol. 2013;304:H260–8.
28. Wijnker PJ, Sequeira V, Foster DB, Li Y, Dos Remedios CG, Murphy AM, Stienen GJ, van der Velden J. Length-dependent activation is modulated by cardiac troponin I bisphosphorylation at Ser23 and Ser24 but not by Thr143 phosphorylation. Am J Physiol Heart Circ Physiol. 2014;306:H1171–81.
29. Salhi HE, Walton SD, Hassel NC, Brundage EA, de Tombe PP, Janssen PM, Davis JP, Biesiadecki BJ. Cardiac troponin I tyrosine 26 phosphorylation decreases myofilament Ca2+ sensitivity and accelerates deactivation. J Mol Cell Cardiol. 2014;76:257–64.
30. Wijnker PJ, Li Y, Zhang P, Foster DB, dos Remedios C, Van Eyk JE, Stienen GJ, Murphy AM, van der Velden J. A novel phosphorylation site, Serine 199, in the C-terminus of cardiac

troponin I regulates calcium sensitivity and susceptibility to calpain-induced proteolysis. J Mol Cell Cardiol. 2015;82:93–103.

31. Hollander Z, Lazárová M, Lam KK, Ignaszewski A, Oudit GY, Dyck JR, Schreiner G, Pauwels J, Chen V, Cohen Freue GV, Ng RT, Wilson-McManus JE, Balshaw R, Tebbutt SJ, McMaster RW, Keown PA, McManus BM; NCE CECR PROOF Prevention of Organ Failure (PROOF) Centre of Excellence. Proteomic biomarkers of recovered heart function. Eur J Heart Fail. 2014;16(5):551–9. doi: 10.1002/ejhf.65. Epub 2014 Feb 23.
32. Hollander Z, Dai DL, Putko BN, Yogasundaram H, Wilson-McManus JE, Thompson RB, Khan A, West ML, McManus BM, Oudit GY. Gender-specific plasma proteomic biomarkers in patients with Anderson-Fabry disease. Eur J Heart Fail. 2015;17:291–300.
33. Helms AS, Davis FM, Coleman D, Bartolone SN, Glazier AA, Pagani F, Yob JM, Sadayappan S, Pedersen E, Lyons R, Westfall MV, Jones R, Russell MW, Day SM. Sarcomere mutation-specific expression patterns in human hypertrophic cardiomyopathy. Circ Cardiovasc Genet. 2014;7: 434–43.
34. Kim PD, Ramirez-Correa GA, Murphy AM. Truncated protein MyBP-C allelic specific expression in hypertrophic cardiomyopathy by Multiple Reaction Monitoring Mass Spectrometry Eastern Society of Pediatric Research abstract presentation. E-PAS. 2015;418:2015. (http://www.abstracts2view.com/pasall/view.php?nu=PAS15L1_418).
35. Gundry RL, Fu Q, Jelinek CA, Van Eyk JE, Cotter RJ. Investigation of an albumin-enriched fraction of human serum and its albuminome. Proteomics Clin Appl. 2007;1:73–88.

Chapter 10
Label-Free Quantification by Data Independent Acquisition Mass Spectrometry to Map Cardiovascular Proteomes

Sarah J. Parker, Ronald J. Holewinski, Irina Tchernyshyov, Vidya Venkatraman, Laurie Parker, and Jennifer E. Van Eyk

Abstract The large-scale identification and quantification of proteins by liquid chromatography mass spectrometry (LC MS) can be achieved by at least three general methods, categorized into targeted, data independent (DIA), and data dependent (DDA) acquisition modes. Each acquisition strategy has its own set of benefits and drawbacks, and the methods serve complementary purposes for the study of protein quantification in biological samples. While not specific to research in cardiovascular physiology, a long-standing but recently popularized proteomic approach, termed Data Independent Acquisition Mass Spectrometry (DIA-MS), promises unique strengths to complement and extend the existing capabilities of traditional "discovery" proteomic profiling by combining development of a peptide library and DIA-MS. In this chapter we will provide background on the DIA-MS technique, highlighting its fundamental differences relative to other mass spectrometry methods, and discuss important considerations for researchers interested in implementing this technique for their proteomic experiments.

Keywords Data independent acquisition mass spectrometry • Label-free • Quantitation • Peptide library • Targeted analysis • Untargeted analysis • Proteomics

S.J. Parker, PhD (✉) • R.J. Holewinski, PhD • I. Tchernyshyov, MSc • V. Venkatraman, MS
The Advanced Clinical Biosystems Research Institute, Heart Institute,
Cedars Sinai Medical Center, Los Angeles, CA, USA
e-mail: sparker43@gmail.com

L. Parker, PhD
Department of Biochemistry, Molecular Biology, and Biophysics,
University of Minnesota, Minneapolis, MN, USA

J.E. Van Eyk, PhD, FAHA, FISHR
The Advanced Clinical Biosystems Research Institute, Heart Institute,
Cedars-Sinai Medical Center, Los Angeles, CA, USA

Department of Medicine, Johns Hopkins University School of Medicine,
Baltimore, MD, USA

G. Agnetti et al. (eds.), *Manual of Cardiovascular Proteomics*,
DOI 10.1007/978-3-319-31828-8_10

Introduction

The traditional approach for large-scale proteomic screening has been to employ data dependent acquisition mass spectrometry (DDA-MS) also called shotgun MS, to sample peptides derived from a complex protein specimen (e.g., cardiac tissue homogenate) as they elute from a high performance liquid chromatography (HPLC) column in line with the mass spectrometer. In this conventional acquisition method, the mass spectrometer first performs a survey scan (i.e. MS1) to measure the maximum number of peptides possible at a given sampling time point in the chromatogram, and then selects a pre-defined number of those peptide ions for fragmentation (i.e. MS/MS or MS2) which allows for the peptide amino acid sequence to be deciphered by similarity to theoretical spectra predicted from a tryptic digest of the genome/proteome databases and then rolled up to identify and estimate expression of the protein from which they were originally derived (Fig. 10.1). The DDA-MS approach has generated phenomenal advances in the descriptive and quantitative analysis of protein composition in a multitude of biological samples spanning simple single-celled organisms to complex disease tissues. While it has been very successful and established the foundation of discovery proteomics, there are limitations to the DDA-MS approach that continue to hamper its analytical depth and reproducibility.

Several factors contribute to these limitations. The speed at which MS instruments can complete MS/MS scans has been a limiting factor suppressing the number of peptides that will be fragmented and identified within a complex sample. Recent calculations estimate that while over 100,000 different peptide species can be observed in MS1 data from a complex sample, fewer than 20 % of these peptides are generally ever selected for fragmentation and MS/MS analysis to be subsequently sequenced and identified [1]. Further, the randomness inherent in the

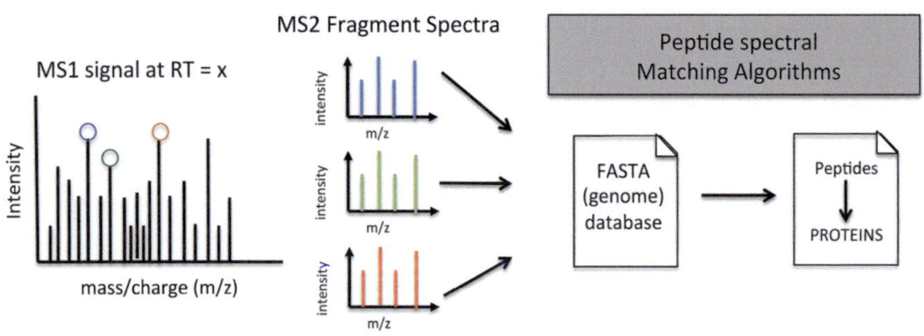

Fig. 10.1 Basic schematic of Data Dependent Acquisition Mass Spectrometry (DDA-MS) approach to proteome screening. In DDA-MS, all co-eluting peptide ions at a given point in the instrument acquisition cycle and HPLC retention time (*RT*) are first surveyed in an MS1 scan, and then a subset (depicted in *blue, orange*, and *green circles*) of ions are selected and shuttled into a collision cell and subsequent detector for fragmentation and MS/MS acquisition. Only the ions selected for fragmentation can be searched against a sequence database to match peptide fragment spectra to likely peptide sequences. Peptide sequences can then be used for protein inferences

stochastic selection of peptides for MS/MS (especially for peptides with relatively lower abundance, despite having adequate signal strength for selection) contributes to up to a 30 % discrepancy in the peptides identified between technical replicates of the exact same complex sample [2]. This becomes a particularly frustrating limitation for comparative quantitative analyses of proteomes between multiple biological conditions, given that in principle these peptides should be detectable, and is especially challenging for large-scale quantitative experiments comparing proteomes across hundreds of biological samples. Even with some of the most elaborate isotopic labeling designs that enable several experimental samples to be mixed and analyzed in the same MS run (e.g. iTRAQ, TMT and multiplexed SILAC strategies), irreproducibility between LC-MS runs across an experiment reduces the number of peptides and proteins that are detected in a sufficient number of biological replicates to be quantitatively compared [3]. Recent advancements in DDA-MS data analysis, especially MS1 filtering of precursor ions with chromatographic alignment to allow extraction of peptide precursor intensities across all replicates of an experimental dataset even in cases where MS/MS data were not triggered in every sample, can substantially improve completeness of the quantitative data matrix [4]. Nevertheless, sampling stochasticity remains a fundamental issue, and especially affects iTRAQ and TMT labeling experiments since the labels enable quantification at the MS/MS level, which requires selection and fragmentation of the labeled peptide ions.

Alternatively to DDA-MS, targeted acquisition MS approaches (e.g. selected reaction monitoring, SRM, or multiple reaction monitoring, MRM) can be employed to maximize sampling of peptide analytes of interest but it is to a limited number of analytes/peptides. In these approaches, the acquisition method is designed to specifically fragment only pre-selected MS1 ions, and to collect MS/MS data on key fragment ions (termed "transitions") as precise reporters of the particular peptides of interest. Targeted acquisition MS can be performed with essentially any type of mass spectrometer, but is best suited to triple quadrupole-type instruments that enable narrow m/z selection of precursor ions via the first quadrupole, fragmentation in the second quadrupole, and m/z selection and measurement of the key fragment ions in the third quadrupole. Because this focus on certain analytes ensures their selection for fragmentation, targeted acquisition MS is not subject to the stochasticity limitations of discovery DDA-MS. It also provides high sensitivity and signal to noise, allowing detection and quantitation of peptide ions at much lower levels than in an untargeted DDA-MS experiment. These two features have made targeted acquisition MS the method most commonly employed for application of proteomics in clinical labs. However, targeted acquisition MS is limited in scope per MS run—the MS cycle time required to select, fragment, and detect fragments from individual peptide analytes, combined with the progression of the chromatographic timescale, mean that at most a few hundred proteins can be quantified in a given targeted run. Compared to the 1000–2000 proteins that can be identified in a well-developed DDA-MS experiment, targeted acquisition MS is much less comprehensive for profiling the breadth of the proteome that should be accessible from a typical sample. Targeted assay methods also require extensive development and optimization for each analyte of interest relative to either DDA-MS or DIA-MS.

As a complementary approach that bridges the gaps between traditional DDA-MS discovery and targeted acquisition MS, emerging analytical and bioinformatic strategies are fueling new enthusiasm for Data Independent Acquisition MS (DIA-MS), due to its powerful capabilities for quantitative unbiased sampling of all observable peptide ions in a sample that facilitates more reproducible detection of peptides across the samples of an experimental dataset. While the implementation of DIA-MS is not a new strategy for proteome quantification, recent analytical approaches have built off of both DDA-MS and targeted-acquisition MS concepts to make the raw data derived from DIA-MS more readily accessible for large-scale quantitative proteome mapping. In this chapter, we will survey the fundamentals of DIA-MS, from design of the acquisition methods to the different strategies for analyzing the data to identify and quantify proteins.

Essentials of Peptide Identification and Quantification by Data Independent Acquisition Mass Spectrometry

DIA-MS involves designing different strategies for both mass spectrometer acquisition and post-acquisition data analysis than either DDA-MS or targeted MS. The use of DIA-MS for proteome profiling is not new to the field [5, 6], however recent developments in analytical strategies, MS instrumentation and software tools [7, 8] are fueling increased interest in the approach. The unique strategy to DIA-MS, is that rather than selecting one peptide at a time for MS/MS analysis, all precursor peptide ions falling within pre-designated m/z windows (e.g., typically in the range of 4–25 m/z) are systematically fragmented as the instrument cycles through a broad m/z range (Fig. 10.2a). Thus, fragment spectra are collected for all observable peptide ions (referred to in some implementations as MS/MS^{all}, Fig. 10.2b), and the analytical challenge requires the de-convolution of the chimeric and complex MS/MS scan data generated from co-eluting peptides that are collectively fragmented within a given mass window at each sampling point across the HPLC gradient (Fig. 10.2b). There are currently two popular approaches for deciphering peptide species in DIA-MS data, categorized simply as targeted analysis, which borrows elements from targeted acquisition MS approaches such as MRM, and untargeted analysis, which incorporates aspects of DDA-MS approaches such as MS/MS spectral matching to proteome database entries.

Targeted Analysis of DIA-MS Data

In its most commonly implemented form, the targeted analysis of DIA-MS data utilizes a 'peptide assay library' to define a set of distinguishing characteristics for a specific peptide (see below) that are then used to extract fragment ion signals from

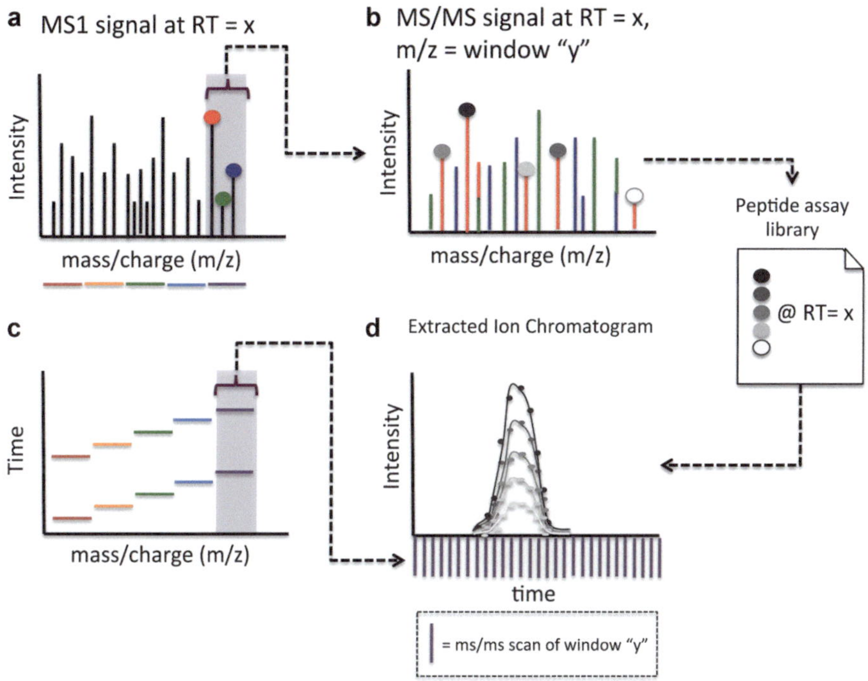

Fig. 10.2 Basic schematic of Data Independent Acquisition Mass Spectrometry (DIA-MS) with targeted data analysis. In DIA-MS, the instrument systematically selects all co-eluting peptide ions within pre-designated mass windows (depicted as *red, orange, green, blue*, and *purple lines* on the x-axis in (**a**) and (**c**). Co-eluting peptide mixtures of a selected mass range are fragmented in the collision cell and detected in complex and chimeric MS/MS spectra (**b**). The instrument systematically cycles through each pre-designated mass window across the entire chromatogram (**c**), with acquisition methods balanced with the HPLC settings to ideally capture a large number of MS/MS observations of a given peptide across its elution profile (**d**). In the targeted analysis strategy, specific peptide fragment masses are queried against the DIA-MS data using a peptide spectral library and an extracted ion chromatogram is generated for a given set of peptide fragments within a designated window of their predicted chromatographic retention time. The resulting 'peak group' extracted ion chromatograms (XICs) are then scored against decoy peak group XICs using various metrics to distinguish confident peptide identifications from false positives

the complex MS/MS^{all} spectra for identification and quantification of peptides and hence, their corresponding proteins. The peptides in the assay library are defined in terms of (1) precursor mass, (2) a handful of characteristic fragment or 'product' masses produced in the collision cell prior to MS/MS detection (i.e., product ions), (3) the expected retention time of the peptide on a given chromatographic set up (i.e., C18 reverse-phase liquid chromatography column), and (4) the expected "fingerprint" of relative intensities of peptide fragments used in the assay as they are generated by a particular mode of fragmentation (e.g., high energy collision [HCD], collision induced dissociation [CID]). An example entry for a peptide from a DIA-MS assay library is shown in Table 10.1.

Table 10.1 Example of selected information typically included in a peptide assay library for DIA-MS experiments

Uniprot ID	Peptide name	Fragment	Precursor Z	Product Z	Precursor Mz	Product Mz	RT_norm	Library intensity
P97351	AC (UniMod:4) QSIYPLHDVFVR	$y9^2$	3	2	568.9557	573.3087	58.2	3100.8
P97351	AC (UniMod:4) QSIYPLHDVFVR	y5	3	1	568.9557	635.3511	58.2	2325.6
P97351	AC (UniMod:4) QSIYPLHDVFVR	y4	3	1	568.9557	520.3242	58.2	2131.8
P97351	AC (UniMod:4) QSIYPLHDVFVR	$y11^2$	3	2	568.9557	673.3668	58.2	1744.2
P97351	AC (UniMod:4) QSIYPLHDVFVR	y6	3	1	568.9557	772.4100	58.2	1550.4

Thus, the peptide assay library defines peptides in terms of precursor ions and specific product ions, which is similar to the strategy used to design MRM assays for quantification of peptides by targeted acquisition MS. A key distinction between MRM and DIA-MS, however, is that since MRM requires targeted acquisition, the data files only contain the precursor/product ion information for the pre-selected peptide analytes of interest—whereas with DIA-MS, thousands of precursor/product ion sets are recorded in the data. The targeting comes after data is obtained and is at the level of the bioinformatics. In discovery mode, the number of molecules targeted is massively expanded but the concepts are the same for 1 or 5000 proteins, and even if an early targeted analysis of the files only involves a limited number of proteins in a peptide assay library, the files can be subsequently re-interrogated at any time with expanded or new peptide assay libraries that incorporate additional precursor/product ion targets [9]. In other words, it is not necessary that the peptide assay library be defined in advance of DIA-MS data generation, and additional peptides/proteins of interest can be added to a peptide assay library and the data reanalyzed later to address new hypotheses, providing unprecedented scope for potential future use of "digitally archived" sample sets or "proteome maps" that can be referred back to in order to ask new biological questions [10].

The end result of targeted extraction of peptide library data from a DIA-MS file is a set of extracted ion chromatograms for the product ions used to define each peptide in the assay library, collectively termed 'peak groups' (Fig. 10.2d). To evaluate and select correct peptide identifications in an automated fashion, the peptide 'peak groups' extracted from a DIA-MS file are scored by software tools according to multiple criteria, and the scores of 'target' peak groups from the library are modeled against the scores generated from a set of 'decoy' peak groups. The peak group scoring and decoy modeling methods differ slightly between the various software packages used for targeted DIA-MS data analysis. A summary of recently released software for DIA-MS is provided in Table 10.2, and a more comprehensive review of DIA-MS software dating back to some of the earliest implementations of the strategy can be found in Bilbao et al. [6].

In general, most scoring criteria include (1) the accuracy of the peptide masses extracted, (2) how close to the predicted retention time the peak group was observed in the chromatographic gradient (peptide retention times), (3) whether the relative intensities of peptide fragments (MS2) extracted match their expected 'fingerprint', (4) whether the apex of each extracted fragment peak lines up with the apices other fragments of that peptide (i.e. precision of apparent co-elution of the peptides and matching transition fragments), and (5) the signal-to-noise ratio of the extracted peaks (ensure peak is above background). Different software programs employ additional scoring metrics designed to confidently assign peptide identity to an extracted peak group. The end product of the targeted DIA-MS analysis is a list comprised of each peptide fragment extracted from the library, along with its extracted intensity value (typically as area under the curve for the extracted ion chromatogram), and information on the false discovery rate (FDR) associated with the peak group

Table 10.2 Selection of recently released software packages for DIA-MS analysis

Software	Availability	Platform	Instrument compatibility	URL
Skyline	Open source	Windows	Any	https://brendanx-uw1.gs.washington.edu/labkey/project/home/software/Skyline/begin.view?
Peak view	Licensed (SCIEX)	Windows	SCIEX TripleTOF	http://sciex.com/products/software/peakview-software
Spectronaut	Licensed[a] (Biognosys)	Windows	Any	https://shop.biognosys.ch/spectronaut
SWATH Prophet	Open source (TPP)	Any	Any	http://tools.proteomecenter.org/wiki/index.php?title=Software:SWATHProphet&redirect=no
Open SWATH	Open source (open MS)	Any[b]	Any	http://www.openswath.org/
DIA Umpire	Open source	Any	Any	http://diaumpire.sourceforge.net/

[a]Free for academic use

[b]Seems to run most efficiently on linux and Apple OS

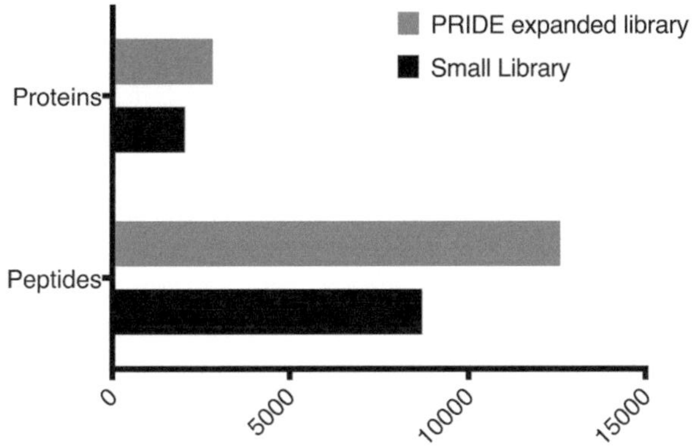

Fig. 10.3 Fractionation and expansion of peptide assay libraries increases depth of DIA-MS analysis. Comparison of peptides and proteins identified in the exact same DIA-MS data file following targeted analysis against a simple library composed of a single DDA-MS injection of cell lysate (small library) versus targeted analysis against an expanded library of the in-house generated DDA-MS file combined with DDA-MS data from fractionated peptide samples that were downloaded from the "PRIDE" mass spectrometry data repository (PRIDE expanded library) (Figure reproduced with permission from Parker et al. [12])

score assigned to a given peptide. As will be discussed below, transition intensity values can then be aggregated to generate peptide level intensities, and peptide intensities aggregated to indicate protein level abundance.

The ability of targeted data analysis to de-convolute and assign amino acid sequences and identities to the full complement of observable peptide ions and fragment species from the complex spectral data generated from DIA-MS depends on having reliable criteria defining the characteristics of the peptides of interest (i.e. the peptide assay library). While the data file from DIA-MS analysis theoretically contains precursor and fragment ion information for all peptides whose intensities fall within the dynamic range of the instrument, their identities will only be accessible if pre-defined fingerprints for extraction are available. Intuitively, the depth of proteome coverage achieved using targeted analysis of DIA-MS data is dependent on the comprehensiveness of the peptide assay library used to extract peptide peak groups for identification. Peptide assay libraries can be obtained by generating them in-house with DDA-MS [11], downloading assay libraries from public data repositories, or a combination of both [12]. Extensive fractionation of complex peptide samples prior to DDA-MS, as well as merging publically available spectral data with locally generated libraries will maximize depth of proteome coverage and will ultimately increase the total number of peptides and proteins quantified in subsequent DIA-MS analyses of unfractionated samples (Fig. 10.3).

There are several other important factors to consider when building peptide assay libraries for targeted DIA-MS analysis, whether from publicly available data or sample-specific, locally generated DDA-MS datasets. For one, it is important that the DDA-MS data used to construct the peptide assay library be generated either on the same type of instrument (and preferably same HPLC method, for instance C18 reverse phase) as for the DIA-MS experiment or at least on an instrument operating in a mode that will generate similar peptide fragmentation patterns to that of the DIA-MS instrument (e.g., Thermo QExactive with HCD fragmentation or SCIEX TripleTOF 5600) [13]. Additionally, extra stringency in verifying the accuracy of peptide spectral matches at the DDA-MS stage should be used so that only the spectral data from high confidence peptide sequence matches are recorded into the peptide assay library. It is also important to normalize (i.e. align) chromatographic retention time (RT) for peptides within the library in a way that allows accurate prediction of peptide elution time in a given DIA-MS experiment regardless of whether the chromatography is matched between library and DIA-MS experiments. Even under the best conditions, peptide retention times are rarely precisely maintained from run to run in LC/MS, especially in nanoLC which is subject to fluid dynamics issues that come from the extraordinarily low flow rates involved. This can create slight retention time differences between replicates, as well as between the original DDA-MS runs for peptide assay library generation and DIA-MS runs, even when performed on the same column and instrument. Retention times in peptide assay libraries can be normalized using either externally spiked reference peptides [14] or internal RT reference peptides such as those identified and used as described in Parker et al. [12]. Finally, while in many cases peptides with post-translational modifications (PTMs) will be readily distinguished in DIA-MS experiments because either their precursor mass is shifted into a different DIA MS1 window or their retention time is shifted substantially enough to be easily identified as a separate analyte, in some circumstances, such as peptides with multiple possible phosphosites or PTMs of very small mass (e.g., the 1 da mass shift due to citrullination), it is necessary to ensure that the y and b ions capable of distinguishing between PTM "isoforms" of a peptide are reliably recorded in the assay library and used for targeted data analysis of the DIA-MS file.

With targeted analysis of DIA-MS data, the extracted peptide identification and quantification is limited in terms of both scale and quality by the composition of the assay library. Larger, more comprehensive libraries can yield greater depth of DIA-MS proteome coverage. As public data repositories become more and more comprehensive, it will also be useful to curate the performance of different fragment ions for DIA-MS analysis in different organisms and tissue types, and identify the best performing fragment sets for a given peptide on a given background proteome. The dependence of targeted DIA-MS analysis on *a priori* knowledge of peptide targets arguably creates a question about whether the technique qualifies as a truly 'discovery' method for proteome screening. However, as discussed above, the capability for iterative re-analysis of DIA-MS data as new targets become of interest or become measurable is a unique strength to this approach, and allows for hypotheses to be generated from initial screens of the data against generic libraries and then follow up hypotheses to be tested with more experiment-specific assay lists without the need to re-acquire the MS data per se (e.g., see Gillet et al, [7]).

Untargeted Analysis of DIA-MS Data

The untargeted analysis of DIA-MS data fundamentally differs from targeted analysis in that no *a priori* established library of peptide targets is required. Instead, patterns of spectral data in the DIA-MS files themselves are used to infer peptide information and match co-eluting sets of precursor and fragment masses to peptide amino acid sequences (Fig. 10.4). Currently, the most commonly used software for the untargeted analysis of DIA-MS data is the open source program "DIA Umpire" [8]. In untargeted analysis by DIA Umpire, a likely peptide ion is first identified from MS1 data based on characteristic profiles of peptide isotope envelopes. The MS1 peak data are then lined up with MS/MS data within an appropriate DIA acquisition window, and all of the fragment masses that co-elute along the precise retention time profile together, as well as with a given MS1 peak, are grouped into 'pseudo spectra' comprising the MS1 and MS/MS information representing a set of potential peptide fragments. These 'pseudo spectra' are then searched against a FASTA database in the same way that a DDA-MS file is searched and scored. The intensity data for peptide sequences matched from 'pseudo-spectra' are then extracted from the DIA-MS file and used to determine the relative abundance of that

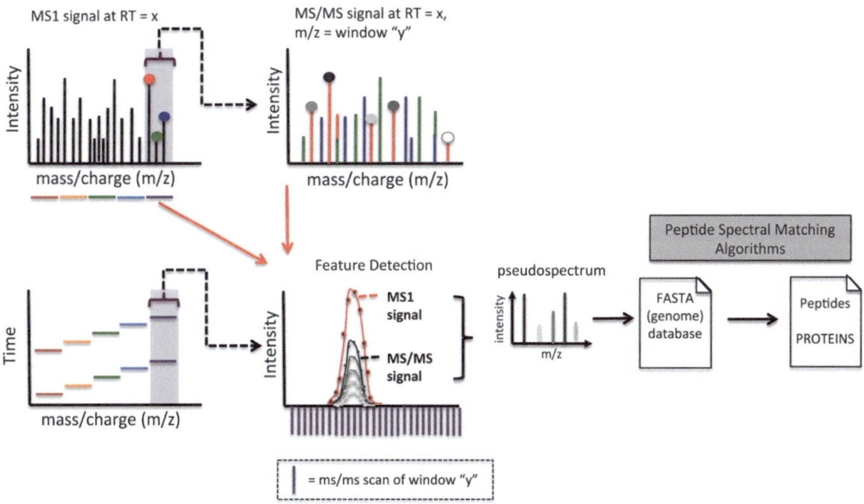

Fig. 10.4 Schematic depicting the general concept behind untargeted analysis of data independent acquisition mass spectrometry data. With untargeted DIA-MS analysis, MS1 and MS/MS spectra are carefully aligned to identify perfectly co-occurring peak groups. These peak groups are then compiled into clean 'pseudo-spectra' that can be searched against sequence databases to match fragments to peptide sequences and their corresponding proteins in the same way DDA-MS data are analyzed

peptide in that sample. Interestingly, the performance of untargeted DIA-MS analysis may be best when the peptide matches from the 'pseudo spectra' are assembled into a peptide assay library and the corresponding peak groups extracted, scored, and quantified in a follow-up targeted re-analysis.

DIA-Umpire and other emerging tools for untargeted DIA data analysis enable systematic acquisition of MS/MS sequence data and subsequent peptide identification without the requirement of a pre-existing assay library, and have extremely powerful implications for discovery-oriented proteomic efforts. Developments in the sensitivity and accuracy of these approaches should continue to improve as DIA-MS proteomic strategies mature over the next several years.

Acquisition Strategies to Maximize Sensitivity and Specificity of Peptide Identification by DIA-MS

DIA-MS performance requires optimization of both data analysis strategies (as discussed above) and acquisition method parameters. In particular, design of appropriate precursor mass window selection parameters is another important factor to consider when optimizing a DIA-MS proteome mapping experiment. The key parameters for consideration are precursor selection window width and ion accumulation times for product ion detection. While smaller window widths and longer dwell times for ion accumulation may be preferable for sensitivity and specificity of the MS/MS signal, this must be balanced against total cycle time to ensure each peptide is observed a sufficient number of times to generate a reliably quantifiable extracted ion chromatogram. Different strategies have emerged to approach an ideal balance between precursor isolation width, ion accumulation time per window, and total cycle time.

One strategy is 'multiplexed acquisition' [15], in which the mass range of potential precursor ions is divided into equally sized small intervals (e.g., 4 m/z). The instrument is set to randomly select a given number of these small mass windows (e.g., 5) to be analyzed in each MS/MS scan. The acquisition method is then designed to scan through the entire mass range within 20 scans, and accumulation times are set to ensure an acceptable total cycle time (e.g., approximately 3–4 s). Thus, while the detector is collecting spectral information on the equivalent amount of ions as would be funneled through a larger, continuous m/z window (e.g., 20 m/z), the randomness and discontinuity of the subset of smaller windows can be harnessed for later mathematical 'de-multiplexing'. The result is that peptide observations are captured for each analyte at a frequency equivalent to that for a simple, continuous and wide (e.g., 20 × 20 m/z) window strategy, but with the capability to mathematically reduce spectral complexity and noise in order to achieve the sensitivity and specificity of substantially smaller precursor window filters (e.g., 4 m/z). The software needed to design a multiplexed DIA-MS acquisition experiment and analyze the resulting data are now built into the open source Skyline program [15, 16].

Another strategy is to employ variable window widths, which aims to optimize sensitivity and specificity of peptide identification by using the density of ions across the m/z range to dictate the size of a precursor isolation window [17]. In this approach, the desired number of isolation windows is typically pre-specified (e.g. 64, 100, etc), and using an initial DDA acquisition file the distribution of precursors masses is determined across the m/z range intended for analysis. The density of peptide ions tends to be highest in particular mass ranges, and those more densely packed m/z ranges are split into much narrower isolation windows than less frequently occurring peptide m/z ranges. However, while increasing the number of windows across an m/z range will enhance specificity of the MS/MS spectra, one must be aware of the effect this will have on data quality and total cycle time as mentioned above. For instance, performing an experiment with 100 variable windows using a cycle time of 30 ms will result in a total cycle time of 3 s, which is sufficient in the majority of cases to give 7–8 points across a peak. However, increasing the number of windows to 200 and keeping the dwell time the same will result in greater specificity but will double the cycle time and reduce the points across the peak to 3–4 per peptide. One can overcome this increase in cycle time by decreasing the dwell time for each window (in this case 15 ms will give a 3 s cycle time for 200 windows) but this will adversely affect the data quality, as 15 ms in most cases will not be sufficient enough time to accumulate enough ions (especially for low abundant peptides) to generate good ion statistics. Therefore, care must be taken in the experimental design of the method, which will also depend on the sample type and its dynamic range and peptide density that is being analyzed and the biological question that is being interrogated.

Proteome-Scale Label Free Quantitation by DIA-MS

Performing DIA-MS on high mass accuracy instruments enables the construction of extracted ion chromatograms (XICs) for both (1) the intact precursor ion prior to fragmentation and (2) the multiple fragments of each peptide analyte generated during MS/MS. The area under the curve of these XICs is the quantitative unit used to estimate peptide and protein abundance in a given sample because it represents the overall intensity of the analyte of interest (Fig. 10.5). The multiple representations of a peptide analyte in the DIA-MS data (MS1 and several MS/MS fragment observations) provide a rich amount of quantitative information for each confidently identified analyte. An ongoing challenge in DIA-MS remains the downstream processing of this quantitative information for optimal estimation of peptide and protein abundances within and between experimental samples. To progress from the simple information of MS/MS signal chromatograms to protein level abundance, several layers of inference must be addressed. Important factors for consideration include (1) normalization of intensity data between separate runs on the mass spectrometer, (2) appropriate aggregation of peptide fragment intensities into an estimate of total peptide intensity, (3) proper aggregation of different peptide intensities

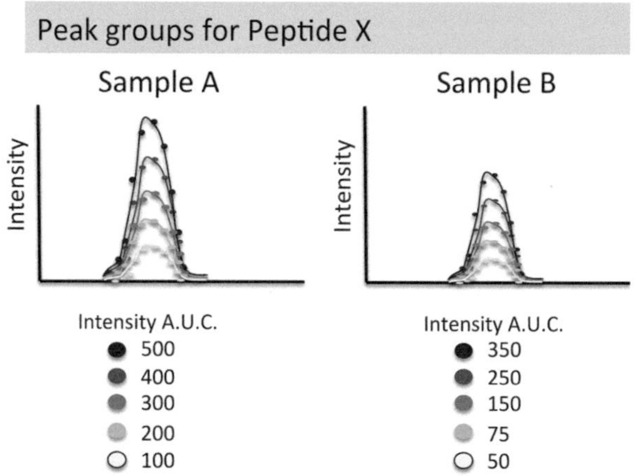

Fig. 10.5 Schematic depicting the peak group extracted ion chromatogram quantification approach in DIA-MS. Chromatograms from each MS/MS fragment of a peptide defined in the assay library are extracted, and the area under the curve (A.U.C.) is calculated as an index of the overall intensity of that fragment in that sample

into an estimate of total protein intensity, [4] statistical modeling of experimental variance in order to identify meaningful differences in protein abundance between experimental groups.

In the final section of this chapter we will summarize some of the issues that a researcher must consider when working through the statistical processing of DIA-MS datasets. Many of these issues are under vigorous debate in the field, and it is likely that no single approach will be appropriate for all experimental designs. To make valid and reliable inferences about the biology in the given system under study, researchers must be intimately familiar with the structure of their data, their experimental design, the sources of variance in their data, and they must also have at least a conceptual understanding of the statistical methods and tools that will be used to calculate and compare protein level abundances.

Data normalization is a critical issue to consider when working with the quantitative output of a DIA-MS experiment both within one run, many runs and between many experiments. Ideally, it is important to know and standardize the mass (i.e. number of total µg) of the complex peptide sample that is loaded onto the LC MS, making sure that an equal amount is loaded for each sample and replicate for any experimental groups within a single data set. However, even when every precaution is taken to ensure equivalent sample mass loaded on column, semi-random fluctuations in instrument performance and other experimental variables can result in inconsistency of overall signal strength from sample to sample – precluding direct, non-normalized comparison of raw intensity values between experimental datasets. Normalization of raw intensity values of each analyte within a sample to a value consistent for the amount of total sample enables a more accurate comparison of relative peptide/protein abundance between samples in the dataset. The different approaches to DIA-MS data normalization include normalization of each analyte

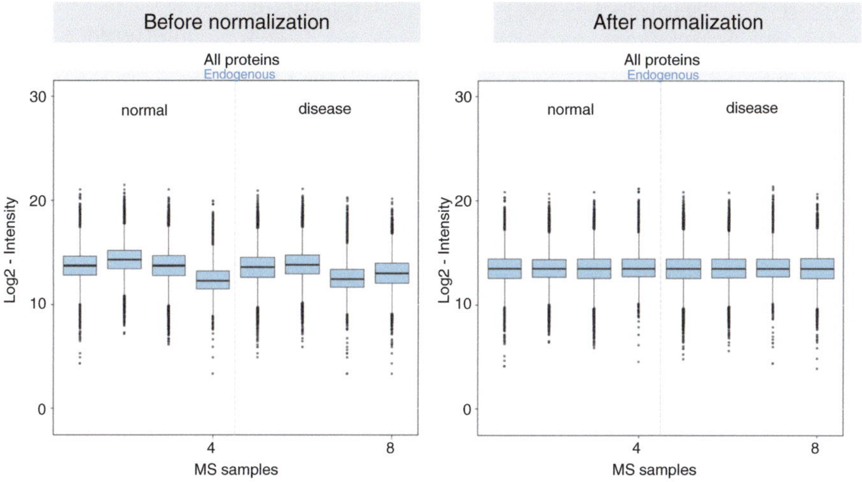

Fig. 10.6 Example demonstrating data normalization in DIA-MS experiments. The overall distribution of protein signal intensities, as calculated from peptide signals, are shown without normalization (*left panel*) and following median normalization (*right panel*)

signal to the median intensity for all analytes of the sample, the 'most likely ratio' normalization that uses fold-change ratios between the replicates and experimental groups of a dataset (see detailed description in [3]), normalization to total signal or total signal in discrete RT windows [18] and normalization to internal standards (e.g., spiked isotopic standards or other reference peptides). Each of these strategies operates under a set of assumptions about the nature of the data, and researchers should carefully consider these assumptions and the validity of a given normalization approach prior to implementing it for their data set. The effectiveness of the normalization should also be verified using quality control plots that visualize signals for all analytes of each sample across the dataset (e.g., Fig. 10.6), ensuring that the overall distributions of normalized analyte signals in each sample are equivalent.

In addition to normalization, another consideration is how to utilize peptide fragment data to best estimate overall peptide intensity within the sample. Current strategies most frequently use the MS/MS signals for quantification of peptides by DIA-MS, because MS/MS data in DIA-MS may be less susceptible to chemical noise and interference (given their somewhat reduced complexity) relative to MS1 signals. This strategy raises the question of how many and which fragment XICs should be used to best approximate peptide intensity for the sample. Short peptide fragments (e.g., y1-3 and b1-3 ions) are often non-specific and subject to interference from other co-eluting peptides, which could skew intensity estimates for a peptide. Accordingly, these more non-specific peptide fragments are usually not ideal for quantification. Random interference in peptide fragment signals can also occur, and is especially problematic if an interfering signal is not uniform across sample replicates and therefore affects the calculated intensity of a peptide fragment

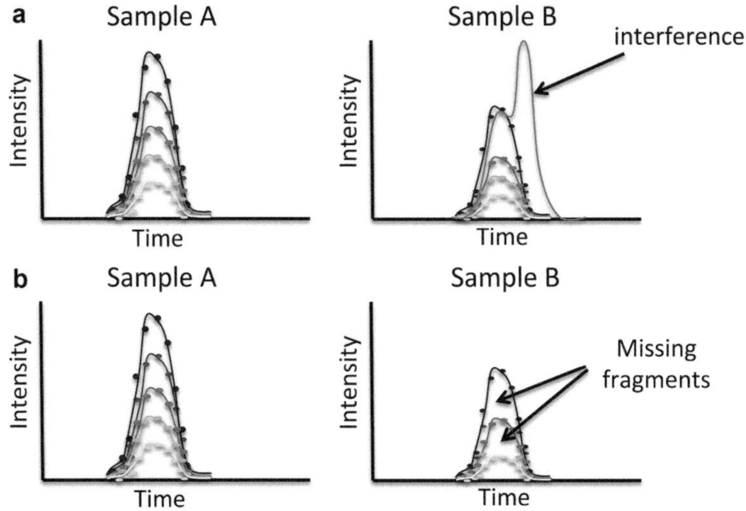

Fig. 10.7 Examples of problematic issues for DIA-MS quantification. Informatic processing of DIA-MS data should include screening for fragment XICs with obvious interferences (**a**) and situations where certain fragments are missing completely due, for instance, to low signal strength or other interference

in one sample disproportionately relative to another (Fig. 10.7a). The simple summing of fragment XIC values to estimate peptide intensity does not account for these interference issues, and could contribute to problematic increases in variance and/or type 1 or type 2 errors when determining proteomic abundance differences between experimental groups. One strategy used to mitigate the influence of random chemical interference is to sum only the top 2 or 3 fragments with the highest intensity values to represent a peptide [19], the assumption being that the highest intensity fragments will be more robust to minor interference than lower intensity fragments. More mathematically sophisticated approaches are also now emerging which include identifying outlier fragments by examining the correlations and intensity pattern between fragments of the same peptide, and either down-weighting or removing outlier fragments in the calculation of overall peptide abundance [18, 20, 21]. Inclusion of the MS1 XIC for the intact peptide can also provide orthogonal quantitative information and aid in the identification and accurate quantification of peptides and their outlier fragments [22].

Recently released statistical processing tools such as the MSSTATS "R" package [20] provide software workflows to process DIA-MS data for peptide or protein-level abundance inference, model random and experimental variance, perform experimental comparisons and significance analysis, and generate graphical visualization of data for further analysis and dissemination. Each offers options for handling some of the major considerations in DIA-MS data analysis discussed above. The comprehensiveness and usability of the "MSSTATS" package is an important advancement for the field, and provides an excellent framework and foundation from which to continue to build and critically test bioinformatic and biostatistics approaches for accurate and reproducible proteome mapping by DIA-MS.

As an additional and exploratory note, the richness of the datasets generated by DIA-MS enables not only careful analysis of fragment data for optimized peptide-level abundance calculation, but also the potential for detecting systematic changes the peptides detected from a given protein. These systematic aberrations may indicate the presence of a novel PTM state or protein isoform in one set of biological samples relative to another. This potential in DIA-MS analysis represents a fascinating capability for proteome analysis and discovery of novel proteoforms unique to specific disease conditions, genetic backgrounds, or biological treatments.

A final issue to consider for quantitative proteomic analysis by DIA-MS is the question of how to select the peptides in the library that will be used for estimating protein level abundance in the full experimental dataset. One possibility is to include all peptides that were observed in a given sample in order to provide protein level intensity, however it may be important to take care to only use peptides observed in all replicates of a dataset for protein level inferences. While in theory the DIA-MS approach results in all observable peptides detected in all replicates of a dataset, in practice issues of signal-to-noise and dynamic range still result in the reliable detection of a peptide in some replicates and/or samples but not others. In these cases it is important to decide whether a potential signal should still be extracted and used as a quantitative value for a peptide even if that peak group failed to generate a score sufficient for confident identification of that peptide in some samples relative to others. This question remains somewhat open in the field and the answer may ultimately depend on the statistical methods used to further process the data. As with the targeted analysis approach, the statistical processing and roll-up of DIA-MS data may borrow somewhat from the targeted field, and strategies for the extremely accurate quantification of peptides and their corresponding proteins by MRM may be appropriately modified and adapted to large-scale screening by DIA-MS. Overall, it should be stated that the approaches and methods for analysis of DIA-MS data are still emerging and rapidly evolving, and require continued development and implementation of powerful algorithms and computational methods to manage variability within the data and accurately capture the similarities and differences in peptide and protein abundance between the biological groups of an experiment.

Summary and Conclusions

Advancements in instrument speed, data acquisition strategies, and data analysis methods are collectively driving the cutting edge towards new boundaries and capabilities for proteome quantification and mapping with mass spectrometry. The recent developments in DIA-MS provide unique advantages that will complement and extend upon the well established and also still rapidly evolving contributions of DDA-MS. Some of the strongest enthusiasm for DIA-MS may be derived from its particular strengths and potential for high-throughput, reproducible and reliable quantitative mapping of proteome data across extremely large numbers of samples. Minimization of instrument-derived stochasticity in peptide identification with DIA-MS over DDA-MS will make even more clear the importance of precision and reproducibility in sample preparation procedures and appropriate experimental design (e.g., blocking, randomization), and researchers setting out to undertake large-scale DIA-MS experiments should consider these factors carefully. Acquisition and analysis methods for DIA-MS proteomics are undergoing constant adaptation and optimization, and this process will continue to improve sensitivity and specificity of the method in coming years. Strategies to summarize the quantitative data extracted by DIA-MS in order to provide the most accurate and informative estimate of peptide and protein abundance are an open area of focus for the field, and there are a number of unanswered questions that must be addressed in ongoing research. Researchers should carefully consider the advantages and limitations of DIA-MS relative to other acquisition strategies and select the most appropriate method to achieve their experimental goals.

References

1. Michalski A, Cox J, Mann M. More than 100,000 detectable peptide species elute in single shotgun proteomics runs but the majority is inaccessible to data-dependent LC-MS/MS. J Proteome Res. 2011;10(4):1785–93.
2. Liu H, Sadygov RG, Yates 3rd JR. A model for random sampling and estimation of relative protein abundance in shotgun proteomics. Anal Chem. 2004;76(14):4193–201.
3. Lambert JP, Ivosev G, Couzens AL, Larsen B, Taipale M, Lin ZY, et al. Mapping differential interactomes by affinity purification coupled with data-independent mass spectrometry acquisition. Nat Methods. 2013;10(12):1239–45. Pubmed Central PMCID: 3882083.
4. Bateman NW, Goulding SP, Shulman NJ, Gadok AK, Szumlinski KK, MacCoss MJ, et al. Maximizing peptide identification events in proteomic workflows using data-dependent acquisition (DDA). Mol Cell Proteomics MCP. 2014;13(1):329–38. Pubmed Central PMCID: 3879624.
5. Venable JD, Dong MQ, Wohlschlegel J, Dillin A, Yates JR. Automated approach for quantitative analysis of complex peptide mixtures from tandem mass spectra. Nat Methods. 2004;1(1):39–45.
6. Bilbao A, Varesio E, Luban J, Strambio-De-Castillia C, Hopfgartner G, Muller M, et al. Processing strategies and software solutions for data-independent acquisition in mass spectrometry. Proteomics. 2015;15(5–6):964–80.
7. Gillet LC, Navarro P, Tate S, Rost H, Selevsek N, Reiter L, et al. Targeted data extraction of the MS/MS spectra generated by data-independent acquisition: a new concept for consistent and accurate proteome analysis. Mol Cell Proteomics MCP. 2012;11(6):O111016717-1-17. Pubmed Central PMCID: 3433915.
8. Tsou CC, Avtonomov D, Larsen B, Tucholska M, Choi H, Gingras AC, et al. DIA-Umpire: comprehensive computational framework for data-independent acquisition proteomics. Nat Methods. 2015;12(3):258–64, 7 p following 64. Pubmed Central PMCID: 4399776.
9. Gillet LC, Navarro P, Tate S, Rost H, Selevsek N, Reiter L, et al. Targeted data extraction of the MS/MS spectra generated by data-independent acquisition: a new concept for consistent and accurate proteome analysis. Mol Cell Proteomics MCP. 2012;11(6):O111 016717. Pubmed Central PMCID: 3433915.
10. Guo T, Kouvonen P, Koh CC, Gillet LC, Wolski WE, Rost HL, et al. Rapid mass spectrometric conversion of tissue biopsy samples into permanent quantitative digital proteome maps. Nat Med. 2015;21(4):407–13. Pubmed Central PMCID: 4390165.

11. Schubert OT, Gillet LC, Collins BC, Navarro P, Rosenberger G, Wolski WE, et al. Building high-quality assay libraries for targeted analysis of SWATH MS data. Nat Protoc. 2015; 10(3):426–41.
12. Parker SJ, Roest H, Rosenberger G, Collins BC, Malstroem L, Amodei D, et al. Identification of a set of conserved eukaryotic internal retention time standards for data-independent acquisition mass spectrometry. Mol Cell Proteomics MCP. 2015;14(10):2800–13.
13. Toprak UH, Gillet LC, Maiolica A, Navarro P, Leitner A, Aebersold R. Conserved peptide fragmentation as a benchmarking tool for mass spectrometers and a discriminating feature for targeted proteomics. Mol Cell Proteomics MCP. 2014;13(8):2056–71. Pubmed Central PMCID: 4125737.
14. Escher C, Reiter L, MacLean B, Ossola R, Herzog F, Chilton J, et al. Using iRT, a normalized retention time for more targeted measurement of peptides. Proteomics. 2012;12(8):1111–21. Pubmed Central PMCID: 3918884.
15. Egertson JD, Kuehn A, Merrihew GE, Bateman NW, MacLean BX, Ting YS, et al. Multiplexed MS/MS for improved data-independent acquisition. Nat Methods. 2013;10(8):744–6. Pubmed Central PMCID: 3881977.
16. MacLean B, Tomazela DM, Shulman N, Chambers M, Finney GL, Frewen B, et al. Skyline: an open source document editor for creating and analyzing targeted proteomics experiments. Bioinformatics. 2010;26(7):966–8. Pubmed Central PMCID: 2844992.
17. Zhang Y, Bilbao A, Bruderer T, Luban J, Strambio-De-Castillia C, Lisacek F, et al. The use of variable Q1 isolation windows improves selectivity in LC-SWATH-MS acquisition. J Proteome Res. 2015;14(10):4359–71.
18. Teo G, Kim S, Tsou CC, Collins B, Gingras AC, Nesvizhskii AI, et al. mapDIA: Preprocessing and statistical analysis of quantitative proteomics data from data independent acquisition mass spectrometry. J Proteomics. 2015;129:108–20. Pubmed Central PMCID: 4630088.
19. Collins BC, Gillet LC, Rosenberger G, Rost HL, Vichalkovski A, Gstaiger M, et al. Quantifying protein interaction dynamics by SWATH mass spectrometry: application to the 14-3-3 system. Nat Methods. 2013;10(12):1246–53.
20. Choi M, Chang CY, Clough T, Broudy D, Killeen T, MacLean B, et al. MSstats: an R package for statistical analysis of quantitative mass spectrometry-based proteomic experiments. Bioinformatics. 2014;30(17):2524–6.
21. Keller A, Bader SL, Shteynberg D, Hood L, Moritz RL. Automated validation of results and removal of fragment Ion interferences in targeted analysis of data-independent acquisition mass spectrometry (MS) using SWATHProphet. Mol Cell Proteomics MCP. 2015;14(5): 1411–8. Pubmed Central PMCID: 4424409.
22. Rardin MJ, Schilling B, Cheng LY, MacLean BX, Sorensen DJ, Sahu AK, et al. MS1 peptide ion intensity chromatograms in MS2 (SWATH) data independent acquisitions. Improving post acquisition analysis of proteomic experiments. Mol Cell Proteomics MCP. 2015;14(9): 2405–19.

Chapter 11
Labeling and Label-Free Shotgun Proteomics Quantification in the Research of Cardiovascular Diseases

Xiaomeng Shen, Shichen Shen, and Jun Qu

Abstract Shotgun proteomics, comprising of labeling and label-free strategies, have expanded rapidly in the past decade and substantially advanced cardiovascular research. Employing liquid chromatography coupled to tandem mass spectrometry as the tool, shotgun proteomics enables unbiased, high-throughput quantitative analysis of proteome-wide alterations in protein abundance, and therefore contributes significantly to the elucidation of the mechanisms underlying cardiovascular diseases (CVD) and the discovery of potential cardiac biomarker candidates. Promising though, further technical improvements are desirable for overcoming technical challenges and ensuring successful application in cardiovascular medicine. Here we elaborate technical aspects of both labeling and label-free shotgun proteomics, including fundamental rationales, limitations, and considerations in CVD proteomics research. Several major challenges and corresponding solutions are also discussed regarding the application of shotgun proteomics in CVD research. Furthermore, we illustrate the applications of shotgun proteomics in cardiovascular research with the emphasis on disease biomarker discovery and mechanism investigation on both subcellular and whole tissue levels, post-translational modification (PTM) characterization.

X. Shen, BS • S. Shen, BS
Department of Biochemistry, University at Buffalo, State University of New York, Buffalo, NY, USA

State of New York Center for Excellence in Bioinformatics and Life Sciences, Buffalo, NY, USA

J. Qu, PhD (✉)
Department of Biochemistry, University at Buffalo, State University of New York, Buffalo, NY, USA

Department of Pharmaceutical Sciences, University at Buffalo, State University of New York, Buffalo, NY, USA

State of New York Center for Excellence in Bioinformatics and Life Sciences, Buffalo, NY, USA
e-mail: junqu@buffalo.edu

G. Agnetti et al. (eds.), *Manual of Cardiovascular Proteomics*,
DOI 10.1007/978-3-319-31828-8_11

Keywords Cardiovascular diseases • Shotgun proteomics • Bottom-up proteomics • LC-MS • Biomarker • Mechanism study • Label-free proteomics • Labeling proteomics • PTM

Introduction: Proteomics and Cardiovascular Disease (CVD)

Cardiovascular disease (CVD) represents one of the top causes for morbidity and mortality among global population, imposing tremendous burdens on public health and medical resources [1]. Under such circumstances, investigations to characterize CVD mechanisms and to discover effective clinical biomarkers for diagnostic, staging and therapeutic purposes are urgently needed, which would significantly ease the difficulties in selection of therapies, patient management, and risk stratification [2]. Conventionally, these studies are performed in a hypothesis-driven paradigm based on predicted or observed targets in clinical or pre-clinical trials, and ligand binding assays (LBA) are often selected as the experimental measure [3]. Despite the great success achieved, such studies turn out to be suboptimal due to their demands for time and labor, as well susceptibility to hard-to-predict biases [4]. In contrast, discovery-based "-omics" methods (e.g. genomics, transcriptomics, proteomics) offer a viable alternative with far more comprehensive and unbiased insights into the molecular basis of cardiac physiological and pathological adaptations in normal and disease conditions, which may significantly promote the research mentioned above. For instance, genomics and transcriptomics approaches have been utilized to generate data with high-throughput and large capacity for CVD research [5]. However, one significant drawback for these strategies is that changes on the messenger RNA level may not be well reflected on the protein level [6, 7], rendering the results obtained inaccurate since proteins are the final executors of biological functions. Proteomics, on the other hand, provides a comprehensive view of proteins on the proteome-wide scale, thus offering vital information on top of genomics and transcriptomics analysis. Furthermore, unlike genomics and transcriptomics, which derive heavily on sequencing, proteomics characterize various facets of a proteome, including protein identify, protein abundance, cellular localization, protein-protein interactions, post-translational modifications (PTM), and protein turnover in specific conditions. While more comprehensive information could be obtained, this also renders proteomics substantially more challenging [8]. In the past decade, to address the technical challenges, new liquid chromatography (LC)-mass spectrometry (MS) technologies, sample preparation techniques and bioinformatics tools have been developed, which remarkably advanced proteomics studies.

Benefiting from the technological advancements, cardiovascular proteomics appears to be a highly dynamic and rapidly developing field, dramatically advancing our knowledge of the complex pathophysiological states in hearts with an enormous potential to rapidly promote thorough understanding of disease mechanisms and to enhance diagnostics. Shotgun proteomics, (i.e. "bottom-up proteomics") the most common type of proteomics strategy employed in cardiovascular proteomics, is conducted by analyzing peptides derived from proteolytic digestion of protein mixtures [9]. Quantitative shotgun proteomics compares the relative protein abundances between case and control proteomes based on quantification at peptide level [10, 11]. This hypothesis-free strategy is frequently applied to identify significantly altered proteins in a system undergoing physiological or pathological changes, which could be either spontaneous or induced. Such a strategy contributes significantly to CVD biomarker discovery and depicts the myocardial "blueprints" underlying pathological conditions such as ischemia and myocardial infarction (MI). Strategies involved in quantitative shotgun proteomics and technical challenges/solutions will be discussed in details in the subsequent sections, including both label-free and labeling methods. The applications of shotgun proteomics in CVD research will also be covered, with an emphasis on the discovery of potential biomarkers, characterization of disease mechanisms in subcellular and whole tissue proteome, and PTM analysis of CVD proteome.

Technical Aspects of Labeled and Label-Free Shotgun Proteomics

In a typical quantitative shotgun proteomics study, multiple proteomes will be compared with each other to identify significantly altered proteins. Identified with little or even no prior knowledge, these proteins could eventually be translated into potential biomarker candidates or key regulatory proteins associated with particular pathophysiological conditions. To this end, researchers have to first determine the quantitative approaches to be used (i.e. labeling or label-free). Peptide samples for LC-MS analysis are then prepared by protein extraction, reduction, alkylation and digestion. Optional protein/peptide fractionation is introduced either before or after protein digestion to increase the depth of the analysis. After LC-MS analysis, peptide/protein identification is fulfilled by comparing the MS/MS spectra generated from fragmented peptides against theoretical MS/MS spectra derived from *in silico* digestion of a specific protein database. Some popular searching engines include SEQUEST, MASCOT, OMSSA, MS-GF+, etc. Quantitative information can be retrieved by inputting the data into specific software depending on the strategies determined at the beginning of the study. A generic workflow of quantitative shotgun proteomics is shown in Fig. 11.1.

Labeling Proteomics

Labeling Strategies

Isotope labeling approaches are ubiquitously employed in quantitative shotgun proteomics. These methods include chemical labeling, isobaric tags labeling, enzymatic labeling and metabolic labeling. Chemical labeling conjugates isotope-coded

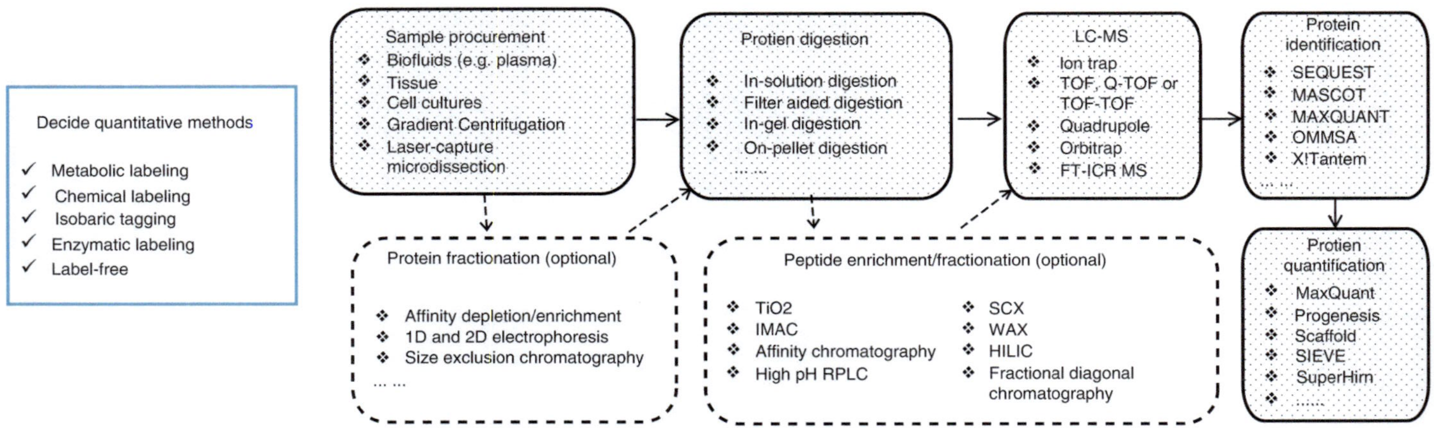

Fig. 11.1 A generic workflow of quantitative proteomics based on shotgun approach. Some examples of techniques are given for each step (Figure is reproduced from Figure 1 of [12])

reagents to reactive groups on the side chains of amino acids or to peptide termini, including isotope-coded affinity tag(ICAT) [13], isotope-coded protein label (ICPL) [14] and dimethyl labeling [15]. Isobaric tags labeling utilizes specifically designed tags comprising of a unique mass reporter, and some examples are Isobaric Tags for Relative and Absolute Quantification(iTRAQ) [16], Tandem Mass Tags(TMT) [17], and Neutron-encoded Mass Signatures(NeuCode) [18]. Enzymatic labeling (i.e. ^{18}O labeling) is a labeling technique in which the carboxyl groups of peptides are labeled with ^{18}O in $H_2{}^{18}O$ [19]. Metabolic labeling methods take advantage of medium or diet enriched in stable isotopes to feed living organisms, in which the isotope-coded amino acids are incorporated into the proteome of organisms via protein synthesis. Two major types of metabolic labeling strategies include Stable Isotope Labeling by Amino Acids in cell culture (SILAC) [20] and Stable Isotope Labeling of Amino acids in Mammals (SILAM) [21]. The conceptual schemes of the four different types of labeling approaches were illustrated in Fig. 11.2. For most of the labeling techniques, the stable isotope labels exhibit considerably similar properties, thus enabling the correction of experimental variability during the labeling step. By sample mixing and prefractionation, in-depth quantitative analysis of samples from multiple conditions could be achieved [22]. Lengqvist et al. have extensively reviewed different types of labeling strategies [23]. In CVD proteomics research, chemical labeling methods are the preferred method owing to their capability of facile labeling of various types of sample proteomes (e.g. tissues and body fluids).

Limitations and Experimental Considerations of Labeling Proteomics

Different labeling methods have their innate pros and cons, depending on the physicochemical features of each type of labeling species. For example, ICAT only reacts with cysteine; hence using ICAT method can significantly reduce sample complexity. However, this strategy eliminates all non-cysteine-containing peptides, which also results in low proteome coverage. Besides, the use of ICAT could cause retention time shift between light and heavy peptides, mainly because of the deuterium labeling [24]. Dimethyl labeling is an easy and less expensive strategy, which introduces 4 Da mass difference by labeling primary amine (i.e. N-terminus and Lys) with dimethyl modification using formaldehyde [15]. Technical hurdles of dimethyl labeling include small mass difference and retention time shift [8]. iTRAQ and TMT compose of three structure elements: a unique mass reporter, a cleavable linker and an amine-reactive group [16, 17]. Labels contain a multiplexed set of reagents with the isobaric masses but different report ions; after labeling, peptides carrying each individual type of label in the mixture represent proteins from samples under certain biological conditions. During MS/MS scan, the linker is fragmented and reporter ions are produced, the intensities of which represent the relative abundance of the peptide from individual samples. To be noticed, these labels reduce data analysis complexity comparing to non-isobaric labeling techniques, since all labeled peptides are co-eluted and co-isolated for MS/MS scan [25]. Both iTRAQ and TMT have relatively large multiplexing capacity (i.e. 8-plex iTRAQ

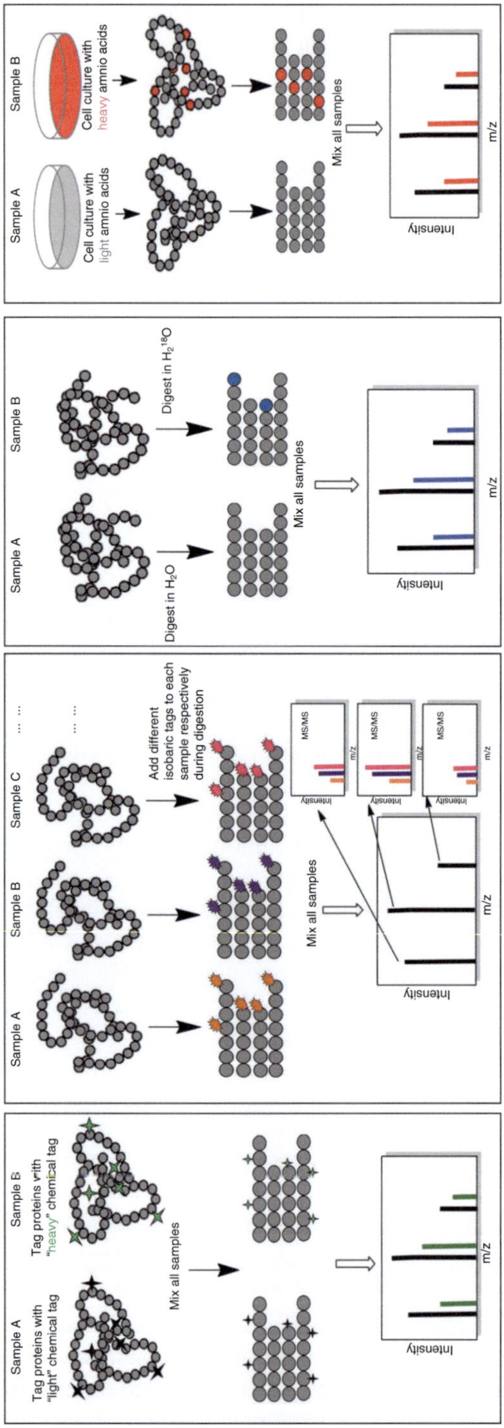

Fig. 11.2 General depiction of various labeling strategies

and 10-plex TMT), which allow the comparison of multiple biological samples simultaneously. So far TMT and iTRAQ are the most commonly used isotope labeling strategies in quantitative proteomics. However, technical drawback still remains for iTRAQ and TMT, including lack of quantification accuracy and interference from coeluting peptides with close m/z [25, 26]. Strategy to overcome these problems includes MS3 analysis, as proposed by Ting et al. [27]. ^{18}O labeling is less widely used probably due to the variability in ^{18}O incorporation and the small mass shift of 2 Da, which may not be enough to separate isotope patterns between the labeled- and non-labeled peptides and thereby rendering data analysis quite challenging [28]. Metabolic labeling is considered to have the highest quantitative accuracy, since the samples can be mixed at protein level prior to digestion, which is likely to introduce the least experimental biases and variation [22]. SILAC and SILAM can be used in cell cultures, model organisms such as mice, drosophila, plants, etc. Nonetheless, metabolic labeling is not as frequently used as the chemical labeling methods, because of the high costs of isotope-enriched reagents and infeasibility of metabolically labeling human subjects. Additionally, as there are very few dividing cell culture systems for ventricular cardiomyocytes, SILAC has limited application in cardiovascular research using cell lines as sample origins [29]. Examples utilizing metabolic labeling for CVD proteomics in animal models include study of cardiac morphogenesis in zebra fish [30] and profiling of mouse heart tissues [31].

Label-Free Proteomics: Ion-Current (IC) and Spectral Counts (SpC)

Quantitative Strategies of Label-Free Proteomics

As an alternative to labeling strategy, label-free approaches abandon sample labeling steps and therefore eliminate several drawbacks of labeling approaches, e.g. cost-prohibitive labeling reagents and uneven labeling efficiency. The fundamental basis of label-free approaches is that LC-MS intensity features and corresponding peptide abundance are linearly correlated [32]. According to the intensity feature chosen for quantification, label-free strategy could be classified into those based on MS1 precursor ion intensity (ion current; IC) [33, 34], those based on the frequency of identified MS2 product ion scans (spectral counting; SpC) [35, 36], and those combining both features [37, 38]. The schematic workflow of label-free quantitative proteomics is shown in Fig. 11.3.

For SpC-based methods, protein quantity is correlated with the frequency of peptide spectral matches (PSM) of a specified protein. Until recently, SpC has been the method-of-choice for label-free approaches in cardiovascular proteomics [39], owing to its conceptual simplicity, easiness in implementation, decent tolerance for variability in sample preparation steps, and suitability for LC-MS analysis under low resolution [40]. Additionally, SpC employs less complicated normalization and

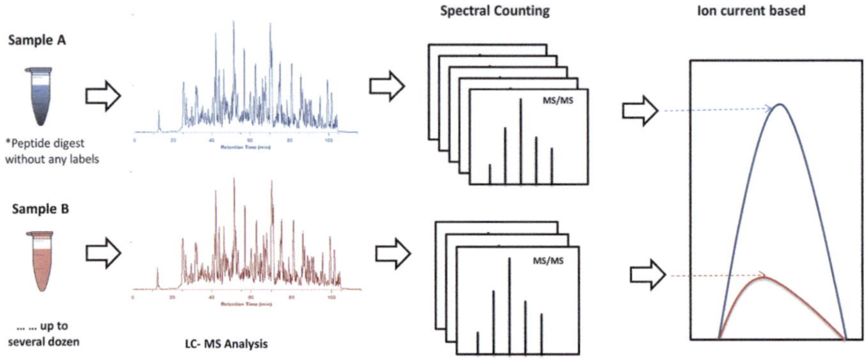

Fig. 11.3 Schematics of label-free quantitative strategies: Spectral Counting (SpC) and ion current(IC) based method. SpC strategy quantifies proteins by frequency of peptide spectral matches of each protein, while IC- based method utilize chromatographic peak areas associated to each protein

statistical analysis than the IC-based approaches [8], e.g. protein abundance index(PAI) [41] and exponentially modified protein abundance index(emPAI) [42] were developed to characterize protein abundance using SpC data. Furthermore, the needs to perform run-to-run alignment of peptide retention time are eliminated for SpC-based methods, which is especially advantageous when the analysis is operated on a low-resolution mass spectrometer or an LC system with poor reproducibility. Consequently, several comparative studies employing low-mass-resolution LC-MS data have made conclusive remarks that for the detection of protein abundance changes in complex samples, SpC surpasses other label-free methods in terms of sensitivity and precision [40, 43].

The procedures of IC-based methods include extraction and integration of all chromatographic peak areas belonging to a given protein [44]. As IC-based methods work more efficiently under high mass resolution, the recent rapid-growing availability of high-resolution MS analyzers such as Fourier Transform-ion Cyclotron Resonance (FTICR) and Orbitrap has dramatically boosted the application of this type of label-free methods [45]. The use of high-resolution analyzers allows extraction of peptide IC within a very narrow m/z range (e.g. <0.02 Th), substantially reducing chemical noises and improving sensitivity as well as specificity of IC-based quantification [46]. When data is acquired from high-resolution LC-MS analysis, IC-based methods are markedly superior to SpC in terms of sensitivity, accuracy, reproducibility and missing data levels [39, 47]. This significant discrepancy can be attributed to several deficiencies of SpC: (1) Dynamic exclusion of precursor ions, a widely practiced feature of mass spectrometers to improve the sequencing of low-abundance peptides, severely undermines quantification by spectral acquisition; (2) MS/MS acquisition of low abundant peptides is often suppressed by peptides of higher abundance; (3) Quantitative counting for lower-abundance proteins/peptides often results in "boundary" counts of "1" and "0", which precludes accurate quantification [43, 48]. In comparison, the IC-based quantification method relies on the measurement of the peak areas of precursor peptides (i.e. using MS1 signals), thus circumventing the above-mentioned problems associated with MS/MS scan.

To be noticed, data-dependent analysis (DDA), in which the MS selects the most abundant ions for product ion scans, are commonly utilized for most MS analysis in label-free proteomics. One limitation of this strategy is that it biases against low abundance peptides, resulting in low dynamic range and inconsistent identification across LC-MS runs [8]. By comparison, data-independent analysis (DIA) acquires MS/MS data based on sequential isolation and fragmentation of defined MS1 window without pre-selection of precursor ions according to their abundance. DIA methods on Q-TOF(i.e. SWATH) [49] or Q-Exactive [50] MS were developed and applied to label-free quantitative proteomics. Though the proteome coverage of DIA methods is to be further improved, the approach constitutes a promising branch of label-free techniques.

Limitations and Experimental Considerations of Label-Free Proteomics

Although the basic idea of label-free strategy is quite straightforward, various factors must be taken into consideration in order to guarantee accurate and precise quantitation. Furthermore, IC-based strategy is technically more demanding than SpC. As no labels are introduced during sample processing and LC-MS analysis, high reproducibility of these steps is required in order for satisfactory outcomes. As recently demonstrated by Nouri-Nigjeh et al., utilization of a long column coupled with large ID trap achieved highly reproducible and comprehensive LC separation, resulting in very small deviation of the retention times of base peaks among 60 consecutive runs (i.e. 4.1–9 % RSD). The high experimental reproducibility notably contributes to the excellent quality of this large-scale label-free proteomics studies [51]. Another important consideration for label-free quantitative proteomics experiments is the development of an optimal data analysis tool for the reliable acquisition of quantitative results. Despite the advances in proteomics software, such a package still remains unavailable. A typical workflow of IC-based proteomics data analysis includes chromatogram alignment, quantitative feature detection and calculation, matching of quantitative features to identified spectrums, normalization, summarization and clean-up. All of these steps require careful evaluation and optimization. Several commercial software packages include Progenesis (Waters), Elucidator (Rosetta Biosoftware) and Expressionist (Genedata). Open-source software packages encompass MaxQuant, OpenMS, SuperHirn, Census and Skyline. For detailed description of proteomics software available, refer to [52].

General Comparison of Labeling and Label-Free Methods

Labeling methods appends universal quantitative references (i.e. internal standards) for all peptide species eligible for labeling and detection, thereby offering accurate quantification of samples involved given that the labeling step is efficient and reproducible enough. The use of different internal standards allows sample mixing and prefractionation/enrichment prior to LC-MS analysis, which considerably reduce sample complexity and dynamic range of protein concentration [23]. Mixing of samples also allow the concomitant analysis of multiple samples in one LC-MS run, significantly shortening the instrument time. Labeling strategy is extremely useful in applications such as PTM quantification, in-depth proteome discovery (e.g. samples subjected to multiple fractionations) and proteome turnover study. However, labeling methods fall short in terms of that it is cost-prohibitive and can only handle small number of replicates due to the limited tag versions [53, 54]. Therefore, labeling proteomics may not be an ideal strategy to investigate a large number of biological samples, as usually required in pharmaceutical and clinical studies. On the other hand, label-free proteomics is capable to deal with a large number of replicates in one set in a cost-effective way and with simpler sample preparations [37, 45], while the data analysis and quality control is challenging, which needs careful optimization and evaluation.

Technical Challenges and Solutions in CVD Proteomics Research

From a technical perspective, an ideal method for shotgun proteomics should provide (*i*) extensive proteomics coverage, (*ii*) achieve protein quantitation with high sensitivity, accuracy and precision, (*iii*) the ability to reliably compare multiple biological replicates without missing values (*iv*) low false-positive biomarker discovery. To meet the abovementioned requirements, researchers need to take specific experimental consideration in the regards of: (1) Sample handling; (2) Sample complexity; (3) Missing value; (4) False positives. This section discusses shotgun proteomics technique with the considerations of above-mentioned aspects, especially experimental procedures applied in CVD proteomics research.

Sample Handling

For shotgun proteomics, exhaustive, efficient and reproducible extraction of proteins from biological samples despite with diverse physical and chemical properties is desired. Preparation of cardiac tissues for proteomics study is particularly challenging, due to the fact that a large portion of myocytes are made up of membranous compartments (e.g. sarcolemma, mitochondrial membrane and sarcoplasmic reticulum), myofilaments and mitochondria [29]. Approximately 80 % of the inner mitochondrial membrane mass is attributed to membrane proteins [55]. Hence, a sample preparation method to effectively disrupt membranous compartments, thoroughly extracting membrane-associating proteins, and adequately cleaning the samples with minimal levels of peptide loss, is urgently needed. The use of strong lysis buffer encompassing several types of detergents to efficiently disrupt membranous compartments and denature proteins has been proved successful in several pieces of cardiovascular proteomics studies [34, 56]. As the importance of subcellular proteomics becomes increasingly recognized in cardiovascular proteomics, the selection of appropriate extraction methods for subcellular compartments becomes a prominent issue. Traditional subcellular isolation technique is based on sucrose gradient sedimentation [57]. Subcellular proteome extraction methods specifically applied to cardiac tissue have also been developed, e.g. the "In sequence" myofilament protein extraction method [58] and the reproducible protocol to purify extracellular matrix from heart tissues [59].

Sample Complexity

To more comprehensively reveal changes in proteome abundance, high proteome coverage, especially for low-abundance proteins playing pivotal regulatory roles in cellular functions, becomes a constant pursuit for shotgun proteomics studies. However, this is always impeded by sample complexity (i.e. the wide dynamic range of protein concentrations in typical proteomes), making the characterization of these low-abundance proteins elusive [46]. For blood plasma and tissues, which are frequently used in cardiovascular proteomics studies, this problem appears to be more pronounced, due to the dynamic range of protein concentrations in these two types of samples (ten orders of magnitude for blood plasma, six for tissues) [60]. This has far exceeded the achievable dynamic range of current shotgun proteomics strategy [61]). To address this problem, protein depletion/equalization and peptide fractionation methods are developed and employed prior to LC-MS/MS analysis to reduce the dynamic range of the proteome [62, 63]. Selective depletion of known high-abundance proteins or equalization of protein concentrations have been demonstrated most useful for plasma or tissue sample analysis. Depletion of abundant proteins such as albumin can be achieved by targeted precipitation using sodium chloride and ethanol [64], disulfide reducing agents [65], or by antibody-based depletion methodologies such as IgY14/Supermix depletion column [66]. Widely practiced though, antibody-based depletion may cause co-depletion of lower-abundance proteins and carry-over [67]; additionally, these methods are often cost-prohibitive and suffer from limited depletion efficiency due to finite antibody binding capacity. Strategy to adjust protein dynamic range using Combinational Peptide Ligand Library(CPLL) serves as a cost-effective approach. CPLL depletes high-abundance proteins while simultaneously enriching low-abundance proteins. The drawback of this method is that a protein must be able to bind one of the ligands in the library in order for retention, which may cause compromised proteome coverage. While these compression procedures of protein concentrations significantly facilitate that reinforcement of protein coverage in blood plasma, quantitative analysis still remains a daunting challenge due to the lingering high complexity and wide dynamic range of the sample [29]. For tissue samples, the dynamic range is also far greater than the achievable one via the majority of quantitative proteomics approaches, while most depletion measures are barely usable [45]. The scenario could be even worse for cardiac tissues, which probably have even wider dynamic range of protein concentrations compared with most of other tissue types [29]. To cope with this problem, peptide fractionation prior to LC-MS/MS analysis is introduced, which could significantly improve comprehensiveness and reduce dynamic range of peptides. Peptides fractionation methods mainly encompass strong anion exchange(SCX) [68], weak anion exchange(WAX) [69], high-pH RPLC separation [70], and zwitterionic(ZIC)-hydrophilic interaction liquid chromatography(HILIC) at median pH [71]. Combination of fractionation separation in multi dimensions has been proved to further enhance proteome coverage [72].

Missing Value

According to literature, shotgun proteomics suffers from poor reproducibility in identification (i.e. overlap of peptide identification across technical replicates is typically <60 %) [73], which leads to high levels of missing values for peptide/protein quantification [38, 50]. As a result, datasets collected from complex biological or clinical samples in a large batch always have high rate of missing data [74]. The missing data represents a major problem for reliable and comprehensive proteomics analysis, especially in large-scale clinical investigations. To meet the increasing needs of comprehensive quantification of proteomes across a large number of biological or clinical replicates, several strategies to address the missing value problem have been proposed: (1) missing value imputation; (2) DIA based strategy; (3) IC-based optimized proteomics workflow. Missing value imputation is a common practice in proteomics data analysis, where researchers use observed values (i.e. lowest or mean) to impute the missing data [75]. Unfortunately, the imputed values do not reflect the real biological variation and may result in inaccurate quantitation. DIA-based method, such as SWATH, is popular, due to its data-independent nature that greatly improve identification reproducibility [49, 50]. However, some limitations of DIA methods also exist: DIA data is more difficult to analyze than other types of proteomics data [76]; a comprehensive spectral library is required. Moreover, the proteomic coverage by DIA is generally poor compared to other methods. IC-based quantitative proteomics workflow employs MS1 signals as quantitative features, which are acquired in a data-independent manner. It has been demonstrated that this strategy has high quantitative reproducibility and very low levels of missing data [39, 45, 51, 77].

False Positives

False-positive discovery of significantly altered proteins is a common yet severe problem for quantitative shotgun proteomics [78], resulting in false biological leads and compromising the validity of the study. Biological and technical variability are the main contributing factors to this problem [78, 79]. Biological variability accumulates when variations among individual subjects are high and/or an insufficient number of biological replicates are employed. This type of variability may be further amplified by pooling all biological replicates for quantification, which is commonly practiced due to technical limitations such as limited multiplexing capacity for labeling techniques, and difficulties in sustaining analytical reproducibility for a long period of time using label-free methods [80, 81]. Technical variability will also exacerbate the false-positives discovery problem via (1) The "multiple hypothesis testing" problems when a large number of inter-group significance test are performed [82] and (2) the susceptibility of sample preparation steps and LC-MS analysis against intrinsic variations and systematic errors, more frequently observed for lower-abundance proteins [79]. Moreover, the noisy signals acquired for lower-abundance peptides may also result in poor reliability for relative quantification [83].

In face of the false-positive discovery issue, various measures have been devised to alleviate biological and technical variability. For instance, the impacts of biological variability may be reduced by involving multiple biological replicates, which can be potentially accommodated by label-free methods as discussed previously. The impacts of technical variability, on the other hand, may be diminished by the introduction of reproducible and quantitative sample preparation/LC-MS procedures, as well as the optimization of data processing procedures. Furthermore, practical measures for the estimation and control of false altered protein discovery rate (FADR) needs to be developed. In shotgun proteomics, the applicability of parametric algorithms such as Benjamini and Hochberg [84] and q-values method [85] are to a large extent limited [82], owing to the fact that the assumptions these algorithms rely on are usually not fulfilled in proteomics data [86, 87]. Evaluation of the FADR empirically enables estimation of false discoveries by experimentally measuring the negative distribution, providing a more reliable alternative to the statistical approaches. Two exemplary FADR control methods utilizing this type of strategy are listed below. An experimental null-based strategy evaluates and controls FADR in proteomics analysis by employing experimental null experiments (i.e. control vs. control experiments), which is described and validated in detail recently [88]. This method provides a straightforward and reliable means for the assessment of false-positive discovery rate, meanwhile facilitating the optimization of proteomic methods for biomarker discovery. The other method estimates FADR by combining the experimentally measured null distributions and Bayesian approach in labeling proteomics strategy [89].

Application of Shotgun Proteomics in CVD Research

Biomarker Discovery in Tissue and Plasma

Feasible to compare protein abundance in multiple proteomes, shotgun proteomics offers a viable approach regarding the discovery of potential biomarker candidates for diagnostic purposes and therapeutic evaluation. Currently there are five established biomarkers used among coronary artery diseases, including (1) cardiac troponin I and T for acute coronary syndromes and myocardial infarction; (2) c-reactive protein for inflammation in atherosclerosis; (3) B-type natriuretic peptides and its

N-terminal form for congestive heart failure. Though fruitful clinical outcomes are obtained from the use of these biomarkers, limitations still remain and hence there are urgent demands to develop novel biomarkers improving risk stratification and management in CVD patient subgroups [90, 91].

Theoretically, samples from human subjects shall be used for biomarker discovery, owing to the fact that the information generated in this way is most clinically-relevant. In reality, however, several major limitations may prevail over the advantages if human samples are selected for biomarker discovery, such as large individual variability [92], as well as difficulties in investigating certain cardiovascular events (e.g. reversible ischemia and hibernating myocardium) in patients [93]. Therefore, several well-established animal disease models serve as the substitutes for human patients in biomarker discovery study. These models include mice [94], rat [95], primate [96] and swine [97].

Tissue Biomarkers

Due to the technical cumbersomeness of plasma proteomics, myocardial tissues are selected as an alternative for biomarker discovery, since proteins may be secreted into circulation under pathophysiological conditions [98]. Myocardial biopsies and post-mortem tissues can both be employed for discovery-based investigation. Hammer et al. characterized inflammatory changes in myocardial biopsies from patients with dilated cardiomyopathy using label-free proteomics [99]. 174 proteins were determined to be altered between 10 patients vs. 7 healthy counterparts, and these proteins were found to be closely involved in mitochondrial and cytoskeleton remodeling. Another label-free analysis was conducted in postmortem myocardial tissues isolated from patients who succumbed to MI [100]. Two proteins, sorbin and SH3 domain-containing protein 2, were proposed to be potential biomarker candidates for early-stage detection of MI. For biomarker discovery using tissues from animal models, there are also a number of reports. For example, Holland et al. investigated the altered proteomes in cardiac tissues from mouse models of Duchenne muscular dystrophy [101]. Using a label-free method, 67 proteins displayed dysregulated levels while drastic changes were observed for 17 proteins, including Ig chains, transferrin, laminin, nidogen and annexin. More recently, proteomics analysis of the ventricle compartment from non-human primates was conducted using TMT labeling and high pH fractionation, in which ~1300 proteins were identified and quantified. A total of 21 proteins showed altered abundance which might be induced by isoproterenol treatment [96].

While in-tissue proteomics analysis has contributed significantly to biomarker discovery, in-plasma analysis may still excel in-tissue one in terms of clinical relevance, since our ultimate objective is to discover circulating disease biomarkers. Therefore, several studies have validated the in-tissue biomarker candidates in plasma samples. For instance, Chugh et al. analyzed tissue samples from a murine model of heart failure for corresponding biomarkers, and 4 potential candidates were verified in both mouse and human blood plasma [102].

Circulating Biomarkers

As mentioned in the technical challenges section, wide dynamic range of protein concentrations [22], as well as the huge biological variability among clinical patients, are the two biggest roadblocks for the discovery of circulating plasma biomarkers. As discussed above, depletion of high-abundance proteins and fractionation techniques can be implemented to alleviate the dynamic range problem. Employing a tandem affinity removal spin column, Jing et al. depleted the top three abundant proteins from plasma, which further led to the identification of several potential biomarker candidates for coronary artery disease in a transgenic mouse model [103]. In order to achieve more comprehensive proteome coverage, the authors also utilized a 2-D LC separation, resulting in the discovery of both known biomarkers and novel candidates [103]. Similar depletion strategy has also been applied for the discovery of biomarker candidates in MI [104] and atherosclerosis patients [105]. Recently, the more powerful IgY14-supermix tandem depletion was introduced, which enabled exhaustive depletion of high- and medium-abundance proteins. This strategy was incorporated into the proteomics workflow by Juhasz et al. used to screen predictive markers for near-term MI [106]. With the established workflow, decent analytical reproducibility as well as the capability of quantifying trivial changes of low-abundance proteins was realized. Chromatographical fractionation can also be applied to reduce the dynamic range of the plasma sample, as reviewed in the technical section. For instance, using combined fractional diagonal chromatography, Mebazaa et al. identified quiescin Q6 as a candidate biomarker for acute heart failure [107].

As an alternative to using whole plasma samples, plasma-derived microvesicles (i.e. microparticle and exosomes) are also an excellent source for the discovery of disease biomarkers [108], in which the problems associated with whole plasma samples could be avoided. Microvesicles enter circulatory system after their release in response to various types of diseases, and mediate cell-to-cell communication under physicochemical conditions [108]. Microvesicles can be enriched from plasma by sequential centrifugation [109]. Proteomics profiling studies on plasma-derived microvesicles have led to the discovery of promising biomarkers implicated in diagnosis of abdominal aortic aneurysms [110], clinically manifest vascular disease [109] and lacunar Infarction [111]. The proteomics strategies employed in these studies are either label-free or iTRAQ.

Mechanism Characterization in Subcellular and Whole Tissue Proteomes

Subcellular Proteomics

In cardiovascular research, shotgun proteomics presents a comprehensive bird's-eye view of the cardiovascular system, which prominently advances our understanding of cardiac pathology, diagnosis, and disease staging. This would be extremely beneficial for the development of therapies against CVD. Previously, shotgun proteomics has been employed for the mechanism interrogation of a number of CVD, e.g. irreversible ischemia [112], diabetic cardiomyopathy [113, 114] congestive heart failure [115], hibernating myocardium [34, 56], aging aorta [116] and cardiac amyloid deposition [117].

Proteomics characterization of subcellular compartments, such as mitochondrion, proteasome, extracellular matrix and myofilaments, has always been of keen interests to researchers given the critical biological relevance underlying pathophysiological changes in these compartments, as well as the sample complexity of the whole tissue proteome. As mentioned above, methods to isolate subcellular fractions include differential centrifugation, immune-based isolation and membrane protein enrichment [118]. Using differential centrifugation and organelle enrichment, Warren et al. successfully isolated proteins from several subcellular compartments (e.g. mitochondria, nuclei, cytoplasmic fractions, microsomes and sarcomere) in rat hearts subjected to regional ischemia, and performed further quantitative analysis of these subcellular proteomes. Based on the results, they concluded that a four-fold increase of the analytical depth was achieved by applying fractionation [119]. Several widely investigated subcellular proteomes in cardiovascular research mainly include mitochondria, extracellular matrix (ECM), and proteasomes, while less-extensively studied ones include myofilaments [120] and membrane proteins [121].

Mitochondrial Proteome Comparative profiling of the mitochondrial proteome represents one of the most popular areas in cardiovascular proteomics research for the last decade [122]. Accounting for about 40 % of the volume of a healthy cardiomyocyte, mitochondria is responsible for ATP generation for contractile function and regulation of programmed cell death (i.e. apoptosis). Accumulating evidences have demonstrated that CVD progression is closely correlated with altered structural and functional features of the mitochondria [123, 124]. Using modified SpC-based method, Zhang et al. quantified protein abundance changes in the mitochondrial proteome subjected to ischemic stress [112]. More recently, Qu et al. performed a comparative study of the mitochondrial proteomes from swine models with hibernating myocardium vs. sham in a relatively large sample capacity ($n=12$ per group). Both IC-based label-free proteomics and 2D-electrophoresis were employed in this study, and it turned out that the results from the two methods correlated well while IC-based label-free method achieved much better proteome coverage [56]. Labeling methods have also been widely utilized. For instance, Julling et al. used iTRAQ to quantitatively characterize the cardiac mitochondrial proteome in streptozotocin-diabetic rats [125] and aged spontaneously hypertensive rats [80]. Mechanisms of type I [113] and type II diabetic cardiomyopathy [114], pressure overload-induced heart failure [115], and atrial fibrillation [126] have also been explored by the application of comparative mitochondrial proteomes. Alterations in mitochondrial proteome elicited by systemic perturbations have been reported in estrogen deficiency [127], diabetic hearts with mPHGPx overexpression [81] and GSK inhibition [128]. Moreover, Kim et al. have studied the dynamics of mitochondrial proteome by supplying mice with deuterium-labeled water. According to their results, the turnover of 458 mitochondrial proteins was characterized in total [129].

Extracellular Matrix (ECM) Proteome The extracellular matrix (ECM) proteome is another extensively-studied subcellular proteome. ECM has emerged as a dynamic component, providing structural support and participating in signaling transductions involved in cardiac injury and remodeling [130]. Investigation of cardiac ECM remains challenging because of the poor solubility of ECM proteins, unpredictable PTMs, propensity to form protein complexes and relatively low abundance of proteins [130]. Current studies attempting to investigate ECM proteome mainly include: (1) ECM proteins isolated or enriched from tissues; (2) cell culture-derived ECM components from cardiac cells (i.e. secretome). In-tissue ECM proteome was first characterized in human aortic samples by Didangelos et al. using SpC-based methods [131]. Similar methodologies have been applied to characterize ECM remodeling process in abdominal aortic aneurysms [132], myocardial ischemia/reperfusion injury [97] and left ventricle subjected to age and SPARC change [133]. Notably, the authors in these studies have developed three-step, reproducible protein extraction protocol to decellurize the tissue and solubilize ECM components, which improved the quality of ECM proteome analysis. Characterization of cell culture-derived ECM proteome also becomes a popular option. For example, using iTRAQ, evaluation of ECM proteins secreted from differentiating cells was performed [59]. More recently, the secretome of rat heart myoblast in response to hypoxia and re-oxygenation stress has also been characterized using both label- free methods and iTRAQ [134].

Proteasome Proteome Several studies have extensively characterized the cardiac proteasome. As proper protein turnover is critical for cardiac homeostasis, proteasomal functions and related cellular pathways are important to CVD occurrence [135]. Ping's group have conducted thorough investigation on cardiac proteasome, including proteome dynamics and functioning of cardiac 19S proteasomes using ^{18}O labeling [136], regulation of acetylation impact myocardial proteasome and its function in diseased heart [137], and more recently, lysine ubiquitination and acetylation of human cardiac 20S proteasomes [138].

Myocardial Tissue Proteomics

Though technically arduous, investigation of the whole cardiac tissue proteome will definitely provide more comprehensive and unbiased insights into the molecular basis of CVD. For example, using label-free methods, Bousette et al. studied the altered proteins in heart tissues from mice subjected to cardiac hypertrophy induced by overexpression of activated calcineurin [94]. Sample prefractionation allowed the elimination of high-abundance contractile proteins in cardiac tissue lysates, and this led to the discovery of 1918 cardiac proteins, in which 290 was found to be dysregulated. In contrast, for comparison, direct analysis of tissue lysates without fractionation resulted in compromised proteome coverage. As shown in a study in 2013, 662 proteins were identified in cardiac tissues subjected to radiation [139]. Recent proteomics studies conducted on whole tissue level include proteomics analysis of left ventricle tissue from hypertensive rats after exercise training [140], laser microdissection sample of amyloid deposits from >100 cases [117] and time-series cardiac tissue samples from murine animal models subjected to myocardial infarction [95].

Characterization of Post Translational Modifications (PTM)

The proteome contains abundant information about PTM, which appears to be an asset that is not accessible by merely investigating either genome or transcriptome. Protein PTM are responsible for the regulation of numerous key molecular pathways and thereby analysis of PTM, presumably in a quantitative manner, will provide highly valuable information for the understanding of disease-related signaling cascades. LC-MS-based shotgun proteomics facilitates PTM analysis with enhanced sensitivity and throughput, thus rendering the global characterization of myocardial PTM possible [141]. Till now, >400 different kinds of PTM have been identified in higher organisms and this number is still accruing [142]. PTM can be identified by mass shifts triggered by the modifications, and the specific amino acid residue where the PTM localizes can be determined by fragment ions with shifted m/z in MS2 spectra. Though the technique is promising, PTM analysis is quite challenging, because modified peptides are in much lower abundance and more difficult to identify from MS spectra than unmodified peptides [143]. Therefore, certain enrichment, fractionation, fragmentation and identification procedures are required to employ in PTM proteomics experiments. Extensively characterized PTM in CVD proteomics encompass phosphorylation, glycosylation, acetylation, oxidative PTM, etc. Of note, more in-depth analysis of PTM is usually accomplished by using subcellular fractions of cardiomyocytes [142].

Phosphorylation Protein phosphorylation by kinases represents one of the most canonical PTM, playing vital roles in a spectrum of signaling pathways associated with myocardial diseases, e.g. contractile function, metabolism, and protein degradation [144]. Thus phosphorylation is the most widely investigated PTM in cardiovascular proteomics [141]. General phosphopeptide enrichment strategies include TiO_2 enrichment, Immobilized Metal Affinity Chromatography (IMAC), the combination of TiO_2 with IMAC or a fraction step (e.g. SCX, SAX, HILIC, ERLIC) prior to LC-MS analysis [8]. The first quantitative phospho-proteomics study in cardiovascular system was performed by Boja et al. on porcine cardiac mitochondria [145], by utilizing TiO_2 enrichment and iTRAQ/HCD to quantify phosphorylated residues. More extensive phosphoproteomics studies have been accomplished in recent years. For instance, Chang et al. enriched phosphopeptides using IMAC, followed by iTRAQ to quantify the temporal changes of global protein phosphorylation in myocardium from pressure-overloaded mice [146]. Their results implicated that Dynamin-related protein 1 might play a potential regulatory role in cardiac hypertrophy rooting from pressure-overload. Scholten et al. employed dimethyl labeling coupled with SCX fractionation and CID/HCD fragmentation for the identification of putative CaMKII targets in cardiac tissues from animals over-expressed with a cardiac-specific CaMKII inhibitor [147]. Lundby et al. sought to identify phosphosites linked with beta-adrenergic receptor signaling, and 670 altered phosphosites were quantified [148]. Furthermore, using TMT labeling plus TiO_2 enrichment, Abdul-Ghani et al. identified 30 differentially expressed phosphoproteins in response to remote ischemic conditioning [149].

Glycosylation N-linked glycosylation links glycan to Asn residues within cell surface and extracellular proteins. There is growing evidence that protein glycosylation plays important roles in mediating normal cardiac functions and survival [150]. Large-scale glycoproteomics profiling is quite challenging due to glycan heterogeneity, which makes identification of glycosylated peptides extremely difficult [8]. Similar with phosphoproteomics analysis, glycoproteomics analysis also requires an enrichment step, which can be realized by methods such as lectin affinity, hydrazide coupling, HILIC, TiO_2 or boronic acid affinity [151]. Initial work of glycosylated peptides analysis in myocardium has been presented by Parker et al. [150]. In their study, peptide samples that were derived from ECM portions were firstly labeled with iTRAQ, enriched using a combined strategy incorporating hydrazide capture, TiO_2 purification and HILIC with and without an ion-pairing agent, then submitted LC-MS/MS analysis. In total 1556 non-redundant N-linked glycosylation sites representing 972 protein groups were identified, while 80 glycosylation sites showed altered abundance. The most recent glycoproteomics analysis was carried out in canine plasma samples, which discovered several dozen of glycosylated sites accounting for the differences between treatment models vs. heart failure models [152]. The authors used solid phase extraction (SPE) to enrich glycopeptides, a similar strategy employed by an earlier study [153].

Other PTM Other PTM studied in global cardiovascular proteomics include S-nitrosylation [154], ubiquinitation, acetylation [138], and citrullination [155]. Via labeling peptides with new isotope coded cysteine thiol-reactive multiplex reagent, cysTMT6, Murray et al. specifically detected and quantified SNO-modifications in human pulmonary arterial endothelia cells [154]. Recently, Fert-Bober et al. characterized citrullination of myofilaments proteins in heart failure for the first time employing SWATH-MS [155].

Concluding Remarks

Owing to the drastic technical advancements, application of shotgun proteomics in CVD investigation becomes increasingly welcomed in recent years. Several examples include: (1) Recent advances in high-resolution MS (e.g. Orbitrap, FTICR) have boosted both proteomic coverage for identification and accuracy for quantitation; (2) A cohort of depletion/equalization, fractionation and subcellular extraction techniques remarkably reduced sample complexity of tissue and plasma samples;

(3) High-plex approaches such as 8-plex iTRAQ [25], 10-plex TMT and 18-plex NeuCode [18] have substantially increased the number of biological replicates that can be analyzed by isotope-labeling methods; (4) DIA-based proteomics strategy provides a promising solution for large-scale, reproducible proteomics analysis.

As discussed in this chapter, in-depth quantitative analysis of plasma and cardiac tissue proteomes, as well as proteome-wide characterization of PTM in the cardiovascular system, are made possible by the innovations of quantitative shotgun proteomics. That being said, a number of issues remain which may markedly hinder progress of cardiovascular proteomics research. Here we highlight two major issues: firstly, a broad gap exists between the identified protein biomarker candidates and clinically applicable biomarkers. The diagnostic utility of the biomarker candidates reported by many shotgun proteomics studies are speculated to be quite limited [90, 91], which may be caused by the lack of a robust pipeline to interface biomarker discovery with clinical validation [79] and lack of large cohorts of well-characterized clinical samples. The combination of shotgun proteomics-based discovery workflow and targeted proteomics-based validation workflow may potentially mitigate this gap [12, 39, 156]. Secondly, CVD proteomics research is still in great need of bioinformatics resources for efficient and high-throughput data analysis, which includes but not limited to: protein database of non-model mammals, reliable and robust procedure to identify and localize PTM, and pathway analysis tools to unravel perplexing molecular signaling. For instance, as the target of CVD proteomics research is primarily the heart, some non-model mammals including pig, canine and rabbit are often employed as preferred experimental animals. However, reviewed, annotated protein databases of these non-model organisms are quite incomplete, which poses technical difficulties for proteomic researchers. Furthermore, PTM study remains challenging due to the lack of robust and reliable procedure to identify and localize medications [143].

In summary, the rapidly evolving proteomics technique has advanced CVD research to a new era, by providing an arsenal of valuable techniques pertinent for the investigation of CVD mechanisms as well as the discovery of potential clinical biomarkers. These new techniques will continue to renew our knowledge on CVD and contribute to the development of better measures for disease management, risk stratification and treatment.

References

1. Yusuf S, Reddy S, Ounpuu S, Anand S. Global burden of cardiovascular diseases: part II: variations in cardiovascular disease by specific ethnic groups and geographic regions and prevention strategies. Circulation. 2001;104(23):2855–64.
2. Cui Z, Dewey S, Gomes AV. Cardioproteomics: advancing the discovery of signaling mechanisms involved in cardiovascular diseases. Am J Cardiovasc Dis. 2011;1(3):274–92.
3. Arab S, Gramolini AO, Ping P, Kislinger T, Stanley B, van Eyk J, et al. Cardiovascular proteomics: tools to develop novel biomarkers and potential applications. J Am Coll Cardiol. 2006;48(9):1733–41.
4. Hoofnagle AN, Wener MH. The fundamental flaws of immunoassays and potential solutions using tandem mass spectrometry. J Immunol Methods. 2009;347(1–2):3–11.

5. Schnabel RB, Baccarelli A, Lin H, Ellinor PT, Benjamin EJ. Next steps in cardiovascular disease genomic research – sequencing, epigenetics, and transcriptomics. Clin Chem. 2012;58(1):113–26.
6. Rugg-Gunn PJ, Cox BJ, Lanner F, Sharma P, Ignatchenko V, McDonald AC, et al. Cell-surface proteomics identifies lineage-specific markers of embryo-derived stem cells. Dev Cell. 2012;22(4):887–901.
7. Cox B, Kotlyar M, Evangelou AI, Ignatchenko V, Ignatchenko A, Whiteley K, et al. Comparative systems biology of human and mouse as a tool to guide the modeling of human placental pathology. Mol Syst Biol. 2009;5:279.
8. Zhang Y, Fonslow BR, Shan B, Baek MC, Yates 3rd JR. Protein analysis by shotgun/bottom-up proteomics. Chem Rev. 2013;113(4):2343–94.
9. Yates 3rd JR. Mass spectrometry and the age of the proteome. J Mass Spectrom. 1998;33(1):1–19.
10. de Godoy LM, Olsen JV, Cox J, Nielsen ML, Hubner NC, Frohlich F, et al. Comprehensive mass-spectrometry-based proteome quantification of haploid versus diploid yeast. Nature. 2008;455(7217):1251–4.
11. Florens L, Washburn MP, Raine JD, Anthony RM, Grainger M, Haynes JD, et al. A proteomic view of the Plasmodium falciparum life cycle. Nature. 2002;419(6906):520–6.
12. Shen X, Young R, Canty JM, Qu J. Quantitative proteomics in cardiovascular research: global and targeted strategies. Proteomics Clin Appl. 2014;8(7–8):488–505.
13. Gygi SP, Rist B, Gerber SA, Turecek F, Gelb MH, Aebersold R. Quantitative analysis of complex protein mixtures using isotope-coded affinity tags. Nat Biotechnol. 1999;17(10):994–9.
14. Schmidt A, Kellermann J, Lottspeich F. A novel strategy for quantitative proteomics using isotope-coded protein labels. Proteomics. 2005;5(1):4–15.
15. Hsu JL, Huang SY, Chow NH, Chen SH. Stable-isotope dimethyl labeling for quantitative proteomics. Anal Chem. 2003;75(24):6843–52.
16. Dayon L, Hainard A, Licker V, Turck N, Kuhn K, Hochstrasser DF, et al. Relative quantification of proteins in human cerebrospinal fluids by MS/MS using 6-plex isobaric tags. Anal Chem. 2008;80(8):2921–31.
17. Rauniyar N, Gao B, McClatchy DB, Yates 3rd JR. Comparison of protein expression ratios observed by sixplex and duplex TMT labeling method. J Proteome Res. 2013;12(2):1031–9.
18. Hebert AS, Merrill AE, Bailey DJ, Still AJ, Westphall MS, Strieter ER, et al. Neutron-encoded mass signatures for multiplexed proteome quantification. Nat Methods. 2013; 10(4):332–4.
19. Reynolds KJ, Yao X, Fenselau C. Proteolytic 18O labeling for comparative proteomics: evaluation of endoprotease Glu-C as the catalytic agent. J Proteome Res. 2002;1(1):27–33.
20. Ong SE, Blagoev B, Kratchmarova I, Kristensen DB, Steen H, Pandey A, et al. Stable isotope labeling by amino acids in cell culture, SILAC, as a simple and accurate approach to expression proteomics. Mol Cell Proteomics. 2002;1(5):376–86.
21. McClatchy DB, Liao L, Park SK, Xu T, Lu B, Yates Iii JR. Differential proteomic analysis of mammalian tissues using SILAM. PLoS One. 2011;6(1):e16039.
22. Mallick P, Kuster B. Proteomics: a pragmatic perspective. Nat Biotechnol. 2010;28(7): 695–709.
23. Lengqvist J, Sandberg A. Stable isotope labeling methods in protein profiling. Methods Mol Biol. 2013;1023:21–51.
24. Regnier FE, Riggs L, Zhang R, Xiong L, Liu P, Chakraborty A, et al. Comparative proteomics based on stable isotope labeling and affinity selection. J Mass Spectrom. 2002;37(2):133–45.
25. Karp NA, Huber W, Sadowski PG, Charles PD, Hester SV, Lilley KS. Addressing accuracy and precision issues in iTRAQ quantitation. Mol Cell Proteomics. 2010;9(9):1885–97.
26. Ow SY, Salim M, Noirel J, Evans C, Rehman I, Wright PC. iTRAQ underestimation in simple and complex mixtures: "the good, the bad and the ugly". J Proteome Res. 2009;8(11): 5347–55.
27. Ting L, Rad R, Gygi SP, Haas W. MS3 eliminates ratio distortion in isobaric multiplexed quantitative proteomics. Nat Methods. 2011;8(11):937–40.

28. Ramos-Fernandez A, Lopez-Ferrer D, Vazquez J. Improved method for differential expression proteomics using trypsin-catalyzed O-18 labeling with a correction for labeling efficiency. Mol Cell Proteomics. 2007;6(7):1274–86.
29. Van Eyk JE. Overview: the maturing of proteomics in cardiovascular research. Circ Res. 2011;108(4):490–8.
30. Konzer A, Ruhs A, Braun H, Jungblut B, Braun T, Kruger M. Stable isotope labeling in zebrafish allows in vivo monitoring of cardiac morphogenesis. Mol Cell Proteomics. 2013;12(6):1502–12.
31. Konzer A, Ruhs A, Braun T, Kruger M. Global protein quantification of mouse heart tissue based on the SILAC mouse. Methods Mol Biol. 2013;1005:39–52.
32. Silva JC, Denny R, Dorschel CA, Gorenstein M, Kass IJ, Li GZ, et al. Quantitative proteomic analysis by accurate mass retention time pairs. Anal Chem. 2005;77(7):2187–200.
33. Gautier V, Mouton-Barbosa E, Bouyssie D, Delcourt N, Beau M, Girard JP, et al. Label-free quantification and shotgun analysis of complex proteomes by one-dimensional SDS-PAGE/NanoLC-MS: evaluation for the large scale analysis of inflammatory human endothelial cells. Mol Cell Proteomics. 2012;11(8):527–39.
34. Duan X, Young R, Straubinger RM, Page B, Cao J, Wang H, et al. A straightforward and highly efficient precipitation/on-pellet digestion procedure coupled with a long gradient nano-LC separation and Orbitrap mass spectrometry for label-free expression profiling of the swine heart mitochondrial proteome. J Proteome Res. 2009;8(6):2838–50.
35. Gao J, Friedrichs MS, Dongre AR, Opiteck GJ. Guidelines for the routine application of the peptide hits technique. J Am Soc Mass Spectrom. 2005;16(8):1231–8.
36. Liu H, Sadygov RG, Yates 3rd JR. A model for random sampling and estimation of relative protein abundance in shotgun proteomics. Anal Chem. 2004;76(14):4193–201.
37. Griffin NM, Yu J, Long F, Oh P, Shore S, Li Y, et al. Label-free, normalized quantification of complex mass spectrometry data for proteomic analysis. Nat Biotechnol. 2010;28(1):83–9.
38. Chen YY, Chambers MC, Li M, Ham AJ, Turner JL, Zhang B, et al. IDPQuantify: combining precursor intensity with spectral counts for protein and peptide quantification. J Proteome Res. 2013;12(9):4111–21.
39. Tu C, Li J, Sheng Q, Zhang M, Qu J. Systematic assessment of survey scan and MS2-based abundance strategies for label-free quantitative proteomics using high-resolution MS data. J Proteome Res. 2014;13(4):2069–79.
40. Asara JM, Christofk HR, Freimark LM, Cantley LC. A label-free quantification method by MS/MS TIC compared to SILAC and spectral counting in a proteomics screen. Proteomics. 2008;8(5):994–9.
41. Cutillas PR, Vanhaesebroeck B. Quantitative profile of five murine core proteomes using label-free functional proteomics. Mol Cell Proteomics. 2007;6(9):1560–73.
42. Ishihama Y, Oda Y, Tabata T, Sato T, Nagasu T, Rappsilber J, et al. Exponentially modified protein abundance index (emPAI) for estimation of absolute protein amount in proteomics by the number of sequenced peptides per protein. Mol Cell Proteomics. 2005;4(9):1265–72.
43. Old WM, Meyer-Arendt K, Aveline-Wolf L, Pierce KG, Mendoza A, Sevinsky JR, et al. Comparison of label-free methods for quantifying human proteins by shotgun proteomics. Mol Cell Proteomics. 2005;4(10):1487–502.
44. Chelius D, Bondarenko PV. Quantitative profiling of proteins in complex mixtures using liquid chromatography and mass spectrometry. J Proteome Res. 2002;1(4):317–23.
45. Tu C, Li J, Jiang X, Sheflin LG, Pfeffer BA, Behringer M, et al. Ion-current-based proteomic profiling of the retina in a rat model of Smith-Lemli-Opitz syndrome. Mol Cell Proteomics. 2013;12(12):3583–98.
46. Yates JR, Ruse CI, Nakorchevsky A. Proteomics by mass spectrometry: approaches, advances, and applications. Annu Rev Biomed Eng. 2009;11:49–79.
47. Matzke MM, Brown JN, Gritsenko MA, Metz TO, Pounds JG, Rodland KD, et al. A comparative analysis of computational approaches to relative protein quantification using peptide peak intensities in label-free LC-MS proteomics experiments. Proteomics. 2013;13(3–4):493–503.
48. Lai X, Wang L, Tang H, Witzmann FA. A novel alignment method and multiple filters for exclusion of unqualified peptides to enhance label-free quantification using peptide intensity in LC-MS/MS. J Proteome Res. 2011;10(10):4799–812.

49. Selevsek N, Chang CY, Gillet LC, Navarro P, Bernhardt OM, Reiter L, et al. Reproducible and consistent quantification of the Saccharomyces cerevisiae proteome by SWATH-mass spectrometry. Mol Cell Proteomics. 2015;14(3):739–49.
50. Bruderer R, Bernhardt OM, Gandhi T, Miladinovic SM, Cheng LY, Messner S, et al. Extending the limits of quantitative proteome profiling with data-independent acquisition and application to acetaminophen-treated three-dimensional liver microtissues. Mol Cell Proteomics. 2015;14(5):1400–10.
51. Nouri-Nigjeh E, Sukumaran S, Tu C, Li J, Shen X, Duan X, et al. Highly multiplexed and reproducible ion-current-based strategy for large-scale quantitative proteomics and the application to protein expression dynamics induced by methylprednisolone in 60 rats. Anal Chem. 2014;86(16):8149–57.
52. Nahnsen S, Bielow C, Reinert K, Kohlbacher O. Tools for label-free peptide quantification. Mol Cell Proteomics. 2013;12(3):549–56.
53. Mueller LN, Brusniak MY, Mani DR, Aebersold R. An assessment of software solutions for the analysis of mass spectrometry based quantitative proteomics data. J Proteome Res. 2008;7(1):51–61.
54. Zhu W, Smith JW, Huang CM. Mass spectrometry-based label-free quantitative proteomics. J Biomed Biotechnol. 2010;2010:840518.
55. Taylor SW, Fahy E, Zhang B, Glenn GM, Warnock DE, Wiley S, et al. Characterization of the human heart mitochondrial proteome. Nat Biotechnol. 2003;21(3):281–6.
56. Qu J, Young R, Page BJ, Shen X, Tata N, Li J, et al. Reproducible ion-current-based approach for 24-plex comparison of the tissue proteomes of hibernating versus normal myocardium in swine models. J Proteome Res. 2014;13(5):2571–84.
57. Huber LA, Pfaller K, Vietor I. Organelle proteomics – implications for subcellular fractionation in proteomics. Circ Res. 2003;92(9):962–8.
58. Arrell DK, Neverova I, Fraser H, Marban E, Van Eyk JE. Proteomic analysis of pharmacologically preconditioned cardiomyocytes reveals novel phosphorylation of myosin light chain 1. Circ Res. 2001;89(6):480–7.
59. Higuchi S, Lin Q, Wang J, Lim TK, Joshi SB, Anand GS, et al. Heart extracellular matrix supports cardiomyocyte differentiation of mouse embryonic stem cells. J Biosci Bioeng. 2013;115(3):320–5.
60. Anderson NL, Anderson NG. The human plasma proteome: history, character, and diagnostic prospects. Mol Cell Proteomics. 2002;1(11):845–67.
61. Tu C, Li J, Young R, Page BJ, Engler F, Halfon MS, et al. Combinatorial peptide ligand library treatment followed by a dual-enzyme, dual-activation approach on a nanoflow liquid chromatography/orbitrap/electron transfer dissociation system for comprehensive analysis of swine plasma proteome. Anal Chem. 2011;83(12):4802–13.
62. Liu T, Qian WJ, Mottaz HM, Gritsenko MA, Norbeck AD, Moore RJ, et al. Evaluation of multiprotein immunoaffinity subtraction for plasma proteomics and candidate biomarker discovery using mass spectrometry. Mol Cell Proteomics. 2006;5(11):2167–74.
63. Washburn MP, Wolters D, Yates JR. Large-scale analysis of the yeast proteome by multidimensional protein identification technology. Nat Biotechnol. 2001;19(3):242–7.
64. Colantonio DA, Dunkinson C, Bovenkamp DE, Van Eyk JE. Effective removal of albumin from serum. Proteomics. 2005;5(15):3831–5.
65. Warder SE, Tucker LA, Strelitzer TJ, McKeegan EM, Meuth JL, Jung PM, et al. Reducing agent-mediated precipitation of high-abundance plasma proteins. Anal Biochem. 2009;387(2):184–93.
66. Shi T, Zhou JY, Gritsenko MA, Hossain M, Camp 2nd DG, Smith RD, et al. IgY14 and SuperMix immunoaffinity separations coupled with liquid chromatography-mass spectrometry for human plasma proteomics biomarker discovery. Methods. 2012;56(2):246–53.
67. Gundry RL, White MY, Nogee J, Tchernyshyov I, Van Eyk JE. Assessment of albumin removal from an immunoaffinity spin column: critical implications for proteomic examination of the albuminome and albumin-depleted samples. Proteomics. 2009;9(7):2021–8.
68. Shen Y, Jacobs JM, Camp 2nd DG, Fang R, Moore RJ, Smith RD, et al. Ultra-high-efficiency strong cation exchange LC/RPLC/MS/MS for high dynamic range characterization of the human plasma proteome. Anal Chem. 2004;76(4):1134–44.

69. Dai J, Wang LS, Wu YB, Sheng QH, Wu JR, Shieh CH, et al. Fully automatic separation and identification of phosphopeptides by continuous pH-gradient anion exchange online coupled with reversed-phase liquid chromatography mass spectrometry. J Proteome Res. 2009;8(1):133–41.
70. Zhou F, Lu Y, Ficarro SB, Webber JT, Marto JA. Nanoflow low pressure high peak capacity single dimension LC-MS/MS platform for high-throughput, in-depth analysis of mammalian proteomes. Anal Chem. 2012;84(11):5133–9.
71. Boersema PJ, Divecha N, Heck AJR, Mohammed S. Evaluation and optimization of ZIC-HILIC-RP as an alternative MudPIT strategy. J Proteome Res. 2007;6(3):937–46.
72. Hao PL, Qian JR, Dutta B, Cheow ESH, Sim KH, Meng W, et al. Enhanced separation and characterization of deamidated peptides with RP-ERLIC-based multidimensional chromatography coupled with tandem mass spectrometry. J Proteome Res. 2012;11(3):1804–11.
73. Tabb DL, Vega-Montoto L, Rudnick PA, Variyath AM, Ham AJ, Bunk DM, et al. Repeatability and reproducibility in proteomic identifications by liquid chromatography-tandem mass spectrometry. J Proteome Res. 2010;9(2):761–76.
74. Weisser H, Nahnsen S, Grossmann J, Nilse L, Quandt A, Brauer H, et al. An automated pipeline for high-throughput label-free quantitative proteomics. J Proteome Res. 2013;12(4):1628–44.
75. Karpievitch YV, Dabney AR, Smith RD. Normalization and missing value imputation for label-free LC-MS analysis. BMC Bioinformatics. 2012;13 Suppl 16:S5.
76. Rost HL, Rosenberger G, Navarro P, Gillet L, Miladinovic SM, Schubert OT, et al. OpenSWATH enables automated, targeted analysis of data-independent acquisition MS data. Nat Biotechnol. 2014;32(3):219–23.
77. Tu C, Sheng Q, Li J, Shen X, Zhang M, Shyr Y, et al. ICan: an optimized ion-current-based quantification procedure with enhanced quantitative accuracy and sensitivity in biomarker discovery. J Proteome Res. 2014;13(12):5888–97.
78. Carr SA, Anderson L. Protein quantitation through targeted mass spectrometry: the way out of biomarker purgatory? Clin Chem. 2008;54(11):1749–52.
79. Rifai N, Gillette MA, Carr SA. Protein biomarker discovery and validation: the long and uncertain path to clinical utility. Nat Biotechnol. 2006;24(8):971–83.
80. Jullig M, Hickey AJ, Chai CC, Skea GL, Middleditch MJ, Costa S, et al. Is the failing heart out of fuel or a worn engine running rich? A study of mitochondria in old spontaneously hypertensive rats. Proteomics. 2008;8(12):2556–72.
81. Baseler WA, Dabkowski ER, Jagannathan R, Thapa D, Nichols CE, Shepherd DL, et al. Reversal of mitochondrial proteomic loss in Type 1 diabetic heart with overexpression of phospholipid hydroperoxide glutathione peroxidase. Am J Physiol Regul Integr Comp Physiol. 2013;304(7):R553–65.
82. Diz AP, Carvajal-Rodriguez A, Skibinski DO. Multiple hypothesis testing in proteomics: a strategy for experimental work. Mol Cell Proteomics. 2011;10(3):M110.004374.
83. Kim YN, Kim HK, Warda M, Kim N, Park WS, Prince Adel B, et al. Toward a better understanding of preeclampsia: comparative proteomic analysis of preeclamptic placentas. Proteomics Clin Appl. 2007;1(12):1625–36.
84. Benjamini YH, Yosef H. Controlling the false discovery rate: a practical and powerful approach to multiple testing. J R Stat Soc Ser B. 1995;57(1):289–300.
85. Storey JD. A direct approach to false discovery rates. J R Stat Soc Ser B. 2002;64(3):479–98.
86. Chang J, Van Remmen H, Ward WF, Regnier FE, Richardson A, Cornell J. Processing of data generated by 2-dimensional gel electrophoresis for statistical analysis: missing data, normalization, and statistics. J Proteome Res. 2004;3(6):1210–8.
87. Cox J, Mann M. MaxQuant enables high peptide identification rates, individualized p.p.b.-range mass accuracies and proteome-wide protein quantification. Nat Biotechnol. 2008;26(12):1367–72.
88. Shen X, Hu Q, Li J, Wang J, Qu J. An experimental null method to guide the development of technical procedures and to control false positive discovery in quantitative proteomics. J Proteome Res. 2015;14:4147–57.

89. Serang O, Cansizoglu AE, Kall L, Steen H, Steen JA. Nonparametric Bayesian evaluation of differential protein quantification. J Proteome Res. 2013;12(10):4556–65.
90. Gerszten RE, Carr SA, Sabatine M. Integration of proteomic-based tools for improved biomarkers of myocardial injury. Clin Chem. 2010;56(2):194–201.
91. Dubois E, Fertin M, Burdese J, Amouyel P, Bauters C, Pinet F. Cardiovascular proteomics: translational studies to develop novel biomarkers in heart failure and left ventricular remodeling. Proteomics Clin Appl. 2011;5(1–2):57–66.
92. McGregor E, Dunn MJ. Proteomics of the heart: unraveling disease. Circ Res. 2006;98(3): 309–21.
93. Canty Jr JM, Suzuki G. Myocardial perfusion and contraction in acute ischemia and chronic ischemic heart disease. J Mol Cell Cardiol. 2012;52(4):822–31.
94. Bousette N, Chugh S, Fong V, Isserlin R, Kim KH, Volchuk A, et al. Constitutively active calcineurin induces cardiac endoplasmic reticulum stress and protects against apoptosis that is mediated by alpha-crystallin-B. Proc Natl Acad Sci U S A. 2010;107(43):18481–6.
95. Li C, Qiu Q, Wang Y, Li P, Xiao C, Wang H, et al. Time course label-free quantitative analysis of cardiac muscles of rats after myocardial infarction. Mol Biosyst. 2014;10:505–13.
96. Song B, Liu Y, Parman T, Liu S, Miller JK, Liu X, et al. Quantitative proteomics for cardiac biomarker discovery using isoproterenol-treated nonhuman primates. J Proteome Res. 2014;13(12):5909–17.
97. Barallobre-Barreiro J, Didangelos A, Schoendube FA, Drozdov I, Yin X, Fernandez-Caggiano M, et al. Proteomics analysis of cardiac extracellular matrix remodeling in a porcine model of ischemia/reperfusion injury. Circulation. 2012;125(6):789–802.
98. Stastna M, Van Eyk JE. Secreted proteins as a fundamental source for biomarker discovery. Proteomics. 2012;12(4–5):722–35.
99. Hammer E, Goritzka M, Ameling S, Darm K, Steil L, Klingel K, et al. Characterization of the human myocardial proteome in inflammatory dilated cardiomyopathy by label-free quantitative shotgun proteomics of heart biopsies. J Proteome Res. 2011;10(5):2161–71.
100. Kakimoto Y, Ito S, Abiru H, Kotani H, Ozeki M, Tamaki K, et al. Sorbin and SH3 domain-containing protein 2 is released from infarcted heart in the very early phase: proteomic analysis of cardiac tissues from patients. J Am Heart Assoc. 2013;2(6):e000565.
101. Holland A, Dowling P, Zweyer M, Swandulla D, Henry M, Clynes M, et al. Proteomic profiling of cardiomyopathic tissue from the aged mdx model of Duchenne muscular dystrophy reveals a drastic decrease in laminin, nidogen and annexin. Proteomics. 2013;13(15):2312–23.
102. Chugh S, Ouzounian M, Lu Z, Mohamed S, Li W, Bousette N, et al. Pilot study identifying myosin heavy chain 7, desmin, insulin-like growth factor 7, and annexin A2 as circulating biomarkers of human heart failure. Proteomics. 2013;13(15):2324–34.
103. Jing L, Parker CE, Seo D, Hines MW, Dicheva N, Yu Y, et al. Discovery of biomarker candidates for coronary artery disease from an APOE-knock out mouse model using iTRAQ-based multiplex quantitative proteomics. Proteomics. 2011;11(14):2763–76.
104. Silbiger VN, Luchessi AD, Hirata RD, Neto LG, Pastorelli CP, Ueda EK, et al. Time course proteomic profiling of human myocardial infarction plasma samples: an approach to new biomarker discovery. Clin Chim Acta Int J Clin Chem. 2011;412(11–12):1086–93.
105. Kristensen LP, Larsen MR, Mickley H, Saaby L, Diederichsen AC, Lambrechtsen J, et al. Plasma proteome profiling of atherosclerotic disease manifestations reveals elevated levels of the cytoskeletal protein vinculin. J Proteomics. 2014;101:141–53.
106. Juhasz P, Lynch M, Sethuraman M, Campbell J, Hines W, Paniagua M, et al. Semi-targeted plasma proteomics discovery workflow utilizing two-stage protein depletion and off-line LC-MALDI MS/MS. J Proteome Res. 2011;10(1):34–45.
107. Mebazaa A, Vanpoucke G, Thomas G, Verleysen K, Cohen-Solal A, Vanderheyden M, et al. Unbiased plasma proteomics for novel diagnostic biomarkers in cardiovascular disease: identification of quiescin Q6 as a candidate biomarker of acutely decompensated heart failure. Eur Heart J. 2012;33(18):2317–24.
108. Thery C, Boussac M, Veron P, Ricciardi-Castagnoli P, Raposo G, Garin J, et al. Proteomic analysis of dendritic cell-derived exosomes: a secreted subcellular compartment distinct from apoptotic vesicles. J Immunol. 2001;166(12):7309–18.

109. Kanhai DA, Visseren FL, van der Graaf Y, Schoneveld AH, Catanzariti LM, Timmers L, et al. Microvesicle protein levels are associated with increased risk for future vascular events and mortality in patients with clinically manifest vascular disease. Int J Cardiol. 2013;168(3): 2358–63.
110. Martinez-Pinna R, Gonzalez de Peredo A, Monsarrat B, Burlet-Schiltz O, Martin-Ventura JL. Label-free quantitative proteomic analysis of human plasma-derived microvesicles to find protein signatures of abdominal aortic aneurysms. Proteomics Clin Appl. 2014;8(7–8):620–5.
111. Datta A, Chen CP, Sze SK. Discovery of prognostic biomarker candidates of lacunar infarction by quantitative proteomics of microvesicles enriched plasma. PLoS One. 2014; 9(4):e94663.
112. Zhang J, Liem DA, Mueller M, Wang Y, Zong C, Deng N, et al. Altered proteome biology of cardiac mitochondria under stress conditions. J Proteome Res. 2008;7(6):2204–14.
113. Baseler WA, Dabkowski ER, Williamson CL, Croston TL, Thapa D, Powell MJ, et al. Proteomic alterations of distinct mitochondrial subpopulations in the type 1 diabetic heart: contribution of protein import dysfunction. Am J Physiol Regul Integr Comp Physiol. 2011;300(2):R186–200.
114. Dabkowski ER, Baseler WA, Williamson CL, Powell M, Razunguzwa TT, Frisbee JC, et al. Mitochondrial dysfunction in the type 2 diabetic heart is associated with alterations in spatially distinct mitochondrial proteomes. Am J Physiol Heart Circ Physiol. 2010;299(2):H529–40.
115. Bugger H, Schwarzer M, Chen D, Schrepper A, Amorim PA, Schoepe M, et al. Proteomic remodelling of mitochondrial oxidative pathways in pressure overload-induced heart failure. Cardiovasc Res. 2010;85(2):376–84.
116. Fu ZM, Wang MY, Everett A, Lakatta E, Van Eyk J. Can proteomics yield insight into aging aorta? Proteomics Clin Appl. 2013;7(7–8):477–89.
117. Larsen BT, Mereuta OM, Dasari S, Fayyaz AU, Theis JD, Vrana JA, et al. Correlation of histomorphologic pattern of cardiac amyloid deposition with amyloid type: a histologic and proteomic analysis of 108 cases. Histopathology. 2015;68:648–56.
118. Sharma P, Cosme J, Gramolini AO. Recent advances in cardiovascular proteomics. J Proteomics. 2013;81:3–14.
119. Warren CM, Geenen DL, Helseth Jr DL, Xu H, Solaro RJ. Sub-proteomic fractionation, iTRAQ, and OFFGEL-LC-MS/MS approaches to cardiac proteomics. J Proteomics. 2010; 73(8):1551–61.
120. Yin X, Cuello F, Mayr U, Hao Z, Hornshaw M, Ehler E, et al. Proteomics analysis of the cardiac myofilament subproteome reveals dynamic alterations in phosphatase subunit distribution. Mol Cell Proteomics. 2010;9(3):497–509.
121. Van Hoof D, Dormeyer W, Braam SR, Passier R, Monshouwer-Kloots J, Ward-van Oostwaard D, et al. Identification of cell surface proteins for antibody-based selection of human embryonic stem cell-derived cardiomyocytes. J Proteome Res. 2010;9(3):1610–8.
122. Gregersen N, Hansen J, Palmfeldt J. Mitochondrial proteomics--a tool for the study of metabolic disorders. J Inherit Metab Dis. 2012;35(4):715–26.
123. McDonald TG, Van Eyk JE. Mitochondrial proteomics. Undercover in the lipid bilayer. Basic Res Cardiol. 2003;98(4):219–27.
124. Edwards AV, White MY, Cordwell SJ. The role of proteomics in clinical cardiovascular biomarker discovery. Mol Cell Proteomics. 2008;7(10):1824–37.
125. Jullig M, Hickey AJ, Middleditch MJ, Crossman DJ, Lee SC, Cooper GJ. Characterization of proteomic changes in cardiac mitochondria in streptozotocin-diabetic rats using iTRAQ isobaric tags. Proteomics Clin Appl. 2007;1(6):565–76.
126. Goudarzi M, Ross MM, Zhou W, Van Meter A, Deng J, Martin LM, et al. Development of a novel proteomic approach for mitochondrial proteomics from cardiac tissue from patients with atrial fibrillation. J Proteome Res. 2011;10(8):3484–92.
127. Lancaster TS, Jefferson SJ, Hunter JC, Lopez V, Van Eyk JE, Lakatta EG, et al. Quantitative proteomic analysis reveals novel mitochondrial targets of estrogen deficiency in the aged female rat heart. Physiol Genomics. 2012;44(20):957–69.

128. Nguyen T, Wong R, Wang G, Gucek M, Steenbergen C, Murphy E. Acute inhibition of GSK causes mitochondrial remodeling. Am J Physiol Heart Circ Physiol. 2012;302(11): H2439–45.
129. Kim TY, Wang D, Kim AK, Lau E, Lin AJ, Liem DA, et al. Metabolic labeling reveals proteome dynamics of mouse mitochondria. Mol Cell Proteomics. 2012;11(12):1586–94.
130. Chang CW, Dalgliesh AJ, Lopez JE, Griffiths LG. Cardiac extracellular matrix proteomics: challenges, techniques, and clinical implications. Proteomics Clin Appl. 2015;10:39–50.
131. Didangelos A, Yin X, Mandal K, Baumert M, Jahangiri M, Mayr M. Proteomics characterization of extracellular space components in the human aorta. Mol Cell Proteomics. 2010;9(9):2048–62.
132. Didangelos A, Yin X, Mandal K, Saje A, Smith A, Xu Q, et al. Extracellular matrix composition and remodeling in human abdominal aortic aneurysms: a proteomics approach. Mol Cell Proteomics. 2011;10(8):M111 008128.
133. de Castro Bras LE, Toba H, Baicu CF, Zile MR, Weintraub ST, Lindsey ML, et al. Age and SPARC change the extracellular matrix composition of the left ventricle. Biomed Res Int. 2014;2014:810562.
134. Li X, Ren Y, Sorokin V, Poh KK, Ho HH, Lee CN, et al. Quantitative profiling of the rat heart myoblast secretome reveals differential responses to hypoxia and re-oxygenation stress. J Proteomics. 2014;98:138–49.
135. Pagan J, Seto T, Pagano M, Cittadini A. Role of the ubiquitin proteasome system in the heart. Circ Res. 2013;112(7):1046–58.
136. Wang D, Zong C, Koag MC, Wang Y, Drews O, Fang C, et al. Proteome dynamics and proteome function of cardiac 19S proteasomes. Mol Cell Proteomics. 2011;10(5):M110 006122.
137. Wang D, Fang C, Zong NC, Liem DA, Cadeiras M, Scruggs SB, et al. Regulation of acetylation restores proteolytic function of diseased myocardium in mouse and human. Mol Cell Proteomics. 2013;12:3793–802.
138. Zong N, Ping P, Lau E, Choi HJ, Ng DC, Meyer D, et al. Lysine ubiquitination and acetylation of human cardiac 20S proteasomes. Proteomics Clin Appl. 2014;8(7–8):590–4.
139. Azimzadeh O, Sievert W, Sarioglu H, Yentrapalli R, Barjaktarovic Z, Sriharshan A, et al. PPAR alpha: a novel radiation target in locally exposed Mus musculus heart revealed by quantitative proteomics. J Proteome Res. 2013;12(6):2700–14.
140. Petriz BA, Almeida JA, Gomes CP, Pereira RW, Murad AM, Franco OL. NanoUPLC/MS(E) proteomic analysis reveals modulation on left ventricle proteome from hypertensive rats after exercise training. J Proteomics. 2015;113:351–65.
141. Edwards AV, Cordwell SJ, White MY. Phosphoproteomic profiling of the myocyte. Circ Cardiovasc Genet. 2011;4(5):575.
142. Agnetti G, Husberg C, Van Eyk JE. Divide and conquer: the application of organelle proteomics to heart failure. Circ Res. 2011;108(4):512–26.
143. Olsen JV, Mann M. Status of large-scale analysis of post-translational modifications by mass spectrometry. Mol Cell Proteomics. 2013;12(12):3444–52.
144. Van Eyk JE. Lessons from old and new kinases. Circ Res. 2004;94(2):135–7.
145. Boja ES, Phillips D, French SA, Harris RA, Balaban RS. Quantitative mitochondrial phosphoproteomics using iTRAQ on an LTQ-Orbitrap with high energy collision dissociation. J Proteome Res. 2009;8(10):4665–75.
146. Chang YW, Chang YT, Wang Q, Lin JJ, Chen YJ, Chen CC. Quantitative phosphoproteomic study of pressure-overloaded mouse heart reveals dynamin-related protein 1 as a modulator of cardiac hypertrophy. Mol Cell Proteomics. 2013;12(11):3094–107.
147. Scholten A, Preisinger C, Corradini E, Bourgonje VJ, Hennrich ML, van Veen TA, et al. Phosphoproteomics study based on in vivo inhibition reveals sites of calmodulin-dependent protein kinase II regulation in the heart. J Am Heart Assoc. 2013;2(4):e000318.
148. Lundby A, Andersen MN, Steffensen AB, Horn H, Kelstrup CD, Francavilla C, et al. In vivo phosphoproteomics analysis reveals the cardiac targets of beta-adrenergic receptor signaling. Sci Signal. 2013;6(278):rs11.

149. Abdul-Ghani S, Heesom KJ, Angelini GD, Suleiman MS. Cardiac phosphoproteomics during remote ischemic preconditioning: a role for the sarcomeric Z-disk proteins. Biomed Res Int. 2014;2014:767812.
150. Parker BL, Palmisano G, Edwards AV, White MY, Engholm-Keller K, Lee A, et al. Quantitative N-linked glycoproteomics of myocardial ischemia and reperfusion injury reveals early remodeling in the extracellular environment. Mol Cell Proteomics. 2011;10(8):M110 006833.
151. Drozdov I, Didangelos A, Yin X, Zampetaki A, Abonnenc M, Murdoch C, et al. Gene network and proteomic analyses of cardiac responses to pathological and physiological stress. Circ Cardiovasc Genet. 2013;6:588–97.
152. Yang S, Chen L, Sun S, Shah P, Yang W, Zhang B, et al. Glycoproteins identified from heart failure and treatment models. Proteomics. 2015;15(2–3):567–79.
153. Tian Y, Koganti T, Yao Z, Cannon P, Shah P, Pietrovito L, et al. Cardiac extracellular proteome profiling and membrane topology analysis using glycoproteomics. Proteomics Clin Appl. 2014;8(7–8):595–602.
154. Murray CI, Uhrigshardt H, O'Meally RN, Cole RN, Van Eyk JE. Identification and quantification of S-nitrosylation by cysteine reactive tandem mass tag switch assay. Mol Cell Proteomics. 2012;11(2):M111 013441.
155. Fert-Bober J, Giles JT, Holewinski RJ, Kirk JA, Uhrigshardt H, Crowgey EL, et al. Citrullination of myofilament proteins in heart failure. Cardiovasc Res. 2015;108:232–42.
156. Fu Q, Schoenhoff FS, Savage WJ, Zhang P, Van Eyk JE. Multiplex assays for biomarker research and clinical application: translational science coming of age. Proteomics Clin Appl. 2010;4(3):271–84.

Chapter 12
Analysis of Proteomic Data

Kai Kammers, D. Brian Foster, and Ingo Ruczinski

Abstract Whether you are a proteomics specialist or simply an end-user of proteomic data, the day will come when you sit down with your dataset, typically a list of proteins or protein clusters whose abundance change in one or more experimental groups. This protein change is often represented as a ratio or fold-change. When the euphoria wears off, the nagging questions set in. How accurate are your data, really? How confident are you in these changes; are they statistically significant? If so, by what statistical test? Are you sure the test is suitable for your data? How would you know? Or perhaps more importantly, as a graduate student, would you spend the next year following up on a proteomic lead? As principal investigator, should you reallocate substantial resources to a new line of enquiry? Given the risk of squandering time and money on false leads or dismissing a nugget that could change existing paradigms, delving more deeply into the principles of robust proteomic analysis, however daunting at first blush, is a good investment.

Keywords Experimental design • Protein quantitation • Differential regulation • Statistical inference • Empirical Bayes • Significance • Multiple comparison correction • Robustness

K. Kammers, PhD • I. Ruczinski, PhD (✉)
Department of Biostatistics, Johns Hopkins Bloomberg School of Public Health, Baltimore, MD, USA
e-mail: kai.kammers@jhu.edu; ingo@jhu.edu; iruczin1@jhu.edu

D.B. Foster, PhD
Laboratory of Cardiovascular Biochemistry, Division of Cardiology, Department of Medicine, The Johns Hopkins University School of Medicine, Baltimore, MD, USA
e-mail: dbrianfoster@jhmi.edu

G. Agnetti et al. (eds.), *Manual of Cardiovascular Proteomics*,
DOI 10.1007/978-3-319-31828-8_12

Introduction

This chapter addresses some basic concepts in experimental design, quantitation, and statistical analysis. They are presented through the lens of labeling-based quantitation using isobaric tags for relative and absolute quantitation (iTRAQ) [1] or tandem mass tags (TMT) [2], though many of the concepts should be applicable in the label-free arena, particularly quantification by precursor ion currents. In section "Quantitation of Relative Protein Abundance," we introduce the median sweep algorithm, which provides a set of principles for robust quantitation, specifically, (1) data transformation (logarithmic), (2) adopting median reporter ion intensity as an estimator of protein abundance, and (3) normalizing each iTRAQ or TMT channel to account for differences in protein load between samples in the multiplex. In section "Statistical Inference," we outline how broad-scope proteomic experiments present an interesting set of challenges for statistical analysis. Namely, proteomics experiments typically elicit long lists of proteins whose abundances are measured (m). Yet because the number of iTRAQ/TMT channels currently is limited to 8 or 10, the number of biological replicates (n) per experimental group tends to be small. This is known as the "high m, low n" problem, for which the often-used 2-sample t-test is sub-optimal. Instead, we introduce the empirical Bayes-modified t-test. In section "Other Practical Considerations," we examine two other more issues that beset large proteomic datasets. The first is data missingness and how the nature of the missingness affects statistical inference. The second is how performing any statistical test for every protein in a large dataset necessarily increases the likelihood that a change in protein abundance may be declared nominally significant ($p<0.05$) by chance alone. We describe how the family-wise error rate (FWER) and the false-discovery rate (FDR) address this multi-hypothesis testing problem differently. Finally, in section "A Case Study," we put it all together and show that for a simulated large proteomic dataset, the statistical power is optimized by the empirical Bayesian method, and demonstrate that modified p-values and FDR-controlled q-values provide a robust rationale for follow-up experiments.

Quantitation of Relative Protein Abundance

Tables of the raw reporter ion intensities associated with each MS^2 mass spectrum and each reporter channel can be exported from many proteomics software packages used to summarize the results of the database search. The "median-sweep" is a simple yet robust algorithm for spectral signal pre-processing prior to statistical downstream analysis [3]. The main goal of these pre-processing steps is to minimize technical or procedural sources of variation in the data, and to maximize chances of extracting biological value from your proteomics experiments. Briefly then, the median sweep entails (a) logarithmic transformation of spectral intensities, (b) summarizing individual protein abundances by taking the median value of the logarithmically transformed intensities for all spectra belonging to that protein, and (c) normalizing relative protein abundances across all channels by median centering. We consider the merits of each step, in turn.

Reporter Ion Intensities and Relative Protein Abundances Are Best Considered on a Logarithmic Scale

When comparing reporter ion intensities and relative protein abundances between groups such as cases and controls, it is easy and convenient to state the findings as fold changes. Saying that a protein is twice as abundant in cases compared to controls is equivalent to saying the protein is half as abundant in controls compared to cases. Ratios such as protein fold changes are, therefore, best considered on the logarithmic scale (base 2 being preferred), since this transformation conveniently allows for statistical inference using fold changes. Note that log(a/b) = log(a) − log(b) = −(log(b) − log(a)) = −log(b/a), which means that differences between log abundances are the same as log fold changes. Statistical approaches like t-tests or ANOVAs use *differences* between groups and the logarithmic transformation allows for such a statistical assessment based on fold changes. For ease of interpretation, the findings are usually reported as fold changes in the literature by simply mapping the data back to the original scale after carrying out the analyses.

The main reason to use a logarithmic transformation in the signal pre-processing is the fact that the variation in the reporter ion intensities typically is a function of ion intensity itself (Fig. 12.1, left). When variance increases with the magnitude of the signal, we say that errors in the data are *multiplicative*. This scenario violates basic assumptions in many statistical tests, which commonly assume that errors are *additive* (i.e. that the variance is not dependent on the signal magnitude), and this can render the statistical inference invalid. However, the problem can typically be resolved by taking the logarithm of the reporter ion intensities (Fig. 12.1, right). In this form, the errors are additive [4] and therefore more tractable statistically.

Median (log_2) Spectral Intensity Provides a Robust Estimate of Protein Abundance

After the logarithmic transformation (Fig. 12.2, row 1), the next step is to estimate the abundance of each protein in each of the reporter channels of the multiplexed experiment. As you might imagine, many research papers have been written on the matter. Which metric best describes the central tendency of individual protein abundance? Is it average spectral intensity? Total spectral intensity? A weighted average? If so, what weightings? Are there peptide-specific factors to be considered that require sophisticated modeling? Fortunately, there is one metric that is as robust as any model and satisfies Occam's razor – the spectral median. Restated, the relative protein abundance (PA) estimate for a particular protein, within a channel (label), is obtained by taking the median value (Fig. 12.2, row 2, open circles) of all median-centered log_2 reporter ion intensities (Fig. 12.2, row 2, black lines), within the channel, assigned to that protein. This is known median summarization. Note that this method ignores information about the peptides to which the spectra were assigned. Empirically, ignoring the peptide information adds additional robustness (Fig. 12.2, row 3), as it reduces artifacts caused by interference.

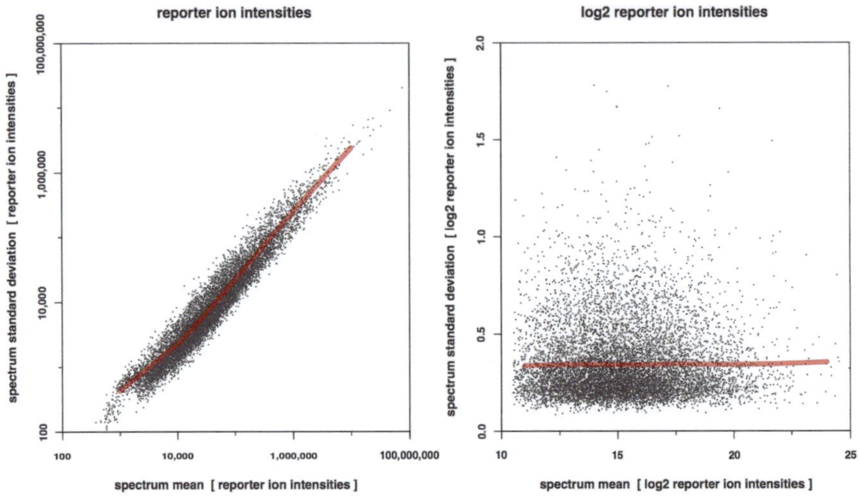

Fig. 12.1 Distributions of Spectral Intensity and Variation. Sample means (x-axis) versus sample standard deviations (y-axis) of reporter ion intensities in 10,000 randomly selected reporter ion spectra, observed in a 10-plex TMT. Means and standard deviations were calculated using the raw reporter ion intensities (*left*), and after using a $\log_2$ transformation of the reporter ion intensities (*right*). A non-parametric smoother was used to highlight the mean-variance relationships (*red lines*)

Median summarization is highly insensitive to outliers; remember the properties of a median dictate that nearly half of the assembled signals may be complete rubbish without unduly interfering with the protein estimate! Therefore, median summarization is likely to be the method of choice for the analysis of minimally fractionated complex peptide samples characterized by non-trivial levels of peptide co-isolation interference. Specifically, co-isolation interference occurs when more than one peptide sequence is fragmented in MS^2 simultaneously, making it difficult to disambiguate the reporter ion signals. In complex samples, depending on the quantitation algorithm, this can be a significant source of protein estimation inaccuracy.

Normalizing the "Protein Load" for Each Labeled Channel Is Essential

There are two fundamental premises that apply to most global-scale proteomic iTRAQ/TMT experiments. The first is that to adequately compare protein levels across experimental groups, an equal amount of protein must be present in each of the channels of the multiplexed experiment. The second assumption is that the experimental treatment causes a relatively small subset of proteins to change significantly (or at least, less than 50 %). It was once thought that channel-to-channel variability could be minimized by taking painstaking care in the determination of

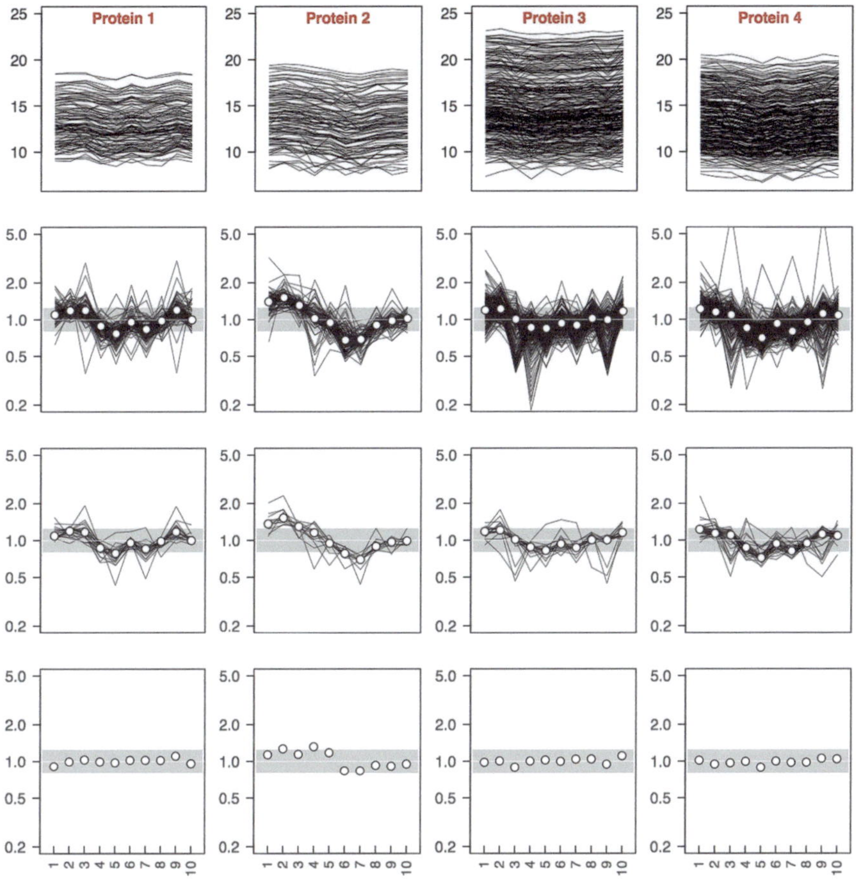

Fig. 12.2 Illustration of the median sweep for the quantitation of relative protein abundances in a 10-plex TMT experiment, simulated based on actual data. The normalization steps are shown for four proteins (*columns 1–4*). Displayed are the absolute reporter ion intensities on the logarithmic scale (base 2) for all spectra detected for the respective proteins (*first row*), the same data after removal of the spectrum median displayed as fold changes (*second row*), the relative peptide abundances calculated as the median of the above row 2 values for each channel and peptide (*third row*), and the relative protein abundances after removing the loading effect (*fourth row*). The gray regions (*rows 2–4*) represent 25 % fold changes in relative protein abundances. The data in row 3 are not used in the quantitation of relative protein abundance, and are only displayed to show the presence of poorly behaved peptides for every protein, highlighting the need for robust estimators such as the median. For proteins 1, 3 and 4, none of the samples shows a fold change larger than 25 %. For protein 2 a fold change of about 50 % is observed when comparing one condition (*channels 1–5*) to the other (*channels 6–10*), possibly indicating a differentially expressed protein (*row 4*). For all proteins, channels 1 and 2, for example, show relatively high intensities and channel 5 shows relatively low intensities (*row 2*), highlighting the need to remove the "loading effect"

protein concentration, subsequent enzymatic digest and peptide labeling phases of sample preparation. This is a tall order, however, given the number of steps involved. In truth, it is impossible to know whether each iTRAQ/TMT channel contains the same amount of protein. In our analysis of sources of variation, protein load issues manifest themselves as a channel effect that must be removed prior to assessment of protein abundances. If there were any doubt of the need to normalize the channels, consider the work of Herbrich et al. [3] who showed that inter-channel variability, both within and between iTRAQ multiplexes, was one of the largest sources of error, *even when the channels consisted of technical sample replicates*.

Fortunately, the data can be normalized. As with PA estimation, there is no shortage of proposed methods to normalize channel data. Some are implemented earlier in the workflow, at the spectral level, e.g. by total channel spectral intensity. However, we advocate normalization by median channel PA. Remember, the effects of outlier data on PA have been minimized previously by median-summarization. Therefore by normalizing to median PA, one is normalizing to an already robust parameter. To be explicit then, channels are normalized by application of "median-centering", that is, subtracting the channel median PA from each of the individual PAs, setting the median PA to zero (Fig. 12.2, row 4). This procedure is repeated for all channels and all multiplexes.

The median sweep borrows from a set of tools that have been applied successfully to the analysis of DNA microarrays for over a decade. When compared, median-summarization performed as well or better in terms of both accuracy and precision than more computationally intensive models of protein abundance quantitation [3]. It is noteworthy that labeled proteomic experiments, in general, do not require internal standards, not even for inter-experiment comparisons. In a head-to-head comparison with a strategy that employed a sample master pool as an internal standard for each individual multiplex, the median sweep generated protein abundances yielded substantially better results [3]. The median sweep algorithm has been implemented in the statistical environment R, and is freely available in source code form under the terms of the GNU General Public License of the Free Software Foundation at www.biostat.jhsph.edu/~kkammers/software/CVproteomics/. It is also available in some of the commercial software packages, for example as one of the options in the Proteome Discoverer software (www.thermoscientific.com).

Statistical Inference

Congratulations. If you've implemented the median sweep, you can be fairly confident that you have prepared your labeled multiplexed data in a manner that minimizes both signal bias and outlier effects and also accounts for differences in the amount of protein analyzed in each labeled channel. The bulk of the variance in your data should now be attributable to the host of fascinating proteins whose abundances are changing between your control and treatment groups. The ensuing questions then seem simple enough. How large is the fold-change in the abundance of protein X, and is it statistically significant or not? But as we will describe, typical proteomics experiments present certain statistical challenges. Luckily, we can again borrow from the field of DNA microarray statistical analysis to address them.

The "High m/Low n" Issue Is a Common Challenge in Proteomics Experiments

Given the speed and sensitivity of modern mass spectrometers, a labeled proteomics experiment may yield thousands of confidently-identified proteins whose abundances are measured (*m*) across experimental groups. Yet because the number of iTRAQ/TMT channels currently is limited to 8 or 10, the number of biological replicates (*n*) per experimental group tends to be small, even when the data from multiple experiments are combined. Thus, many multiplexed studies suffer from the so-called "high *m*/low *n*" problem. The high *m*/low *n* issue is problematic in at least two ways. The low number of biological replicates (*n*) saps statistical power and is discussed hereafter, while the large number of measurements (*m*) ensures that no matter what statistical test is used, one is bound to discover a significant fold-change by chance alone (a false positive). This is covered in section "Other Practical Considerations".

Low n-Values and the 2-Sample t-Test

Standard 2-sample t-tests are frequently used to compare the $\log_2$ relative abundances for each protein (PA) across the conditions of interest. As a refresher, let's assume a balanced experimental design with *n* cases and *n* controls. Let $X_{1p}, \ldots, X_{np}$ and $Y_{1p}, \ldots, Y_{np}$ be the PAs for each identified and quantified protein *p* for the cases and controls, respectively, with corresponding group means $\bar{X}_p$ and $\bar{Y}_p$. Assuming the PAs are independent and normally distributed with equal within-group variability (departures from the normality assumption are actually of no major concern in a *t*-test unless severe outliers are present, Rice, 1995 [6]), the pooled within-group standard deviation is given by

$$s_p = \sqrt{\frac{\sum_i \left(X_{ip} - \bar{X}_p\right)^2 + \sum_i \left(Y_{ip} - \bar{Y}_p\right)^2}{n-2}},$$

and the test statistic for protein *p* is given as

$$t_p = \frac{estimated\ \log_2\ fold\ change}{estimated\ standard\ error} = \frac{\bar{X}_p - \bar{Y}_p}{s_p\sqrt{2/n}}.$$

A p-value for t_p is derived by referring the test statistic to a t-distribution with $d_p = 2 \times n - 2$ degrees of freedom, as the null distribution, and calculating the probability that a random draw from said distribution exceeds t_p in absolute value.

One can see how "low n" influences the denominator of the within-group variability and the estimated standard error, respectively, which in turn affects the assessment of statistical significance of the fold-change. A frequent consequence is that proteins that exhibit large fold-changes are often declared non-significant because of a large sample variance, while at the same time small observed fold changes might be declared statistically significant, because of a small sample variance. In other words, in "low n" scenarios, standard t-tests may yield both more false positives and false negatives.

Introducing Empirical Bayes-Moderated t-Tests for "Low n" Experiments

To stabilize these variance estimates, an empirical Bayes (EB) approach can be used to shrink the observed protein sample variances towards a pooled variance estimate, while allowing for a realistic distribution of the underlying biological variances. Only the general type of the distribution is hereby assumed, and the shape parameters are estimated from the observed data (thus, *empirical* Bayes). This method was first introduced for gene expression studies as LIMMA ("Linear Models for Microarray Data" [7]) and has now been adapted to mass spectrometry-based proteomics experiments [e.g. 8, 9].

Specifically, the EB procedure uses the fact that assuming the $\log_2$ relative abundances for protein p are normally distributed, the sample variance for the proteins s_p^2 follows a scaled χ^2 distribution

$$s_p^2 \mid \sigma_p^2 \sim \frac{\sigma_p^2}{d_p} \times \chi_{d_p}^2,$$

where σ_p^2 denotes the true (unknown) variance, and d_p are the degrees of freedom determined by the experimental design (e.g. $d_p = 2 \times n - 2$ in the above described balanced design). Under the assumption that these protein variances σ_p^2 follow a so-called scaled inverse χ^2 distribution, i.e.

$$\frac{1}{\sigma_p^2} \sim \frac{1}{d_0 \times s_0^2} \times \chi_{d_0}^2,$$

it follows [6] that the protein sample variances s_p^2 are described by a scaled F distribution given by

$$s_p^2 \sim s_0^2 \times F_{d_p, d_0}.$$

Thus, the parameters d_0 and s_0^2 can be estimated from the observed sample variances using maximum likelihood methods, and the posterior for a protein's sample variance is "moderated": the sample variance is shrunk towards the common prior value s_0^2 using

$$s_{p[moderated]}^2 = \frac{d_p \times s_p^2 + d_0 \times s_0^2}{d_p + d_0} = \lambda \times s_p^2 + (1 - \lambda) \times s_0^2, \text{ with } \lambda = \frac{d_p}{d_p + d_0} \in (0,1).$$

This equation shows that the shrinkage amount depends on the relative sizes of the observed and prior degrees of freedom d_p and d_0, respectively, and will be most pronounced for small sample sizes when d_p and thus, λ, are small. The moderated t-statistics are calculated by simply replacing the sample standard deviation by the moderated sample standard deviation in the denominator of the test statistic, i.e.

$$t_{p[moderated]} = \frac{estimated\ log_2\ fold\ change}{estimated\ moderated\ standard\ error} = \frac{\bar{X}_p - \bar{Y}_p}{s_{p[moderated]}\sqrt{2/n}}.$$

Strengths of the Empirical Bayesian Modeling

It is important to note that only the estimated standard errors in the moderated test statistics change, whereas the estimated fold changes remain the same. The p-value is derived by referring this moderated test statistic to a t-distribution with $d_p + d_0$ degrees of freedom. In section "A Case Study," our case study will demonstrate that inference based on moderated test statistics is far more stable and powerful compared to inference based on ordinary t-tests – particularly when the sample size is small. The empirical Bayes procedure allows for pooling information across all proteins within any linear model framework, generating moderated t-statistics and p-values for the contrast(s) of interest, e.g. the fold changes in case–control studies, but also experiments with multiple group comparisons [7]. It also readily extends to settings where the data from multiple iTRAQ or TMT experiments are analyzed [9].

Another strength of this EB approach is that is can be applied to cases in which there are missing data (section "Other Practical Considerations"). Note that missing data, to be expected in any high-throughput mass spectrometry experiment, result in a loss of power to detect differentially expressed proteins, but these data can still be analyzed using linear models with moderated test statistics, and proper type I error control. In particular, the amount of shrinkage of the variance terms depends on the observed experimental degrees of freedom, with more missing data resulting in stronger shrinkage towards the common prior variance s_0^2 (see Kammers et al. [9] for a more detailed discussion). All of this is implemented in the open source limma software package, and should not be an obstacle for the laboratory scientist in his or her data analysis.

Other Practical Considerations

Missing Data

The labeled proteomic workflow has many advantages. It is relatively simple to perform and its multiplex design helps minimize some of the technical variability. That said, it is not without minor complications, one of which is missing data. In an iTRAQ/TMT design, missing data can usually be ascribed to one or two sources.

Within a multiplex, you will find spectra for which a signal is missing from one or a few channels. This probably best explained by the issues in labeling efficiency. Performed properly, the efficiency of peptide labeling with its tags is very high, though not 100 %. This means that there will be a small number of spectra with an incomplete set of reporter intensities. Generally speaking, this missingness tends to correlate with fairly strongly with peptide abundance. In our hands, this accounts for <1 % of spectra and a small number of novel proteins. Though there are a number of ways of interpolating missing data, we prefer to simply omit these spectra from prior to implementing the median sweep.

The second type of missing data arises in experimental designs that involve more than one multiplex run. From the previous chapters, you will be familiar with the fact that iTRAQ/TMT quantitation is an MS^2-based method. To recap, peptides are introduced into the mass spectrometer where the parent mass is measured in MS^1. Masses are then selected or sampled for fragmentation in MS^2 where b- and y- and reporter ions are released and detected. This is called data-dependent acquisition (DDA). DDA is the reason why, even if you run the same complex sample on a mass spectrometer three times, you will identify three modestly different datasets. It is important to note that the number of times a peptide or protein is identified, in itself, is not a reflection on the quality or confidence of the identification, simply a function of sampling. Nevertheless, these run-to-run differences arising from DDA might appear to pose a problem. Does this mean multiplex experiments are intrinsically ill-suited for the analysis of large clinical cohorts? How many proteins could possibly be common to, say, ten 10-plex experiments? Does it matter?

To answer that question it might help first to understand how statisticians classify types of data missingness [10]. Data can be missing completely at random (MCAR), missing at random (MAR), or missing not at random (MNAR). For missing data to be MCAR, the likelihood of missing must be completely independent of all variables in an experiment, both observed and unobserved. This is a rare situation and not relevant to the realm of mass spectrometry-based quantitation. MNAR applies when failure to observe the data is intrinsically related to the nature of the variable being measured, for example when the rate of missingness in proteomic data is inversely related to protein abundance [11]. This kind of missingness poses problems for standard statistical inference, since simply omitting the missing data can cause strong biases in the analysis. Thus, more sophisticated methods such as multiple data imputation are required. MAR on the other hand, perhaps counter-intuitively, describes missingness that is not necessarily random, but can be reasonably accounted for by factors relating to the experiment. In this light, missingness between multiplex runs is strongly related to sampling by the instrument (i.e., the experiment itself), rather than protein or peptide abundance per se (unobserved by definition for all missing data). This is not to suggest that missingness is not influenced at all by low abundance, merely that stochastic sampling within a multiplex experiment is by far the major driver behind missing data. Therefore, missingness between multiplex runs is typically well described under the MAR assumption.

This is important because under the MAR assumption, analyses like the empirical Bayes-modified t-tests can easily be conducted on data pooled from several multiplex runs, regardless of how many times a protein is observed, without fear of introducing undue bias. Note though that the degree of variance shrinkage afforded by EB modeling, and therefore statistical power, drops with the number of experimental degrees of freedom, determined by the n-value. Therefore, inference about a protein observed in a subset of multiplex runs will not be as strong as it would for a protein observed in all runs.

Multiple Comparisons

Just as low *n*-values necessitate special considerations when assessing differential protein abundance, so too, do large numbers of compared measurements (high *m*). Let's consider that for a given statistical test, where we often set the false positive rate nominally at 5 % (calling abundances significantly different if $p<0.05$). But what if we performed the same test on 100 different proteins? We would then expect to identify 5 proteins as differentially regulated by chance when, in truth, they really are not. This is the conundrum of multi-hypothesis testing. In high throughput experiments, practitioners typically employ the family-wise error rate or the false discovery rate control when testing multiple hypotheses, which are different strategies to handle the balance of false positive and false negative identifications.

Use the Family-Wise Error Rate (FWER) to Guard Against Any False Positives

The FWER is the probability of generating at least one false positive among all hypotheses tested, i.e. the probability of incorrectly declaring one or more null proteins differentially expressed. The Bonferroni procedure is the best-known approach to control the FWER, only declaring proteins with nominal p-values less than α/m differentially expressed, where α is the desired FWER (typically, 5 %; Fig. 12.3, blue line), and m is the number of proteins measured in the mass spectrometry experiment. This strong protection against *any* false positives comes at the expense of generating false negatives, that is, a potentially large number of truly differentially expressed proteins are not declared statistically significant (and with small sample sizes, it is common that no protein achieves differential expression significance at the Bonferroni level). This type of error control is appropriate in a setting where false negatives are of no concern to the investigator, and his or her objective is solely to generate a list of proteins he or she firmly believes are differentially expressed (even though that list might be very short or even empty).

Setting the False-Discovery Rate (FDR): Discover More by Accepting Some Risk

More commonly, however, the objective of high-throughput experiments is to detect as many differentially expressed proteins as possible, subject to a tolerable false detection rate. Thus, the FDR is a more desirable parameter in these proteomic settings, devised to control the proportion of false positives *among a set of proteins declared differentially expressed*. The procedure proposed by Benjamini et al. [12] arguably is the best-known procedure to control the FDR. To control FDR at a level δ, one orders the unadjusted p-values $p_1 \leq p_2 \leq \cdots \leq p_m$, and finds the test (i.e. protein) with the highest rank j for which the p-value p_j is less than or equal to $j / m \times \delta$ (Fig. 12.3, red line). The list of declared differentially expressed proteins includes those of rank j or less.

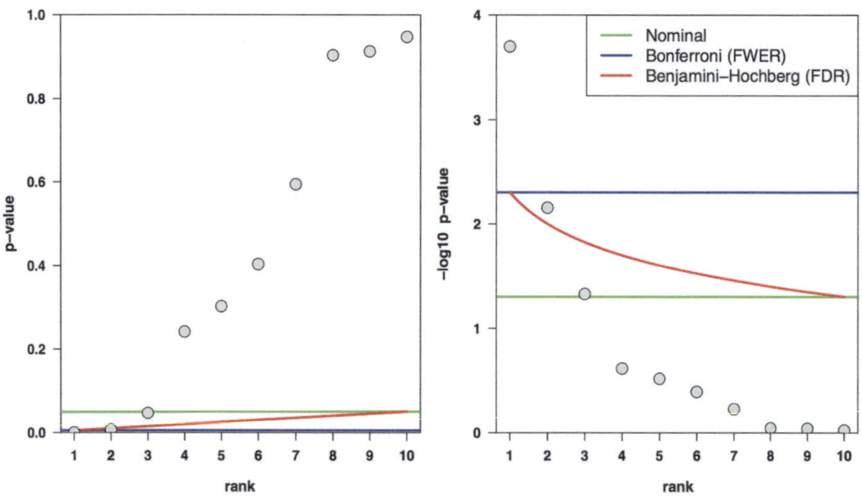

Fig. 12.3 Illustration of three types of error control in a setting with ten hypothesis tests. The p-values are ordered *left to right* from smallest to largest, and shown as actual values (*left panel*) and on the negative $\log_{10}$ scale (*right panel*). Using the nominal type I error level of 5 % to declare proteins differentially expressed (*green line*) yields three significant findings in this example, however, each non-differentially expressed protein has an *a priori* chance of 5 % to be falsely declared significant. In a mass spectrometry experiment with hundreds or thousands of proteins identified and quantified, this yields an unacceptably high rate of false positives. Using the Bonferroni procedure to control the family-wise error rate at 5 % (*blue line*) yields only one significant finding. However, this strong protection against *any* false positives comes at the expense of potentially many false negatives, that is, truly differentially expressed proteins not declared statistically significant. In high-throughput experiments where many proteins might be differentially expressed, this is often not acceptable either. The FDR, devised to control the proportion of false positives among a set of proteins declared differentially expressed, is often a more desirable parameter in these proteomic settings. Here, the procedure proposed by Benjamini and Hochberg is shown (*red line*), yielding 2 significant findings at a 5 % FDR

The Positive FDR or "q-Value"

Storey and Tibshirani (2003) [13] extended the Benjamini and Hochberg FDR to the *positive* FDR, reported as "q-values", which have a similar interpretation for the FDR as p-values have for type-I error control: the q-value for a protein is defined as the minimum FDR that can be attained when calling that protein differentially expressed. As an example: if a protein has a q-value of 0.05, the expected proportion of false positives *among the list of differentially-expressed proteins* is 5 %. The q-values are calculated from the reported p-values, using a monotone transformation after estimating the actual proportion of differentially expressed proteins in the experiment [13]. Surprisingly, q-values are rarely used to assess differential regulation of proteins despite this intuitive and appealing interpretation. It is noteworthy that protein abundance levels are often not independent (i.e., genes are co-expressed), which actually violates one of the assumptions in the calculations of the FDR, since the test statistics and resulting p-values are not independent either. However, they exhibit "clumpy dependence" (i.e. local and finite), and reassuringly, Storey and Tibshirani (2003) [13] have shown that the FDR is controlled despite the violation of the independence assumption, and the estimated q-values conservatively estimate the true q-values.

What Is Your Definition of Differential Abundance?

A popular procedure for assigning proteins as differentially regulated is to consider only proteins that meet some arbitrary thresholds, usually $p<0.05$ and a fold-change larger than 1.5. The rationale stems from the intuitive notion that when a standard *t*-test is used to evaluate a proteomic dataset, low p-values may sometimes arise from marginal fold changes. By imposing the additional criterion of a 1.5-fold change, we are attempting to weed out some of those false positives. Though well-intended, the practice reveals a certain level of cognitive dissonance. On the one hand, we implement a standard *t*-test to lend a study the *imprimatur* of statistical rigor, but then proceed to disregard certain p-values arising from marginal fold-changes because we don't really believe they are likely to be biologically meaningful. How would one then assess the FDR in that scenario?

A more rigorous approach would be to be employ EB-moderated t-tests and positive FDR (q-value) assessment. The empirical Bayes procedure yields moderated p-values that implicitly achieve ends similar to applying thresholds. Specifically, by shrinking the variance, the net effect of EB is to penalize inference from marginal fold-changes while typically boosting the significance of proteins with large fold changes (see the section below and the companion webpage to this chapter). Finally, by combining EB analysis with multi-hypothesis correction, like the q-value, one can obtain a statistically robust list of biologically meaningful protein changes with a known FDR.

A Case Study

In this section, we illustrate the comparative benefits of EB-moderated t-tests and q-values over the standard 2-sample *t*-test, with respect to statistical power and type I error control (true and false positives). We simulated data mimicking a 10-plex TMT experiment with 5 cases and 5 controls, using simulation parameters from actual proteomic experiments (the experimental details and the code are available from the companion website at www.biostat.jhsph.edu/~kkammers/software/CVproteomics/). In brief, $\log_2$ relative abundances were generated for 5000 proteins in 5 cases and 5 controls, with 100 proteins assumed to exhibit a 50 % fold change between cases to controls (and no fold change for the other proteins).

According to the 2-Sample t-Test

Among the 5000 proteins simulated in the single 10-plex TMT experiment, a total of 340 achieved statistical significance using ordinary t-tests with a nominal 5 % type I error level ($p<0.05$), including 252 false positive identifications (roughly 5000×0.05), emphasizing the need for multiple comparisons correction. Under a false discovery rate control of 5 % (here, $q<0.05$), 31 of the initial 340 proteins were declared differentially expressed (comprised of 28 true and 3 false positive identifications; Fig. 12.4, left).

According to the EB-Modified t-Test

Though essentially the same number of proteins (337) was significant at a nominal 5 % level ($p<0.05$) using moderated t-statistics, a substantially larger number of proteins (49) was declared differentially expressed at a FDR of 5 % ($q<0.05$), mostly comprised of true identifications (46 true positives, 3 false positives; Fig. 12.4, right). Therefore compared to the 2-sample *t*-test the EB method yielded 64 % more true positives while the number of false positives was unchanged at the same FDR.

Overall Assessment

To assess EB performance over a range of FDRs, we averaged over 1000 simulations. For all levels of FDR control between 1 and 10 %, the empirical Bayes approach using moderated test statistic produced substantially larger lists of proteins declared differentially expressed, detecting more truly differentially expressed proteins while maintaining proper error control (Fig. 12.5, left). On average, at an FDR of 1 % (i.e. $q<0.01$) only 6 proteins were declared differentially expressed when using ordinary

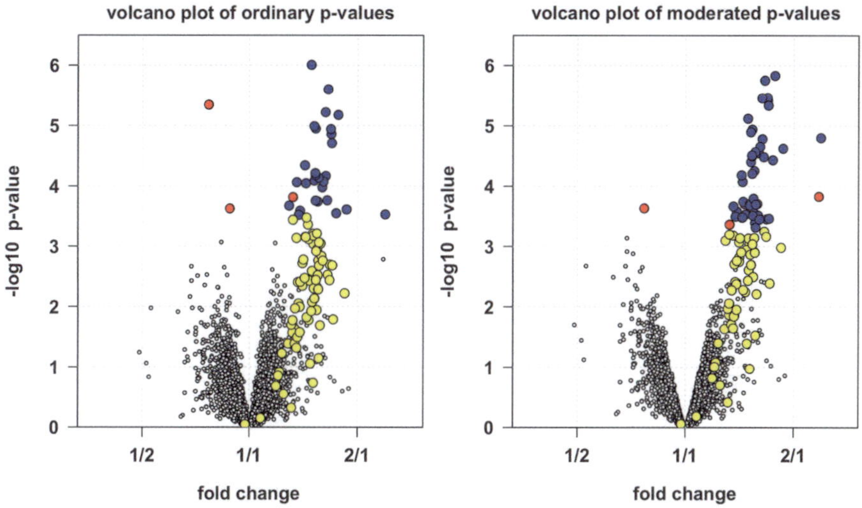

Fig. 12.4 Inference from one simulated 10-plex TMT experiment with five cases and five controls, based on 5000 proteins. A 50 % fold change was simulated for 100 proteins. *Left*: The volcano plot showing the estimated fold changes (x-axis) versus the $-\log_{10}$ p-values (y-axis) for each protein using ordinary 2-sample t-tests. *Right*: The volcano plot from the inference based on the moderated t-statistics. In each panel, proteins with false discovery rate adjusted p-values (i.e. q-values) smaller than 5 % are highlighted in *blue* (true positives) and *red* (false positives). False negatives are highlighted in *yellow*. Only 31 proteins are declared differentially expressed when using ordinary test statistics (28 true positives, 3 false positives), compared to 49 proteins (46 true positives, 3 false positives) declared differentially expressed when moderated test statistics are employed. Thus, pooling information from the distribution of all proteins improves power to detect differentially expressed proteins, and in particular attenuates the statistical significance of proteins with small fold changes and small sample variability

test statistics, compared to 18 proteins declared differentially expressed when moderated test statistics were employed (FDR of 5 %: 32 and 54 proteins, respectively; FDR of 10 %: 50 and 73 proteins). At equal false positive rates on a receiver operating characteristic (ROC) curve, the true positive rate is substantially higher when using moderated test statistics (Fig. 12.5, right). Equivalent results were obtained when fold changes and the number of proteins are varied, and when multiple experiments were analyzed simultaneously in the presence of missing data (see Kammers et al. [9] and the companion webpage to this chapter).

Conclusion

To review, the goal of this chapter is to introduce some of the pitfalls that often compromise or confound relative protein quantitation and statistical analysis. These include, but are not necessarily limited to, the influence of outliers, differences in

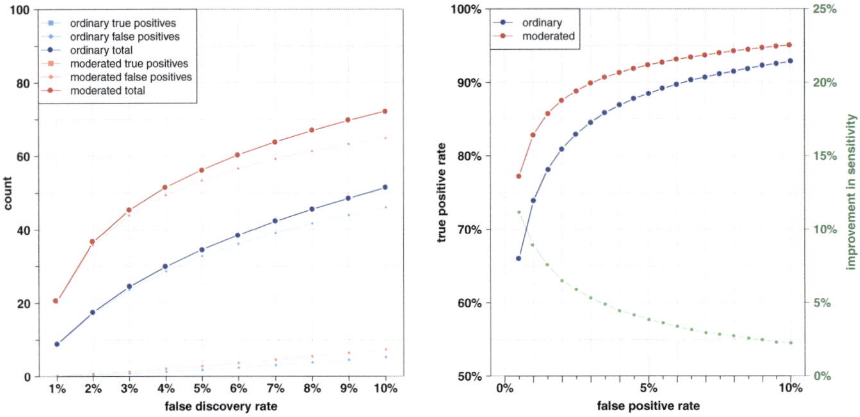

Fig. 12.5 Results for detection rates averaged across 1000 simulations. *Left*: for all levels of false discovery rate control targeted (1–10 %, x-axis), the empirical Bayes approach using moderated test statistics produces larger lists of proteins declared differentially expressed (y-axis), thus detecting more truly differentially expressed proteins while maintaining proper error control. *Right*: the same results presented in a ROC curve. For all false positive rates (0.5–10 %, x-axis), the true positive rate (y-axis, *black*) is substantially higher when using moderated test statistics, yielding a large difference in sensitivity (y-axis, *green*)

protein load across reporter ion channels, as well as false positive and false negative calls of protein regulation. These can be addressed, in turn, by using robust estimation methods, protein normalization, empirical Bayes-modified tests and proper control for multiple-hypothesis testing.

Note though, that following basic principles of good experimental design is imperative for any findings to be trustworthy. For example, it is critically important that under no circumstances should the outcome of interest (such as the case–control status) be confounded with the experiment. For example, if all cases are run in one TMT experiment and all controls are run in another TMT experiment, it is impossible to detect any fold changes between cases and controls. Assume, for simplicity, that the abundance of some protein the in all cases is twice as large as the abundance in the controls. Since quantitation of relative abundance is carried out within each experiment, the estimates for relative abundance would be equal to one in each experiment, masking the fact that this protein is differentially regulated. No amount of statistical wizardry can fix this problem. Sometimes practitioners use reference samples in the experiments to compare abundances across experiments, however that practice is strongly discouraged as it results in higher variability and thus a loss of power, and can easily be avoided with proper experimental design [3].

In our case study, a single 10-plex TMT experiment with 5 cases and controls, we have demonstrated that inference based on moderated test statistics is far more stable and powerful compared to inference based on ordinary test statistics. These findings extend to settings where the data from multiple TMT or iTRAQ experiments are jointly analyzed (see for example [9]), and when more than two groups are contrasted using an Analysis of Variance (ANOVA; see for example [7]).

ANOVAs have also been proposed to analyze all proteins simultaneously, in particular in the context of linear mixed effects models [4, 5], to stabilize variance estimates. However, this assumes the same biological variability for all proteins (an assumption that is clearly not correct), and can be viewed as a statistical procedure that shrinks all variances to one common value. This approach has also been explored in the gene expression literature [14, 15], but eventually abandoned in favor of more realistic empirical-Bayes procedures.

Though analyzing raw reporter signals "from scratch", as we have described, is not terribly difficult (knowledge of Excel and R will generally suffice), many may prefer to stick with commercially-available software or established open-source proteomic analysis workflows to determine protein ratios and p-values. Hopefully, however, you are now in a position to understand the issues at hand and assess the output by answering the following questions for yourself, in light of key principles. How, precisely, does your software package evaluate protein abundance? Has the data been normalized for protein load? If so, how? If you would like to merge multiplex data, does your software allow you to identify and remove batch bias? What statistical test was used in the workflow? What are your criteria for differential protein abundance? And finally, what FDR are you prepared to live with in pursuit of tantalizing new leads? *In proteomicum veritas* (tr. in proteomics, there is truth). Good luck!

References

1. Ross PL, Huang YN, Marchese JN, Williamson B, Parker K, Hattan S, Khainovski N, Pillai S, Dey S, Daniels S, Purkayastha S, Juhasz P, Martin S, Bartlet-Jones M, He F, Jacobson A, Pappin DJ. Multiplexed protein quantitation in saccharomyces cerevisiae using amine-reactive isobaric tagging reagents. Mol Cell Proteomics. 2004;3(12):1154–69.
2. Thompson A, Schäfer J, Kuhn K, Kienle S, Schwarz J, Schmidt G, Neumann T, Hamon C. Tandem mass tags: a novel quantification strategy for comparative analysis of complex protein mixtures by MS/MS. Anal Chem. 2003;75(8):1895–904, 2003.
3. Herbrich SM, Cole RN, West KP Jr, Schulze K, Yager JD, Groopman JD, Christian P, Wu L, O'Meally RN, May DH, McIntosh MW, Ruczinski I. Statistical inference from multiple itraq experiments without using common reference standards. J Proteome Res. 2013;12(2):594–604.
4. Oberg AL, Mahoney DW, Eckel-Passow JE, Malone CJ, Wolfinger RD, Hill EG, Cooper LT, Onuma OK, Spiro C, Therneau TM, Bergen HR. Statistical analysis of relative labeled mass spectrometry data from complex samples using ANOVA. J Proteome Res. 2008;7(1):225–33.
5. Hill EG, Schwacke JH, Comte-Walters S, Slate EH, Oberg AL, Eckel-Passow JE, Therneau TM, Schey KL. A statistical model for iTRAQ data analysis. J Proteome Res. 2008;7(8): 3091–101.
6. Rice J. Mathematical statistics and data analysis. 2nd ed. Boston: Duxbury Press; 1995.
7. Smith GK. Linear models and empirical Bayes methods for assessing differential expression in microarray experiments. Stat Appl Genet Mol Biol. 2004;3(1).
8. Orellana CA, Marcellin E, Schulz BL, Nouwens AS, Gray PP, Nielsen LK. High antibody producing Chinese hamster ovary cells upregulate intracellular protein transport and glutathione synthesis. J Proteome Res. 2015;14(2):609–18.
9. Kammers K, Cole RN, Tiengwe C, Ruczinski I. Detecting significant changes in protein abundance. EuPA Open Proteom. 2015;7:11–9.

10. Little RJA, Rubin DB. Statistical analysis with missing data. New York: John Wiley & Sons; 1987.
11. Wang P, Tang H, Zhang H, Whiteaker J, Paulovich AG, Mcintosh M. Normalization regarding non-random missing values in high-throughput mass spectrometry data. Pac Symp Biocomput. 2006;315–26.
12. Benjamini Y, Hochberg Y. Controlling the false discovery rate: a practical and powerful approach to multiple testing. J R Stat Soc Series B. 1995;57(1):289–300.
13. Storey JD, Tibishrani R. Statistical significance for genomewide studies. Proc Natl Acad Sci U S A. 2003;100(16):9440–5.
14. Kerr MK, Martin M, Churchill GA. Analysis of variance for gene expression microarray data. J Comput Biol. 2000;7(6):819–37.
15. Kerr MK, Churchill GA. Experimental design for gene expression microarrays. Biostatistics. 2001;2(2):183–201.

Chapter 13
Post-translational Modifications in the Cardiovascular Proteome

Christopher I. Murray, Heaseung Sophia Chung, Kyriakos N. Papanicolaou, D. Brian Foster, and Mark J. Kohr

Abstract The analysis of post-translational modifications is critical for understanding the regulation of protein function in the heart. These small, often charged, groups are added to a protein's structure to modulate its activity, localization or associations. The development of proteomic technologies has greatly improved the identification and subsequent characterization of these modifications. However, due to the complex nature of the cardiovascular proteome, a particular post-translational modification may represent only a tiny fraction of the milieu. Additionally, some modifications are too labile for mass spectrometry analysis. To address these limitations, a variety of techniques and strategies have been developed to specifically target and improve the detection of these modifications. In the following chapter, we review the challenges and solutions to identifying several prominent post-translational modifications in the cardiovascular system.

C.I. Murray, PhD (✉)
Department of Anesthesiology, Pharmacology and Therapeutics, University of British Columbia, Vancouver, BC, Canada

Life Sciences Centre,
2350 Health Science Mall, Room 2.303, Vancouver, BC V6T 1Z3, Canada
e-mail: cmurra17@mail.ubc.ca

H.S. Chung, MS
Department of Biological Chemistry, Johns Hopkins University School of Medicine, Baltimore, MD 21218, USA

K.N. Papanicolaou, PhD
Division of Cardiology, Department of Medicine, Johns Hopkins University, Johns Hopkins School of Medicine, Baltimore, MD 21218, USA

D.B. Foster, PhD
Laboratory of Cardiovascular Biochemistry, Division of Cardiology, Department of Medicine, The Johns Hopkins University School of Medicine, Baltimore, MD, USA

M.J. Kohr, PhD (✉)
Department of Environmental Health Sciences, Bloomberg School of Public Health, Johns Hopkins University, 615 N. Wolfe Street, Baltimore, MD 21205, USA
e-mail: mkohr1@jhu.edu

G. Agnetti et al. (eds.), *Manual of Cardiovascular Proteomics*,
DOI 10.1007/978-3-319-31828-8_13

Keywords Post-translational modification • Mass spectrometry • Proteomics • Phosphorylation • O-GlcNAcylation • Acetylation • S-nitrosylation • Biotin switch assay • BEMAD • IMAC

Introduction

A full understanding of the cardiovascular proteome would not be complete without investigation of its post-translational modifications (PTM). These alterations to a protein's sequence are not directly encoded in the genome but affect the protein after it has been expressed. Most post-translational modifications can arise in two basic ways: the enzyme catalyzed addition of a chemical group, usually as part of a signaling pathway, or by spontaneous interaction with a reactive molecule [1]. In general, enzyme catalyzed modifications are the result of evolutionarily refined signaling pathways while spontaneous modifications tend to be more adventitious in nature. These protein modifications can result in a conformational change, affecting activity, interactions or subcellular localization. Well-studied modifications include phosphorylation, glycosylation, acetylation, and various oxidations, which are discussed below; however, to date over 300 different PTMs have been identified each with its own regulatory potential. The net effect of these modifications in the cell has been appreciated as a powerful mechanism in both healthy and disease states.

In the heart, PTMs have been found to regulate many critical processes including calcium handling, force generation, energy production, the development of heart failure and cardioprotection following ischemia reperfusion injury, among others [2, 3]. A classic example of this regulation can be found in the phosphorylation of phospholamban [4]. In its unphosphorylated state, phospholamban is an inhibitor of the sarcoplasmic reticulum (SR) Ca^{2+}-ATPase (SERCA), which sequesters contraction-triggering Ca^{2+} from the cytoplasm to the SR during diastole. Following beta-adrenergic stimulation, phospholamban is phosphorylated by protein kinase A at serine 16, which removes its inhibitory effect on SERCA enhancing uptake of Ca^{2+} to the SR [5]. The subsequent loading of the SR with Ca^{2+} results in an increased release for the next heart beat. The overall effect of this modification is to boost both relaxation and contractility, improving cardiac output.

PTMs have also been found to have deleterious effects on the heart [2, 6]. In the cardiovascular system, changes in the oxidative balance can affect many aspects of cellular physiology. Depending on the magnitude, fluctuations in the cell's production of reactive oxygen species can regulate normal metabolic processes, activate protective mechanisms, or be cytotoxic. Production of these reactive species can result in a variety of reversible and irreversible oxidative protein modifications. In general, harmful oxidative damage results from the increased presence of irreversible modifications while reversible modifications are more likely to be redox-signaling events [7], although reversible modifications can signal both adaptive and maladaptive cardiac responses [8]. Increases in oxidative stress have been linked to the development of heart disease, including cardiac hypertrophy, ischemia/reperfusion injury and heart failure [6, 9]. In particular, the contractile myofilament proteins appear to be particularly sensitive to oxidative modifications. S-nitrosation of regulatory protein tropomyosin has been linked to heart failure [10]. In addition, contractile dysfunction has been associated with the oxidation of tropomyosin in both ischemia reperfusion injury and microembolism [11, 12].

Over the past three decades significant effort has been expended to systematically identify, map and characterize the various PTMs in physiological and pathophysiological states [1, 13–16]. The increased power and scale of proteomic technologies has been very valuable in this endeavor, significantly increasing the number of identified sites. In the following chapter, we introduce and review several state-of-the-art approaches for preparing, enriching, and identifying several different post-translational modifications: phosphorylation, O-GlcNAcylation, acetylation, and oxidative cysteine modifications. While this list is by no means comprehensive, it provides an introduction to the considerations necessary for any proteomic PTM study.

Strategies for Post-translational Modification Identification

Phosphorylation

Protein phosphorylation is the most common post-translational modification (PTM) on serine (~90 %), threonine (~10 %), and tyrosine (~ <1 %) residues [17], which has been best studied due to its key roles in regulating protein structure and function. Since phosphorylation has been found to effect critical cellular processes such as protein–protein interactions and cell signaling, much energy has been focused on identifying, quantifying and monitoring the temporal changes in phosphorylation in a variety of biological systems. However, due to the complexity of the phosphoproteome and the dynamics of phosphorylation, it has been challenging to detect and monitor this modification.

Over the past few years, mass spectrometry (MS) based methods have emerged introducing powerful tools to identify and quantify phosphorylated sites. These approaches have allowed the global profiling of phosphorylated serine, threonine, and tyrosine residues in a biological sample of interest. The following section describes important considerations for the experimental design of large-scale characterization of phosphorylation. The major steps for phosphoproteomics workflow are cell fractionation, enrichment and analysis via liquid chromatography (LC) coupled with tandem MS. Each of these steps can be further modified to serve the purpose of an individual study [18].

Considerations for Sample Preparation

Phosphoproteins represent only a small portion of the total proteome and thus samples for phosphoproteomic analysis are enriched from the unmodified peptides. For this reason, the sample size of the starting material is very large, typically in the range of 1–10 mg [19] and even >20 mg in some studies [18]. However, with the development of efficient enrichment techniques and highly-sensitive mass spectrometers, it has become possible to start with a few hundred micrograms of protein [20].

Cell and tissue harvesting should be performed rapidly at a low temperature or followed by snap freezing. During cell lysis or tissue homogenization, proper protease inhibitors and phosphatase inhibitors are required to prevent artificial proteolysis and preserve protein phosphorylation status. Once soluble lysates are extracted, intact proteins are digested into peptides using trypsin or alternate proteases such as Glu-C, Lys-N, Lys-C, Asp-N or chymotrypsin in combination with trypsin to increase protein sequence coverage, followed by fractionation and enrichment for MS analysis. Alternatively, the protein mixtures can be fractionized using a gel-based method, then digested and enriched prior to MS identification.

Phospho-Enrichment Strategies

Liquid Chromatographic Fractionation

Since millions of peptides are potentially generated after proteolytic digest, phosphopeptides are nearly impossible to detect in the whole cell lysate or tissue sample. Thus, fractionation by liquid chromatographic strategies such as strong cation exchange chromatography (SCX), strong anion exchange chromatography (SAX) or hydrophilic interaction liquid chromatography (HILIC) are required to reduce sample complexity, coupled with further enrichment of phosphopeptides. Different fractionation strategies before or after enrichment have been studied to enhance the selectivity of phosphopeptides [21–23].

SCX is one of the most powerful approaches to separate phosphopeptides. At low pH (<2.6), a typical peptide has a net charge of +2, whereas an identical peptide that is phosphorylated would have +1 by the negatively charged phosphate group [24]. SCX separates phosphorylated peptides from nonphosphorylated peptides utilizing this charge difference. Using a SCX column, more negatively charged phosphopeptides are eluted earlier than more positively charged non-phosphopeptides by increasing ionic strength in the mobile phase. The multi-phosphorylated peptides, which have a net zero or even negative charge, can be observed in the flow-through fraction (unbound). Gygi and coworkers [25] and Macek et al. [24] observed highly enriched phosphopeptides in the +1 SCX fractions and flow-through fraction.

HILIC has been reported to be a promising method in large scale proteomics [23]. It separates phosphopeptides based on more hydrophilic nature of phosphopeptides compared to other peptides: more hydrophilic peptides are eluted in the later fraction by gradually decreasing the organic content of the mobile phase [23, 26].

Enrichment

In order to separate low abundance phosphopeptides from unmodified peptides, a selective enrichment technique is required to increase the relative abundance of phosphopeptides above the threshold for detection by MS. The most common techniques for enrichment are immobilized metal affinity chromatography (IMAC) [27, 28] and metal oxide affinity chromatography (MOAC) [29].

IMAC separates phosphopeptides based on the high affinity coordination of phosphate groups to metal ions such as Fe^{3+}, Zn^{2+}, and Ga^{3+}[30]. A variety of different IMAC resins are commercially available in column or magnetic bead formats, which show different efficiencies and specificities [19]. However, the main limitation of the IMAC method is the nonspecific binding of acidic peptides.

Recently, MOAC-based methods have also been used for phosphopeptide enrichment in large-scale proteomic analysis [29, 31, 32]. Among these, titanium dioxide (TiO_2) has emerged as the most common MOAC-based phosphopeptide enrichment method [30]. The principle of this method is similar to IMAC, thus TiO_2 technique is considered to be interchangeable with IMAC depending on the feature of the samples (each method has a different bias and selectivity). The advantages of this technique are the shorter preparation time and increased capacity compared to IMAC [30]. The nonspecific binding of acidic nonphosphorylated peptides can be reduced using a quenching agent, such as 2,5-dihydroxybenzoic acid [33] during MOAC enrichment. Commonly, the phosphopeptides fractionated by SCX chromatography are enriched by IMAC or TiO_2 strategy prior to MS characterization.

One of the antibody-based methods, immunopurification (IP) using immobilized anti-phosphotyrosine antibodies, is also a well-established enrichment approach in phosphoproteomics and has been reported to be highly specific for targeting tyrosine kinase signaling [19]. However, the specificity of immunoaffinity chromatography using anti-phosphoserine/threonine antibody was demonstrated to be less efficient compared to an anti-phosphotyrosine antibody [19, 34].

MS Acquisition and Data Analysis

MS Acquisition

Due to the high complexity of phosphopeptide mixtures, phosphopeptides are usually separated initially on a LC column that is coupled to MS. The effluent from the LC column is directly electrosprayed and analyzed in mass spectrometers, which measure the mass-to-charge (m/z) ratio and intensity in a MS1 spectrum [24]. Then, the mass spectrometer dissociates the peptides, which results in fragment ions in a so-called MS2 (MS/MS) spectrum [24]. Several MS acquisition strategies have been developed due to the low abundance characteristic and poor fragmentation patterns of phosphopeptides in complex protein digests.

Collision induced dissociation (CID) is the standard fragmentation technique to induce fragmentation for the analysis of phosphopeptides. However, considerable neutral loss (NL) of phosphoric acid (from serine and threonine) from the phosphopeptide ion is observed in the CID process, which results in uninformative fragment ions in MS2 and thus, a major challenge in phosphopeptide identification. To overcome this problem, the NL ion from an MS2 spectrum can be further subjected to an additional CID to generate a so-called MS3 spectrum (MS/MS/MS). This has improved fragment information and has resulted in better identification of phosphopeptides.

As an alternative fragmentation mode, electron transfer dissociation (ETD) or electron capture dissociation (ECD) can be utilized. ECD/ETD utilizes a low fragmentation energy that preserves the phosphorylation group to greatly improve site assignment [35, 36]. While CID is more sensitive for phosphopeptide identification and thus, is the preferable workflow for cases with limited sample amounts [36, 37], ETD yields more confident phosphorylation site assignments [36, 38] and has a great potential in global analyses of peptides with multiple charges (3+ and higher) and multiple phosphorylation sites [24]. Complementary application of CID and ETD modes have shown good performance in large-scale phosphoproteome analyses [39, 40], for example, by combining CID with ETD for doubly charged peptides and for more highly charged peptides, respectively [41]. Recently, newly developed high energy collision dissociation (HCD) provided sufficient fragmentation of phosphopeptides to interpret the peptide sequence without MS3 fragmentation [19].

Data Analysis

As it is often difficult to validate phosphopeptides and localize (assign) the phosphorylation site, especially for multiple phosphorylated peptides, manual assessment of each spectrum is necessary. However, this is not a feasible method for large-scale phosphoproteomic experiments since it is very time consuming. To overcome this issue, bioinformatics tools have been developed. For phosphopeptide validation, by selecting the appropriate features and corresponding criteria, NL in MS2 spectrum during CID can be used as a marker to confirm the phosphopeptides [36, 42] and the combined information from both MS2 and MS3 can improve the validation of the phosphopeptide. Recently, a classification filtering strategy using different filtering for four classes of phosphopeptides was developed to identify the phosphopeptide [36, 43]. For ETD based-phosphoproteomic experiments, standard search engines (Sequest, OMSSA, X!Tandem, or Protein Prospector) have been reported to generate good results [44].

To localize phosphorylation sites, algorithms with probability based scoring strategies such as Ascore or MSQuant have been developed to localize the phosphorylation sites. In these algorithms, the spectrum of the peptide is compared to the theoretical spectra and the localization score for each (possible) sequence position is calculated based on the probability, providing a statistical means for assigning individual phosphorylation sites within the peptide sequences [24, 36]. The phosphosites are classified according to this localization score [45] and a cutoff criterion is set to evaluate the confidence of the localization.

Conclusion

In this section, the basic workflow of shotgun phosphoproteomics analysis was described. Furthermore, quantification approaches of the phosphorylation can be incorporated into the described standard workflow. While further improvement of step-specific enrichment, the sensitivity and scan speed of the mass spectrometer, and the bioinformatics tools may improve the accuracy of phosphoproteomic analysis, the current proteomic strategy is still a very powerful tool and has its advantages in providing a high-throughput identification of protein phosphorylation, which enables a global perspective in understanding signaling and regulation of cellular functions in a broad range of biological research topics.

O-GlcNAcylation

Another competitor for the hydroxyl groups of Ser/Thr is the O-linked β-D-N-acetylglucosamine modification (O-GlcNAcylation). This modification was first identified in 1984 by Torres and Hart [46] as a small monosaccharide that is not elongated to complex sugar structures like N-linked glycosylations. O-GlucNAc is reversible, highly dynamic and in some cases, can act in a reciprocal relationship with phosphorylation. Unlike phosphorylation, which is regulated by numerous kinases and phosphatases, only two enzymes control the addition and removal of O-GlcNAc: O-GlcNAc transferase (OGT) and β-D-N-acetylglucosaminidase (O-GlcNAcase). While O-GlcNAcylation is most prevalent in metazoans, it has been found in all mammalian tissues, primarily on nuclear, cytoplasmic and mitochondrial proteins [47].

O-GlcNAcylation plays an important role in a variety of biological processes effecting the regulation of protein expression, signaling and stress response [47, 48]. Levels of O-GlcNAc have been linked to several chronic metabolic diseases including diabetes [49, 50], cancer [51, 52], and neurodegenerative disorders [53, 54]. In the heart, O-GlcNAc is thought to be one of the signals of acute stress. Levels have been found to increase with oxidative stress [55, 56]. O-GlcNAc also appears to be involved in cardioprotection in ischemia/reperfusion injury since it was found to be correlated with reduced infarct size [55]. This modification has also been found to modify proteins of the contractile myofilament proteins, thereby regulating function [57, 58]. O-GlcNAc signaling activity has also been found to increase in hypertrophy and heart failure [59].

Considerations for Sample Preparation

Like phosphorylation, O-GlcNAcylation represents a relatively small proportion of the total proteome. Key O-GlcNAc modifications have been found on regulatory transcriptions factors and other proteins in the cell that are low in abundance [48]. To account for this, it is often necessary to have a large amount of starting material. Some studies have utilized 1–30 mg of starting material although targeted enrichment and modern mass spectrometry techniques have reduced this requirement somewhat [60–63]. In addition to having enough starting material, it is important to take steps to preserve the sample's complement of O-GlcNAc modified sites. O-GlcNAcase (OGA) is the only mammalian enzyme known to remove O-GlcNAc residues. It can be specifically inhibited by PUGNAc [64], Thiamet-G [65] or others. These inhibitors can be added to cell culture or administered *in vivo* to prevent loss of O-GlcNAcylaton prior to homogenization of cells or tissues [66]. Several of the enrichment options discussed below also require removal of N-linked glycosylation, which can be done by treatment with the Peptide-N-Glycosidase F (PNGase F). This prevents cross-reaction with the other class of carbohydrate post-translational modifications.

Another important consideration is the choice of enzyme for digestion. Trypsin is the gold standard in proteomic work; however, several functionally important O-GlcNAc sites have been identified in Ser/Thr clusters with few Lys/Arg residues [47]. In these cases, digestion with trypsin alone would not produce a viable peptide for identifications. Chemical cutters, such as cyanogen bromide, can be used in combination with other proteases Asp-N, and Glu-C to increase the sequence coverage and improve the number of sites identified.

Enrichment Strategies

Due to relatively low levels of O-GlcNAc in most cells; it is necessary to perform a pre-enrichment step prior to mass spectrometry analysis. One technique that has been used successfully exploits the affinity of lectins for glycoproteins. For example, wheat germ agglutinin (WGA) will bind O-GlcNAcylated peptides using WGA-based lectin affinity chromatography although affinity and specificity can be a concern [63, 67, 68]. Also, O-GlcNAc specific antibodies have been used to enrich modified proteins and peptides for MS analysis [69, 70].

An alternate and more specific strategy has emerged utilizing an azide modified GlcNAc (UDP-GlcNAz) [71]. This modified sugar can be supplemented to cultured cells or fed to an organism and the glycan biosynthetic pathway via OGT will incorporate the UDP-GlcNAz into protein targets. Once labeled, these GlcNAz-modified proteins can be covalently derivatized by click chemistry with various biochemical probes at the site of the modified sugar [62, 72, 73]. The most common probe is biotin, which allows for high affinity capture of GlcNAz modified peptides with streptavidin. Since UDP-GlcNAz is a common substrate for many glycol-modifications, it is important to pretreat samples with PNGase F prior to the click reaction to prevent the enrichment of many other glycosylated peptides.

MS Acquisition

Once O-GlcNAc modified peptides have been enriched, one of the challenges in identifying the modified sites is that the glycosidic bond is weaker than the bonds in the peptide backbone and is much more likely to fragment during high-energy

CID [74, 75]. Since the O-GlcNAc group is preferentially lost, it is very difficult to determine the site of modification [47]. An alternative to CID has been the ECD/ETD fragmentation technique [35]. ECD/ETD uses a lower fragmentation energy that does not break the glycosidic bond, leaving the modification intact for analysis. This approach has greatly improved the direct identification of O-GlcNAcylated sites [76–78]. However, it should be noted that ECD/ETD is a less robust technique than CID and does produce fewer overall peptide identifications.

While the introduction of ECD/ETD has improved identifications of peptides carrying an intact sugar, a more popular approach has been a replacement strategy. β-elimination followed by Michael addition with dithiothreitol (BEMAD) is a procedure where the O-GlcNAc modifications are removed by mild β-elimination and the formally modified amino acid residue is labeled with DTT (see Fig. 13.1) [79, 80]. This replaces the labile glycosidic bond with a more stable thiol that can be fragmented by CID. One of the advantages of this approach is that isopotically heavy and light DTT can be used in the Micheal addition to allow relative quantitative analysis of each modified site [81]. Also, the BEMAD approach has been used in combination with UDP-GlcNAz metabolic labeling to first selectively enrich and then derivatize the O-GlcNAc modifications to improve site-specific identifications [57, 58]. It should be noted that in some cases, mild β-elimination can react with phosphorylation or other glycosylation sites. To prevent mis-assignments, it is advantageous to pre-treat samples with alkaline phosphatase and PNGase F to remove these modifications.

Conclusion

These types of approaches have been used to successfully identify hundreds of O-GlcNAc modifications in cardiac tissue. While these improvements have been substantial, they are by no means exhaustive. Further targeted investigations are likely to reveal many additional regulatory sites in physiological and pathophysiological states.

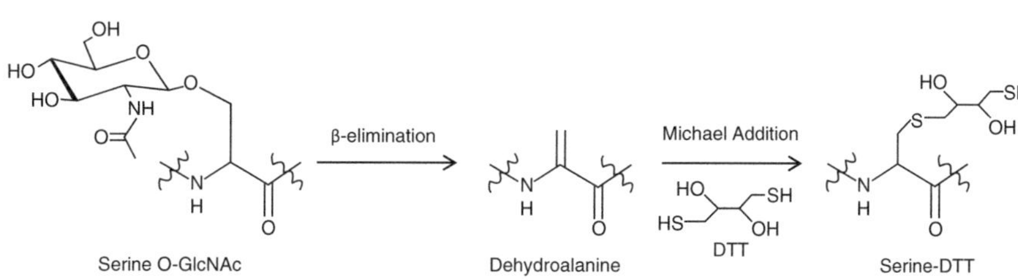

Fig. 13.1 β-elimination and Michael addition with DTT (BEMAD) reaction for replacing O-GlcNAc residue with a more mass spectrometry favorable group. DTT can also substituted for a thiol reactive biotin group

Acetylation and Other Metabogenic Modifications of Lysine

Lysine Acetylation

Students of the biological sciences are likely familiar with acetylation that caps the N-termini of nearly half of all encoded proteins. First reported over 50 years ago [82], the addition of an acetyl (CH_3CO) moiety provides stability to the protein, making it resistant to proteolysis [83]. Not long thereafter, a second more labile form of protein acetylation was found on the ε-amino group of internal lysine residues of histones [84]. The transfer of an acetyl group from acetyl-CoA to the ε-amino group neutralizes lysine's positive charge weakening a histone's interaction with DNA. This is a critical step in chromatin remodeling and gene regulation and is therefore tightly regulated by enzymes: the histone acetyltransferases (HATs) and deacetylases (HDACs). These enzymes have garnered much attention in the field of cardiac biology as they have been implicated in the gene regulatory programs underlying both exercise- and disease-induced cardiac hypertrophy [85].

Non-Enzymatic Factors that Influence Acetylation: Acetyl-CoA and NAD+

In recent years, it has become apparent that acetylation and deacetylation extend beyond histones and the nucleus, thanks in large measure, to advances in mass spectrometry and the development of acetylation-specific antibodies. Lysine "acetylomes" have now been compiled from bacteria and yeast to fruit flies and mammals [86–88]. In prokaryotes, acetylation tends to target metabolic enzymes and plays an important role in determining overall metabolic activity in response to changes in the type of nutrient source or availability [89].

The extent of protein acetylation in mammalian cells is likely influenced by levels Acetyl-CoA or at least the Acetyl-CoA/CoA ratio. Acetyl-CoA is the culminating metabolite of both fatty acid β oxidation and glycolysis, whose product pyruvate is converted to Acetyl-CoA by the pyruvate dehydrogenase complex. It is the primary source of fuel for the tricarboxylic acid cycle. Though made in the mitochondria, it can be exported to the cytoplasm. In the fed state, acetyl-CoA levels accrue, and decline in upon fasting or longer-term caloric restriction.

It is also noteworthy that some acetylation sites are substrates for a second class of deacetylase, the sirtuins, whose activity is also dependent on a metabolite cofactor NAD^+. NAD^+ may be synthesized *de novo* from dietary tryptophan and nicotinate, or reclaimed through a salvage pathway from nicotinamide, which is the primary pathway in mammals. The cell maintains two discrete pools of NAD^+, one mitochondrial pool and a cytosolic pool that is continuous with the nucleus. In the healthy heart, total NAD^+ pools do not fluctuate much. However, NAD^+ along with NADH forms an important redox couple in both compartments. Therefore the NAD^+/NADH ratio can be altered by feeding status and, in the case of the heart, changes in workload (i.e. heart rate).

Strategy for Identification of Lysine Acetylation

Considerations for Sample Preparation

As with other PTMs, the acetylation state of proteins varies broadly in terms of stoichiometry, and preserving the acetylation state over the course of sample preparation, is a top priority. Fortunately, acetylation of lysine is a chemically stable modification, provided that care is taken to minimize the activity of enzymes that add and remove acetyl groups in your cell lysate or tissue homogenate. Therefore, the age-old adage of using ice-cold extraction buffer applies yet again, as it reduces the catalytic activity of acetyltransferases and deacetylases. Activity is further minimized through the use of inhibitors that target the two broad classes of deacetylase, the HDACs and the sirtuins. To preserve the acetylome for proteomic studies, it is often advisable to use inhibitors whose specificity is fairly broad. Trichostatin A (1 μM) and sodium butyrate (5 mM) would be an example of a broad specificity HDAC inhibitors, whereas sirtuins are effectively inhibited by inclusion of nicotinamide (5 mM). Finally, unless one is performing organellar enrichment, it is often desirable to perform a protein precipitation step as quickly as possible after cell lysis. Precipitation methods such as ice-cold trichloroacetic acid precipitation or methanol/chloroform/water extraction simultaneously denature enzymes that would alter acetylation, while simultaneously removing necessary cofactors such as NAD^+. Proteins can then be resolubilized in a strong denaturant such as urea and diluted to a concentration amenable to enzymatic digestion (e.g. 1.5–2 M urea), typically by trypsin.

Acetyl-Peptide Enrichment

Like other PTMs, lysine acetylation stoichiometry varies greatly and the probability of acetyl-peptide identification, from a complex mixture of peptides, is enhanced greatly by their enrichment prior to MS. The single most important advance on this front has been the development of specific antibodies that recognize the acetylated lysine. Specifically, the antibodies are conjugated to a matrix such as agarose for the purposes of immuno-enrichment. These antibody-resin conjugates are now available from several manufacturers. Enrichment is most often performed following enzymatic digestion and peptide purification by C18 reversed-phase solid phase extraction (SPE). The amount of antibody-resin used depends on the conjugation density of the antibody (i.e. nmolAb/mg resin) and depends on the manufacturer. Typically SPE-purified peptides are dissolved in a simple buffer with a pH between 7.5 and 8. Ionic strength should not greatly impact the antibody/acetyl-peptide interaction but low ionic strength may result in greater non-specific adsorption of unacetylated peptides to the resin, and therefore the degree of acetylpeptide enrichment. Therefore, phosphate-buffered saline and Tris-buffered saline buffers are often used. Incubation of the peptides with the antibody-resin is typically often conducted at 4 °C. After incubation, antibody resin by low-speed centrifugation and the supernatant, containing unacetylated peptides, is removed. The resin is resuspended in binding buffer and re-centrifuged to remove residual supernatant and remove nonspecifically adsorbed peptides. The procedure is repeated twice more before switching to distilled deionized water for the final washes. Acetyl-peptide elution is accomplished by using a low pH solution. Protocols differ here. Elution with glycine (pH 3) is common if the eluted peptides are re-purified by SPE. Otherwise, acetyl-peptides may be eluted with 0.1 % formic acid, in which case, they are ready for LC-MS/MS.

Detection of Acetylated Peptides by Mass Spectrometry

Detection of acetylpeptides is relatively straightforward. By comparison to phosphorylation and O-GlcNacylation, the acetyl-lysine linkage is highly stable. Acetylated lysine changes the peptide mass by 42 Da, and can be readily identified on b- and y-ions yielded by CID and HCD fragmentation.

Beyond Lysine Acetylation: New Metabogenic Modifications of Lysine

Sources of Low Molecular Weight Acyl-CoA Species

If lysine acetylation arises by reaction proteins with acetyl-CoA, does it follow that proteins may be modified at lysine by other acyl-CoA species? Indeed, it does. Within the last 10 years newly characterized acyl modifications of lysine include succinylation, malonylation, glutarylation, propionylation, butyrylation and crotonylation [90–95] stemming from the reaction of proteins with succinyl-, malonyl, glutaryl-, propionyl-CoA, butyryl- and crotonyl-CoA respectively. The modifications obviously differ from acetylation with respect to the mass of the acyl moiety appended to lysine. However, the dicarboxylates, such as succinyl, malonyl, and glutaryl moieties also impart a negative charge to the modified site at physiological pH.

The Acyl-CoA metabolites arise primarily from fatty acid and amino acid catabolic pathways. For instance, butyryl-CoA is the penultimate (4-carbon) acyl-CoA species generated by beta oxidation of even-number carbon chains. Beta oxidation of odd-chain fatty acids ultimately yields propionyl-CoA. Propionyl-CoA, in turn, is metabolized to succinyl-CoA in the mitochondria. Propionyl-CoA and succinyl-CoA, are also downstream metabolites in valine, isoleucine and methionine metabolism, whereas glutaryl- and crotonyl-CoA belong to lysine and tryptophan catabolic pathways. Only malonyl-CoA is generated by a biosynthetic reaction, namely the carboxylation of acetyl-CoA by acetyl-CoA carboxylase.

Enrichment and Detection of Lysine Acyl Peptides

Advances in this field can be attributed to yeoman's work by Yingming Zhao and colleagues. The original discovery of the modifications involved synthesis of acylated peptides derived from the histones, to serve as standards, then searching for spectra that matched those standards. As in the case of acetylation, broad-scale identification of the novel lysine acylation has subsequently been greatly facilitated through the generation of modification-specific antibodies. Acyl-lysine bearing peptides are best prepared and enriched essentially as described in the section **Strategy for Identification of Lysine Acetylation**.

The modifications mirror acetylation in terms of stability in the mass spectrometer, and while dicarboxylate modifications introduce a negative charge at physiological pH, they are fully protonated in low pH MS workflows. The modifications can, therefore, be distinguished by the masses they impart to the modified peptide. Propionylation, crotonylation, butyrylation, malonylation, succinylation and glutarylation confer peptide mass shifts of 56, 68, 70, 86, 100 and 114 Da respectively [90–95].

Conclusion

The identification of new acylation modifications on internal lysines has flourished within the present decade. The building blocks for these modifications, the various acyl-CoAs, operate at a unique node, coupling cellular metabolism to protein function. Our understanding of the complex relationships between the cell's metabolic state, the quantitative and qualitative characteristics of the modifications and their impact on protein function is crude at this stage. More work is needed to understand the implications of dysregulated protein acylation in the context of cardiac dysfunction. There is no question that proteomics will continue to provide exciting new discoveries, as to the nature of these modifications, their favored targets, and their impact in diverse cellular functions.

Cysteine Modifications

Nitroso-Redox Equilibrium

The nitroso-redox equilibrium is emerging as a crucial regulator of many processes that govern normal cellular physiology and pathophysiology [96]. This equilibrium is characterized by a delicate balance that exists between the production of reactive nitrogen and reactive oxygen species, which are maintained at low levels through the actions of cellular anti-oxidant defenses. At low levels, reactive nitrogen and reactive oxygen species play important roles as signaling molecules and second messengers, in part, via post-translational protein modification [97–99]. There is a number of different amino acids that are targeted for redox-dependent post-translational modification, including methionine oxidation, tyrosine nitration, and lysine or histidine carbonylation, but this section will focus exclusively on cysteine thiol modifications. Reactive nitrogen and reactive oxygen species typically signal at low levels through reversible cysteine thiol modifications, including *S*-nitrosylation, *S*-sulfhydration, and *S*-glutathionylation [6]. However, the overproduction of reactive nitrogen and/or reactive oxygen species can induce nitrosative or oxidative stress, a common hallmark of many human diseases, including ischemic heart disease, diabetes, and stroke [100–102]. Nitrosative and/or oxidative stress typically results in the higher order oxidation of cysteine thiols, and these modifications include disulfide adducts, sulfenylation, sulfinic acid, and sulfonic acid. The widespread oxidation of cysteine thiols can lead to a nitroso-redox imbalance, which may alter cellular signaling, induce cellular dysfunction and possibly lead to cell death.

Sources of Reactive Nitrogen and Reactive Oxygen Species

Nitric oxide is one of the primary reactive nitrogen species in the cell, and is produced through the action of nitric oxide synthase [99, 103]. The non-enzymatic reduction of nitrite is another potential source of nitric oxide, although this typically only occurs under conditions of low pH as with ischemia [104]. Typical reactive oxygen species include superoxide and hydrogen peroxide, as well as stronger oxidants like peroxynitrite and hydroxyl radicals [105]. Under physiological conditions, the generation of reactive oxygen species occurs via mitochondrial respiration. Under pathological conditions, however, the upregulation of enzymatic electron donors like NADPH oxidase and xanthine oxidase can greatly enhance the production of reactive oxygen species; uncoupled nitric oxide synthase is another potential source.

Redox-Based Cysteine Thiol Modifications

Reversible Cysteine Thiol Modifications

Many reversible cysteine thiol modifications play important roles as signaling molecules for reactive nitrogen and reactive oxygen species. The effects of these modifications on target proteins are diverse, with potential impacts on protein function and stability, protein-protein interaction, and protein localization [6, 99]. These modifications may also protect the modified cysteine thiol from additional oxidative modification [106–109]. *S*-nitrosylation is one of the most well-described redox-dependent cysteine thiol modifications. This modification is specific and reversible, and results from the covalent addition of a nitric oxide moiety to a cysteine thiol (RS-NO). *S*-glutathionylation is another reversible modification that is more stable than *S*-nitrosylation, and results from the addition of glutathione to a cysteine thiol

(RS-SG). *S*-sulfhydration is also a stable and reversible modification that results from the interaction between hydrogen sulfide and a cysteine thiol, leading to the formation of a hydropersulfide moiety (RS-SH). Increased reactive oxygen species production can lead to the further oxidation of cysteine thiols, resulting in the formation of disulfide adducts and sulfenylation. Sulfenylation or sulfenic acid commonly results from the oxidation of cysteine thiols by hydrogen peroxide (RS-OH). This modification was initially thought to be detrimental to cellular function, but recent evidence suggests that the highly reactive and labile nature of this modification may allow it to serve as a critical intermediate for the formation of other oxidative modifications, including disulfide bonds. Disulfide bonds (RS-SR) are important for the maintenance of protein structure and/or complex formation, and recent studies suggest that disulfide adducts may also impact protein function [110, 111].

Methods to Assess Reversible Cysteine Thiol Modifications

To characterize the role of redox-dependent cysteine thiol modifications in health and disease, a number of proteomic and mass spectrometry-based approaches have been developed to identify modified proteins and modification sites. Many of these approaches take advantage of reducing agents that target specific redox-dependent cysteine thiol modifications, as well as a large number of cysteine-reactive compounds and labels. These approaches range from the use of switch assays with isobaric mass tags for labile modifications to antibody-based enrichment strategies for more stably modified proteins. These strategies are critical for assessing the role of redox-dependent modifications in the regulation of protein function, and for determining spatial localization, target specificity, and mechanisms for the addition and removal of reversible modifications.

S-nitrosylation

The labile nature of *S*-nitrosylation has often made this a difficult modification to evaluate, as *S*-nitrosylation can easily be lost during sample preparation, but a number of different approaches have been developed to identify specific modification sites and assess the biological function of *S*-nitrosylation in cellular physiology. These strategies incorporate various measures to prevent the degradation of *S*-nitrosylation.

Direct Measure with Mass Spectrometry

A small number of studies have successfully utilized mass spectrometry to directly identify *S*-nitrosylated cysteine residues. Unfortunately, these studies were limited to the use of purified proteins [112, 113], and the identification of *S*-nitrosylated proteins in complex samples has not been reported. This approach remains a challenge because *S*-nitrosylation tends to degrade upon exposure to the common acidic conditions utilized for liquid chromatography tandem mass spectrometry (LC-MS/MS). The low level of endogenous *S*-nitrosylation is also a confounding factor [114].

Biotin Switch Assay

The most common approach for examining protein *S*-nitrosylation is the biotin switch assay [115]. This assay circumvents lability issues by replacing modified cysteine residues with a pyridyldithiol biotin (biotin-HPDP) label (see Fig. 13.2). With this approach, free cysteine thiols are blocked with a methylthiolating agent, such as methyl methanethiosulfonate (MMTS). *S*-nitrosylated cysteine residues are then preferentially reduced with ascorbate and labeled with biotin-HPDP. At this point, biotinylated proteins can be examined via western blot. Alternatively, biotinylated proteins can be enriched using streptavidin resin for subsequent LC-MS/MS analysis. Since the inception of the biotin switch assay, many variations of this approach have been developed. These variations include the use of fluorescent labels (i.e., DyLight maleimide, CyDye maleimide) in tandem with two-dimensional gel electrophoresis [116–119]. The fluorescent basis of this approach provides for accurate quantitation, as well as a visual representation of protein *S*-nitrosylation differences between samples. Fluorescent spots can also be picked and modified proteins identified via LC-MS/MS. However, this approach is not effective for the identification of specific modification sites [120]. Fortunately, there are a number of high-throughput approaches that have since been developed for the specific identification of *S*-nitrosylation sites, including *S*-nitrosylation-resin-assisted capture (SNO-RAC) [107, 120, 121]. SNO-RAC utilizes a thiol-binding resin (i.e., thiopropyl sepharose) instead of biotin-HPDP to capture and enrich for *S*-nitrosylated protein targets. Specific modification sites can then be identified using LC-MS/MS, and quantified via label-free peptide analysis. Cysteine-reactive isotope-coded affinity tags [122] and tandem mass tags [123, 124] have also been incorporated into the modified biotin switch approach for the identification and quantitation of specific modification sites using LC-MS/MS. These cysteine-reactive labels confer the advantage of multiple isobaric tags (isotope coded affinity tag: 2-plex; tandem mass tag: 6-plex) with reporter ions in defined regions of the mass spectra. These labels have also been used to assess *S*-nitrosylation occupancy, or the percentage of a given cysteine residue that is modified via *S*-nitrosylation [114]. In this case, free cysteine thiols are blocked with one isobaric label, while *S*-nitrosylated cysteine thiols are reduced and labeled with a second isobaric label. Additional variations of the biotin switch assay include *S*-nitrosylation Site Identification (SNOSID) [125], the HIS-TAG switch method [126], and *S*-nitrosothiol capture (SNOCAP) [127]. As

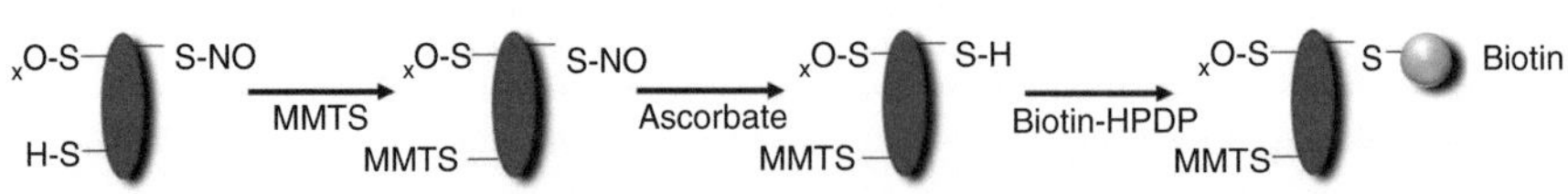

Fig. 13.2 Biotin switch assay for examining protein *S*-nitrosylation. *S-H* free cysteine thiol, *S-NO* *S*-nitrosylated cysteine thiol, *S-O$_x$* oxidized cysteine thiol (other than S-NO)

with any approach, the biotin switch has inherent limitations, namely due to the potential of ascorbate to non-specifically reduce other cysteine thiol modifications [128–130]. Although this claim has been refuted [131, 132], appropriate positive and negative controls should always be used with this approach.

Mercuric Chloride

An alternative to the biotin switch assay utilizes the reaction between *S*-nitrosylation and phenylmercury compounds to enrich for *S*-nitrosylated proteins [133]. With this approach, free cysteine thiols are blocked with MMTS, and ρ-amino-phenylmercuric acetate conjugated to agarose beads is used to capture *S*-nitrosylated proteins. Following the release of captured peptides, cysteine thiols are then oxidized to sulfonic acid to facilitate the site-specific identification of modified cysteine residues using LC-MS/MS. This approach avoids the use of ascorbate, but appropriate controls should still be used to account for the specificity of the phenylmercury compound.

Antibody Enrichment

There are a number of commercial antibodies for the detection of protein *S*-nitrosylation, which may allow for antibody-based enrichment and identification via LC-MS/MS. However, many of these antibodies were raised against *S*-nitrosylated bovine serum albumin, so antibody specificity may be a potential issue with this approach.

S-glutathionylation

S-glutathionylation is a relatively stable redox-dependent modification in comparison to more labile modifications like *S*-nitrosylation. The primary high-throughput approach for examining *S*-glutathionylation uses a very similar workflow to that described for the biotin switch assay, with several notable exceptions [134–136]. *S*-nitrosylation is first removed via UV-induced photolysis, followed by blocking with *N*-ethylmaleimide; *N*-ethylmaleimide is used instead of MMTS in order to form an irreversible linkage that cannot be reduced in the later steps of this protocol. *S*-glutathionylation is then reduced with recombinant glutatredoxin-1, which preferentially targets mixed disulfides. Previously modified residues are then enriched using phenylmercury capture and identified using LC-MS/MS [136]. In theory, glutaredoxin-1 can be used to examine *S*-glutathionylation in tandem with many of the approaches previously described for *S*-nitrosylation, including fluorescent maleimide labels for gel electrophoresis and cysteine-reactive tandem mass tags for site identification and quantification. There are also a number of commercial antibodies for the detection of protein *S*-glutathionylation, which may allow for antibody-based enrichment and identification via LC-MS/MS.

S-sulfhydration

S-sulfhydration is one of the more recently described redox-dependent modifications, and to date, only a small number of *S*-sulfhydrated proteins and modification sites have been identified. A modified version of the biotin switch assay was initially developed to identify *S*-sulfhydrated protein targets [137]. With this approach, free cysteine thiols are blocked with MMTS. Biotin-HPDP is then added to react directly with *S*-sulfhydrated residues; ascorbate and other reducing agents are not used with this approach. Biotinylated proteins can then be enriched via streptavidin beads and analyzed using LC-MS/MS. A reverse labeling approach has also been developed whereby free and *S*-sulfhydrated thiols are both labeled with a fluorescent maleimide tag [138]. The labeled protein is then subjected to reduction with dithiothreitol (DTT), which will only reduce the labeled *S*-sulfhydrated cysteine thiols. The decrease in fluorescence intensity is reported to correspond the level of *S*-sulfhydration for a given protein target.

Disulfide Bonds

The identification of disulfide bonds can be complex, namely due to the involvement of two cysteine thiols within the same protein (intra-protein disulfide) or between different proteins (inter-protein disulfide). Inter-protein disulfides are commonly examined with gel-based approaches, including diagonal gel electrophoresis [139–141]. Modified proteins can then be extracted from the gel and identified using LC-MS/MS. Intra-protein disulfides can be examined using compounds such as dibromobimane, which selectively modifies vicinal thiols [142]. The remaining non-vicinal thiols are reduced and blocked (i.e., with *N*-ethylmaleimide). Subsequently, dibromobimane-labeled thiols can be reduced, labeled, and enriched for LC-MS/MS analysis [142]. Additional approaches use LC-MS/MS to identify disulfide bonds by comparing the masses of reduced and non-reduced protein digests which are proteolyzed with intact disulfide bonds [143], or with partial reduction and partial alkylation strategies [144].

Sulfenylation

Sulfenylation is a highly reactive and unstable modification that can be easily lost during sample preparation. However, despite the labile nature of this modification, a number of approaches have been developed to examine sulfenylation.

The first approach utilizes a modified version of the biotin switch assay, with the exception that the sulfenic acid-specific reducing agent arsenite is substituted for ascorbate. Following the arsenite-mediated reduction of sulfenic acid, cysteine thiols can be labeled with biotin-maleimide and examined via western blot, or enriched with streptavidin resin and analyzed using LC-MS/MS [145]. Isotope coded affinity tags have also been substituted for biotin-maleimide with this approach [146, 147]. A second approach involves the use of dimedone and analogs thereof, which convert sulfenylation into a more stable derivative [136, 148–151]. Dimedone-labeled proteins can then be enriched and analyzed via LC-MS/MS. An anti-dimedone sulfenic acid antibody is also commercially available.

Irreversible Cysteine Thiol Modifications

Sulfinic acid (RS-O_2H) and sulfonic acid (RS-O_3H) result from the further oxidation of sulfenic acid. These modifications are extremely stable and considered to be irreversible by normal cellular mechanisms. As such, these modifications are commonly used as markers of oxidative stress [152], although recent evidence suggests that these modifications may also play an important role in cellular signaling [153].

The irreversible nature of sulfinic and sulfonic acid rules out the possibility of using the reduction/labeling/capture approach described for other redox-dependent cysteine thiol modifications. However, the extreme stability of these modifications is highly conducive to the direct detection of sulfinic [154, 155] and sulfonic acid [152, 156] via LC-MS/MS. A number of commercial antibodies are also available for examining sulfinic and sulfonic acid, leaving open the possibility for antibody-based enrichment strategies.

Conclusion

There are a number of different proteomic and mass spectrometry-based approaches that have been developed for examining redox-dependent cysteine thiol modifications. Many of these approaches are emerging and adaptable, and can be used to examine additional redox-dependent modifications by simple reagent substitution. When selecting a specific approach for a particular study, it is important to consider the specific experimental objectives (i.e., targeted approach vs. high-throughput approach). A multi-faceted approach utilizing multiple strategies is likely to yield more robust results, avoid the common pitfalls of each of the individual approaches, and reduce the probability of false-positive identifications. However, as with all approaches, suitable positive and negative controls should always be utilized. Non-specific reducing agents such as tris(2-carboxyethyl)phosphine hydrochloride (TCEP) and DTT are useful for reducing a large majority of the redox-dependent cysteine thiol modifications described previously in order to assesses global changes in cysteine thiol oxidation, but the ultimate challenge moving forward will lie in the identification of compounds and strategies to specifically reduce individual cysteine thiol modifications. These targeted strategies will allow for additional and expanded mechanistic studies examining the dynamic nature of cysteine thiol modifications in health and disease.

Overall Conclusions

Post-translational modifications are a central aspect of the dynamic cardiovascular proteome. The temporal and spatial regulation of protein function by modifications can alter tissue function or signal disease. Using the tools of advanced proteomic analysis it has been possible to systemically identify and characterize these modifications with the goal of understanding their role in regulation. In addition to mechanistic insights obtained from these studies, this information has great clinical potential. Identifying critical post-translational modifications can be used to develop new and more effective pharmaceuticals. More recently, proteomic analysis and characterization of these modifications has been combined with powerful diagnostic approaches such as multiple reaction monitoring or data independent acquisition techniques. These approaches can screen large numbers of clinical samples for biomarker discovery or to stage the progression of disease. Incorporating PTM data into these advanced diagnostic platforms will better reflect the dynamic regulation involved in pathophysiological conditions.

References

1. Mann M, Jensen ON. Proteomic analysis of post-translational modifications. Nat Biotechnol. 2003;21(3):255–61.
2. Liddy KA, White MY, Cordwell SJ. Functional decorations: post-translational modifications and heart disease delineated by targeted proteomics. Genome Med [Internet]. 2013;5(2):20. Available from: http://www.ncbi.nlm.nih.gov/pubmed/23445784.
3. Smith LE, White MY. The role of post-translational modifications in acute and chronic cardiovascular disease. Proteomics Clin Appl [Internet]. 2014;8(7–8):506–21. Available from: http://www.ncbi.nlm.nih.gov/pubmed/24961403.
4. James P, Inui M, Tada M, Chiesi M, Carafoli E. Nature and site of phospholamban regulation of the Ca^{2+} pump of sarcoplasmic reticulum. Nature. 1989;342(6245):90–2.
5. Wegener A, Simmerman H, Lindemann J, Jones L. Phospholamban phosphorylation in intact ventricles. Phosphorylation of serine 16 and threonine 17 in response to beta-adrenergic stimulation. J Biol Chem [Internet]. 1989;264(19):11468. Available from: http://www.jbc.org/content/264/19/11468.abstract.
6. Chung HS, Wang SB, Venkatraman V, Murray CI, Van Eyk JE. Cysteine oxidative posttranslational modifications: emerging regulation in the cardiovascular system. Circ Res. 2013;112(2):382–92.
7. Buchanan BB, Balmer Y. Redox regulation: a broadening horizon. Annu Rev Plant Biol. 2005;56:187–220.
8. Bae YS, Oh H, Rhee SG, Yoo YD. Regulation of reactive oxygen species generation in cell signaling. Mol Cells [Internet]. 2011;32(6):491–509. Available from: http://www.ncbi.nlm.nih.gov/pmc/articles/PMC3887685/.

9. Shao D, Oka S, Brady CD, Haendeler J, Eaton P, Sadoshima J. Redox modification of cell signaling in the cardiovascular system. J Mol Cell Cardiol [Internet]. 2012;52(3):550–8. Available from: http://www.pubmedcentral.nih.gov/articlerender.fcgi?artid=3237398&tool=pmcentrez&rendertype=abstract.
10. Canton M, Menazza S, Sheeran FL, Polverino De Laureto P, Di Lisa F, Pepe S. Oxidation of myofibrillar proteins in human heart failure. J Am Coll Cardiol [Internet]. 2011;57(3):300–9. Available from: http://dx.doi.org/10.1016/j.jacc.2010.06.058.
11. Canton M, Neverova I, Menabò R, Van Eyk J, Di Lisa F. Evidence of myofibrillar protein oxidation induced by postischemic reperfusion in isolated rat hearts. Am J Physiol Heart Circ Physiol. 2004;286(3):H870–7.
12. Canton M, Skyschally A, Menabò R, Boengler K, Gres P, Schulz R, et al. Oxidative modification of tropomyosin and myocardial dysfunction following coronary microembolization. Eur Heart J [Internet]. 2006;27(7):875–81. Available from: http://www.ncbi.nlm.nih.gov/pubmed/16434410.
13. Zhao Y, Jensen ON. Modification-specific proteomics: strategies for characterization of post-translational modifications using enrichment techniques. Proteomics. 2009;9(20):4632–41.
14. Huang J, Wang F, Ye M, Zou H. Enrichment and separation techniques for large-scale proteomics analysis of the protein post-translational modifications. J Chromatogr A [Internet]. 2014;1372C:1–17. Available from: http://www.sciencedirect.com/science/article/pii/S0021967314017208.
15. Mertins P, Qiao JW, Patel J, Udeshi ND, Clauser KR, Mani DR, et al. Integrated proteomic analysis of post-translational modifications by serial enrichment. Nat Methods [Internet]. 2013;10(7):634–7. Available from: http://www.pubmedcentral.nih.gov/articlerender.fcgi?artid=3943163&tool=pmcentrez&rendertype=abstract.
16. Murray CI, Van Eyk JE. Chasing cysteine oxidative modifications: proteomic tools for characterizing cysteine redox status. Circ Cardiovasc Genet. 2012;5(5):1–10.
17. Seet BT, Dikic I, Zhou M-M, Pawson T. Reading protein modifications with interaction domains. Nat Rev Mol Cell Biol. 2006;7(7):473–83.
18. Roux PP, Thibault P. The coming of age of phosphoproteomics – from large data sets to inference of protein functions. Mol Cell Proteomics [Internet]. 2013;12(12):3453–64. Available from: http://www.ncbi.nlm.nih.gov/pubmed/24037665.
19. Sun Z, Hamilton KL, Reardon KF. Phosphoproteomics and molecular cardiology: techniques, applications and challenges. J Mol Cell Cardiol. 2012;53(3):354–68.
20. Hoffert JD, Knepper MA. Taking aim at shotgun phosphoproteomics. Anal Biochem. 2008;375(1):1–10.
21. Ficarro SB, Zhang Y, Carrasco-Alfonso MJ, Garg B, Adelmant G, Webber JT, et al. Online nanoflow multidimensional fractionation for high efficiency phosphopeptide analysis. Mol Cell Proteomics. 2011;10(11):O111.011064.
22. Villén J, Gygi SP. The SCX/IMAC enrichment approach for global phosphorylation analysis by mass spectrometry. Nat Protoc. 2008;3(10):1630–8.
23. McNulty DE, Annan RS. Hydrophilic interaction chromatography reduces the complexity of the phosphoproteome and improves global phosphopeptide isolation and detection. Mol Cell Proteomics. 2008;7(5):971–80.
24. Macek B, Mann M, Olsen JV. Global and site-specific quantitative phosphoproteomics: principles and applications. Annu Rev Pharmacol Toxicol. 2009;49:199–221.
25. Beausoleil SA, Jedrychowski M, Schwartz D, Elias JE, Villén J, Li J, et al. Large-scale characterization of HeLa cell nuclear phosphoproteins. Proc Natl Acad Sci U S A. 2004;101(33):12130–5.
26. Alpert AJ. Hydrophilic-interaction chromatography for the separation of peptides, nucleic acids and other polar compounds. J Chromatogr. 1990;499:177–96.
27. Nühse T, Yu K, Salomon A. Isolation of phosphopeptides by immobilized metal ion affinity chromatography. Curr Protoc Mol Biol. 2007;Chapter 18:Unit 18.13. pp. 1–23.
28. Andersson L, Porath J. Isolation of phosphoproteins by immobilized metal (Fe3+) affinity chromatography. Anal Biochem. 1986;154(1):250–4.
29. Pinkse MWH, Uitto PM, Hilhorst MJ, Ooms B, Heck AJR. Selective isolation at the femtomole level of phosphopeptides from proteolytic digests using 2D-NanoLC-ESI-MS/MS and titanium oxide precolumns. Anal Chem. 2004;76(14):3935–43.

30. Nita-Lazar A, Saito-Benz H, White FM. Quantitative phosphoproteomics by mass spectrometry: past, present, and future. Proteomics. 2008;8(21):4433–43.
31. Aryal UK, Ross ARS. Enrichment and analysis of phosphopeptides under different experimental conditions using titanium dioxide affinity chromatography and mass spectrometry. Rapid Commun Mass Spectrom. 2010;24(2):219–31.
32. Sano A, Nakamura H. Titania as a chemo-affinity support for the column-switching HPLC analysis of phosphopeptides: application to the characterization of phosphorylation sites in proteins by combination with protease digestion and electrospray ionization mass spectrometry. Anal Sci. 2004;20(5):861–4.
33. Larsen MR, Thingholm TE, Jensen ON, Roepstorff P, Jørgensen TJD. Highly selective enrichment of phosphorylated peptides from peptide mixtures using titanium dioxide microcolumns. Mol Cell Proteomics. 2005;4(7):873–86.
34. Collins MO, Yu L, Choudhary JS. Analysis of protein phosphorylation on a proteome-scale. Proteomics. 2007;7(16):2751–68.
35. Syka JEP, Coon JJ, Schroeder MJ, Shabanowitz J, Hunt DF. Peptide and protein sequence analysis by electron transfer dissociation mass spectrometry. Proc Natl Acad Sci U S A. 2004;101(26):9528–33.
36. Wang F, Song C, Cheng K, Jiang X, Ye M, Zou H. Perspectives of comprehensive phosphoproteome analysis using shotgun strategy. Anal Chem. 2011;83(21):8078–85.
37. Stingl C, Ihling C, Ammerer G, Sinz A, Mechtler K. Application of different fragmentation techniques for the analysis of phosphopeptides using a hybrid linear ion trap-FTICR mass spectrometer. Biochim Biophys Acta Pro Proteomics. 2006;1764(12):1842–52.
38. Beck F, Lewandrowski U, Wiltfang M, Feldmann I, Geiger J, Sickmann A, et al. The good, the bad, the ugly: validating the mass spectrometric analysis of modified peptides. Proteomics. 2011;11(6):1099–109.
39. Molina H, Horn DM, Tang N, Mathivanan S, Pandey A. Global proteomic profiling of phosphopeptides using electron transfer dissociation tandem mass spectrometry. Proc Natl Acad Sci U S A. 2007;104(7):2199–204.
40. Swaney DL, Wenger CD, Thomson JA, Coon JJ. Human embryonic stem cell phosphoproteome revealed by electron transfer dissociation tandem mass spectrometry. Proc Natl Acad Sci U S A. 2009;106(4):995–1000.
41. Good DM, Wirtala M, McAlister GC, Coon JJ. Performance characteristics of electron transfer dissociation mass spectrometry. Mol Cell Proteomics. 2007;6(11):1942–51.
42. Martin DMA, Nett IRE, Vandermoere F, Barber JD, Morrice NA, Ferguson MAJ. Prophossi: automating expert validation of phosphopeptide-spectrum matches from tandem mass spectrometry. Bioinformatics. 2010;26(17):2153–9.
43. Jiang X, Ye M, Han G, Dong X, Zou H. Classification filtering strategy to improve the coverage and sensitivity of phosphoproteome analysis. Anal Chem. 2010;82(14):6168–75.
44. Bakalarski CE, Haas W, Dephoure NE, Gygi SP. The effects of mass accuracy, data acquisition speed, and search algorithm choice on peptide identification rates in phosphoproteomics. Anal Bioanal Chem. 2007;389(5):1409–19.
45. Olsen JV, Blagoev B, Gnad F, Macek B, Kumar C, Mortensen P, et al. Global, in vivo, and site-specific phosphorylation dynamics in signaling networks. Cell. 2006;127(3):635–48.
46. Torres CR, Hart GW. Topography and polypeptide distribution of terminal N-acetylglucosamine residues on the surfaces of intact lymphocytes. Evidence for O-linked GlcNAc. J Biol Chem. 1984;259(5):3308–17.
47. Ma J, Hart GW. O-GlcNAc profiling: from proteins to proteomes. Clin Proteomics [Internet]. 2014;11(1):8. Available from: http://www.pubmedcentral.nih.gov/articlerender.fcgi?artid=4015695&tool=pmcentrez&rendertype=abstract.
48. Ngoh GA, Facundo HT, Zafir A, Jones SP. O-GlcNAc signaling in the cardiovascular system. Circ Res. 2010;107(2):171–85.
49. Issad T, Masson E, Pagesy P. O-GlcNAc modification, insulin signaling and diabetic complications. Diabetes Metab [Internet]. 2010;36(6 Pt 1):423–35. Available from: http://dx.doi.org/10.1016/j.diabet.2010.09.001.

50. Vaidyanathan K, Wells L. Multiple tissue-specific roles for the O-GlcNAc post-translational modification in the induction of and complications arising from type II diabetes. J Biol Chem. 2014;289(50):34466–71.
51. Slawson C, Hart GW. O-GlcNAc signalling: implications for cancer cell biology. Nat Rev Cancer [Internet]. 2011;11(9):678–84. Available from: http://dx.doi.org/10.1038/nrc3114.
52. Zhang X, Ma L, Qi J, Shan H, Yu G, Gu Y. MAPK/ERK signaling pathway-induced hyper-O-GlcNAcylation enhances cancer malignancy. Mol Cell Biochem. 2015;410(1–2):101–10.
53. Lazarus BD, Love DC, Hanover JA. O-GlcNAc cycling: implications for neurodegenerative disorders. Int J Biochem Cell Biol [Internet]. 2009;41(11):2134–46. Available from: http://www.pubmedcentral.nih.gov/articlerender.fcgi?artid=2757632&tool=pmcentrez&rendertype=abstract.
54. Wang P, Lazarus B, Forsythe M, Love D, Krause M, Hanover J. O-GlcNAc cycling mutants modulate proteotoxicity in Caenorhabditis elegans models of human neurodegenerative diseases. Proc Natl Acad Sci U S A. 2012;109(43):17669–74.
55. Jones SP, Zachara NE, Ngoh GA, Hill BG, Teshima Y, Bhatnagar A, et al. Cardioprotection by N-acetylglucosamine linkage to cellular proteins. Circulation. 2008;117(9):1172–82.
56. Champattanachai V, Marchase RB, Chatham JC. Glucosamine protects neonatal cardiomyocytes from ischemia-reperfusion injury via increased protein-associated O-GlcNAc. Am J Physiol Cell Physiol [Internet]. 2007;292(1):C178–87. Available from: http://www.ncbi.nlm.nih.gov/pubmed/16899550.
57. Ramirez-Correa GA, Jin W, Wang Z, Zhong X, Gao WD, Dias WB, et al. O-linked GlcNAc modification of cardiac myofilament proteins: a novel regulator of myocardial contractile function. Circ Res. 2008;103:1354–8.
58. Hédou J, Bastide B, Page A, Michalski J-C, Morelle W. Mapping of O-linked beta-N-acetylglucosamine modification sites in key contractile proteins of rat skeletal muscle. Proteomics [Internet]. 2009;9(8):2139–48. Available from: http://www.ncbi.nlm.nih.gov/pubmed/19322778.
59. Lunde IG, Aronsen JM, Kvaløy H, Qvigstad E, Sjaastad I, Tønnessen T, et al. Cardiac O-GlcNAc signaling is increased in hypertrophy and heart failure. Physiol Genomics. 2012;44(2):162–72.
60. Alfaro JF, Gong C-X, Monroe ME, Aldrich JT, Clauss TRW, Purvine SO, et al. Tandem mass spectrometry identifies many mouse brain O-GlcNAcylated proteins including EGF domain-specific O-GlcNAc transferase targets. Proc Natl Acad Sci U S A [Internet]. 2012;109(19):7280–5. Available from: http://www.pubmedcentral.nih.gov/articlerender.fcgi?artid=3358849&tool=pmcentrez&rendertype=abstract.
61. Trinidad JC, Barkan DT, Gulledge BF, Thalhammer A, Sali A, Schoepfer R, et al. Global identification and characterization of both O-GlcNAcylation and phosphorylation at the murine synapse. Mol Cell Proteomics. 2012;11(8):215–29.
62. Hahne H, Sobotzki N, Nyberg T, Helm D, Borodkin VS, van Aalten DMF, et al. Proteome wide purification and identification of O-GlcNAc-modified proteins using click chemistry and mass spectrometry. J Proteome Res [Internet]. 2013;12(2):927–36. Available from: http://www.ncbi.nlm.nih.gov/pubmed/23301498\n, http://dx.doi.org/10.1021/pr300967y.
63. Myers SA, Panning B, Burlingame AL. Polycomb repressive complex 2 is necessary for the normal site-specific O-GlcNAc distribution in mouse embryonic stem cells. Proc Natl Acad Sci U S A. 2011;108(23):9490–5.
64. Horsch M, Hoesch L, Vasella A, Rast DM. N-acetylglucosaminono-1,5-lactone oxime and the corresponding (phenylcarbamoyl)oxime. Novel and potent inhibitors of beta-N-acetylglucosaminidase. Eur J Biochem. 1991;197(3):815–8.
65. Yuzwa SA, Macauley MS, Heinonen JE, Shan X, Dennis RJ, He Y, et al. A potent mechanism-inspired O-GlcNAcase inhibitor that blocks phosphorylation of tau in vivo. Nat Chem Biol. 2008;4(8):483–90.
66. Ramirez-Correa GA, Ferrando IM, Hart G, Murphy A. Detection of O-GlcNAc modifications on cardiac myofilament proteins. Methods Mol Biol [Internet]. 2013;1005:157–68. Available from: http://www.ncbi.nlm.nih.gov/pubmed/23606256.

67. Vosseller K, Trinidad JC, Chalkley RJ, Specht CG, Thalhammer A, Lynn AJ, et al. O-linked N-acetylglucosamine proteomics of postsynaptic density preparations using lectin weak affinity chromatography and mass spectrometry. Mol Cell Proteomics. 2006;5(5):923–34.
68. Chalkley RJ, Thalhammer A, Schoepfer R, Burlingame AL. Identification of protein O-GlcNAcylation sites using electron transfer dissociation mass spectrometry on native peptides. Proc Natl Acad Sci U S A [Internet]. 2009;106(22):8894–9. Available from: http://www.ncbi.nlm.nih.gov/entrez/query.fcgi?cmd=Retrieve&db=PubMed&dopt=Citation&list_uids=19458039.
69. Teo CF, Ingale S, Wolfert MA, Elsayed GA, Nöt LG, Chatham JC, et al. Glycopeptide-specific monoclonal antibodies suggest new roles for O-GlcNAc. Nat Chem Biol [Internet]. 2010;6(5):338–43. Available from: http://dx.doi.org/10.1038/nchembio.338.
70. Zhao P, Viner R, Teo CF, Boons GJ, Horn D, Wells L. Combining high-energy C-trap dissociation and electron transfer dissociation for protein O-GlcNAc modification site assignment. J Proteome Res. 2011;10(9):4088–104.
71. Vocadlo DJ, Hang HC, Kim E-J, Hanover JA, Bertozzi CR. A chemical approach for identifying O-GlcNAc-modified proteins in cells. Proc Natl Acad Sci U S A. 2003;100(Track II):9116–21.
72. Gurcel C, Vercoutter-Edouart A-S, Fonbonne C, Mortuaire M, Salvador A, Michalski J-C, et al. Identification of new O-GlcNAc modified proteins using a click-chemistry-based tagging. Anal Bioanal Chem [Internet]. 2008;390(8):2089–97. Available from: http://www.ncbi.nlm.nih.gov/pubmed/18369606.
73. Parker BL, Gupta P, Cordwell SJ, Larsen MR, Palmisano G. Purification and identification of O-GlcNAc-modified peptides using phosphate-based alkyne CLICK chemistry in combination with titanium dioxide chromatography and mass spectrometry. J Proteome Res [Internet]. 2011;10(4):1449–58. Available from: http://www.ncbi.nlm.nih.gov/pubmed/21158410.
74. Chalkley RJ, Burlingame AL. Identification of novel sites of O-N-acetylglucosamine modification of serum response factor using quadrupole time-of-flight mass spectrometry. Mol Cell Proteomics. 2003;2(3):182–90.
75. Yuzwa SA, Yadav AK, Skorobogatko Y, Clark T, Vosseller K, Vocadlo DJ. Mapping O-GlcNAc modification sites on tau and generation of a site-specific O-GlcNAc tau antibody. Amino Acids. 2011;40(3):857–68.
76. Myers SA, Daou S, Affar EB, Burlingame A. Electron transfer dissociation (ETD): the mass spectrometric breakthrough essential for O-GlcNAc protein site assignments-a study of the O-GlcNAcylated protein host cell factor C1. Proteomics [Internet]. 2013;13(6):982–91. Available from: http://www.pubmedcentral.nih.gov/articlerender.fcgi?artid=3988289&tool=pmcentrez&rendertype=abstract.
77. Klein AL, Berkaw MN, Buse MG, Ball LE. O-linked N-acetylglucosamine modification of insulin receptor substrate-1 occurs in close proximity to multiple SH2 domain binding motifs. Mol Cell Proteomics [Internet]. 2009;8(12):2733–45. Available from: http://www.ncbi.nlm.nih.gov/pubmed/19671924\n, http://www.mcponline.org/content/8/12/2733.full.pdf.
78. Berk J, Maitra S, Dawdy A, Shabanowitz J, Hunt DF, Wilson K. O-GlcNAc regulates emerin binding to BAF in a chromatin- and lamin B-enriched "niche". J Biol Chem. 2013;288:30192–209.
79. Greis KD, Hayes BK, Comer FI, Kirk M, Barnes S, Lowary TL, et al. Selective detection and site-analysis of O-GlcNAc-modified glycopeptides by beta-elimination and tandem electrospray mass spectrometry. Anal Biochem. 1996;234(1):38–49.
80. Wells L, Vosseller K, Cole RN, Cronshaw JM, Matunis MJ, Hart GW. Mapping sites of O-GlcNAc modification using affinity tags for serine and threonine post-translational modifications. Mol Cell Proteomics. 2002;1(10):791–804.
81. Vosseller K, Hansen KC, Chalkley RJ, Trinidad JC, Wells L, Hart GW, et al. Quantitative analysis of both protein expression and serine/threonine post-translational modifications through stable isotope labeling with dithiothreitol. Proteomics. 2005;5(2):388–98.
82. Phillips DMP. The presence of Acetyl Groups in Collagen. Biochem J. 1965;87(258):258–63.

83. Hershko A, Heller H, Eytan E, Kaklij G, Rose IA. Role of the alpha-amino group of protein in ubiquitin-mediated protein breakdown. Proc Natl Acad Sci U S A. 1984;81(22):7021–5.
84. Gershey EL, Vidali G, Allfrey VG. Chemical studies of histone acetylation: the occurrence of ε-N-acetyllysine in the f2a1 histone. J Biol Chem [Internet]. 1968;243(19):5018–22. Available from: http://www.jbc.org/content/243/19/5018.abstract.
85. Backs J, Olson EN. Control of cardiac growth by histone acetylation/deacetylation. Circ Res. 2006;98(1):15–24.
86. Weinert BT, Wagner SA, Horn H, Henriksen P, Liu WR, Olsen JV, et al. Proteome-wide mapping of the Drosophila acetylome demonstrates a high degree of conservation of lysine acetylation. Sci Signal. 2011;4(183):ra48.
87. Henriksen P, Wagner SA, Weinert BT, Sharma S, Bacinskaja G, Rehman M, et al. Proteome-wide analysis of lysine acetylation suggests its broad regulatory scope in Saccharomyces cerevisiae. Mol Cell Proteomics. 2012;11(11):1510–22.
88. Zhang J, Sprung R, Pei J, Tan X, Kim S, Zhu H, et al. Lysine acetylation is a highly abundant and evolutionarily conserved modification in Escherichia coli. Mol Cell Proteomics. 2009;8(2):215–25.
89. Wang Q, Zhang Y, Yang C, Xiong H, Lin Y, Yao J, et al. Acetylation of metabolic enzymes coordinates carbon source utilization and metabolic flux. Science. 2010;327(5968):1004–7.
90. Du J, Zhou Y, Su X, Yu JJ, Khan S, Jiang H, et al. Sirt5 is a NAD-dependent protein lysine demalonylase and desuccinylase. Science. 2011;334(6057):806–9.
91. Peng C, Lu Z, Xie Z, Cheng Z, Chen Y, Tan M, et al. The first identification of lysine malonylation substrates and its regulatory enzyme. Mol Cell Proteomics. 2011;10(12):M111.012658.
92. Tan M, Luo H, Lee S, Jin F, Yang JS, Montellier E, et al. Identification of 67 histone marks and histone lysine crotonylation as a new type of histone modification. Cell. 2011;146(6):1016–28.
93. Zhang Z, Tan M, Xie Z, Dai L, Chen Y, Zhao Y. Identification of lysine succinylation as a new post-translational modification. Nat Chem Biol. 2011;7(1):58–63.
94. Tan M, Peng C, Anderson KA, Chhoy P, Xie Z, Dai L, et al. Lysine glutarylation is a protein posttranslational modification regulated by SIRT5. Cell Metab. 2014;19(4):605–17.
95. Chen Y, Sprung R, Tang Y, Ball H, Sangras B, Kim SC, et al. Lysine propionylation and butyrylation are novel post-translational modifications in histones. Mol Cell Proteomics. 2007;6(5):812–9.
96. Hare JM, Stamler JS. NO/redox disequilibrium in the failing heart and cardiovascular system. J Clin Invest. 2005;115(3):509–17.
97. Finkel T. Signal transduction by reactive oxygen species. J Cell Biol. 2011;194(1):7–15.
98. Holmström KM, Finkel T. Cellular mechanisms and physiological consequences of redox-dependent signalling. Nat Rev Mol Cell Biol [Internet]. 2014;15(6):411–21. Available from: http://www.ncbi.nlm.nih.gov/pubmed/24854789.
99. Weidinger A, Kozlov A. Biological activities of reactive oxygen and nitrogen species: oxidative stress versus signal transduction. Biomolecules [Internet]. 2015;5(2):472–84. Available from: http://www.mdpi.com/2218-273X/5/2/472/.
100. Dhalla NS, Golfman L, Takeda S, Takeda N, Nagano M. Evidence for the role of oxidative stress in acute ischemic heart disease: a brief review. Can J Cardiol. 1999;15(5):587–93.
101. Victorino VJ, Mencalha AL, Panis C. Post-translational modifications disclose a dual role for redox stress in cardiovascular pathophysiology. Life Sci [Internet]. Elsevier Inc.; 2015;129:42–7. Available from: http://linkinghub.elsevier.com/retrieve/pii/S0024320514009291.
102. Varga ZV, Giricz Z, Liaudet L, Haskó G, Ferdinandy P, Pacher P. Interplay of oxidative, nitrosative/nitrative stress, inflammation, cell death and autophagy in diabetic cardiomyopathy. Biochimica et Biophysica Acta Mol Basis Dis. 2014;1852(2):232–42.
103. Ziolo MT, Kohr MJ, Wang H. Nitric oxide signaling and the regulation of myocardial function. J Mol Cell Cardiol. 2008;45(5):625–32.
104. Dezfulian C, Raat N, Shiva S, Gladwin MT. Role of the anion nitrite in ischemia-reperfusion cytoprotection and therapeutics. Cardiovasc Res. 2007;75(2):327–38.

105. Zweier JL, Talukder MAH. The role of oxidants and free radicals in reperfusion injury. Cardiovasc Res. 2006;70(2):181–90.
106. Kohr MJ, Evangelista AM, Ferlito M, Steenbergen C, Murphy E. S-nitrosylation of TRIM72 at cysteine 144 is critical for protection against oxidation-induced protein degradation and cell death. J Mol Cell Cardiol. 2014;69:67–74.
107. Kohr MJ, Sun J, Aponte A, Wang G, Gucek M, Murphy E, et al. Simultaneous measurement of protein oxidation and S-nitrosylation during preconditioning and ischemia/reperfusion injury with resin-assisted capture. Circ Res. 2011;108(4):418–26.
108. Chen YY, Chu HM, Pan KT, Teng CH, Wang DL, Wang AHJ, et al. Cysteine S-nitrosylation protects protein-tyrosine phosphatase 1B against oxidation-induced permanent inactivation. J Biol Chem. 2008;283(50):35265–72.
109. Roos G, Messens J. Protein sulfenic acid formation: from cellular damage to redox regulation. Free Radic Biol Med. 2011;51(2):314–26.
110. Wang S-B, Foster DB, Rucker J, O'Rourke B, Kass DA, Van Eyk JE. Redox regulation of mitochondrial ATP synthase: implications for cardiac resynchronization therapy. Circ Res. 2011;109(7):750–7.
111. Prysyazhna O, Eaton P. Redox regulation of cGMP-dependent protein kinase Iα in the cardiovascular system. Front Pharmacol [Internet]. 2015;6(July):1–8. Available from: http://journal.frontiersin.org/Article/10.3389/fphar.2015.00139/abstract.
112. Wang Y, Liu T, Wu C, Li H. A strategy for direct identification of protein S-nitrosylation sites by quadrupole time-of-flight mass spectrometry. J Am Soc Mass Spectrom. 2008; 19(9):1353–60.
113. Mirza UA, Chait BT, Lander HM. Monitoring reactions of nitric oxide with peptides and proteins by electrospray ionization-mass spectrometry. J Biol Chem. 1995;270(29): 17185–8.
114. Kohr MJ, Aponte A, Sun J, Gucek M, Steenbergen C, Murphy E. Measurement of S-nitrosylation occupancy in the myocardium with cysteine-reactive tandem mass tags: short communication. Circ Res. 2012;111(10):1308–12.
115. Jaffrey SR, Snyder SH. The biotin switch method for the detection of S-nitrosylated proteins. Sci STKE. 2001;2001(86):11.
116. Lin J, Steenbergen C, Murphy E, Sun J. Estrogen receptor-β activation results in s-nitrosylation of proteins involved in cardioprotection. Circulation. 2009;120(3):245–54.
117. Sun J, Kohr MJ, Nguyen T, Aponte AM, Connelly PS, Esfahani SG, et al. Disruption of caveolae blocks ischemic preconditioning-mediated S-nitrosylation of mitochondrial proteins. Antioxid Redox Signal. 2012;16(1):45–56.
118. Sun J, Morgan M, Shen RF, Steenbergen C, Murphy E. Preconditioning results in S-nitrosylation of proteins involved in regulation of mitochondrial energetics and calcium transport. Circ Res. 2007;101(11):1155–63.
119. Santhanam L, Gucek M, Brown TR, Mansharamani M, Ryoo S, Lemmon CA, et al. Selective fluorescent labeling of S-nitrosothiols (S-FLOS): a novel method for studying S-nitrosation. Nitric Oxide Biol Chem. 2008;19(3):295–302.
120. Kohr MJ, Aponte AM, Sun J, Wang G, Murphy E, Gucek M, et al. Characterization of potential S-nitrosylation sites in the myocardium. Am J Physiol Heart Circ Physiol. 2011;300(4):H1327–35.
121. Forrester MT, Thompson JW, Foster MW, Nogueira L, Moseley MA, Stamler JS. Proteomic analysis of S-nitrosylation and denitrosylation by resin-assisted capture. Nat Biotechnol. 2009;27(6):557–9.
122. Gygi SP, Rist B, Gerber SA, Turecek F, Gelb MH, Aebersold R. Quantitative analysis of complex protein mixtures using isotope-coded affinity tags. Nat Biotechnol. 1999;17(10):994–9.
123. Murray CI, Uhrigshardt H, O'Meally RN, Cole RN, Van Eyk JE. Identification and quantification of S-nitrosylation by cysteine reactive tandem mass tag switch assay. Mol Cell Proteomics. 2012;11(2):M111.013441.

124. Pan K-T, Chen Y-Y, Pu T-H, Chao Y-S, Yang C-Y, Bomgarden RD, et al. Mass spectrometry-based quantitative proteomics for dissecting multiplexed redox cysteine modifications in nitric oxide-protected cardiomyocyte under hypoxia. Antioxid Redox Signal [Internet]. 2014;20(9):1365–81. Available from: http://www.pubmedcentral.nih.gov/articlerender.fcgi?artid=3936484&tool=pmcentrez&rendertype=abstract.
125. Hao G, Derakhshan B, Shi L, Campagne F, Gross SS. SNOSID, a proteomic method for identification of cysteine S-nitrosylation sites in complex protein mixtures. Proc Natl Acad Sci U S A. 2006;103(4):1012–7.
126. Camerini S, Polci ML, Restuccia U, Usuelli V, Malgaroli A, Bachi A. A novel approach to identify proteins modified by nitric oxide: the HIS-TAG switch method. J Proteome Res. 2007;6(8):3224–31.
127. Paige JS, Xu G, Stancevic B, Jaffrey SR. Nitrosothiol reactivity profiling identifies S-nitrosylated proteins with unexpected stability. Chem Biol. 2008;15(12):1307–16.
128. Landino LM, Koumas MT, Mason CE, Alston JA. Ascorbic acid reduction of microtubule protein disulfides and its relevance to protein S-nitrosylation assays. Biochem Biophys Res Commun. 2006;340(2):347–52.
129. Huang B, Chen C. An ascorbate-dependent artifact that interferes with the interpretation of the biotin switch assay. Free Radic Biol Med. 2006;41(4):562–7.
130. Giustarini D, Dalle-Donne I, Colombo R, Milzani A, Rossi R. Is ascorbate able to reduce disulfide bridges? A cautionary note. Nitric Oxide Biol Chem. 2008;19(3):252–8.
131. Forrester MT, Foster MW, Benhar M, Stamler JS. Detection of protein S-nitrosylation with the biotin-switch technique. Free Radic Biol Med. 2009;46(2):119–26.
132. Forrester MT, Foster MW, Stamler JS. Assessment and application of the biotin switch technique for examining protein S-nitrosylation under conditions of pharmacologically induced oxidative stress. J Biol Chem. 2007;282(19):13977–83.
133. Doulias P-T, Greene JL, Greco TM, Tenopoulou M, Seeholzer SH, Dunbrack RL, et al. Structural profiling of endogenous S-nitrosocysteine residues reveals unique features that accommodate diverse mechanisms for protein S-nitrosylation. Proc Natl Acad Sci U S A. 2010;107(39):16958–63.
134. Lind C, Gerdes R, Hamnell Y, Schuppe-Koistinen I, Von Löwenhielm HB, Holmgren A, et al. Identification of S-glutathionylated cellular proteins during oxidative stress and constitutive metabolism by affinity purification and proteomic analysis. Arch Biochem Biophys. 2002;406(2):229–40.
135. Hamnell-Pamment Y, Lind C, Palmberg C, Bergman T, Cotgreave IA. Determination of site-specificity of S-glutathionylated cellular proteins. Biochem Biophys Res Commun. 2005;332(2):362–9.
136. Gould NS, Evans P, Martínez-Acedo P, Marino SM, Gladyshev VN, Carroll KS, et al. Site-specific proteomic mapping identifies selectively modified regulatory cysteine residues in functionally distinct protein networks. Chem Biol [Internet]. 2015;22(7):965–75. Available from: http://linkinghub.elsevier.com/retrieve/pii/S1074552115002367.
137. Mustafa AK, Gadalla MM, Sen N, Kim S, Mu W, Gazi SK, et al. H2S signals through protein S-sulfhydration. Sci Signal. 2009;2(96):ra72.
138. Sen N, Paul BD, Gadalla MM, Mustafa AK, Sen T, Xu R, et al. Hydrogen sulfide-linked sulfhydration of NF-κB mediates its antiapoptotic actions. Mol Cell. 2012;45(1):13–24.
139. Brown JR, Hartley BS. Location of disulphide bridges by diagonal paper electrophoresis. The disulphide bridges of bovine chymotrypsinogen A. Biochem J. 1966;101(1):214–28.
140. Sommer A, Traut RR. Diagonal polyacrylamide-dodecyl sulfate gel electrophoresis for the identification of ribosomal proteins crosslinked with methyl-4-mercaptobutyrimidate. Proc Natl Acad Sci U S A. 1974;71(10):3946–50.
141. Avner BS, Shioura KM, Scruggs SB, Grachoff M, Geenen DL, Helseth DL, et al. Myocardial infarction in mice alters sarcomeric function via post-translational protein modification. Mol Cell Biochem. 2012;363(1–2):203–15.

142. Charles R, Jayawardhana T, Eaton P. Gel-based methods in redox proteomics. Biochimica et Biophysica Acta Gen Sub. 2014;1840(2):830–7.
143. Gorman JJ, Wallis TP, Pitt JJ. Protein disulfide bond determination by mass spectrometry. Mass Spectrom Rev. 2002;21(3):183–216.
144. Tsai PL, Chen SF, Huang SY. Mass spectrometry-based strategies for protein disulfide bond identification. Rev Anal Chem. 2013;32(4):257–68.
145. Saurin AT, Neubert H, Brennan JP, Eaton P. Widespread sulfenic acid formation in tissues in response to hydrogen peroxide. Proc Natl Acad Sci U S A. 2004;101(52):17982–7.
146. Sethuraman M, McComb ME, Heibeck T, Costello CE, Cohen RA. Isotope-coded affinity tag approach to identify and quantify oxidant-sensitive protein thiols. Mol Cell Proteomics. 2004;3(3):273–8.
147. Sethuraman M, McComb ME, Huang H, Huang S, Heibeck T, Costello CE, et al. Isotope-coded affinity tag (ICAT) approach to redox proteomics: identification and quantitation of oxidant-sensitive cysteine thiols in complex protein mixtures. J Proteome Res. 2004;3(6):1228–33.
148. Charles RL, Schröder E, May G, Free P, Gaffney PRJ, Wait R, et al. Protein sulfenation as a redox sensor: proteomics studies using a novel biotinylated dimedone analogue. Mol Cell Proteomics. 2007;6(9):1473–84.
149. Reddie KG, Seo YH, Muse Iii WB, Leonard SE, Carroll KS. A chemical approach for detecting sulfenic acid-modified proteins in living cells. Mol Biosyst. 2008;4(6):521–31.
150. Leonard SE, Reddie KG, Carroll KS. Mining the thiol proteome for sulfenic acid modifications reveals new targets for oxidation in cells. ACS Chem Biol. 2009;4(9):783–99.
151. Paulsen CE, Truong TH, Garcia FJ, Homann A, Gupta V, Leonard SE, et al. Peroxide-dependent sulfenylation of the EGFR catalytic site enhances kinase activity. Nat Chem Biol. 2011;8(1):57–64.
152. Paulech J, Liddy KA, Engholm-Keller K, White MY, Cordwell SJ. Global analysis of myocardial peptides containing cysteines with irreversible sulfinic and sulfonic Acid post-translational modifications. Mol Cell Proteomics. 2015;14:609–20.
153. Canet-Avilés RM, Wilson MA, Miller DW, Ahmad R, McLendon C, Bandyopadhyay S, et al. The Parkinson's disease protein DJ-1 is neuroprotective due to cysteine-sulfinic acid-driven mitochondrial localization. Proc Natl Acad Sci U S A. 2004;101(24):9103–8.
154. Wagner E, Luche S, Penna L, Chevallet M, Van Dorsselaer A, Leize-Wagner E, et al. A method for detection of overoxidation of cysteines: peroxiredoxins are oxidized in vivo at the active-site cysteine during oxidative stress. Biochem J. 2002;366(Pt 3):777–85.
155. Kinumi T, Shimomae Y, Arakawa R, Tatsu Y, Shigeri Y, Yumoto N, et al. Effective detection of peptides containing cysteine sulfonic acid using matrix-assisted laser desorption/ionization and laser desorption/ionization on porous silicon mass spectrometry. J Mass Spectrom. 2006;41(1):103–12.
156. Medzihradszky KF, Guan S, Maltby DA, Burlingame AL. Sulfopeptide fragmentation in electron-capture and electron-transfer dissociation. J Am Soc Mass Spectrom. 2007;18 (9):1617–24.

Chapter 14
Proteomic Network Systems Analysis

D. Kent Arrell and Andre Terzic

Abstract Proteomics and other high throughput technologies generate extensive molecular lists, the scope of which renders their accurate interpretation a daunting task. Thus, generalizable approaches by which to extract insight from this complexity are indispensable. Network systems biology principles and their application offer a modular, interchangeable data analytics pipeline by which to collate, integrate, and prioritize such datasets. By understanding the basis and utility of various organizing and interpretive profiling elements including ontological classification, functional enrichment and over-representation algorithms, and combining these with pathway analysis resources and the versatile tools and applications of complex network analysis, an applied network systems approach yields actionable insights into tackling the biology underlying high throughput data. Providing a framework to proteomic newcomers and experienced practitioners alike, we here outline data analytic approaches and provide concrete examples of the pairing of network systems prognostication with informed follow-up, through application of complementary physiological experimentation to validate proteomic observations in cardiovascular health and disease.

Keywords Bioinformatics • Cardiac • Cardiovascular • Complex network analysis • Heart disease • Network biology • Protein • Proteome • Systems biology

Introduction

The proteomics field incorporates a diverse array of methods and approaches by which to examine consequences of posttranscriptional and posttranslational effects on protein abundance, structure, modifications, and interactions. Many such elements are detailed in this manual, with chapters addressing various applications of

D.K. Arrell, PhD • A. Terzic, MD, PhD
Center for Regenerative Medicine and Marriott Heart Disease Research Program,
Division of Cardiovascular Diseases, Department of Medicine,
Department of Molecular Pharmacology and Experimental Therapeutics,
and Department of Medical Genetics, Mayo Clinic,
Rochester, MN, USA
e-mail: arrell.kent@mayo.edu; terzic.andre@mayo.edu

G. Agnetti et al. (eds.), *Manual of Cardiovascular Proteomics*,
DOI 10.1007/978-3-319-31828-8_14

proteomic methodology, experimental design, data acquisition, and analysis of subcellular compartments and specific protein modifications to further our understanding of cardiovascular health and disease [1]. Regardless of this plethora of experimental and technical approaches, all proteomic studies share a common denominator: their output. Every study yields a compilation of proteins, either as simple lists of identities or subsets partitioned by differential expression or modification in response to the biological question under investigation. The key to extracting further insight from this output is the implementation of a generalized approach to data organization and interpretation.

From a logistical standpoint, this commonality – a list of proteins – suggests that it should be possible for any proteomic dataset to be examined in a similar manner. Unlike reductionist molecular approaches, where detailed functional analysis is conducted on a protein-by-protein basis, a more expansive strategy is required to account for the sheer volume of data. Indeed, while proteomic methods may be used to examine individual or small numbers of proteins, they are designed for and applied primarily to large scale analysis of entire proteomes or isolated subproteomes (Table 14.1), such that they now typically yield lists of several hundred to upwards of a thousand proteins in a single report. As one might anticipate, extracting insight from hundreds of proteins simultaneously is not facile, and can be perceived as altogether unmanageable. These lists can be simplified by narrowing the focus to a few choice proteins, such as those exhibiting the greatest extent of change, residing within a particular organelle, or executing a particular biological function. This offers advantages of reduced complexity and the potential to pinpoint functions germane to the topic of interest, but subjective protein exclusion during downstream analysis invariably leads to information loss. Moreover, care must be taken to ensure that data reduction decisions are not influenced by selection bias, whether intentional (e.g. confirmation bias [2], expectation bias [3]) or not. Finally, acquiring a mountain of data and ignoring all but the peak is patently counterproductive to an experimental rationale predicated on the capacity to conduct high throughput profiling. As we have emphasized previously [4], information reduction strategies may result in overlooking critical functional interactions and mechanistically important processes associated with and evident only upon examination of the complete dataset. In making these assessments, awareness of proteins not changing in a system is often just as valuable as detecting those that are altered. Thus, a judicious proteomic data analytics strategy should yield the same advantages – reduced information complexity and provision of functional insights – while remaining free of selection bias and ensuring inclusivity of all detected proteins [4, 5].

Network systems biology (Table 14.1) principles offer an attractive approach for comprehensive data analysis. Systems based strategies facilitate objective organization, prioritization, and integration of proteomic and other big data in their entirety, regardless of abundance, scope or complexity. To this end, a suite of bioinformatic and computational applications designed specifically to interrogate high throughput data is currently available, and in conjunction with protein databases that provide ease of access for molecular information retrieval, they harbor the capacity to bring clarity and cohesion to proteomic output [4–6]. Various applications and tools can

Table 14.1 Proteomic network systems analysis glossary

Subproteome	Subset of the proteome, often grouped by a shared cellular function, alteration by or dependency on a specific biological process, or components of a particular extracellular or intracellular organelle or compartment
Systems biology	Analytical high throughput data-inclusive approach to investigating and modeling relationships among a system's components in order to understand and predict emergent properties
Network or Interactome	Representation of pairwise collections of interactions and relationships, known as a graph in mathematics, which exhibits emergent properties based on their topological structure
Network systems biology	Complex network theory applied to the analysis of molecular interaction networks, including evaluation and assessment of their composition and topology to guide systems biological interpretation
Node	Individual subunits that comprise a graph or network via their cumulative interactions. In network biology, one or more of proteins, genes, metabolites, and endogenous small molecules or drugs are typically represented as nodes
Edge	An interaction or relationship between two nodes of a network. In network biology this might represent a physical complex of protein subunits, or a regulatory, genetic, or signaling effect, visualized as a line connecting interacting nodes. Based on the relationship, edges may also be portrayed as undirected or directed
Degree distribution	Node degree, the number of connections each node possesses to other nodes in a network, can be used to convey a network's degree distribution, which represents the probability distribution of the degrees of all network nodes
Scale-free topology	Non-stochastic power law distribution of node connectivity characteristic of biological networks, the non-random structure of which influences subsequent emergent network properties such as functional robustness
Hub	Network node that is much more highly connected (i.e. a greater degree) than would be expected to occur at random
Clustering coefficient	Extent to which nodes within a network cluster together as a measure of neighborhood relatedness, where a node's clustering coefficient is defined as the proportion of nodes connecting to it that also connect to each other
Bridging node	Node bridging the shortest direct path between two other nodes or modules within a network, often forming the shortest path between a high proportion of node pairs, leading to an increased likelihood of functional importance due to its position as a network bottleneck
Network motif	Patterns of complex network node-edge connections that are observed more frequently than expected relative to randomly assembled networks
Network module	Highly interconnected collection of nodes that nevertheless exhibits few interactions with the remaining network, often comprised of nodes with a shared function such as subunits of a protein complex or members of a metabolic or signaling pathway

(continued)

Table 14.1 (continued)

Functional robustness	Network property of functional resilience when subjected to random removal or inhibition of a single node, based on the low connectivity of most nodes and the high degree of a limited proportion of nodes
Path length	Distance between two nodes in a network based on the minimum number of edges required to connect them. Average path length of scale-free networks is extremely small, meaning very few steps are required to connect any two nodes
Network diameter	Greatest path length required to connect any two nodes of a network
Metaboproteome	Subproteome comprised of and involving proteins supporting cell metabolism

be arranged and organized to suit particular projects and data sources, but in general there are four elements that form the basis of a proteomic network systems approach. The first component is ontological classification, an initial compartmentalization for partitioning high-throughput proteomic data into discrete biological categories. This serves to reduce complexity of an initial list of proteins, and enables assessment of the relative frequency or infrequency of occurrence for particular functional categories within acquired data relative to a specified reference set or between experimental cohorts. The second component, pathway analysis, extends categorical assignment by evaluating proteomic findings in the context of biological pathways. In this regard, data is superimposed onto canonical pathways and functional annotations, providing further evidence of enrichment properties and establishing connections between distinct elements of the measured dataset. Such connections, retrieved from pathway analysis resources or via stand-alone molecular interaction applications, are vital for the third component, complex network analysis. Networks, comprised of proteins connected by their collective functional and structural interactions, position proteins within the context of their local interaction neighborhood. Network composition, topology, and positional relevance of specific proteins provide value-added properties for data interpretation extending beyond what can be achieved with pathway analysis alone, and these can be exploited for hypothesis generation to assist in developing validation experiments to explain mechanistic underpinnings of initial proteomic measurements. A fourth and final optional component is systems modeling. If sufficient functional data is obtained relating to specific components, pathways, or network elements, it may be possible to develop mathematical or computational models to explain or predict functional outcomes on the basis of experimental proteomic findings. This particular element requires both extensive additional information to model the system under investigation as well as expertise in calculus and mathematics to design and implement. To simplify this introduction of proteomic network systems analysis for beginners, modeling will thus not be discussed here. The first three components, meanwhile, can be undertaken with only an initial list of output proteins and their related expression values, without any requisite expertise in standard tools of the trade. For those interested in further pursuing systems modeling, a recent description of fundamental concepts in its comprehension and application for cardiovascular proteomics is available [6].

Bioinformatic and computational network systems analytic approaches are prognostic in nature, providing hypothesis generating predictions requiring subsequent validation of underlying biological effects for the observed proteomic output. To avoid pitfalls of overestimating the impact of predictions or of statistical overfitting of high throughput data, predictive analytics should therefore be paired with complementary experimental data to ensure hypothesis validation. Ideally, then, prognosis should be actionable, and if predictions are valid, verifiable. This chapter provides a structured description of network systems biology procedures by which to address the daunting task of organizing and interpreting proteomic output, for generation of "actionable prognostication", and provides discrete examples of how information arising from these tools is paired with complementary physiological data to validate proteomic observations in cardiovascular research. The conceptual approach to proteomic network systems biology described herein extends just as readily to the analysis of genomic, transcriptomic, and metabolomic data. Thus, these organization and prioritization principles prove equally versatile for interrogation and interpretation of data from other high throughput molecular profiling methods, including strategies that integrate multi-level -omics datasets.

Ontological Classification, Functional Enrichment and Over-Representation

Current large-scale proteomic studies generate datasets exceedingly difficult to comprehend or interpret without initial data reduction or clustering. This is due to a combination of sheer size, i.e. the number of proteins, and an even greater complexity imposed by their associated biological functions and processes. In this regard, shared biological properties can and do serve as a useful starting point for data comprehension. As noted, when this process is conducted selectively to pare down an extensive list and focus immediately on a protein subset rather than the entire dataset, inclusivity is bypassed for simplicity and subsequent analysis is compromised by user bias. A more objective rationale involves collation of all detected proteins using extant biological information, with prioritization based on subsequent assessment and interrogation of the complete dataset. Even selection of statistical cutoffs should not be arbitrary, but rather applied as a reasonable fold change and statistical test with sufficient power to clearly establish a difference between experimental cohorts [7]. Implementing this strategy is aided by knowledge of the Gene Ontology (GO) [8], organized and structured specifically to document gene and protein properties, and awareness of where to find and access databases containing GO information.

In the field of proteomics, the UniProt Knowledge Base [9] (UniProtKB, accessible at www.uniprot.org) is an established protein sequence repository for assignment of spectral data acquired during mass spectrometry. Its current iteration emerged from a consortium combining several previous protein databases, including

Swiss-Prot, the Translated European Molecular Biology Laboratory, the Protein Information Resource and, more recently, the International Protein Index. Besides serving as a mass spectrometry resource, UniProtKB incorporates a wealth of additional information on various protein properties and characteristics, making it a useful starting point for in-depth dataset analysis [9]. Each entry cross-references with hyperlinks to a wealth of protein database resources listed under as many as 15 sub-categories, depending upon extent of protein characterization, and includes all available GO information [10]. As a result, initial ontology classification can be achieved by simply parsing GO data from UniProtKB, enabling data reduction to cluster proteins without necessitating in-depth or formal knowledge of the GO resource. In this way, proteins may be classified and grouped by a specific function (e.g. kinase, oxidoreductase), by particular biological processes to which they contribute (e.g. glycolysis, muscle contraction), where each protein might execute a unique function while working together as a collective within a particular pathway or as members of a multi-protein complex, and by their localization within one or more discrete cellular components (e.g. mitochondrion, nucleus).

It is precisely these concepts – molecular function, biological process, and cellular component – upon which the GO structure is based [11–16]. These are considered root terms forming 'domains' within the controlled vocabulary set out by the GO Consortium [8], and thus all other GO terms fall into one of these three domains. Each GO classification is unique, but together they form a loosely hierarchical structure, whereby more specialized 'child' terms link to one or multiple more generalized 'parent' terms. As such, the structure of related GO terms can be portrayed or described as a graph, or network, where every GO term serves as a node, interrelationships between pairs of GO terms form edges connecting them (Table 14.1), and edges in turn form nested connections within the hierarchy, the various elements of which fan out from their respective domain root terms (Fig. 14.1).

Comprehension of this structure is advantageous for individuals involved in high throughput research, as GO categorization is now applied to most cardiovascular proteomic studies [4, 6]. Clustering by shared functional properties would be straightforward if all proteins were defined by single GO terms. However, well characterized proteins are often assigned multiple GO associations, either as nested molecular functions of increasing specificity, as a result of participation in multiple biological processes, or a combination of the two, whereas a small proportion of proteins lack any GO term due to unknown function. Thus, a single protein may be included in and defined by multiple GO categories simultaneously (Fig. 14.1), or appear in none at all.

Following dataset assignment of protein GO designations, it is then beneficial to determine their categorical frequency of occurrence. On its own, the frequency at which a particular GO term appears in a dataset is somewhat meaningless. It must be interpreted in the context of an established benchmark such as the known proteome for a species or tissue of interest, or the full extent of proteins detected within the constraints of the experimental technology being applied, such as the complete set of proteins present on a chip array [12–19]. Statistical assessment is carried out by a hypergeometric distribution (Fig. 14.1), defining the probability of whether GO terms appear more or less frequently in an experimental dataset than would be

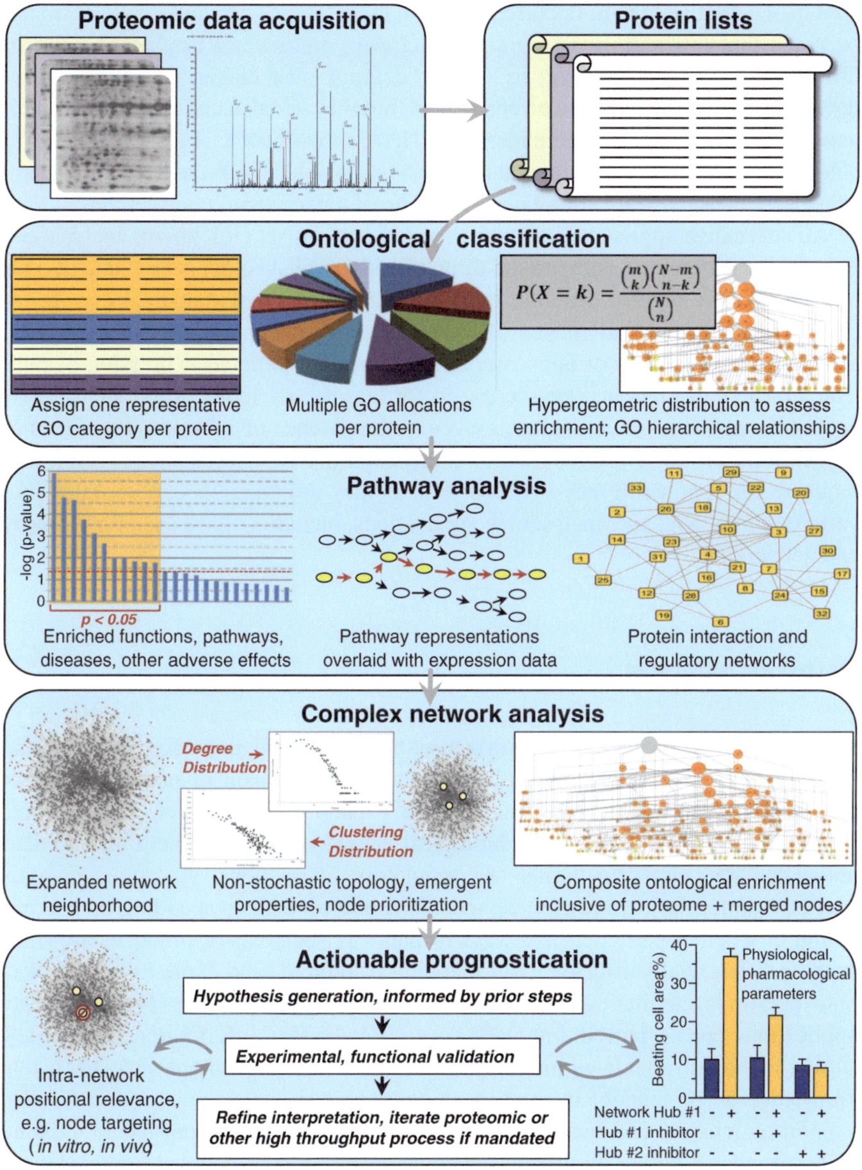

Fig. 14.1 Proteomic network systems analysis. Systems comprehension requires moving beyond simple lists by undertaking their organization, clustering and prioritization to reduce information complexity and extract functional insights into the biology underlying acquired proteomic data. Subjecting protein lists to ontological classification, pathway analysis, and complex network analysis, interpretation can be used to generate actionable hypotheses for functional validation. Shown as a simplified workflow, these elements do not necessarily follow a prescribed sequence, but instead serve as interchangeable components to be included as needed, or repeated multiple times at different stages of an analysis when warranted. Indeed, the entire process may be cyclical or iterative, as experimental validation and insights gleaned from an initial analysis may be further refined by subsequent proteomic or other high throughput data acquisition and a successive round of systems analysis. Abbreviation: *GO* Gene Ontology

anticipated relative to their occurrence within the reference benchmark set. It should be noted that overrepresentation and enrichment metrics are useful not only for differential expression analyses in terms of defining what classes of upregulated or downregulated proteins are overrepresented, but they are also applicable for examination of a simple list of protein identities. Here, enrichment analysis can be used to determine whether an experimental methodology is selective as desired or intended, or perhaps biased towards or against detection of a particular subproteome.

An alternative approach known as gene (or protein) set enrichment analysis can be used for differential comparison to previously published datasets [20–22]. This method takes a slightly different tack, avoiding prescribed statistical cutoffs for comparison, and instead makes use of rank ordered expression, which is then assessed for correlation or anti-correlation to the rank ordered expression of published data. This approach can be useful for teasing out subtle ontological differences between experimental groups, even in the absence of substantial numbers of significantly differing proteins. Fortunately, expertise in mathematics is not a prerequisite for conducting any of these tests as they are typically incorporated into various commercial and open source applications, including those used for pathway and network analysis.

Pathway Analysis

Efforts to extend proteomic categorization beyond the relative occurrence of, and assignment to, particular GO classifications are the domain of pathway analysis applications. Pathway algorithms facilitate expanded examination of proteomic data in the context of established biological pathways, protein functions, and their associated structural, functional and regulatory interactions. Proteomic datasets may be mapped across one or more specific canonical pathways, aiding in determination of whether particular pathways or pathway segments or branches are similarly or differentially affected, with mapping of proteins and expression level data onto pathways facilitating their visualization and representation. Many pathway applications enable protein annotation, using embedded information with details similar to those available within the UniProtKB, and tie-ins or hyperlinks to current knowledge on proteins of interest. As a consequence, many such algorithms also support enrichment and overrepresentation analysis, some making use of existing GO molecular function, biological process, and cellular component nomenclature, while other applications construct and implement platform-specific ontology terms and classifiers [4]. Finally, some pathway resources focus entirely on protein interactions, while others include interaction network generation as one element of a suite of functions, such as that offered by commercial pathway tools, e.g. MetaCore and Ingenuity Pathways Analysis (IPA). Network generation in these applications can then be further scrutinized for functional annotation enrichment or canonical pathway overrepresentation in the context of a broader network neighborhood in a fashion similar to that which is carried out on initial proteomic input. Beyond an understanding of what pathway analysis entails, other primary issues that beginners are faced with are where to find these repositories, potential costs involved in their use, if any, and specified data formats, if required, all with the overarching consideration of what beneficial attributes are present and desirable in specific pathway analysis algorithms.

Even at first glance, it is evident that the extent of information available in individual pathway analysis resources varies greatly between applications. In part, this may be due to the fact that many pathway databases arose from investigator-generated data accumulation. Thus these databases may relate to a specific research area of interest, sometimes with a focus on only a limited set of pathways or a discrete protein property – such as protein-protein interactions – or an emphasis on data from only a single species or particular organelle. Other contributing factors are the sheer number of repositories, and their broad applicability to tackle the variability of biological questions under examination. The latest update of Pathguide, the largest online compendium of biological pathway and molecular interaction related resources, now lists nearly 550 different pathway applications [23]. This number has almost doubled in the past 3 years [4], signifying the tremendous growth in this field. Pathguide lists eight distinct pathway application categories: protein-protein interactions; metabolic pathways; signaling pathways; pathway diagrams; transcription factor/gene regulatory networks; protein-compound interactions; genetic interaction networks; protein sequence focused databases; and 17 separate resources listed under the category of 'Other'. This site provides detailed information about each resource, accessibility (cost and current availability), and whether they adhere to or accommodate specific bioinformatics language standards, e.g. BioPAX [24], which was designed to enable integration, exchange, visualization and analysis of biological pathway data. Many of these databases are freely accessible or free for use by academics, although the more comprehensive resources are typically commercial entities with license requirements that can be cost prohibitive to some investigators. In general, these resources map current biological knowledge to known pathways rather than serving as inference tools to predict theoretical interactions or novel biological outcomes. Some applications are attractive for use because of their broad applicability. For example, the MetaCore pathway database and the IPA Ingenuity Pathways Knowledge Base harbor suites of functions, accounting for these resources being listed in 6 and 5 Pathguide categories, respectively, making them popular choices for cardiovascular proteomic pathway analysis [11–14, 25–42]. Although less versatile, more specialized applications often prove highly desirable when matched to user-specific needs, for instance, exploiting mitochondrial protein interaction databases for bioenergetics research. For those with little knowledge of this bioinformatic field, Pathguide is an excellent source of information for an informed decision on selecting pathway analysis applications that optimally align with experimental needs.

Once a pathway analysis platform is selected, implementation may take many forms. A quick overview of IPA search parameters provides useful considerations in this regard. Input data can range from a number of high throughput experimental procedures, including proteomic, mRNA, microRNA, or metabolomic profiling, as well as RNA-Seq and other next generation sequencing experiments. To enable diverse inputs, this application accepts over 20 different types of molecule identifier. As a consequence, this provides users with the potential to combine data from multiple high-throughput sources, channeling them through a single location for concomitant interrogation under identical bioinformatic parameters. This facilitates proteomic data integration with high throughput information spanning multiple regulatory levels, increasing the potential for comprehensive insight into cellular function [43]. A number of user-specified parameters are then applied to define how broad or refined an analysis is desired. For proteomic analysis, protein identities are generally submitted together with expression level data such as fold change, log ratios, or P-values to set prescribed cutoffs for differential expression between experimental groups. Subsets of upregulated and downregulated proteins can then be examined in isolation, or together as a complete differentially expressed cohort. The scope and stringency of input functional relevance is also user controlled, such as breadth of species data to interrogate, whether direct and indirect biological relationships are acceptable, and whether these interactions must be documented relationships only or if predicted interactions are also acceptable.

Pathway analysis output, such as that obtained with IPA and MetaCore, contains enrichment analysis functions highlighting specific functional annotations, canonical pathways, diseases, or other adverse effects overrepresented within the examined dataset (Fig. 14.1). Images representing signaling pathways, metabolic pathways, or other clusters of interest such as protein complexes, may then be opened for inspection, typically with expression data for constituent proteins overlaid on the image for ease of visual representation. Additional predictive elements are now being incorporated to enhance these pathway diagrams. IPA, for example, recently introduced tools designed to predict upstream regulatory effectors with the greatest likelihood of explaining observed input data, including predicted regulator activation states based on observed protein expression values. Moreover, generated pathways overlaid with expression data can now be used to infer whether expression of other known elements within the pathway might also be altered despite not being detected during initial proteomic analysis. Finally, comprehensive pathway analysis algorithms also generate protein interaction and regulatory networks (Fig. 14.1) [4], which can be tailored by settings for maximum network size, number of networks, and whether network nodes are limited to proteins and genes only, or if their composition may be expanded to include other bioactive molecules such as drugs, endogenous chemicals, and metabolites.

What must be kept in mind is that pathway analysis outputs are inferred biological consequences arising from or explaining input proteins and their observed expression values. As these are predictions and not mechanistic explanations, it is best to view pathway analysis as an interpretative tool facilitating hypothesis generation [4]. Ideally, these hypotheses are then tested and validated by experimental follow-up. With the realization that pathway analysis is operating from the standpoint of partial information, and the knowledge that quality and reliability of supported interactions and relationships gleaned from the literature by these applications can be highly variable, the quality of which are not readily apparent without in depth analyses of all relevant literature, it is best to approach any results with a healthy dose of skepticism by designing and executing validation experiments whenever possible. Continuous refinement of pathway analysis applications improves as data acquisition increases, but the most convincing systems proteomics studies will always complement predictive analytics with supportive experimental validation.

Pathway analysis algorithms also harbor limitations with respect to generated protein and gene interaction networks [4]. While these networks can be evaluated for enrichment and overrepresentation in a manner similar to that of an unconnected protein or gene dataset that served as initial input, pathway applications are not designed to characterize additional characteristics such as network topology or structure, which imparts additional emergent properties of relevance for particular nodes within the network. Moreover, pathway analysis networks are often intended for visual esthetics rather than functional interpretation, so these applications tend to have upper bounds in their capacity to assemble large networks. As proteomic datasets continue to increase in magnitude, this limitation becomes more problematic for network-oriented biological interpretation. To properly exploit network structure and composition for proteomic systems analysis, it is therefore essential to move beyond pathway network applications and make use of dedicated network analysis tools. Comprehension of some basic principles of complex network analysis, including those that confer value-added properties for systems analysis, facilitates their use for visualizing and interpreting proteomic data.

Complex Network Analysis

What, exactly, is a network? As noted for the hierarchical structure of the complete assemblage of GO terms, a network, or graph, is a collection of nodes, each connecting to one or more additional nodes in a pairwise manner (Table 14.1). In proteomics, then, a network serves as a mathematical representation of known or predicted biological relationships between collections of proteins. Nodes or vertices designate the proteins, while any relationship between two proteins is represented by an edge, or line, connecting the two nodes (Table 14.1). The number of edges connecting a node to other nodes in the network is defined as the first node's degree (Table 14.1). Networks are now understood to assemble into nonrandom structures, where most nodes within the network contain very few connections to other nodes, and thus have a small degree, whereas a much smaller proportion of nodes have many connections, or a large degree.

Once believed to be randomly arranged in terms of connectivity, over the past 15 years this non-stochastic connectivity tendency in biological networks has become better understood. It is now well established that this arrangement of biological network degree distribution (Table 14.1, Fig. 14.1) approximates a power law, leading to a characteristic network topology that is termed scale-free (Table 14.1) [44]. Nodes of extremely high degree are defined as hubs (Table 14.1), and their extensive connectivity is often reflected in these nodes being critical for network functionality.

Another useful network parameter is termed the clustering coefficient (Table 14.1). This is a property of secondary interactions within a network [45], as the clustering coefficient for a particular node indicates the proportion of nodes linking to it that also connect to each other. In other words, this measure defines how interconnected are a node's nearest neighbors. Tightly clustered groups of proteins, which often share similar functional attributes or serve as partners in a multi-protein structural complex, create local regions of high clustering coefficient nodes in a network, and in turn, clusters of clusters can be observed in extremely large networks, such that the network forms a hierarchical structure. Nodes that bridge two or more regions of high clustering within a network are known as bridging nodes (Table 14.1). Due to their position spanning large numbers of nodes on either side, they form a conduit as the shortest path between an inordinately high proportion of node pairs within the network. Therefore, bridging nodes often are also critical to overall network function, like hubs, despite typically being of rather limited degree, unlike hubs. Network non-stochasticity also imparts other emergent properties of biological relevance beyond that of hubs and bridging nodes, such as network structural motifs, modularity, and functional robustness (Table 14.1) [46, 47].

From these rather esoteric descriptions, it may not be readily apparent how networks are useful for representation of proteomic data. Proteins carry out the vast majority of functions within cells, doing so not in isolation but rather in concert with a plethora of other proteins and macromolecules, as components in structural or regulatory interactions, or as part of signaling or metabolic cascades. Accordingly, arrangement of these interactions in the form of a biological network serves as a rational means of assembling complex data in a functional, coherent format. Once generated, biological interaction networks can be evaluated on the basis of their composition via ontological and functional enrichment analysis, and on the basis of network topology or structure, both in terms of its overall architecture as well as by mathematical measures identifying nodes with positions of prominence throughout the network, e.g. hubs and bridging nodes (Table 14.1, Fig. 14.1) [46, 47]. Because proteomic networks are also non-stochastic, regardless of network size or scale, they possess predictable structural characteristics that can be useful for functional interrogation and hypothesis generation. Importantly, such network topology-dependent traits are not readily apparent when their constituent proteins are instead arranged only as lists.

Methods used to generate networks from biological data can differ widely, depending on the source of information used to define interactions, and on underlying presumptions used to evaluate what properties constitute an edge to connect two nodes. When produced in conjunction with pathway analysis, proteomic interaction networks are most often generated using current biological knowledge to establish connectivity. Such networks typically include structural, regulatory, and signaling based interactions, comprising both direct and indirect relationships between nodes. If only a particular subset of interactions is warranted or desired for network assembly, such as protein-protein interactions, these can be assembled directly from the literature or by *de novo* experimental data acquisition. There are also statistically guided methods for network construction, connecting nodes on the basis of co-expression or correlation [48], or by reverse engineering from expression dynamics using *ab initio* methods [49], although these methods require greater detail regarding data input for modeling and prediction of network interactions than is typically available from proteomic studies. Thus, networks derived from pathway analysis and the accumulated knowledge archived therein is currently the most applicable methodology available for proteomic systems analysis.

Examining protein networks from both a structure and function standpoint requires an understanding of dedicated network analysis and visualization applications. A prominent example is Cytoscape, developed by a multi-institute consortium as an open-source, stand-alone tool for network visualization and evaluation of network structural properties, which over time has added the capacity to access protein and gene data directly from other repositories, thus enhancing network comprehension in biological contexts [50]. As it is open-source, users and developers are welcome to create and contribute new peripherals, or apps, to further advance Cytoscape's utility for network interpretation and interrogation. Newcomers to complex network analysis will appreciate the fact that neither bioinformatic proficiency nor expertise in network biology are required to begin using Cytoscape, although its maturation and broad appeal has led to a plethora of tools and applications that require a substantial commitment to accurately comprehend and exploit to their full potential.

Numerous options exist for visualization of networks constructed in or imported into Cytoscape. Besides the basic requirement for a list of pairwise interactions, additional attributes can be uploaded and appended, such as protein expression data, which can then be superimposed as node or edge attributes or to network layout to organize networks visually on the basis of expression information. Thus, extent and direction of biological change, i.e. up- or down-regulation, can be conveyed by color, shape, or size of nodes and edges, with a variety of optional layouts available to assist researchers in emphasizing particular network elements or properties [11–14]. For example, the layout of nodes in one network can be applied to another in order to co-localize nodes shared by both networks in the same relative spatial regions, enhancing the visual capacity to compare and contrast related networks [14]. Once attributes are applied to network nodes, this information can also be exploited to enable layout co-localization by regional clustering of nodes sharing one or more common attributes [12–14]. Unlike pathway analysis network tools, these functions can be achieved in Cytoscape without concern for upper bounds on network size or structural complexity.

Beyond visualization properties, substantial effort has been devoted to Cytoscape analysis, interrogation and interpretation tools to examine network structure and network functional characteristics. Network Analyzer [51] was developed soon after the introduction of Cytoscape, and was such a popular app for network topology analysis that it is now fully integrated as a standard tool on all new platform downloads. Analyzer assesses a variety of topological elements for both directed and undirected Cytoscape networks, including but not limited to number of nodes and edges, number of connected components, degree distribution, clustering coefficient, average path lengths between pairs of nodes, and network diameter (Table 14.1) [51]. These network topological attributes are not addressed by pathway analysis network functions, making network specific applications a valuable addition to the systems proteomics repertoire.

From a network functional enrichment standpoint, the most popular Cytoscape app is the Biological Network Gene Ontology (BiNGO) tool [52]. It was designed to acquire current external resources of the GO and apply them internally to fulfill network GO analysis within Cytoscape. Thus, BiNGO interprets biological network overrepresentation across the breadth of GO domains—biological process, cellular component, and molecular function—by comparison to the entirety of a species-specific reference set, a process similar to enrichment analysis tools in pathway algorithms. BiNGO output can be displayed in two ways. The first is as a significance ranked spreadsheet of terms defined within a particular GO domain, with domain of choice and significance threshold pre-defined in user settings. The second is as a nested hierarchical ontology network representing GO terms as nodes, parent-child relationships as edges, node coloring graded in relation to the presence and extent of statistical significance, and node size scaled to the proportion of initial network nodes mapping to each GO network term. Ultimately, BiNGO generates an ontology network defining the functional attributes of its parent molecular network (Fig. 14.1) [52].

There are now over 250 unique Cytoscape apps designed for specific functions, including but not limited to, import of networks and their attributes, network inference, analysis of existing networks, enrichment and ontology analysis, systems biology, comparison between networks, and communication and scripting applications. Cytoscape has become increasingly popular for cardiovascular proteomic network analysis [11–14, 17, 25] due to the litany of contributors building it into a comprehensive program addressing almost every network-oriented concept imaginable. While described here extensively to outline dedicated network platform applications, Cytoscape is not the only useful network analysis program available, and readers are encouraged to investigate other network visualization and analysis tools. Similar to the Pathguide repository of pathway analysis applications, Graph Visualization Software References formerly served as a single site resource providing information on several dozen network analysis algorithms to enlighten and guide software selection [53], but unfortunately a recent search indicates that this database appears to be no longer available. At this time, no comparable resource is available to guide newcomers to appropriate tools for network analysis, but investigators are encouraged to seek out and apply network associated applications to maximize proteomic systems-oriented data analysis.

Putting the Components Together – Actionable Prognostication with Experimental Validation

Unlike traditional reductionist approaches where hypotheses are formulated and subsequently investigated by applying various molecular biology techniques, often with an emphasis on characterizing function of only a single protein or biological pathway, proteomic and other high throughput techniques are often applied without a preconceived notion of what may or will be detected or discovered. Indeed, the biology may not even be sufficiently well understood to formulate actionable hypotheses until after such data is first analyzed and interpreted. Accordingly, high throughput analyses are often viewed or approached from a different scientific standpoint, wherein data analysis serves as the hypothesis generating step that must subsequently be validated (Fig. 14.1). Thus, acquired proteomic data does not typically serve as a final answer in and of itself. Instead, delivering on the promise of proteomic data often requires the application of actionable prognostication.

A case in point is proteomic comprehension of the cardiac implications of ATP-sensitive K^+ (K_{ATP}) channel deficiency, caused by absence of the *KCNJ11*-encoded Kir6.2 pore forming subunit of the channel multi-subunit protein complex [54]. Functional consequence of genetic knockout in cardiac myocytes is a loss of K^+ conductance across the cell membrane, but K_{ATP} channel activity influences far more, modulating membrane potential-dependent cellular metabolism much like a rheostat, adjusting function to match cellular energy demands [55–58]. Even though K_{ATP} channelopathies are implicated in human cardiac disease [59–61], consequences of channel deficiency predisposing to disease vulnerability escaped broader molecular comprehension, mandating proteomic systems interrogation of channel dysfunction in various contexts [12–15]. In the *KCNJ11*-knockout, rather than simply being a case of presence *versus* absence of a single protein, proteomic analysis determined that, even in the absence of superimposed cardiac stress, more than 100 proteins were significantly altered in response to chronic K_{ATP} channel deficiency [13]. Taking this list of proteins through a network systems analysis is particularly revealing for comprehension of the underlying mechanistic consequences of channel dysfunction.

Initial ontological classification (**Step 1 of what to do with your proteomic list**) indicated that a little over 60 % of differentially expressed proteins could be categorized as having direct involvement in metabolic function, whereas the remainder participated in a variety of other cellular processes, including proteolysis, chaperones, cytostructure, oxidoreductases, transcription or translation, and regulation of cell signaling [13]. This abundance of metabolic connections is consistent with the known impact of the channel as a metabolic rheostat, the prominence of which was reinforced by IPA functional ontology classification (**Step 2 of what to do with your proteomic list**) [13]. Moreover, metabolic relevance was further strengthened by subsequent complex network ontology enrichment analysis (**Step 2 applied to output acquired from Step 3 of what to do with your proteomic list**). In this regard, BiNGO analysis was conducted within Cytoscape to define overrepresented biological processes – one of the three primary GO domains – associated with the expanded K_{ATP} channel–dependent metaboproteome (Table 14.1) network derived from the remodeled proteome. The resultant BiNGO ontology network comprised nearly 1,000 distinct GO terms associating with the parental molecular network, yet only 55 of these were significantly overrepresented [13]. Moreover, every one of the 55 was a metabolic process, collectively forming a highly nested ontology network within a limited number of broader metabolic functions. Primarily overrepresented were GO terms involved in glycolysis, as well as tricarboxylic acid cycle, fatty acid, and other substrate metabolism branches, along with some degree of protein catabolism enrichment [13].

These parameters provide a sense of altered proteome functional attributes, but not of the biological consequences of proteome remodeling. Pathway analysis was thus also applied in a complementary manner to predict potential adverse effects, yielding actionable insight into the implications arising from and consistent with the altered proteome. Here, "cardiovascular disease" was significantly overrepresented at the level of the proteome and, even more extensively, at the interactome (Table 14.1) level integrating all proteome changes in their broader network neighborhood [13]. Experimental evidence supporting susceptibility of the K_{ATP} channel deficient cohort to cardiovascular disease was evident in measures of cardiac mass, cardiac function, and survivorship in response to increasing levels of imposed cardiac stress [13]. Thus, proteomic network systems analysis here incorporated actionable experimental evaluation, validating the predicted functional consequences of observed proteome remodeling (**Step 4, actionable prognostication with experimental validation**).

Similar systems approaches have also been applied to understand proteomic consequences of K_{ATP} channel deficiency in the setting of superimposed cardiac stress [12, 14]. For example, prediction of overrepresented adverse effects facilitated experimental follow-up in a model of deoxycorticosteroid and salt-induced hypertension, where pathway analysis of proteomic data predicted three adverse effects related exclusively to cardiac function – cardiac damage, cardiac enlargement, and cardiac fibrosis [12]. Each effect was subsequently confirmed by assessment of cardiac output, measurement of heart-to-body-weight ratios, and evaluation of collagen deposition, respectively, validating predicted detrimental cardiac effects of K_{ATP} channel-dependent proteome remodeling in response to physiological stress [12]. Pathway analysis adverse effect screening also proved instrumental in evaluating consequences of proteome remodeling in cardiomyopathy and the structural and functional remodeling mediated by the response to stem cell therapy in cardiomyopathic hearts [14]. Here, proteome changes associated with cardiomyopathy were subjected to pathway analysis, with *in silico* prediction of both enrichment of cardiac disease as well as several cardiac adverse effects, which were greatly ameliorated or completely absent when evaluating the stem cell treated cardiomyopathic proteome [14]. Prognostication was validated by a range of echocardiographic metrics and anatomical measurements, confirming predicted deleterious structural and functional outcomes of disease and their improvement following cell mediated therapy [14].

Further supporting network approaches, a distinct benefit of extending systems analysis to complex networks is their potential to provide value-added elements to evaluate implications of the proteomic data in a network-oriented context. Network structure, i.e. topology, can be exploited on the basis of identification and targeted inhibition of network hubs. While most nodes in scale-free networks possess a low degree and can be removed or inhibited without great risk of leading to a loss in network functionality, i.e. robustness, the reverse is also true, wherein targeting highly connected nodes can be exploited to evaluate whether a predicted network function or other emergent property is dependent on network integrity mediated by their hubs [62]. For instance, such connectivity properties suggest that inhibition of one or more primary hubs of an endodermal secretome network might prevent its potentiating effect on cardiac differentiation [11]. This prediction was reinforced by *in silico* modulation of composite network generation, whereby prioritization of cardiovascular development predicted for the network was demoted after arbitrarily removing the most highly connected hub from pathway analysis input data [11]. Indeed, this was demonstrated functionally when application of pharmacological inhibitors of the two most highly connected nodes each abolished the cardiac potentiation effect mediated by the secretome. This included the primary hub that was detected during initial proteomic analysis as well as the secondary hub that was only incorporated during network generation but was noticeably absent from the proteomic data [11]. Thus, network topology assessment has the potential to provide value-added emergent properties for hypothesis generation of intra-network positional relevance. Moreover, even though network generation increases overall molecular complexity by adding more proteins to the initial proteomic list, in doing so it also yields further potentially relevant candidates for systems evaluation that may be critical contributors to the underlying biology that were nevertheless overlooked during initial proteomic screening [4, 5, 11].

Conclusion

Continued technological advances, with improved instrument sensitivity and resolution combined with expanded, more detailed protein databases, will lead to increasingly larger proteomic datasets, each harboring tremendous biological intricacy. Network systems analysis strategies will therefore become ever more critical for proteomic and other high throughput data deconvolution [4, 63]. Herein, guidance is provided on generalized analytic approaches by which to systematically organize, cluster, and prioritize proteomic datasets in their entirety to reduce information complexity while simultaneously yielding functional insights. An important qualifier is that ontological classification, and enrichment, pathway, and complex network analyses may be considered as interchangeable modular components of a network systems approach. Rather than applying each as a standalone topic or steps that must be adhered to in a prescribed order, these elements may be arranged flexibly and used in a variety of ways, as required to address a specific biological question. Indeed, the same step may even be repeated multiple times at different stages in an analysis. For instance, enrichment analysis can be conducted on initial ontological categories, during pathway analysis, or on final output networks [6], potentially revealing shifts in focus at successive points in an analysis. When actionable hypotheses are generated via systems analysis and examined experimentally, refinement or modification of the initial interpretation may potentially mandate an additional round of high throughput data acquisition. Thus, network systems analysis can also be viewed as an iterative process, with a cyclical transition from high throughput data to interrogation, followed by experimentation to validate or refine interpretation, ultimately guiding subsequent decisions on additional proteomic or other high throughput experiments (Fig. 14.1) [5, 6]. Comprehension of the basis and utility of these organizing and interpretive principles provides a foundation for their application, preparing students, proteomic practitioners, and clinicians alike for effective application of basic and translational proteomic network medicine to further our understanding of cardiovascular health and disease.

Acknowledgements Dr. Arrell is supported by a Ruth L. Kirschstein National Research Service Award from the National Institutes of Health. Dr. Terzic holds the Marriott Family Professorship in Cardiovascular Research and is the Michael S. and Mary Sue Shannon Director, Mayo Clinic Center for Regenerative Medicine. This work was supported by the National Institutes of Health (T32HL07111), Fondation Leducq, Marriott Heart Disease Research Program, and Mayo Clinic Center for Regenerative Medicine.

References

1. Agnetti G, Lindsey ML, Foster DB, editors. Manual of cardiovascular proteomics. Cham, Switzerland: Springer; 2016.
2. Oswald ME, Grosjean S. Confirmation Bias. In: Pohl RF, editor. Cognitive illusions: a handbook on fallacies and biases in thinking, judgement and memory. Hove, UK, and New York, USA: Psychology Press; 2004. p. 79–96.
3. Jeng M. A selected history of expectation bias in physics. Am J Phys. 2006;74:578–83.
4. Arrell DK, Terzic A. Systems proteomics for translational network medicine. Circ Cardiovasc Genet. 2012;5:o8–16.
5. Arrell DK, Terzic A. Network systems biology for drug discovery. Clin Pharmacol Ther. 2010;88:120–5.
6. Lindsey ML, Mayr M, Gomes AV, Delles C, Arrell DK, Murphy AM, Lange RA, Costello CE, Jin YF, Laskowitz DT, Sam F, Terzic A, Van Eyk J, Srinivas PR. Transformative impact of proteomics on cardiovascular health and disease: a scientific statement from the American Heart Association. Circulation. 2015;132:852–72.
7. Levin Y. The role of statistical power analysis in quantitative proteomics. Proteomics. 2011;11:2565–7.
8. Ashburner M, Ball CA, Blake JA, Botstein D, Butler H, Cherry JM, Davis AP, Dolinski K, Dwight SS, Eppig JT, Harris MA, Hill DP, Issel-Tarver L, Kasarskis A, Lewis S, Matese JC, Richardson JE, Ringwald M, Rubin GM, Sherlock G. Gene Ontology: tool for the unification of biology. The Gene Ontology Consortium. Nat Genet. 2000;25:25–9.
9. Pundir S, Magrane M, Martin MJ, O'Donovan C; UniProt Consortium. Searching and navigating UniProt databases. Curr Protoc Bioinformatics. 2015;50:1.27.1–1.27.10.
10. Huntley RP, Sawford T, Mutowo-Meullenet P, Shypitsyna A, Bonilla C, Martin MJ, O'Donovan C. The GOA database: Gene Ontology annotation updates for 2015. Nucleic Acids Res. 2015;43(Database issue):D1057–63.
11. Arrell DK, Niederländer NJ, Faustino RS, Behfar A, Terzic A. Cardioinductive network guiding stem cell differentiation revealed by proteomic cartography of tumor necrosis factor alpha-primed endodermal secretome. Stem Cells. 2008;26:387–400.
12. Zlatkovic J, Arrell DK, Kane GC, Miki T, Seino S, Terzic A. Proteomic profiling of K_{ATP} channel-deficient hypertensive heart maps risk for maladaptive cardiomyopathic outcome. Proteomics. 2009;9:1314–25.

13. Arrell DK, Zlatkovic J, Kane GC, Yamada S, Terzic A. ATP-sensitive K^+ channel knockout induces cardiac proteome remodeling predictive of heart disease susceptibility. J Proteome Res. 2009;8:4823–34.
14. Zlatkovic-Lindor J, Arrell DK, Yamada S, Nelson TJ, Terzic A. ATP-sensitive K^+ channel-deficient dilated cardiomyopathy proteome remodeled by embryonic stem cell therapy. Stem Cells. 2010;28:1355–67.
15. Arrell DK, Zlatkovic-Lindor J, Yamada S, Terzic A. K_{ATP} channel-dependent metaboproteome decoded: systems approaches to heart failure prediction, diagnosis, and therapy. Cardiovasc Res. 2011;90:258–66.
16. Foster DB, Liu T, Rucker J, O'Meally RN, Devine LR, Cole RN, O'Rourke B. The cardiac acetyl-lysine proteome. PLoS One. 2013;8:e67513.
17. Isserlin R, Merico D, Alikhani-Koupaei R, Gramolini A, Bader GD, Emili A. Pathway analysis of dilated cardiomyopathy using global proteomic profiling and enrichment maps. Proteomics. 2010;10:1316–27.
18. Monte E, Mouillesseaux K, Chen H, Kimball T, Ren S, Wang Y, Chen JN, Vondriska TM, Franklin S. Systems proteomics of cardiac chromatin identifies nucleolin as a regulator of growth and cellular plasticity in cardiomyocytes. Am J Physiol Heart Circ Physiol. 2013;305:H1624–38.
19. Drozdov I, Didangelos A, Yin X, Zampetaki A, Abonnenc M, Murdoch C, Zhang M, Ouzounis CA, Mayr M, Tsoka S, Shah AM. Gene network and proteomic analyses of cardiac responses to pathological and physiological stress. Circ Cardiovasc Genet. 2013;6:588–97.
20. Subramanian A, Tamayo P, Mootha VK, Mukherjee S, Ebert BL, Gillette MA, Paulovich A, Pomeroy SL, Golub TR, Lander ES, Mesirov JP. Gene set enrichment analysis: a knowledge-based approach for interpreting genome-wide expression profiles. Proc Natl Acad Sci U S A. 2005;102:15545–50.
21. Cha S, Imielinski MB, Rejtar T, Richardson EA, Thakur D, Sgroi DC, Karger BL. In situ proteomic analysis of human breast cancer epithelial cells using laser capture microdissection: annotation by protein set enrichment analysis and gene ontology. Mol Cell Proteomics. 2010;9:2529–44.
22. Lavallée-Adam M, Rauniyar N, McClatchy DB, Yates JR 3rd. PSEA-Quant: a protein set enrichment analysis on label-free and label-based protein quantification data. J Proteome Res. 2014;13:5496–509.
23. Bader GD, Cary MP, Sander C. Pathguide: a pathway resource list. Nucleic Acids Res. 2006;34(Database issue):D504–6.
24. Demir E, Cary MP, Paley S, Fukuda K, Lemer C, Vastrik I, Wu G, D'Eustachio P, Schaefer C, Luciano J, Schacherer F, Martinez-Flores I, Hu Z, Jimenez-Jacinto V, Joshi-Tope G, Kandasamy K, Lopez-Fuentes AC, Mi H, Pichler E, Rodchenkov I, Splendiani A, Tkachev S, Zucker J, Gopinath G, Rajasimha H, Ramakrishnan R, Shah I, Syed M, Anwar N, Babur O, Blinov M, Brauner E, Corwin D, Donaldson S, Gibbons F, Goldberg R, Hornbeck P, Luna A, Murray-Rust P, Neumann E, Ruebenacker O, Samwald M, van Iersel M, Wimalaratne S, Allen K, Braun B, Whirl-Carrillo M, Cheung KH, Dahlquist K, Finney A, Gillespie M, Glass E, Gong L, Haw R, Honig M, Hubaut O, Kane D, Krupa S, Kutmon M, Leonard J, Marks D, Merberg D, Petri V, Pico A, Ravenscroft D, Ren L, Shah N, Sunshine M, Tang R, Whaley R, Letovksy S, Buetow KH, Rzhetsky A, Schachter V, Sobral BS, Dogrusoz U, McWeeney S, Aladjem M, Birney E, Collado-Vides J, Goto S, Hucka M, Le Novère N, Maltsev N, Pandey A, Thomas P, Wingender E, Karp PD, Sander C, Bader GD. The BioPAX community standard for pathway data sharing. Nat Biotechnol. 2010;28:935–42.
25. Behfar A, Perez-Terzic C, Faustino RS, Arrell DK, Hodgson DM, Yamada S, Puceat M, Niederländer N, Alekseev AE, Zingman LV, Terzic A. Cardiopoietic programming of embryonic stem cells for tumor-free heart repair. J Exp Med. 2007;204:405–20.
26. Fu Z, Wang M, Gucek M, Zhang J, Wu J, Jiang L, Monticone RE, Khazan B, Telljohann R, Mattison J, Sheng S, Cole RN, Spinetti G, Pintus G, Liu L, Kolodgie FD, Virmani R, Spurgeon H, Ingram DK, Everett AD, Lakatta EG, Van Eyk JE. Milk fat globule protein epidermal

growth factor-8: a pivotal relay element within the angiotensin II and monocyte chemoattractant protein-1 signaling cascade mediating vascular smooth muscle cells invasion. Circ Res. 2009;104:1337–46.
27. Mayr M, Zampetaki A, Sidibe A, Mayr U, Yin X, De Souza AI, Chung YL, Madhu B, Quax PH, Hu Y, Griffiths JR, Xu Q. Proteomic and metabolomic analysis of smooth muscle cells derived from the arterial media and adventitial progenitors of apolipoprotein E-deficient mice. Circ Res. 2008;102:1046–56.
28. Grant JE, Bradshaw AD, Schwacke JH, Baicu CF, Zile MR, Schey KL. Quantification of protein expression changes in the aging left ventricle of Rattus norvegicus. J Proteome Res. 2009;8:4252–63.
29. Li Z, Liu N, Zhang LS, Gong K, Cai Y, Gao W, Liu Z, Liu S, Han Q, Zhang Y. Proteomic profiling reveals comprehensive insights into adrenergic receptor-mediated hypertrophy in neonatal rat cardiomyocytes. Proteomics Clin Appl. 2009;3:1407–21.
30. Mayr M, Liem D, Zhang J, Li X, Avliyakulov NK, Yang JI, Young G, Vondriska TM, Ladroue C, Madhu B, Griffiths JR, Gomes A, Xu Q, Ping P. Proteomic and metabolomic analysis of cardioprotection: interplay between protein kinase C epsilon and delta in regulating glucose metabolism of murine hearts. J Mol Cell Cardiol. 2009;46:268–77.
31. Agnetti G, Kaludercic N, Kane LA, Elliott ST, Guo Y, Chakir K, Samantapudi D, Paolocci N, Tomaselli GF, Kass DA, Van Eyk JE. Modulation of mitochondrial proteome and improved mitochondrial function by biventricular pacing of dyssynchronous failing hearts. Circ Cardiovasc Genet. 2010;3:78–87.
32. Bugger H, Schwarzer M, Chen D, Schrepper A, Amorim PA, Schoepe M, Nguyen TD, Mohr FW, Khalimonchuk O, Weimer BC, Doenst T. Proteomic remodelling of mitochondrial oxidative pathways in pressure overload-induced heart failure. Cardiovasc Res. 2010;85:376–84.
33. Farina A, Chambery A, Esposito S, Agozzino L, Cotrufo M, Della Corte A, Parente A. Proteomic profiling of medial degeneration in human ascending aorta. Clin Biochem. 2010;43:387–96.
34. Parguiña AF, Fernández Parguiña A, Grigorian-Shamajian L, Agra RM, Teijeira-Fernández E, Rosa I, Alonso J, Viñuela-Roldán JE, Seoane A, González-Juanatey JR, García A. Proteins involved in platelet signaling are differentially regulated in acute coronary syndrome: a proteomic study. PLoS One. 2010;5:e13404.
35. Angel PM, Nusinow D, Brown CB, Violette K, Barnett JV, Zhang B, Baldwin HS, Caprioli RM. Networked-based characterization of extracellular matrix proteins from adult mouse pulmonary and aortic valves. J Proteome Res. 2011;10:812–23.
36. Fogle RL, Hollenbeak CS, Stanley BA, Vary TC, Kimball SR, Lynch CJ. Functional proteomic analysis reveals sex-dependent differences in structural and energy-producing myocardial proteins in rat model of alcoholic cardiomyopathy. Physiol Genomics. 2011;43:346–56.
37. Hammer E, Goritzka M, Ameling S, Darm K, Steil L, Klingel K, Trimpert C, Herda LR, Dörr M, Kroemer HK, Kandolf R, Staudt A, Felix SB, Völker U. Characterization of the human myocardial proteome in inflammatory dilated cardiomyopathy by label-free quantitative shotgun proteomics of heart biopsies. J Proteome Res. 2011;10:2161–71.
38. Hanzu FA, Palomo M, Kalko SG, Parrizas M, Garaulet M, Escolar G, Gomis R, Diaz-Ricart M. Translational evidence of endothelial damage in obese individuals: inflammatory and prothrombotic responses. J Thromb Haemost. 2011;9:1236–45.
39. Qi YX, Jiang J, Jiang XH, Wang XD, Ji SY, Han Y, Long DK, Shen BR, Yan ZQ, Chien S, Jiang ZL. PDGF-BB and TGF-β1 on cross-talk between endothelial and smooth muscle cells in vascular remodeling induced by low shear stress. Proc Natl Acad Sci U S A. 2011;108:1908–13.
40. Li W, Rong R, Zhao S, Zhu X, Zhang K, Xiong X, Yu X, Cui Q, Li S, Chen L, Cai J, Du J. Proteomic analysis of metabolic, cytoskeletal and stress response proteins in human heart failure. J Cell Mol Med. 2012;16:59–71.
41. de la Cuesta F, Barderas MG, Calvo E, Zubiri I, Maroto AS, Darde VM, Martin-Rojas T, Gil-Dones F, Posada-Ayala M, Tejerina T, Lopez JA, Vivanco F, Alvarez-Llamas G. Secretome

analysis of atherosclerotic and non-atherosclerotic arteries reveals dynamic extracellular remodeling during pathogenesis. J Proteomics. 2012;75:2960–71.
42. Sun Z, Biela LM, Hamilton KL, Reardon KF. Concentration-dependent effects of the soy phytoestrogen genistein on the proteome of cultured cardiomyocytes. J Proteomics. 2012;75:3592–604.
43. Lu R, Markowetz F, Unwin RD, Leek JT, Airoldi EM, MacArthur BD, Lachmann A, Rozov R, Ma'ayan A, Boyer LA, Troyanskaya OG, Whetton AD, Lemischka IR. Systems-level dynamic analyses of fate change in murine embryonic stem cells. Nature. 2009;462:358–62.
44. Barabási A-L, Albert R. Emergence of scaling in random networks. Science. 1999;286: 509–12.
45. Arrell DK, Terzic A. Interpreting networks in systems biology. Clin Pharmacol Ther. 2013;93:389–92.
46. Barabási A-L, Oltvai ZN. Network biology: understanding the cell's functional organization. Nat Rev Genet. 2004;5:101–13.
47. Newman MEJ. The structure and function of complex networks. SIAM Rev. 2003;45: 167–256.
48. Dewey FE, Perez MV, Wheeler MT, Watt C, Spin J, Langfelder P, Horvath S, Hannenhalli S, Cappola TP, Ashley EA. Gene coexpression network topology of cardiac development, hypertrophy, and failure. Circ Cardiovasc Genet. 2011;4:26–35.
49. Khalil I, Brewer MA, Neyarapally T, Runowicz CD. The potential of biologic network models in understanding the etiopathogenesis of ovarian cancer. Gynecol Oncol. 2010;116:282–5.
50. Cline MS, Smoot M, Cerami E, Kuchinsky A, Landys N, Workman C, Christmas R, Avila-Campilo I, Creech M, Gross B, Hanspers K, Isserlin R, Kelley R, Killcoyne S, Lotia S, Maere S, Morris J, Ono K, Pavlovic V, Pico AR, Vailaya A, Wang PL, Adler A, Conklin BR, Hood L, Kuiper M, Sander C, Schmulevich I, Schwikowski B, Warner GJ, Ideker T, Bader GD. Integration of biological networks and gene expression data using Cytoscape. Nat Protoc. 2007;2:2366–82.
51. Assenov Y, Ramírez F, Schelhorn SE, Lengauer T, Albrecht M. Computing topological parameters of biological networks. Bioinformatics. 2008;24:282–4.
52. Maere S, Heymans K, Kuiper M. BiNGO: a Cytoscape plugin to assess overrepresentation of gene ontology categories in biological networks. Bioinformatics. 2005;21:3448–9.
53. Pinaud B, Kuntz P. GVSR: an on-line guide for choosing a graph visualization software. In: Brandes U, Cornelsen S, editors. Graph drawing: 18th international symposium, GD 2010, lecture notes in computer science. Berlin: Springer; 2011. p. 400–1.
54. Miki T, Nagashima K, Tashiro F, Kotake K, Yoshitomi H, Tamamoto A, Gonoi T, Iwanaga T, Miyazaki J, Seino S. Defective insulin secretion and enhanced insulin action in K_{ATP} channel-deficient mice. Proc Natl Acad Sci U S A. 1998;95:10402–6.
55. Zingman LV, Alekseev AE, Bienengraeber M, Hodgson DM, Karger AB, Dzeja PP, Terzic A. Signaling in channel/enzyme multimers: ATPase transitions in SUR module gate ATP-sensitive K^+ conductance. Neuron. 2001;31:233–45.
56. Liu XK, Yamada S, Kane GC, Alekseev AE, Hodgson DM, O'Cochlain DF, Jahangir A, Miki T, Seino S, Terzic A. Genetic disruption of Kir6.2, the pore-forming subunit of ATP-sensitive K^+ channel, predisposes to catecholamine induced ventricular dysrhythmia. Diabetes. 2004;53: S165–8.
57. Selivanov VA, Alekseev AE, Hodgson DM, Dzeja PP, Terzic A. Nucleotide-gated K_{ATP} channels integrated with creatine and adenylate kinases: amplification, tuning and sensing of energetic signals in the compartmentalized cellular environment. Mol Cell Biochem. 2004;256–7: 243–56.
58. Gumina RJ, O'Cochlain DF, Kurtz CE, Bast PH, Pucar D, Mishra P, Miki T, Seino S, Macura S, Terzic A. K_{ATP} channel knockout worsens myocardial calcium stress load in vivo and impairs recovery in stunned heart. Am J Physiol Heart Circ Physiol. 2007;292:H1706–13.
59. Yamada S, Kane GC, Behfar A, Liu XK, Dyer RB, Faustino RS, Miki T, Seino S, Terzic A. Protection conferred by myocardial ATP-sensitive K^+ channels in pressure overload induced

congestive heart failure revealed in KCNJ11 Kir6.2-null mutant. J Physiol. 2006;577: 1053–65.
60. Bienengraeber M, Olson TM, Selivanov VA, Kathmann EC, O'Cochlain DF, Gao F, Karger AB, Ballew JD, Hodgson DM, Zingman LV, Pang Y-P, Alekseev AE, Terzic A. ABCC9 mutations identified in human dilated cardiomyopathy disrupt catalytic K_{ATP} channel gating. Nat Genet. 2004;36:382–7.
61. Olson TM, Alekseev AE, Moreau C, Liu XK, Zingman LV, Miki T, Seino S, Asirvatham SJ, Jahangir A, Terzic A. K_{ATP} channel mutation confers risk for vein of Marshall adrenergic atrial fibrillation. Nat Clin Pract Cardiovasc Med. 2007;4:110–6.
62. Albert R, Jeong H, Barabási A-L. Error and attack tolerance of complex networks. Nature. 2000;406:378–82.
63. Loscalzo J, Barabási A-L. Systems biology and the future of medicine. Wiley Interdiscip Rev Syst Biol Med. 2011;3:619–27.

Chapter 15
Sensing and Remembering Cellular States Through Chromatin

Shanxi Jiang and Thomas M. Vondriska

Abstract Chromatin is the means by which the same genome encodes multiple cells: it enables orderly development, normal physiology and, when it goes haywire, malfunctioning chromatin is a hallmark of disease. In the cardiovascular system, the epigenomic features controlling gene expression have recently become the focus of intense research. This chapter examines the principles of chromatin structure, details their regulation and identifies areas of rapid development in our understanding of how the genome is packaged. Also explored are the recent observations indicating that deranged epigenomic features on a genome-wide scale may underpin various cardiovascular diseases.

Keywords Heart • Vasculature • Epigenetics • Epigenomics • Genomics • Proteomics • Transcription

Introduction

Around two billion years ago, natural selection for a strategy of information storage that utilized RNA and protein to package DNA presaged the evolution of multicellularity. To get the same DNA substrate to produce different outcomes, the method or conditions of extracting information has to change. Chromatin accomplishes this task in all plants and animals: comprised of histone proteins and DNA, chromatin is the structural form of the genome in vivo, compacting the enormous chromosome molecules for storage in the nucleus. That plants and animals can achieve the spectacular range of appearance and function observed in the natural world is the result of the ability to produce highly specialized cells and organs. This specialization is achieved through chromatin and its regulation, which enables

S. Jiang, BS • T.M. Vondriska, PhD (✉)
Department of Anesthesiology, David Geffen School of Medicine at UCLA,
Los Angeles, CA, USA
e-mail: shanxi.jiang@gmail.com; tvondriska@mednet.ucla.edu

G. Agnetti et al. (eds.), *Manual of Cardiovascular Proteomics*,
DOI 10.1007/978-3-319-31828-8_15

the same genomes to orchestrate diverse phenotypes throughout development and into adulthood.

The fundamental unit of chromatin is the nucleosome, consisting of two copies of four histone proteins (named H2A, H2B, H3 and H4) and entwined by 147 base pairs of DNA [1]. Interesting relationships have emerged relating primary DNA sequence to the binding of nucleosomes across a genome: however, it is now recognized that a diverse range of regulatory mechanisms control where nucleosomes reside, how they combine with each other to form higher order structures and the resulting accessibility (or lack thereof) for transcriptional machinery to interact with and express a gene. Broadly construed, DNA can be either accessible for transcription, or euchromatic, or inaccessible for transcription, heterochromatic [2]. As depicted in Fig 15.1, the landscape of chromatin features combine to facilitate or prevent gene transcription.

It is now commonly accepted that chromatin patterns are cell and developmental stage specific, underlying transcriptome changes that enable phenotype specification. Individual nucleosomes can be modified by swapping the histone variants that comprise them, by post-translational modification of those histones, by ATP-dependent processes that reorganize groups of nucleosomes in response to environmental stress or developmental cue and by alterations to the DNA, cytosine methylation in particular. Furthermore, RNA and protein can combine with nucleosomes to form higher

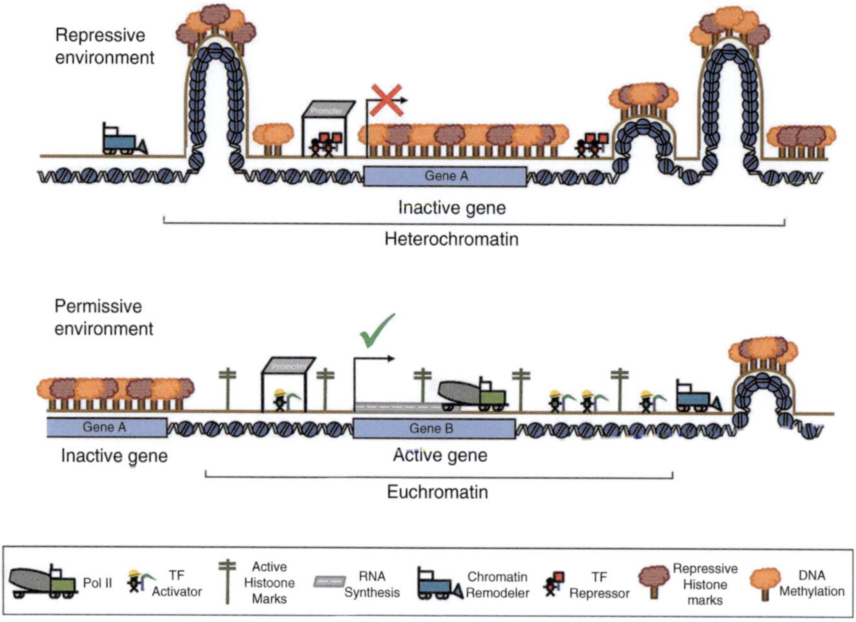

Fig 15.1 Chromatin landscaping. At individual loci, a balance of features promoting or inhibiting chromatin accessibility determines whether transcription occurs. As described in the text, these features decorate individual nucleosomes and combine to influence gene expression and chromatin structure on the scale of genes, chromatin territories and the entire genome

order chromatin structures, compacting or relaxing sub-chromosomal regions. With the development of the next generation sequencing techniques, there has been an explosion in the analysis of how different proteins bind to the genome and the combinatorial patterns of protein binding that specify chromatin structure and thereby transcriptional behavior at individual loci [3]. A major challenge for biology and medicine is now to decode the logic of chromatin regulation, to enable targeting of gene expression programs for therapeutic modulation during disease as well as to use epigenomic information for improved patient stratification and diagnoses.

This chapter reviews the current understanding of chromatin regulation in disease with a particular emphasis on the cardiovascular system, describing how genetics and environment are integrated in the epigenome, ultimately controlling disease susceptibility and progression.

Basics of Chromatin Structure

Histone Variants

Although they all contain two copies each of four histone family proteins, nucleosomes are not homogeneous protein complexes. Mammalian genomes harbor dozens of histone variant genes that, when expressed, can combine to specify a range of chromatin features. In the heart, proteomic mass spectrometry has been used to quantify histone variants [4], investigating how these variants change with disease [5]. Much of what is known about the role of individual histone variants comes from loss of function studies in animal models.

Replacing histone H3, centromere protein A (CENP-A) epigenetically defines centromere organization through a process that involves RNA interference [6]. In addition to CENP-A, there are several more centromeric histone associated proteins, including CENP-B, CENP-C and CENP-T, that maintain the function and structure of CENP-A and promote centromere formation during cell division [7]. In the setting of some cancers, overexpression of CENP-A results in enrichment at non-centromeric sites, thereby altering chromatin structure as part of the disease pathogenesis [8].

Another histone H3 variant is H3.3, which is highly conserved, with only four amino acid differences from the canonical H3 in eukaryotes [9]. Localization of histone H3.3 corresponds to transcriptionally active chromatin regions with the highest turnover rate at RNA polymerase II (RNAP II) binding sites and transcription start and end sites, indicating its function in transcription initiation and termination [10]. Furthermore, the presence of H3.3 regulates repressive histone marks (e.g. H3K27me3) [11] and histone variants (e.g. H2A.Z) [12]. *De novo* synthesized H3.3 replaces the canonical histone H3 and remodels donor nuclear chromatin for gene activation during oocyte reprogramming and knockdown of H3.3 disrupts the reprogramming process [13]. Complete deletion of H3.3 embryonic lethal due in part to p53 activation [14].

The N-terminal MacroH2A is 64 % homologous to the canonical histone H2A and contains a carboxyl-terminal ~ 30 kDa macrodomain [15]. MacroH2A had long been thought to regulate gene silencing, however recent chromatin immunoprecipitation (ChIP) plus DNA sequencing findings indicate that macroH2A localizes to areas of both repressive and active chromatin, having its effects on transcription by inhibiting activator binding sites (to cause repression) or inhibiting repressor sites (to cause activation) [16].

Histone variant H2AX, constituting between 2 and 25 % of total H2A protein by gel image quantification in mammals [17], has been functionally implicated in mitotic/meiotic division, stem cell development and aging [18]. Recently, it has been reported that H2AX deposition can be serve as an epigenetic mark for quality control of induced pluripotent stem cells [19]. Phosphorylation of H2AX serves as a marker of DNA damage.

H2A.B is unstable at the protein level and distant in terms of sequence homology compared with other histone H2A variants, with ~40–50 % sequence identity [20, 21]. H2A.B-containing nucleosomes wrap only 116–130 bp of DNA (rather than the conventional 147 bp) and as a result transiently associate with the genome during processes of DNA replication and repair [21]. Genome-wide analysis showed that H2A.B correlates with DNA methylation (in some scenarios a repressive marker itself) and facilitates methylation related to transcription elongation, suggesting a positive role in regulating gene expression [22].

Histone H2A.Z is a highly conserved H2A variant with about 60 % homology to canonical histone H2A, demonstrating both active and repressive transcriptional regulation. Recent findings show that H2A.Z associated with gene coding regions as well as 3' and 5' ends of genes, facilitates cryptic, antisense transcription and RNAP II regulated transcription [23–25]. H2A.Z works with Nanog, a key transcription factor for stem cell identity, in regulating pluripotency/reprogramming and serves as a biomarker for asymmetrically self-renewing cells [26, 27]. Increased protein levels of H2A.X cause cardiac hypertrophy, whereas knockdown of H2A.Z prevents pathologic cardiomyocyte growth [28].

Excluded from the nucleosome core, the last family of histone variants, known as linker histones, plays an intriguing and incompletely understood role in controlling how nucleosomes interact with each other. The lysine-rich linker histone (also referred as H1 and H5) contains a highly conserved central globular domain, a short amino-terminal tail and a long carboxyl-terminal domain, which are vital to its role in higher-order chromatin structure [29]. The globular domain of chicken linker histone H5 and Drosophila linker histone H1 have been shown to possess off-dyad and on-dyad binding respectively, perhaps due to differences in key amino acids between the proteins [30]. Linker histones can interact with both core histones and other proteins. Through mass spectrometry and microscopy, Histone H1 has been found to interact with H2A C-terminus and proteins that are involved in rDNA chromatin structure, rRNA processing and mRNA splicing [31–33]. Furthermore, knockout of all three H1 isoforms in mouse embryonic stem (ES) cells revealed roles for the protein family in gene silencing and nucleosome organization [33]. The main modification of linker histones is phosphorylation by cyclin-dependent kinases

(CDK), which is lowest in G1 and highest in G2 and Mitosis [34]. Infrared spectroscopy of phosphorylated linker histone with CDK2 showed the induction of β-structure that may result in chromatin condensation [35]. Impaired linker histones dynamics can trigger multiple diseases including cancer [36].

Notable to the heart, triple knockout of H1cH1dH1e of E9.5 showed pericardial expansion [37]. At the mRNA level, these triple knockout mice did not show alterations in cardiac specific transcripts such as Nkx2.5 and alpha myosin heavy chain [38]. In an adult model of cardiac hypertrophy, global epigenome remodeling involves changes in histone stoichiometry at the protein level: following pressure overload, the ratio of linker to core histones was decreased as compared to the healthy heart, suggesting a more permissive transcriptional environment. This interpretation is supported by genome-wide transcriptome changes in cardiac hypertrophy, which would necessitate relaxed chromatin at multiple loci. Lastly, this change was associated with a global shift of histone post-translational modifications favoring euchromatin over heterochromatin [5].

High mobility group proteins, originally shown to be structural components of chromatin in the 1970s [39] have more recently been observed to contribute to cardiac gene expression in a locus-specific manner, facilitating higher order chromatin structure [5]. These proteins may act similar to linker histones, facilitating higher order structure of chromatin by binding bent DNA and enabling compaction of tracts of nucleosomes.

Chromatin Domains

While there is widespread agreement on the importance of the nucleosome as functional unit of chromatin and its modulation by chromatin remodelers (discussed in detail below), the principles of genome organization beyond this scale are decidedly more nebulous. Let's estimate there are perhaps 20 million nucleosomes in a given nucleus and around 3 billion base pairs of DNA: how this molecular morass is organized for reproducible, timely access and repackaging is of key importance, and indeed has been the focus of intense investigation. Early observations of in vitro reconstituted DNA and histones revealed the formation of the titular '30 nm fiber', in which the linker histone teams up with several nucleosomes to form intermediate domains of packaging. More recently, 10 nm substructures have been identified, again comprised of nucleosomes plus linker histones, and suggested to be the functional units of both euchromatin and heterochromatin [40]. The folding of this 10 nm nucleosome fiber has been demonstrated to be irregular and gathered in heterogeneous groups, leading to variable chromatin structure, and challenging the long held view about a higher order 30 nm fiber secondary structure [41]. It has been found that chromatin secondary structure is affected by many factors, such as linker histones, length of linker DNA and thus spacing of nucleosomes, histone variant and histone/DNA modifications [42, 43]. In this active area of investigation, novel techniques [42] are continually being brought to bear on the question of whether there is a finite intermediate structure of chromatin, larger than a nucleosome and smaller than the chromosome.

For decades it has been recognized that the genome segments into non-random chromosome territories which may play a role (although the direction of a causal arrow is unclear in this relationship) in transcriptional programs [44]. Recent advances in next generation sequencing have expanded and textured this model in interesting ways. First, innovation in chromosome capture technologies have enabled multiple studies [45] into the endogenous structure of the mammalian genome, revealing folding principles, providing higher resolution to aforementioned chromosomal territories and establishing the fractal, self-repeating structure adopted by the cell's most complex multi-molecular complex. As further resolution has been achieved with these studies, topologically associated domains (TADs) have been described, accounting for ~90 % of genomic structure in mouse ES cells genome and averaging 880 kb in size [46]. The boundary regions of TADs are enriched for CTCF, housekeeping genes and short interspersed elements, but not histone modifications such as enhancer-related H3K4me1 and heterochromatin-related H3K9me3 [46]. TAD boundaries tend to be conserved between cell types, however within these boundaries, the histone modifications and chromatin structural proteins that decorate chromatin impose upon it transcriptional phenotypes, including all variations of active and silenced chromatin [47]. Thus TADs appear to be a structure within which epigenomic modifiers specify the transcriptome. Likewise, TADs correspond to eukaryotic replication-timing reprogram, translocating from nuclear interior during DNA synthesis (active transcription) to nuclear periphery during later replication (repressive transcription) [48]. Deletion or disruption of TADs by CRISPR/Cas genome editing leads to altered gene expression and *de novo* enhancer-promoter interactions [49].

Open questions in chromatin structure include: To what extent is our ability to discern intermediate chromatin features limited by the fixation and sequencing protocols currently in vogue? If TADs are shared between cell types, what is the role of global chromatin structure in cell type specific transcriptomes and phenotypes? What is the impact of genetic variation, which has been all but ignored in epigenomic studies to date, on chromatin structure?

Remodeling Chromatin for Development and Disease

To accommodate multiple cellular transcriptomes, chromatin accessibility has evolved to be highly dynamic throughout normal organismal development. This plasticity is also exploited in disease to enact abnormal transcriptional programs. Recent advances in understanding how epigenomic memories are created, remembered, erased—and, to extend the metaphor—in some cases hallucinated, has advanced our understanding of the basic biology of gene expression and shifted our understanding of disease to include aberrant chromatin structure and function.

Histone Tail Modification

While the bulk of the histone protein mass is ordered within the core nucleosome particle, each of the four variants wears a capricious amino terminus that, resisting fixed structure, is exposed to the nuclear milieu, solicitous of molecular interaction and post-translational modification. The ability of enzymes to modify histone tails to influence transcription in vivo has been known for nearly 40 years [50, 51], but recent advances in mass spectrometry has exploded the list of modifications documented on histone tails to include virtually every known type of post-translational modification, numbering in some cases over 100 modifications on nucleosomes from a single cell type and with most but not all occurring on the soluble tails [52]. Methylation (active or repressive transcription) and acetylation (active transcription) represent two major classes and coordinate with each other [53]. Other histone modifications include phosphorylation, ubiquitination, SUMOylation and crotonylation. Through computational methods, the relationship between histone modifications and gene expression has been explored [54].

Some of these principles for histone modification-dependent regulation of gene expression and phenotype have been tested in the heart. Methylation and acetylation of histones with their associated histone modifiers (such as acetyltransferase, histone deacetylases and histone methyltransferases) have been particularly well explored [55, 56]. Other modifications, such as phosphorylation of histone H3 is involved in cardiac hypertrophy through transcriptional elongation [57]. Moreover, hyper-acetylation has been shown to change gene expression through RNA alternative slicing, thus affecting cardiac cell growth [58]. Stimulation of SUMOylation may exert a protective function on heart [59]. Genome wide analyses have identified cooperative functions of active (H3K9ac, H3K27ac, H3K4me3, and H3K79me2) and repressive (H3K9me2, H3K9me3, and H3K27me3) histone marks [60]. Using data from the ENCODE Project, Roadmap Epigenomics and several other studies, researchers found that E11.5 active enhancers can be accurately predicted by three dimensional analysis of genome-wide H3K27ac and H3K4me1 across developmental time, between tissues within an organism and for the corresponding tissue within species [61]. An algorithm named histoneHMM (hidden markov model) has been used to predict genomic regions that affect cardiac hypertrophy, mostly focused on H3K27me3 [62]. Moreover, histone methylation levels in the heart have been found to be regulated mostly in trans, prominently for H3K4me3 [63].

ATP-Dependent Remodelers

Where nucleosomes reside along the genome affects gene expression and is thus a highly regulated process. One of the most direct mechanisms to influence nucleosome positioning in an active, stimulus responsive manner is through ATP-dependent chromatin remodeling enzymes, which are usually subdivided into four groups: SWI/SNF (switching defective/sucrose non-fermenting), ISWI (imitation switch), INO80 (inositol requiring 80) and CHD (chromodoman, helicase, DNA binding). These groups are identified by a specific, highly conserved ATPase that belongs to the SF2 helicase superfamily and can alter histone-DNA interactions through a process that consumes ATP. These proteins influence transcription as well as chromatin structure by nucleosome translocation [64], nucleosome (whole nucleosome/H2A-H2B dimers) eviction [65] and histone variant exchange (e.g. H2A.Z) [66].

In the heart, most studies have focused on SWI/SNF in the setting of development and hypertrophy, yet all four groups have been explored in some manner [55]. Chromodomain-helicase-DNA-binding protein 7 (CHD7; from the CHD family) and its mutation has been associated with CHARGE syndrome, revealing its novel function in calcium excitation-contraction coupling [67]. Probably the most well studied family is that regulated by Brg1, a member of the BAF complex, which has been shown to be critical for fetal gene activation (myosin heavy chain switching in particular) in the mouse heart following stress [68].

DNA Methylation

Not all things chromatin occur on proteins. The first, and perhaps only truly *epigenetic* (as defined by transgenerational heritability of acquired features) mark, DNA methylation is defined as the addition of a methyl group (CH_3) to the C5 position of cytosine and usually occurs within the major groove of DNA at CpG dinucleotides The methyl group is transferred by DNA methyltransferase (DNMT) family with DNMT1 functioning as the maintenance, and DNMT3a/b as the *de novo*, DNA methyltransferase [69]. DNA methylation mostly happens in the promoter region, gene body and less in intergenic regions. Across the genome, DNA methylation and CpG density follow a bimodal distribution, with high methylation level in CpG-poor regions (CpG depletion) and low methylation level in CpG-rich (typically entails $>50\,\%$ GC) region that are commonly defined as CpG islands (CGIs) [70]. CGIs are located mainly in promoter regions and their methylation can initiate vigorous, long-term transcription repression such as X-inactivation [1].

Although DNA methylation usually associated with repressive transcription, studies show that DNA methylation, especially in gene bodies, is altered during transcriptional elongation [71]. Another long held view—that DNA methylation is highly stable—is also being reconsidered in light of the finding of active DNA demethylation (higher in gene bodies) through base excision repair proteins such as Ten-eleven translocation (TET) [72]. Interesting, DNA methylation has been regarded as a regulator of alternative splicing possibly by influencing chromatin structure, thus affecting RNAP II recruitment and binding of heterochromatin protein 1 [73].

In the heart, DNA methylation influences cardiomyopathy and heart development. In congenital heart diseases, hypermethylation of MSX1 and GATA4 has been found [74]. In Tetralogy of Fallot, certain genes such as EGFR and TBX5 have shown significant differences in methylation status compared to unaffected individuals [75]. In patients with dilated cardiomyopathy, aberrant DNA methylation has been found in lymphocyte antigen 75 and adenosine receptor A2A, which has been further confirmed in zebrafish [76]. Also, three angiogenesis-related genes (AMOTL2, ARHGAP24 and PECAM1) have been identified that exhibited altered methylation status [77]. During development, only a small fraction of genes showed aberrant methylation: this subset, however, were highly related to cardiac specific processes when comparing developmental day E11.5 to E14.5 [78]. When comparing adult to developing mouse heart, DNA methylation is increased in active enhancers [79]. When comparing developing, mature and diseased cardiomyocytes, DNA methylation is quite dynamic, supporting a role for this modification in promoting, or responding to, condition-specific gene expression [80]. Cardiomyocytes treated with endothelin-1 (a hypertrophic agonist) showed increased DNA methylation; conversely, inhibition of DNA methylation rescued the norepinephrine-induced hypertrophy [81, 82]. Therefore, alteration of DNA methylation has been proposed as a novel therapeutic target in the heart, although the mechanisms of action, and principal targets, remain to be determined.

RNA-Based Mechanisms

Widespread application of next generation RNA sequencing technologies have dramatically increased the portion of the genome that is understood to be transcribed in a given cell type, now estimated at ~75 %. The beguiling nature of this observation is that most of these newly identified transcripts do not appear to be messenger RNAs: instead, they belong to a class of RNAs called, perhaps misleadingly, non-coding RNAs (ncRNAs) which may code small peptides/proteins, may function independently as RNA scaffolds for various cellular processes or may in fact be, as was previously thought, transcriptional noise. ncRNAs can be divided into small RNA (<200 nucleotides) and long non coding RNA (lncRNA, >200 nucleotides) [83]. Small RNAs are thought to modulate heterochromatin and gene silencing with the help of Argonaut [84]. They may also participate in alternative splicing and transcription together with other epigenetic regulation such as DNA methylation and histone modifications [85]. lncRNAs connect chromatin loci with chromatin remodelers, transcription factors and other RNAs both in cis and trans [86, 87] by binding modules nestled in their secondary structure, leading to changes of chromatin structure and nuclear organization [88]. Besides their close relationship to silenced chromatin, different classes of lncRNAs alternatively contribute to active transcription, especially transcriptional enhancer element RNA (eRNA) that binds the Mediator complex [89]. Natural antisense transcripts (NAT), another class of lncRNA, are read from the opposite strand of the mRNAs that they regulate in a complementary, cis manner. Alternations in lncRNA levels have been implicated in cancer, skeletal defects, embryogenesis abnormalities and brain defects [90].

Roles for lncRNAs in heart development, differentiation, and disease have recently emerged. Depletion of the lncRNA *Braveheart* revealed its role in cardiovascular lineage commitment by activating MesP1 (a master transcription factor expressed in multipotent cardiac progenitor) and interacting with SUZ12 (a component of polycomb repressive complex 2(PRC2)) [91]. Targeted homozygous deletion of *Fendrr* in mouse, another lncRNA, resulted in embryonic lethality and PRC2 reduction, leading to decreased H3K27me3 and increased H3K4me3 in the promoter regions of target genes [92]. Restoration of repressed myosin heavy-chain-associated lncRNA transcripts (abbreviated as *Myheart*) in the setting of pressure overload hypertrophy protects the heart from cardiomyopathy by interacting with and inhibiting Brg1 [93]. Myocardial infarction associated transcript (*Miat*), a potential risk factor of myocardial infarction, was discovered through a case–control association study of single nucleotide polymorphism markers [94]. The relationship of microRNAs to heart and vascular diseases, has been extensively reviewed [95, 96].

circRNA, a novel class of ncRNA, is formed by backsplicing and features a covalently joined loop structure without free 3′ and 5′ ends [97]. Circular Antisense Noncoding RNA in the INK4 Locus (ANRIL) is the first cardiovascular related circRNA that correlated with INK4/ARF expression and atherosclerosis risk [98]. Using a statistical method, circular Sodium/Calcium Exchanger (NCX1) was found to increase more rapidly during fetal heart development than its linear version [99]. Recently, using human umbilical vein endothelial cells, researchers showed that circZNF292 is regulated by hypoxia and displays proangiogenic activity [100], although this family of RNAs remains largely unexplored in the cardiovascular system.

New Epigenomic Techniques

Sequencing-Based Techniques

For histone variants, restriction endonuclease digestion of chromatin coupled to deep sequencing (RED-seq) offers an unbiased and sensitive method to study chromatin accessibility in nucleosome depleted regions, within nucleosome arrays and between different histone variants [101]. RED-seq is performed on permeabilized cells using restriction endonuclease digestion, DNA extraction with unbiased sonication, two separate steps of ligation of linkers, PCR amplification and sequencing. For tertiary structures such as TADs, Hi-C, that studies in situ DNA-DNA contacts, has been applied [46]. Sub-TAD structure, however, has been explored with the help of higher resolution 5C [102]. Apart from DNA-DNA interactions in shaping the three-dimensional arrangement of chromatin, the function of specific proteins, especially transcription factors, can be identified through chromatin interaction analysis with paired end tag (ChIA-PET) that involves immunoprecipitation [103]. ChIA-PET can investigate specific proteins that modulate genome organization through formaldehyde crosslinking, DNA sonication, ChIP enrichment, followed by ligation and sequencing.

For examining the occupancy profile of nucleosomes along chromatin, multiple techniques have emerged. For open chromatin, transposase-accessible chromatin using sequencing (ATAC-seq) features a simple and sensitive two-step protocol to explore the nucleosome landscape [104]. Without fixation, permeabilized gDNA is ligated with the help of Tn5 transposase. After purification, open chromatin is amplified and sequenced. Another novel approach named methidiumpropyl-EDTA sequencing (MPE-seq) provides a sensitive method for detection the upstream open chromatin region of active promoters' transcription start sites, which can be combined with micrococcal nuclease-sequencing (MNase-seq) to generate a detailed readout of chromatin structure and regulation [105]. MPE-seq uses ferrous iron to generate DNA breakage followed by sequencing with minimal bias. For histone modifications and transcription factor binding sites, ChIP followed by qPCR or sequencing has been extensively used to map the proteins or histone modifications bound to DNA with cell fixation, DNA sonication and sequencing [106]. Reduced representation bisulfite sequencing (RRBS) is a commonly used method for DNA methylation (cytosine methylation and hydroxylmethylation) analysis. RRBS uses bisulfite to convert unmodified cytosines to uracil, thereby revealing methylation distribution. A new technique named methyl-sequencing has been developed that has the ability to detect the 5-formylcytosine that is resistant to conversion in RRBS, combined with comprehensive methylation detection level by using cytosine-methylated universal adapters [107]. For detecting the hydroxylmethylation alone, Tet-assisted bidulfite sequencing (TAB-seq) [108] and oxidative bisulfite sequencing (oxBS-seq) [109] can be employed. TAB-seq uses Tet to convert methylcytosine to carbomethylcytosine, whereas oxBS-seq applies oxidization to convert hydroxymethylcytosine to formylcytosine. For transcriptome and alternative splicing, RNA sequencing remains to be the top choice. For the relationship between lncRNA and chromatin, chromatin isolation by RNA purification (ChIRP) sequencing can be used [110]. ChIRP maps the genomic localization of a known ncRNA by antibody pull down. Besides RNAs, nascent transcripts that bind to RNAP II can be detected through methods such as nascent transcript sequencing (NET-seq) [111].

Through whole genome sequencing, specific genes that may affect the congenital heart diseases, dilated cardiomyopathy, arrhythmia and other heart-related diseases have been detected [112]. For changes in coding sequences, exome sequencing has emerged as a powerful choice due to lower cost and simpler interpretation in contrast with whole genome sequencing. Many heart diseases, such as congenital heart diseases [113], myocardial infarction [114], and coronary heart disease [115] have been explored with these approaches. Similar to RNA sequencing, DNA methylation sequencing has been compared with normal and diseased conditions, revealing aberrant methylation in genes associated with dilated cardiomyopathy [76], cardiac fibrosis [116], and congenital heart diseases [78]. Numerous proteins have been shown, by ChIP-seq experiments, to alter their association with the genome during heart disease, development and regeneration [93, 117–126].

Microscopy

Using microscopy to visualize chromatin structure has tantalizing appeal, but limitations abound with regard to resolution. Light microscopy has a resolution around 200–300 nm and employs florescent probes to visualize living cells. Certain techniques, such as fluorescence recovery after photobleaching (FRAP) and fluorescence in situ hybridization, can be coupled with light microscopy. For super-resolution microscopy, besides the antibody-based methods which have been applied to examine chromatin structure in cardiac myocytes during disease [127], new technique such as TALENs [128] and CRISPR/Cas9 [129] will enable the endogenous gene labeling for both proteins and DNA sequences that will give more reliable results that can be tailored to a specific genetic locus. Other readouts of genomic organization including RNA polymerase II, histone modifications and nascent RNA have been investigated [130, 131]. Imaging of the core histone H2B has also been used to reveal the spatial organization of chromatin fiber [132]. Cryogenic electron microscopy, with around 10 nm resolution, has been used to study chromatin structure, revealing the formation of higher-order chromatin features influenced by the presence of H1 [42].

In the heart, super-resolution microscopy, especially its application in T-tubule and calcium signaling, has been reviewed [133]. Label-free microscopy, generating light signals based on molecules' photo-physiology, has been applied to reveal cardiac-vascular interactions [134]. Intravital microscopy, which is applied to living tissues, has been used to investigate the beating heart with single cell resolution during cardiac ischemia [135] and monocyte circulation during myocardial infarction [136]. Atomic force microscopy, a type of scanning probe microscopy with resolution ranging from 0.1 to 100 μm, allows 3D structural assessment in living cells. This method has been used to detect sarcomere lengthening [137] and valve leaflet stiffness [138]. Different kinds of electron microscopy have enabled the identification of myocyte morphology [139], ultrastructure of the intercalated disc [140], morphology of mitochondria [141] and the interaction of telocytes and myocytes [142]. Frontiers for microscopic analysis of cardiac chromatin will be breached by novel labeling reagents as well as by the application of new imaging techniques, such as 3D super resolution and light sheet microscopy.

Concluding Remarks and Perspective

Chromatin is the substrate of cellular memory—it is the way cells, and therefore organs, remember what they are. The last decade of research in genomics, epigenomics and transcriptomics have revolutionized our understanding of the mechanisms through which these different tiers of biological information interact. Despite this progress in the basic science realm, this holistic approach to biology is challenging for translation: progress requires that we come to utilize cellular networks in similar terms as the EKG…to interpret a readout from 'omics measurements akin to how chest auscultation is employed as an integral part of clinical decision making. This advance requires novel approaches to analyzing big data and, most importantly, investigations of epigenomic regulation and epigenomic susceptibility directly in human populations.

Acknowledgements We thank Dr. Manuel Rosa Garrido for help with the figure and all members of the Vondriska laboratory for helpful discussions. Research in the Vondriska laboratory is supported by the National Heart, Lung and Blood Institute of the NIH, the American Heart Association, Thermo Fisher Scientific and the Department of Anesthesiology in the David Geffen School of Medicine at UCLA.

References

1. Jaenisch R, Bird A. Epigenetic regulation of gene expression: how the genome integrates intrinsic and environmental signals. Nat Genet. 2003;33(Suppl):245–54.
2. Kouzarides T. Chromatin modifications and their function. Cell. 2007;128:693–705.
3. Soon WW, Hariharan M, Snyder MP. High-throughput sequencing for biology and medicine. Mol Syst Biol. 2013;9:640.
4. Franklin S, Zhang MJ, Chen H, Paulsson AK, Mitchell-Jordan SA, Li Y, Ping P, Vondriska TM. Specialized compartments of cardiac nuclei exhibit distinct proteomic anatomy. Mol Cell Proteomics MCP. 2011;10:M110.000703.
5. Franklin S, Chen H, Mitchell-Jordan S, Ren S, Wang Y, Vondriska TM. Quantitative analysis of the chromatin proteome in disease reveals remodeling principles and identifies high mobility group protein b2 as a regulator of hypertrophic growth. Mol Cell Proteomics MCP. 2012;11:M111.014258.
6. Black BE, Cleveland DW. Epigenetic centromere propagation and the nature of cenp-a nucleosomes. Cell. 2011;144:471–9.
7. Logsdon GA, Barrey EJ, Bassett EA, DeNizio JE, Guo LY, Panchenko T, Dawicki-McKenna JM, Heun P, Black BE. Both tails and the centromere targeting domain of cenp-a are required for centromere establishment. J Cell Biol. 2015;208:521–31.
8. Lacoste N, Woolfe A, Tachiwana H, Garea AV, Barth T, Cantaloube S, Kurumizaka H, Imhof A, Almouzni G. Mislocalization of the centromeric histone variant cenh3/cenp-a in human cells depends on the chaperone daxx. Mol Cell. 2014;53:631–44.
9. Elsaesser SJ, Goldberg AD, Allis CD. New functions for an old variant: no substitute for histone h3.3. Curr Opin Genet Dev. 2010;20:110–7.
10. Ha M, Kraushaar DC, Zhao K. Genome-wide analysis of h3.3 dissociation reveals high nucleosome turnover at distal regulatory regions of embryonic stem cells. Epigenetics Chromatin. 2014;7:38.
11. Weth O, Paprotka C, Gunther K, Schulte A, Baierl M, Leers J, Galjart N, Renkawitz R. Ctcf induces histone variant incorporation, erases the h3k27me3 histone mark and opens chromatin. Nucleic Acids Res. 2014;42:11941–51.
12. Chen P, Wang Y, Li G. Dynamics of histone variant h3.3 and its coregulation with h2a.Z at enhancers and promoters. Nucleus. 2014;5:21–7.
13. Wen D, Banaszynski LA, Liu Y, Geng F, Noh KM, Xiang J, Elemento O, Rosenwaks Z, Allis CD, Rafii S. Histone variant h3.3 is an essential maternal factor for oocyte reprogramming. Proc Natl Acad Sci U S A. 2014;111:7325–30.
14. Jang CW, Shibata Y, Starmer J, Yee D, Magnuson T. Histone h3.3 maintains genome integrity during mammalian development. Genes Dev. 2015;29:1377–92.

15. Pehrson JR, Fuji RN. Evolutionary conservation of histone macroh2a subtypes and domains. Nucleic Acids Res. 1998;26:2837–42.
16. Lavigne MD, Vatsellas G, Polyzos A, Mantouvalou E, Sianidis G, Maraziotis I, Agelopoulos M, Thanos D. Composite macroh2a/nrf-1 nucleosomes suppress noise and generate robustness in gene expression. Cell Rep. 2015;11:1090–101.
17. Rogakou EP, Pilch DR, Orr AH, Ivanova VS, Bonner WM. DNA double-stranded breaks induce histone h2ax phosphorylation on serine 139. J Biol Chem. 1998;273:5858–68.
18. Turinetto V, Giachino C. Multiple facets of histone variant h2ax: a DNA double-strand-break marker with several biological functions. Nucleic Acids Res. 2015;43:2489–98.
19. Wu T, Liu Y, Wen D, Tseng Z, Tahmasian M, Zhong M, Rafii S, Stadtfeld M, Hochedlinger K, Xiao A. Histone variant h2a.X deposition pattern serves as a functional epigenetic mark for distinguishing the developmental potentials of ipscs. Cell Stem Cell. 2014;15:281–94.
20. Chadwick BP, Willard HF. A novel chromatin protein, distantly related to histone h2a, is largely excluded from the inactive x chromosome. J Cell Biol. 2001;152:375–84.
21. Arimura Y, Kimura H, Oda T, Sato K, Osakabe A, Tachiwana H, Sato Y, Kinugasa Y, Ikura T, Sugiyama M, Sato M, Kurumizaka H. Structural basis of a nucleosome containing histone h2a.B/h2a.Bbd that transiently associates with reorganized chromatin. Sci Rep. 2013;3:3510.
22. Chen Y, Chen Q, McEachin RC, Cavalcoli JD, Yu X. H2a.B facilitates transcription elongation at methylated cpg loci. Genome Res. 2014;24:570–9.
23. Gu M, Naiyachit Y, Wood TJ, Millar CB. H2a.Z marks antisense promoters and has positive effects on antisense transcript levels in budding yeast. BMC Genomics. 2015;16:99.
24. Jeronimo C, Watanabe S, Kaplan CD, Peterson CL, Robert F. The histone chaperones fact and spt6 restrict h2a.Z from intragenic locations. Mol Cell. 2015;58:1113–23.
25. Weber CM, Ramachandran S, Henikoff S. Nucleosomes are context-specific, h2a.Z--modulated barriers to rna polymerase. Mol Cell. 2014;53:819–30.
26. Wang J, Qiao M, He Q, Shi R, Loh SJ, Stanton LW, Wu M. Pluripotency activity of nanog requires biochemical stabilization by variant histone protein h2a.Z. Stem Cells. 2015;33:2126–34.
27. Huh YH, Noh M, Burden FR, Chen JC, Winkler DA, Sherley JL. Sparse feature selection identifies h2a.Z as a novel, pattern-specific biomarker for asymmetrically self-renewing distributed stem cells. Stem Cell Res. 2015;14:144–54.
28. Chen IY, Lypowy J, Pain J, Sayed D, Grinberg S, Alcendor RR, Sadoshima J, Abdellatif M. Histone h2a.Z is essential for cardiac myocyte hypertrophy but opposed by silent information regulator 2alpha. J Biol Chem. 2006;281:19369–77.
29. Hartman PG, Chapman GE, Moss T, Bradbury EM. Studies on the role and mode of operation of the very-lysine-rich histone h1 in eukaryote chromatin. The three structural regions of the histone h1 molecule. Eur J Biochem/FEBS. 1977;77:45–51.
30. Zhou BR, Jiang J, Feng H, Ghirlando R, Xiao TS, Bai Y. Structural mechanisms of nucleosome recognition by linker histones. Mol Cell. 2015;59:628–38.
31. Vogler C, Huber C, Waldmann T, Ettig R, Braun L, Izzo A, Daujat S, Chassignet I, Lopez-Contreras AJ, Fernandez-Capetillo O, Dundr M, Rippe K, Langst G, Schneider R. Histone h2a c-terminus regulates chromatin dynamics, remodeling, and histone h1 binding. PLoS Genet. 2010;6:e1001234.
32. Shukla MS, Syed SH, Goutte-Gattat D, Richard JL, Montel F, Hamiche A, Travers A, Faivre-Moskalenko C, Bednar J, Hayes JJ, Angelov D, Dimitrov S. The docking domain of histone h2a is required for h1 binding and rsc-mediated nucleosome remodeling. Nucleic Acids Res. 2011;39:2559–70.
33. Szerlong HJ, Herman JA, Krause CM, DeLuca JG, Skoultchi A, Winger QA, Prenni JE, Hansen JC. Proteomic characterization of the nucleolar linker histone h1 interaction network. J Mol Biol. 2015;427:2056–71.
34. Talasz H, Helliger W, Puschendorf B, Lindner H. In vivo phosphorylation of histone h1 variants during the cell cycle. Biochemistry. 1996;35:1761–7.

35. Lopez R, Sarg B, Lindner H, Bartolome S, Ponte I, Suau P, Roque A. Linker histone partial phosphorylation: effects on secondary structure and chromatin condensation. Nucleic Acids Res. 2015;43:4463–76.
36. Chen Y, Hoover ME, Dang X, Shomo AA, Guan X, Marshall AG, Freitas MA, Young NL. Quantitative mass spectrometry reveals that intact histone h1 phosphorylations are variant specific and exhibit single molecule hierarchical dependence. Mol Cell Proteomics MCP. 2015;15:813–33.
37. Fan Y, Nikitina T, Morin-Kensicki EM, Zhao J, Magnuson TR, Woodcock CL, Skoultchi AI. H1 linker histones are essential for mouse development and affect nucleosome spacing in vivo. Mol Cell Biol. 2003;23:4559–72.
38. Zhang Y, Cooke M, Panjwani S, Cao K, Krauth B, Ho PY, Medrzycki M, Berhe DT, Pan C, McDevitt TC, Fan Y. Histone h1 depletion impairs embryonic stem cell differentiation. PLoS Genet. 2012;8:e1002691.
39. Vidali G, Boffa LC, Allfrey VG. Selective release of chromosomal proteins during limited dnaase 1 digestion of avian erythrocyte chromatin. Cell. 1977;12:409–15.
40. Fussner E, Strauss M, Djuric U, Li R, Ahmed K, Hart M, Ellis J, Bazett-Jones DP. Open and closed domains in the mouse genome are configured as 10-nm chromatin fibres. EMBO Rep. 2012;13:992–6.
41. Maeshima K, Imai R, Tamura S, Nozaki T. Chromatin as dynamic 10-nm fibers. Chromosoma. 2014;123:225–37.
42. Song F, Chen P, Sun D, Wang M, Dong L, Liang D, Xu RM, Zhu P, Li G. Cryo-em study of the chromatin fiber reveals a double helix twisted by tetranucleosomal units. Science. 2014;344:376–80.
43. Collepardo-Guevara R, Schlick T. Chromatin fiber polymorphism triggered by variations of DNA linker lengths. Proc Natl Acad Sci U S A. 2014;111:8061–6.
44. Cremer T, Cremer C. Chromosome territories, nuclear architecture and gene regulation in mammalian cells. Nat Rev Genet. 2001;2:292–301.
45. Dekker J, Marti-Renom MA, Mirny LA. Exploring the three-dimensional organization of genomes: interpreting chromatin interaction data. Nat Rev Genet. 2013;14:390–403.
46. Dixon JR, Selvaraj S, Yue F, Kim A, Li Y, Shen Y, Hu M, Liu JS, Ren B. Topological domains in mammalian genomes identified by analysis of chromatin interactions. Nature. 2012;485:376–80.
47. Ciabrelli F, Cavalli G. Chromatin-driven behavior of topologically associating domains. J Mol Biol. 2015;427:608–25.
48. Pope BD, Ryba T, Dileep V, Yue F, Wu W, Denas O, Vera DL, Wang Y, Hansen RS, Canfield TK, Thurman RE, Cheng Y, Gulsoy G, Dennis JH, Snyder MP, Stamatoyannopoulos JA, Taylor J, Hardison RC, Kahveci T, Ren B, Gilbert DM. Topologically associating domains are stable units of replication-timing regulation. Nature. 2014;515:402–5.
49. Lupianez DG, Kraft K, Heinrich V, Krawitz P, Brancati F, Klopocki E, Horn D, Kayserili H, Opitz JM, Laxova R, Santos-Simarro F, Gilbert-Dussardier B, Wittler L, Borschiwer M, Haas SA, Osterwalder M, Franke M, Timmermann B, Hecht J, Spielmann M, Visel A, Mundlos S. Disruptions of topological chromatin domains cause pathogenic rewiring of gene-enhancer interactions. Cell. 2015;161:1012–25.
50. Travis GH, Colavito-Shepanski M, Grunstein M. Extensive purification and characterization of chromatin-bound histone acetyltransferase from saccharomyces cerevisiae. J Biol Chem. 1984;259:14406–12.
51. Allfrey VG. In chromatin and chromosome structure. New York: Academic; 1977.
52. Zhao Y, Garcia BA. Comprehensive catalog of currently documented histone modifications. Cold Spring Harb Perspect Biol. 2015;7:a025064.
53. Rada-Iglesias A, Bajpai R, Swigut T, Brugmann SA, Flynn RA, Wysocka J. A unique chromatin signature uncovers early developmental enhancers in humans. Nature. 2011;470:279–83.

54. Karlic R, Chung HR, Lasserre J, Vlahovicek K, Vingron M. Histone modification levels are predictive for gene expression. Proc Natl Acad Sci U S A. 2010;107:2926–31.
55. Gillette TG, Hill JA. Readers, writers, and erasers: chromatin as the whiteboard of heart disease. Circ Res. 2015;116:1245–53.
56. McKinsey TA, Zhang CL, Olson EN. Control of muscle development by dueling hats and hdacs. Curr Opin Genet Dev. 2001;11:497–504.
57. Awad S, Al-Haffar KM, Marashly Q, Quijada P, Kunhi M, Al-Yacoub N, Wade FS, Mohammed SF, Al-Dayel F, Sutherland G, Assiri A, Sussman M, Bers D, Al-Habeeb W, Poizat C. Control of histone h3 phosphorylation by camkiidelta in response to haemodynamic cardiac stress. J Pathol. 2015;235:606–18.
58. Sharma A, Nguyen H, Geng C, Hinman MN, Luo G, Lou H. Calcium-mediated histone modifications regulate alternative splicing in cardiomyocytes. Proc Natl Acad Sci U S A. 2014;111:E4920–8.
59. Blakeslee WW, Wysoczynski CL, Fritz KS, Nyborg JK, Churchill ME, McKinsey TA. Class i hdac inhibition stimulates cardiac protein sumoylation through a post-translational mechanism. Cell Signal. 2014;26:2912–20.
60. Papait R, Cattaneo P, Kunderfranco P, Greco C, Carullo P, Guffanti A, Vigano V, Stirparo GG, Latronico MV, Hasenfuss G, Chen J, Condorelli G. Genome-wide analysis of histone marks identifying an epigenetic signature of promoters and enhancers underlying cardiac hypertrophy. Proc Natl Acad Sci U S A. 2013;110:20164–9.
61. Capra JA. Extrapolating histone marks across developmental stages, tissues, and species: an enhancer prediction case study. BMC Genomics. 2015;16:104.
62. Heinig M, Colome-Tatche M, Taudt A, Rintisch C, Schafer S, Pravenec M, Hubner N, Vingron M, Johannes F. Histonehmm: differential analysis of histone modifications with broad genomic footprints. BMC Bioinf. 2015;16:60.
63. Rintisch C, Heinig M, Bauerfeind A, Schafer S, Mieth C, Patone G, Hummel O, Chen W, Cook S, Cuppen E, Colome-Tatche M, Johannes F, Jansen RC, Neil H, Werner M, Pravenec M, Vingron M, Hubner N. Natural variation of histone modification and its impact on gene expression in the rat genome. Genome Res. 2014;24:942–53.
64. Whitehouse I, Flaus A, Cairns BR, White MF, Workman JL, Owen-Hughes T. Nucleosome mobilization catalysed by the yeast swi/snf complex. Nature. 1999;400:784–7.
65. Dechassa ML, Sabri A, Pondugula S, Kassabov SR, Chatterjee N, Kladde MP, Bartholomew B. Swi/snf has intrinsic nucleosome disassembly activity that is dependent on adjacent nucleosomes. Mol Cell. 2010;38:590–602.
66. Mizuguchi G, Shen X, Landry J, Wu WH, Sen S, Wu C. Atp-driven exchange of histone h2az variant catalyzed by swr1 chromatin remodeling complex. Science. 2004;303:343–8.
67. Payne S, Burney MJ, McCue K, Popal N, Davidson SM, Anderson RH, Scambler PJ. A critical role for the chromatin remodeller chd7 in anterior mesoderm during cardiovascular development. Dev Biol. 2015;405:82–95.
68. Hang CT, Yang J, Han P, Cheng HL, Shang C, Ashley E, Zhou B, Chang CP. Chromatin regulation by brg1 underlies heart muscle development and disease. Nature. 2010;466:62–7.
69. Okano M, Bell DW, Haber DA, Li E. DNA methyltransferases dnmt3a and dnmt3b are essential for de novo methylation and mammalian development. Cell. 1999;99:247–57.
70. Deaton AM, Bird A. Cpg islands and the regulation of transcription. Genes Dev. 2011;25:1010–22.
71. Jones PA. Functions of DNA methylation: Islands, start sites, gene bodies and beyond. Nat Rev Genet. 2012;13:484–92.
72. Tahiliani M, Koh KP, Shen Y, Pastor WA, Bandukwala H, Brudno Y, Agarwal S, Iyer LM, Liu DR, Aravind L, Rao A. Conversion of 5-methylcytosine to 5-hydroxymethylcytosine in mammalian DNA by mll partner tet1. Science. 2009;324:930–5.
73. Lev Maor G, Yearim A, Ast G. The alternative role of DNA methylation in splicing regulation. Trends Genet TIG. 2015;31:274–80.
74. Serra-Juhe C, Cusco I, Homs A, Flores R, Toran N, Perez-Jurado LA. DNA methylation abnormalities in congenital heart disease. Epigenetics. 2015;10:167–77.

75. Sheng W, Qian Y, Zhang P, Wu Y, Wang H, Ma X, Chen L, Ma D, Huang G. Association of promoter methylation statuses of congenital heart defect candidate genes with tetralogy of fallot. J Transl Med. 2014;12:31.
76. Haas J, Frese KS, Park YJ, Keller A, Vogel B, Lindroth AM, Weichenhan D, Franke J, Fischer S, Bauer A, Marquart S, Sedaghat-Hamedani F, Kayvanpour E, Kohler D, Wolf NM, Hassel S, Nietsch R, Wieland T, Ehlermann P, Schultz JH, Dosch A, Mereles D, Hardt S, Backs J, Hoheisel JD, Plass C, Katus HA, Meder B. Alterations in cardiac DNA methylation in human dilated cardiomyopathy. EMBO Mol Med. 2013;5:413–29.
77. Movassagh M, Choy MK, Goddard M, Bennett MR, Down TA, Foo RS. Differential DNA methylation correlates with differential expression of angiogenic factors in human heart failure. PLoS One. 2010;5:e8564.
78. Chamberlain AA, Lin M, Lister RL, Maslov AA, Wang Y, Suzuki M, Wu B, Greally JM, Zheng D, Zhou B. DNA methylation is developmentally regulated for genes essential for cardiogenesis. J Am Heart Assoc. 2014;3:e000976.
79. Hon GC, Rajagopal N, Shen Y, McCleary DF, Yue F, Dang MD, Ren B. Epigenetic memory at embryonic enhancers identified in DNA methylation maps from adult mouse tissues. Nat Genet. 2013;45:1198–206.
80. Gilsbach R, Preissl S, Gruning BA, Schnick T, Burger L, Benes V, Wurch A, Bonisch U, Gunther S, Backofen R, Fleischmann BK, Schubeler D, Hein L. Dynamic DNA methylation orchestrates cardiomyocyte development, maturation and disease. Nat Commun. 2014;5:5288.
81. Paradis A, Xiao D, Zhou J, Zhang L. Endothelin-1 promotes cardiomyocyte terminal differentiation in the developing heart via heightened DNA methylation. Int J Med Sci. 2014;11:373–80.
82. Xiao D, Dasgupta C, Chen M, Zhang K, Buchholz J, Xu Z, Zhang L. Inhibition of DNA methylation reverses norepinephrine-induced cardiac hypertrophy in rats. Cardiovasc Res. 2014;101:373–82.
83. Cech TR, Steitz JA. The noncoding rna revolution-trashing old rules to forge new ones. Cell. 2014;157:77–94.
84. Su H, Trombly MI, Chen J, Wang X. Essential and overlapping functions for mammalian argonautes in microrna silencing. Genes Dev. 2009;23:304–17.
85. Li LC. Chromatin remodeling by the small rna machinery in mammalian cells. Epigenetics. 2014;9:45–52.
86. Zhang B, Arun G, Mao YS, Lazar Z, Hung G, Bhattacharjee G, Xiao X, Booth CJ, Wu J, Zhang C, Spector DL. The lncrna malat1 is dispensable for mouse development but its transcription plays a cis-regulatory role in the adult. Cell Rep. 2012;2:111–23.
87. Gupta RA, Shah N, Wang KC, Kim J, Horlings HM, Wong DJ, Tsai MC, Hung T, Argani P, Rinn JL, Wang Y, Brzoska P, Kong B, Li R, West RB, van de Vijver MJ, Sukumar S, Chang HY. Long non-coding rna hotair reprograms chromatin state to promote cancer metastasis. Nature. 2010;464:1071–6.
88. Moazed D. Small rnas in transcriptional gene silencing and genome defence. Nature. 2009;457:413–20.
89. Lam MT, Li W, Rosenfeld MG, Glass CK. Enhancer rnas and regulated transcriptional programs. Trends Biochem Sci. 2014;39:170–82.
90. Li L, Chang HY. Physiological roles of long noncoding rnas: insight from knockout mice. Trends Cell Biol. 2014;24:594–602.
91. Klattenhoff CA, Scheuermann JC, Surface LE, Bradley RK, Fields PA, Steinhauser ML, Ding H, Butty VL, Torrey L, Haas S, Abo R, Tabebordbar M, Lee RT, Burge CB, Boyer LA. Braveheart, a long noncoding rna required for cardiovascular lineage commitment. Cell. 2013;152:570–83.
92. Grote P, Wittler L, Hendrix D, Koch F, Wahrisch S, Beisaw A, Macura K, Blass G, Kellis M, Werber M, Herrmann BG. The tissue-specific lncrna fendrr is an essential regulator of heart and body wall development in the mouse. Dev Cell. 2013;24:206–14.
93. Han P, Li W, Lin CH, Yang J, Shang C, Nurnberg ST, Jin KK, Xu W, Lin CY, Lin CJ, Xiong Y, Chien HC, Zhou B, Ashley E, Bernstein D, Chen PS, Chen HS, Quertermous T, Chang

CP. A long noncoding rna protects the heart from pathological hypertrophy. Nature. 2014;514:102–6.
94. Ishii N, Ozaki K, Sato H, Mizuno H, Saito S, Takahashi A, Miyamoto Y, Ikegawa S, Kamatani N, Hori M, Saito S, Nakamura Y, Tanaka T. Identification of a novel non-coding rna, miat, that confers risk of myocardial infarction. J Hum Genet. 2006;51:1087–99.
95. Boon RA, Dimmeler S. Micrornas in myocardial infarction. Nat Rev Cardiol. 2015;12:135–42.
96. Dorn 2nd GW. Decoding the cardiac message: the 2011 thomas w. Smith memorial lecture. Circ Res. 2012;110:755–63.
97. Jeck WR, Sharpless NE. Detecting and characterizing circular rnas. Nat Biotechnol. 2014;32:453–61.
98. Burd CE, Jeck WR, Liu Y, Sanoff HK, Wang Z, Sharpless NE. Expression of linear and novel circular forms of an ink4/arf-associated non-coding rna correlates with atherosclerosis risk. PLoS Genet. 2010;6:e1001233.
99. Szabo L, Morey R, Palpant NJ, Wang PL, Afari N, Jiang C, Parast MM, Murry CE, Laurent LC, Salzman J. Statistically based splicing detection reveals neural enrichment and tissue-specific induction of circular rna during human fetal development. Genome Biol. 2015;16:126.
100. Boeckel JN, Jae N, Heumuller AW, Chen W, Boon RA, Stellos K, Zeiher AM, John D, Uchida S, Dimmeler S. Identification and characterization of hypoxia-regulated endothelial circular rna. Circ Res. 2015;117:884–90.
101. Chen PB, Zhu LJ, Hainer SJ, McCannell KN, Fazzio TG. Unbiased chromatin accessibility profiling by red-seq uncovers unique features of nucleosome variants in vivo. BMC Genomics. 2014;15:1104.
102. Nora EP, Lajoie BR, Schulz EG, Giorgetti L, Okamoto I, Servant N, Piolot T, van Berkum NL, Meisig J, Sedat J, Gribnau J, Barillot E, Bluthgen N, Dekker J, Heard E. Spatial partitioning of the regulatory landscape of the x-inactivation centre. Nature. 2012;485:381–5.
103. Fullwood MJ, Liu MH, Pan YF, Liu J, Xu H, Mohamed YB, Orlov YL, Velkov S, Ho A, Mei PH, Chew EG, Huang PY, Welboren WJ, Han Y, Ooi HS, Ariyaratne PN, Vega VB, Luo Y, Tan PY, Choy PY, Wansa KD, Zhao B, Lim KS, Leow SC, Yow JS, Joseph R, Li H, Desai KV, Thomsen JS, Lee YK, Karuturi RK, Herve T, Bourque G, Stunnenberg HG, Ruan X, Cacheux-Rataboul V, Sung WK, Liu ET, Wei CL, Cheung E, Ruan Y. An oestrogen-receptor-alpha-bound human chromatin interactome. Nature. 2009;462:58–64.
104. Buenrostro JD, Giresi PG, Zaba LC, Chang HY, Greenleaf WJ. Transposition of native chromatin for fast and sensitive epigenomic profiling of open chromatin, DNA-binding proteins and nucleosome position. Nat Methods. 2013;10:1213–8.
105. Ishii H, Kadonaga JT, Ren B. Mpe-seq, a new method for the genome-wide analysis of chromatin structure. Proc Natl Acad Sci U S A. 2015;112:E3457–65.
106. Landt SG, Marinov GK, Kundaje A, Kheradpour P, Pauli F, Batzoglou S, Bernstein BE, Bickel P, Brown JB, Cayting P, Chen Y, DeSalvo G, Epstein C, Fisher-Aylor KI, Euskirchen G, Gerstein M, Gertz J, Hartemink AJ, Hoffman MM, Iyer VR, Jung YL, Karmakar S, Kellis M, Kharchenko PV, Li Q, Liu T, Liu XS, Ma L, Milosavljevic A, Myers RM, Park PJ, Pazin MJ, Perry MD, Raha D, Reddy TE, Rozowsky J, Shoresh N, Sidow A, Slattery M, Stamatoyannopoulos JA, Tolstorukov MY, White KP, Xi S, Farnham PJ, Lieb JD, Wold BJ, Snyder M. Chip-seq guidelines and practices of the encode and modencode consortia. Genome Res. 2012;22:1813–31.
107. Urich MA, Nery JR, Lister R, Schmitz RJ, Ecker JR. Methylc-seq library preparation for base-resolution whole-genome bisulfite sequencing. Nat Protoc. 2015;10:475–83.
108. Booth MJ, Branco MR, Ficz G, Oxley D, Krueger F, Reik W, Balasubramanian S. Quantitative sequencing of 5-methylcytosine and 5-hydroxymethylcytosine at single-base resolution. Science. 2012;336:934–7.
109. Booth MJ, Marsico G, Bachman M, Beraldi D, Balasubramanian S. Quantitative sequencing of 5-formylcytosine in DNA at single-base resolution. Nat Chem. 2014;6:435–40.
110. Chu C, Qu K, Zhong FL, Artandi SE, Chang HY. Genomic maps of long noncoding rna occupancy reveal principles of rna-chromatin interactions. Mol Cell. 2011;44:667–78.

111. Churchman LS, Weissman JS. Nascent transcript sequencing visualizes transcription at nucleotide resolution. Nature. 2011;469:368–73.
112. Churko JM, Mantalas GL, Snyder MP, Wu JC. Overview of high throughput sequencing technologies to elucidate molecular pathways in cardiovascular diseases. Circ Res. 2013;112:1613–23.
113. Li Y, Klena NT, Gabriel GC, Liu X, Kim AJ, Lemke K, Chen Y, Chatterjee B, Devine W, Damerla RR, Chang C, Yagi H, San Agustin JT, Thahir M, Anderton S, Lawhead C, Vescovi A, Pratt H, Morgan J, Haynes L, Smith CL, Eppig JT, Reinholdt L, Francis R, Leatherbury L, Ganapathiraju MK, Tobita K, Pazour GJ, Lo CW. Global genetic analysis in mice unveils central role for cilia in congenital heart disease. Nature. 2015;521:520–4.
114. Do R, Stitziel NO, Won HH, Jorgensen AB, Duga S, Angelica Merlini P, Kiezun A, Farrall M, Goel A, Zuk O, Guella I, Asselta R, Lange LA, Peloso GM, Auer PL, Project NES, Girelli D, Martinelli N, Farlow DN, DePristo MA, Roberts R, Stewart AF, Saleheen D, Danesh J, Epstein SE, Sivapalaratnam S, Hovingh GK, Kastelein JJ, Samani NJ, Schunkert H, Erdmann J, Shah SH, Kraus WE, Davies R, Nikpay M, Johansen CT, Wang J, Hegele RA, Hechter E, Marz W, Kleber ME, Huang J, Johnson AD, Li M, Burke GL, Gross M, Liu Y, Assimes TL, Heiss G, Lange EM, Folsom AR, Taylor HA, Olivieri O, Hamsten A, Clarke R, Reilly DF, Yin W, Rivas MA, Donnelly P, Rossouw JE, Psaty BM, Herrington DM, Wilson JG, Rich SS, Bamshad MJ, Tracy RP, Cupples LA, Rader DJ, Reilly MP, Spertus JA, Cresci S, Hartiala J, Tang WH, Hazen SL, Allayee H, Reiner AP, Carlson CS, Kooperberg C, Jackson RD, Boerwinkle E, Lander ES, Schwartz SM, Siscovick DS, McPherson R, Tybjaerg-Hansen A, Abecasis GR, Watkins H, Nickerson DA, Ardissino D, Sunyaev SR, O'Donnell CJ, Altshuler D, Gabriel S, Kathiresan S. Exome sequencing identifies rare ldlr and apoa5 alleles conferring risk for myocardial infarction. Nature. 2015;518:102–6.
115. Myocardial Infarction Genetics Consortium I, Stitziel NO, Won HH, Morrison AC, Peloso GM, Do R, Lange LA, Fontanillas P, Gupta N, Duga S, Goel A, Farrall M, Saleheen D, Ferrario P, Konig I, Asselta R, Merlini PA, Marziliano N, Notarangelo MF, Schick U, Auer P, Assimes TL, Reilly M, Wilensky R, Rader DJ, Hovingh GK, Meitinger T, Kessler T, Kastrati A, Laugwitz KL, Siscovick D, Rotter JI, Hazen SL, Tracy R, Cresci S, Spertus J, Jackson R, Schwartz SM, Natarajan P, Crosby J, Muzny D, Ballantyne C, Rich SS, O'Donnell CJ, Abecasis G, Sunyaev S, Nickerson DA, Buring JE, Ridker PM, Chasman DI, Austin E, Ye Z, Kullo IJ, Weeke PE, Shaffer CM, Bastarache LA, Denny JC, Roden DM, Palmer C, Deloukas P, Lin DY, Tang ZZ, Erdmann J, Schunkert H, Danesh J, Marrugat J, Elosua R, Ardissino D, McPherson R, Watkins H, Reiner AP, Wilson JG, Altshuler D, Gibbs RA, Lander ES, Boerwinkle E, Gabriel S, Kathiresan S. Inactivating mutations in npc1l1 and protection from coronary heart disease. N Engl J Med. 2014;371:2072–82.
116. Tao H, Yang JJ, Shi KH, Deng ZY, Li J. DNA methylation in cardiac fibrosis: new advances and perspectives. Toxicology. 2014;323:125–9.
117. Zhao JY, Yang XY, Shi KH, Sun SN, Hou J, Ye ZZ, Wang J, Duan WY, Qiao B, Chen YJ, Shen HB, Huang GY, Jin L, Wang HY. A functional variant in the cystathionine beta-synthase gene promoter significantly reduces congenital heart disease susceptibility in a han chinese population. Cell Res. 2012;23:242–53.
118. Sanchez-Castro M, Gordon CT, Petit F, Nord AS, Callier P, Andrieux J, Guerin P, Pichon O, David A, Abadie V, Bonnet D, Visel A, Pennacchio LA, Amiel J, Lyonnet S, Le Caignec C. Congenital heart defects in patients with deletions upstream of sox9. Hum Mutat. 2013;34:1628–31.
119. Morikawa Y, Zhang M, Heallen T, Leach J, Tao G, Xiao Y, Bai Y, Li W, Willerson JT, Martin JF. Actin cytoskeletal remodeling with protrusion formation is essential for heart regeneration in hippo-deficient mice. Sci Signal. 2015;8:ra41.
120. Heallen T, Zhang M, Wang J, Bonilla-Claudio M, Klysik E, Johnson RL, Martin JF. Hippo pathway inhibits wnt signaling to restrain cardiomyocyte proliferation and heart size. Science. 2011;332:458–61.

121. Huang GN, Thatcher JE, McAnally J, Kong Y, Qi X, Tan W, DiMaio JM, Amatruda JF, Gerard RD, Hill JA, Bassel-Duby R, Olson EN. C/ebp transcription factors mediate epicardial activation during heart development and injury. Science. 2012;338:1599–603.
122. Takeuchi JK, Lou X, Alexander JM, Sugizaki H, Delgado-Olguin P, Holloway AK, Mori AD, Wylie JN, Munson C, Zhu Y, Zhou YQ, Yeh RF, Henkelman RM, Harvey RP, Metzger D, Chambon P, Stainier DY, Pollard KS, Scott IC, Bruneau BG. Chromatin remodelling complex dosage modulates transcription factor function in heart development. Nat Commun. 2011;2:187.
123. Lei I, Gao X, Sham MH, Wang Z. Swi/snf protein component baf250a regulates cardiac progenitor cell differentiation by modulating chromatin accessibility during second heart field development. J Biol Chem. 2012;287:24255–62.
124. He A, Ma Q, Cao J, von Gise A, Zhou P, Xie H, Zhang B, Hsing M, Christodoulou DC, Cahan P, Daley GQ, Kong SW, Orkin SH, Seidman CE, Seidman JG, Pu WT. Polycomb repressive complex 2 regulates normal development of the mouse heart. Circ Res. 2012;110:406–15.
125. Anand P, Brown JD, Lin CY, Qi J, Zhang R, Artero PC, Alaiti MA, Bullard J, Alazem K, Margulies KB, Cappola TP, Lemieux M, Plutzky J, Bradner JE, Haldar SM. Bet bromodomains mediate transcriptional pause release in heart failure. Cell. 2013;154:569–82.
126. He A, Kong SW, Ma Q, Pu WT. Co-occupancy by multiple cardiac transcription factors identifies transcriptional enhancers active in heart. Proc Natl Acad Sci U S A. 2011;108:5632–7.
127. Mitchell-Jordan S, Chen H, Franklin S, Stefani E, Bentolila LA, Vondriska TM. Features of endogenous cardiomyocyte chromatin revealed by super-resolution sted microscopy. J Mol Cell Cardiol. 2012;53:552–8.
128. Wood AJ, Lo TW, Zeitler B, Pickle CS, Ralston EJ, Lee AH, Amora R, Miller JC, Leung E, Meng X, Zhang L, Rebar EJ, Gregory PD, Urnov FD, Meyer BJ. Targeted genome editing across species using zfns and talens. Science. 2011;333:307.
129. Hsu PD, Lander ES, Zhang F. Development and applications of crispr-cas9 for genome engineering. Cell. 2014;157:1262–78.
130. Ghamari A, van de Corput MP, Thongjuea S, van Cappellen WA, van Ijcken W, van Haren J, Soler E, Eick D, Lenhard B, Grosveld FG. In vivo live imaging of rna polymerase ii transcription factories in primary cells. Genes Dev. 2013;27:767–77.
131. Wang Y, Maharana S, Wang MD, Shivashankar GV. Super-resolution microscopy reveals decondensed chromatin structure at transcription sites. Sci Rep. 2014;4:4477.
132. Ricci MA, Manzo C, Garcia-Parajo MF, Lakadamyali M, Cosma MP. Chromatin fibers are formed by heterogeneous groups of nucleosomes in vivo. Cell. 2015;160:1145–58.
133. Kohl T, Westphal V, Hell SW, Lehnart SE. Superresolution microscopy in heart – cardiac nanoscopy. J Mol Cell Cardiol. 2013;58:13–21.
134. Bhat S, Ohn J, Liebling M. Motion-based structure separation for label-free, high-speed, 3d cardiac microscopy. IEEE Trans Image Process Publ IEEE Signal Process Soc. 2012;21:3638–47.
135. Lee S, Vinegoni C, Feruglio PF, Fexon L, Gorbatov R, Pivoravov M, Sbarbati A, Nahrendorf M, Weissleder R. Real-time in vivo imaging of the beating mouse heart at microscopic resolution. Nat Commun. 2012;3:1054.
136. Jung K, Kim P, Leuschner F, Gorbatov R, Kim JK, Ueno T, Nahrendorf M, Yun SH. Endoscopic time-lapse imaging of immune cells in infarcted mouse hearts. Circ Res. 2013;112:891–9.
137. Nance ME, Whitfield JT, Zhu Y, Gibson AK, Hanft LM, Campbell KS, Meininger GA, McDonald KS, Segal SS, Domeier TL. Attenuated sarcomere lengthening of the aged murine left ventricle observed using two-photon fluorescence microscopy. Am J Physiol Heart Circ Physiol. 2015;309:H918–25.
138. Sewell-Loftin MK, Brown CB, Baldwin HS, Merryman WD. A novel technique for quantifying mouse heart valve leaflet stiffness with atomic force microscopy. J Heart Valve Dis. 2012;21:513–20.

139. Pauza DH, Rysevaite K, Inokaitis H, Jokubauskas M, Pauza AG, Brack KE, Pauziene N. Innervation of sinoatrial nodal cardiomyocytes in mouse. A combined approach using immunofluorescent and electron microscopy. J Mol Cell Cardiol. 2014;75:188–97.
140. Leo-Macias A, Liang FX, Delmar M. Ultrastructure of the intercellular space in adult murine ventricle revealed by quantitative tomographic electron microscopy. Cardiovasc Res. 2015;107:442–52.
141. Fujioka H, Tandler B, Hoppel CL. Mitochondrial division in rat cardiomyocytes: an electron microscope study. Anat Rec. 2012;295:1455–61.
142. Gherghiceanu M, Popescu LM. Heterocellular communication in the heart: electron tomography of telocyte-myocyte junctions. J Cell Mol Med. 2011;15:1005–11.

Chapter 16
Synergizing Proteomic and Metabolomic Data to Study Cardiovascular Systems

Jennifer E. Gilda, Clifford D.L. Folmes, Jenice X. Cheah, Tuleagh Innes-Gawn, Merry L. Lindsey, and Aldrin V. Gomes

Abstract Investigation of biological systems requires an understanding of the crosstalk between complex regulatory processes and how disturbances in these processes contribute to the development of a disease phenotype. While proteomic studies have significantly advanced our understanding of the types and relative amounts of proteins in complex samples, these analyses are now being complemented by additional -omic platforms. For example, global metabolic investigations are increasingly leveraged to determine the underlying mechanisms of cardiovascular diseases. These investigations allow the determination and relative quantification of metabolites in complex samples. As our ability to analyze and quantify large experimental proteomic and metabolomic data sets continues to improve, combining these data sets allows for the identification of pathways and sub-pathways that would not be detected if either analytical method was used in isolation. In this book chapter, we discuss how to design a cardiovascular metabolomic experiment and how to utilize combined proteomic and metabolomic data for a more comprehensive exam-

J.E. Gilda, BS • J.X. Cheah • T. Innes-Gawn
Department of Neurobiology, Physiology, and Behavior, University of California, Davis, Davis, CA 95616, USA
e-mail: jegilda@ucdavis.edu

C.D.L. Folmes, PhD
Division of Cardiovascular Diseases, Department of Medicine, Center for Regenerative Medicine, Mayo Clinic, Rochester, MN, USA

M.L. Lindsey, PhD
Department of Physiology and Biophysics, Mississippi Center for Heart Research, University of Mississippi Medical Center, Jackson, MS 39216, USA

Research Service, G.V. (Sonny) Montgomery Veterans Affairs Medical Center, Jackson, MS 39216, USA

A.V. Gomes (✉)
Department of Physiology and Membrane Biology, University of California, Davis, CA 95616, USA

Department of Neurobiology, Physiology, and Behavior, University of California, Davis, Davis, CA 95616, USA

G. Agnetti et al. (eds.), *Manual of Cardiovascular Proteomics*,
DOI 10.1007/978-3-319-31828-8_16

ination of biological systems. In the near future, improved software to manage the integration of large datasets and development of new bioinformatics tools will help to harness the potential of these large datasets and make integrative –omics more accessible.

Keywords Proteomics • Cardiovascular • Mass spectrometry • Metabolomics • Sample preparation • Study design • Targeted analysis • Untargeted analysis

Introduction

Metabolic Data Complements Proteomic Data

While metabolomics, or metabonomics, has only recently emerged as a field of investigation [1, 2], metabolic studies have spanned millennia, with observations pertaining to "honey urine" in diabetics dating back to 1500 BC [3]. Traditionally, investigators have utilized a targeted approach to examine metabolites within specific pathways based upon an a priori knowledge of putative metabolic alterations associated with a biological or disease state. The ultimate goal of systems biology is to fully integrate the complex biological data obtained from different -omic platforms, including proteomics, metabolomics, and transcriptomics [4]. The goal of whole tissue or cell metabolomics is to determine all metabolites (30–1500 Da) present in the tissue or cell at the time of evaluation. Metabolite profiling has become a comprehensive method to identify and quantify hundreds of metabolites across a variety of tissues and cell types. The use of metabolomics in biomarker discovery is increasing in frequency, particularly for cardiovascular diseases [5], and has been utilized to discover critical metabolites in serum or plasma (e.g., branched-chain amino acids [BCAAs], acylcarnitines, purines) that contribute to underlying mechanisms of cardiovascular disease (CVD) [6]. As one example, metabolite profiling of serum revealed differences in pathways associated with purines, carnitine, and betaine as gender differences in patients with myocardial infarction [7]. While the number of studies integrating the parallel analysis of protein expression and metabolite expression is still relatively small compared to genomics and even proteomics, the results from these studies highlight the benefits of incorporating a variety of data sets [8–10]. For example, Mayr et al. utilized a combined proteomic and metabolomic approach to investigate metabolic processes in persistent atrial fibrillation [9]. A combined approached has also been used to investigate congestive heart failure, myocardial hibernation, and the role of PKCδ in vascular smooth muscle cells [8, 11, 12]. These benefits include the increased confidence in identification of altered signaling pathways, increased information on these pathways is revealed, and the identification of affected signaling pathways that were not identified with either technique alone.

Recent Advances in Metabolomics Studies.

Recent developments in the two main techniques used in metabolomics, nuclear magnetic resonance (NMR) and mass spectrometry (MS), have led to significantly greater knowledge and coverage of the metabolome than was previously possible [13]. In NMR spectroscopy and MS, approaches have been developed to address spectra that were poorly aligned between analyses as a result of shifts due to variations in experimental conditions or performance of the instrument [14–17]. Small shifts in spectral alignment can be overcome using bin-based approaches while larger shifts are overcome by peak alignment approaches [14, 17]. Some spectral shifts can be minimized by changes in sample preparation, such as pH buffering of samples for NMR analysis.

Significant improvements have been made in the sensitivity and resolution of mass spectrometers. MS is usually coupled with chromatography to enhance the mass resolving capabilities, with the majority of studies still relying on liquid chromatography (LC) or gas chromatography (GC) based separations; however, high resolution mass spectrometers are now available such that a chromatography step is no longer necessary for low to moderately complex samples. Direct infusion or flow injection without chromatography can be used to introduce samples, which requires less time, leading to higher throughput [18, 19] but also has the disadvantage of inducing ion suppression. The balance among parameters needs to be considered when optimizing protocols.

The increasing knowledge on metabolic data available to researchers has allowed them to perform pathway analysis. Metabolites in a metabolic pathway are linked by a series of chemical reactions carried out by an enzyme or set of enzymes. Knowledge of metabolic pathways enables researchers to utilize pathway-based approaches, and many pathway analysis tools are now available to researchers, such as Cytoscape and MetaboAnalyst. Pathway analysis allows the researcher to assess the effect of changes in metabolite levels on specific pathways [13]. The development of large-scale databases has also made this possible [18].

Study Design

Importance of Controls

Like other -omic technologies, proper quality control is critical for reliable proteomic and metabolomic results. Quality control issues can arise due to the sample collection and preparation method, as well as the analytical method employed to identify and quantify metabolites. Appropriate quality controls, including experimental runs without sample (blank controls), randomized sample runs, and inclusion of external calibration standards in samples should be incorporated into metabolomics experiments [20] and are essential for routine, large-scale, untargeted metabolomics investigations. External calibration standards are especially important to reduce analytical variation in metabolomics experiments, as these standards allow signal correction between runs. In addition, it is important to assess the coverage and quantification levels of metabolites over time to ensure that the instrument is functioning optimally.

Designing a Proteomics Experiment

Adequate and appropriate experimental design of a proteomic experiment is crucial, and the chapter by Lindsey et al. published in this volume, describes how to design a successful cardiovascular proteomics experiment [21]. In that chapter, authors discuss important considerations for sample preparation for proteomic experiments, protein fractionation, protein MS, peptide and protein identification, and statistical evaluation of proteomic experiments. Another good resource for designing a successful cardiovascular proteomics experiment is the recent Statement on the transformative impact of proteomics published by the American Heart Association [22].

Designing a Metabolomics Experiment

The initial consideration when designing a metabolomics experiment is to determine the number of metabolites that are expected to be quantified, as the number and chemical properties of these metabolites will define the experimental design and analytical approach. This distinction separates metabolomics experiments into two broad classes: (a) untargeted analysis whereby the goal is to quantify as many metabolites as possible to provide a global and unbiased representation of the metabolome and (b) targeted analysis whereby a specific set of metabolites, ranging from tens to hundreds of metabolites, that reside within a small number of related intermediary metabolic pathways is quantified [19, 23]. Each approach presents unique advantages and disadvantages. Untargeted analysis allows the detection of a wide range of metabolites utilizing either a single analytical platform (GC-MS, LC-MS, capillary electrophoresis (CE)-MS, or NMR), or a combination of these analytical platforms to obtain a comprehensive profile of the metabolome, while targeted analysis allows qualitative and quantitative investigations of one or a few chemically similar metabolites. While the untargeted approach provides broader coverage and the opportunity to discover novel metabolites and pathways, it requires advanced statistical analysis to process and interpret the large datasets and is limited by identification and characterization of unannotated metabolites, which can be time consuming and challenging [24]. In contrast, targeted analysis reduces complexity by examining a limited set of metabolites with better quantitation by including internal standards and requires less time to process and analyze the raw data

[24]. Researchers can combine untargeted and targeted metabolomics approaches to identify information that the other approach alone could not, and gain a better idea of how a specific pathway or area of interest is affected under different conditions [25, 26]. The following section provides an overview of some of the major considerations when designing a metabolomics experiment. More detailed step-by-step protocols have been published elsewhere [27–31].

Sample Collection and Processing for Metabolomic Experiments

Sample preparation is a critical component of the metabolomics workflow, with the ultimate goal of producing a metabolic profile that is representative of the *in vivo* biological status at the time of sampling. The composition and reproducibility of the metabolite profile are largely dependent on these pre-analytical steps, consisting of sample collection/quenching and metabolite extraction.

Sample Collection and Quenching

Many metabolites have high turnover rates and are extremely labile (e.g., glycolytic intermediates and phosphates content can change on the order of milliseconds), necessitating quenching of metabolism immediately upon sampling to conserve a metabolite profile that is representative of the time of collection. The quenching processes must balance the ability to rapidly inhibit metabolic enzyme activity to prevent changes in the metabolite pool, while limiting the possible leakage of intracellular metabolites, which can occur due to cell membrane permeabilization during hard quenching protocols. The specific protocols are dependent on sample type, but largely consist of rapid cooling using liquid nitrogen or cold organic solvents. Some recent studies have also utilized rapid heating [32], although this technique must still be validated in large-scale metabolomic studies.

For tissue sampling, tissues are isolated, quickly perfused or rinsed with Krebs-Henseleit buffer to remove contaminating blood and immediately freeze-clamped using Wollenberger clamps precooled to the temperature of liquid nitrogen and placed directly in liquid nitrogen. Subsequently, tissues are ground when still frozen using mortar and pestle or ball mill that has been cooled with liquid nitrogen, with care taken to prevent thawing of the samples [28]. For blood samples, serum is collected by allowing the sample to clot naturally on ice. Plasma is collected by mixing with an anticoagulant followed by centrifugation to remove the clot or blood cells and platelets, respectively. Lithium heparin is typically utilized for metabolomics studies, as other anticoagulants such as EDTA and citrate can interfere with subsequent metabolite profiling [27]. It is best for samples to be aliquoted to avoid freeze/thaw cycles, rapidly frozen in liquid nitrogen, and stored at −80 °C until analysis.

For cells in culture, metabolites can be quantified from both the cells (metabolomic fingerprinting) and conditioned cell culture media (metabolomic footprinting) [33].

For adherently grown cells, media can be harvested, centrifuged to remove cells and debris, aliquoted and frozen in liquid nitrogen. The cells can be washed with buffered saline and frozen directly in liquid nitrogen or alternatively directly quenched, scraped and extracted using ice-cold organic solvents [34]. Although cell detachment to collect adherent cells is commonly utilized, this has potential drawbacks including inducing metabolite leakage [35]. Suspension cultured cells can be quenched directly using precooled solutions such as 60 % methanol with 0.85 % ammonium bicarbonate [36] and collected either by cold centrifugation or filtration [34].

Metabolite Extraction

For global metabolite profiling approaches, the extraction protocol should efficiently isolate as many metabolites as possible in an unbiased and non-destructive manner, while remaining compatible with downstream analytical approaches. Due to the wide variation in chemical properties of metabolites, each extraction protocol will favor isolation of specific metabolite classes. Therefore, it may be necessary to extract samples with several protocols in order to maximize the coverage obtained. Polar metabolites are typically extracted using a combination of water and organic solvents such as methanol or acetonitrile that simultaneously precipitate proteins and extract metabolites. Alternatively, acidic extractions using perchloric acid can be utilized to extract metabolites, yet require an additional step to neutralize the supernatant with potassium carbonate. Although many different extraction protocols have been utilized for polar metabolites, a combination of methanol and water extracts the greatest range of metabolites [37]. Extraction of lipophilic metabolites can be enhanced using a methanol/chloroform/water extraction, with the upper methanol/water layer containing polar metabolites and the lower chloroform layer containing the lipophilic metabolites. The efficiency of metabolite extraction can be increased by promoting tissue/cell lysis mechanically using a homogenizer, through sonication, or by repeated freeze-thaw cycles. For polar metabolites, samples are lyophilized or solvents removed using a speed vacuum concentrator and stored at −80 °C until analysis.

Analytical Techniques

A range of analytical platforms has been utilized to target individual classes of metabolites including enzymatic analysis, electrochemical detectors, ultraviolet–visible (UV–VIS) spectroscopy, infrared and Raman spectroscopy, NMR spectroscopy and MS. Due to their ability to detect a wide range of chemical compounds with high specificity and reproducibility coupled with advances in instrumentation, compound databases and bioinformatics tools, NMR and chromatography-coupled MS have become the main analytical platforms for metabolomics. Since metabolites have diverse physical and chemical properties, no single platform is capable of quantifying the whole metabolome. However, multiplexing of analytical platforms, such as NMR and MS, which provide partially overlapping yet distinct metabolite coverage, ultimately increases the metabolite coverage towards quantifying the entire metabolome.

Mass Spectrometry

Beyond the well-established role for MS in proteomics experiments, MS also represents an major analytical platform for metabolomics due to its high sensitivity (picomole to femtomole), ability to quantify a wide range of metabolites and provide specific chemical information [38]. Although direct MS analysis of crude mixtures has been utilized, MS is typically coupled with chromatographic separation such as gas chromatography following chemical derivatization or liquid chromatography for pre-separation to improve overall metabolite coverage [38]. With limited metabolite overlap, these complementary approaches are often utilized together to increase metabolite coverage.

GC offers consistent, reproducible, stable, and high-resolution separation. When coupled with electron impact ionization (EI), time-of-flight or quadrupole mass analyzers offer a robust platform that can reliably detect hundreds of metabolites in serum or plasma [27]. Combining reproducible GC retention times/indices with consistent and characteristic EI fragmentation patterns enables compound identification across instruments through comparison to mass spectral libraries such as FiehnLib [39] or the National Institute of Standards and Technologies database [40]. The major limitation of GC is that it requires volatile and thermally stable analytes, criteria which few metabolite classes meet. These limitations can be overcome through chemical derivatization, typically consisting of an oximation step followed by trimethylsilylation, which enables the detection of many classes of metabolites within central carbon metabolism [27]. The drawback of such preprocessing approaches is that they can induce artifacts, increase sample to sample variability, and may lead to loss of metabolite classes that are not amenable to derivatization [27, 41–43].

LC-MS has become one of the major analytical platforms for global untargeted metabolomics due to the ability to detect thousands of metabolite features within a biological sample without the requirement for chemical derivatization [44]. The majority of applications utilize either reverse-phase or hydrophilic interaction chromatography, which provide complementary metabolite profiles, coupled with high resolution and high mass accuracy instruments such as quadrupole time-of-flight and Orbitrap mass spectrometers. Electrospray ionization is typically utilized, and profiles are often obtained in both positive and negative ion mode to further increase the metabolome coverage, as specific metabolites are preferentially detected in one mode versus the other. The major limitation and bottleneck in untargeted LC-MS based metabolomics is metabolite identification. Due to the difficulty in predicting metabolite fragmentation patterns for tandem MS, metabolomics data are often acquired in MS mode, in contrast to proteomics where predictable fragment patterns are used to identify peptides [44]. Further impairing metabolite identification, retention time and mass spectra are not reproducible across systems; hence comparison of accurate masses to metabolite databases such as METLIN [45] only produce putative identities that must be compared to retention times and MS/MS data of pure compounds run on the same instrument with the same parameters [31]. Indeed, complete identification of metabolite features in biological samples is currently not possible, as many features do not match to specific metabolites found in databases, thus requiring de novo metabolite identification [46]. Alternatively, LC-MS/MS with selected reaction monitoring has been utilized for targeted analysis of many metabolites, including carboxylic acids, amino acids, and nucleotides, which represent critical constituents of central carbon, nucleotide and amino acid metabolism [47, 48].

Nuclear Magnetic Resonance Spectroscopy

NMR spectroscopy of single compounds or complex mixtures provides detailed information about molecular structure in a non-destructive way and has been utilized in biological systems to examine macromolecule structure and function, image *in vivo* anatomical structures and study energy metabolism and bioenergetics [49]. While the theoretical basis of NMR spectroscopy is well established, it is beyond the scope of this chapter, and the reader is directed to an excellent overview that has been previously published [49]. Early metabolic applications of proton NMR date back to 1977 when a handful of metabolites were assigned and quantified in a suspension of red blood cells [50], with this technique quickly being adapted to study the metabolite composition of body fluids [51]. Advances in NMR technology, such as development of cryogenic probes and increasing instrument magnetic field strength, has improved both sensitivity and peak dispersion and enabled the simultaneous detection of hydrogen-containing metabolites in a complex mixture with a limit of detection in the nanomolar range [52]. NMR requires no pre-separation of metabolites, with sample preparation limited to addition of deuterium oxide (D_2O) as a magnetic field lock signal for the instrument, a reference standard such as 3-trimethylsilylpropionic acid (TSP) or 2,2-dimethyl-2-sila-pentane-5-sulfonate sodium salt (DSS), an internal standard of known concentration for absolute quantitation and buffering, as chemical shifts of some metabolites are pH-dependent. Conventionally, samples have been analyzed using 5 mm or capillary NMR tubes; however large-scale studies now utilize robotic liquid handling and flow injection methods to perform analysis in a high throughput manner. Using high-resolution spectrometers, metabolites can be detected using acquisition times on the order of several minutes with standard NMR pulse sequences that include water pre-saturation to suppress the large water signal observed in biological samples (for details of pulse sequences please see [28]). Alternatively, metabolites can also be quantified in intact tissues without pretreatment using magic angle spinning NMR spectroscopy, although metabolite resolution is reduced compared to tissue extracts [53].

Metabolite identification remains one of the bottlenecks for NMR-based metabolomics studies [54]. NMR spectral peaks are initially assigned by comparison with spectra of pure metabolites acquired using a similar solvent system and analytical conditions (e.g., pH, temperature, or pulse sequence). A number of open access databases, including the Human Metabolome Database [55], Biological Magnetic Resonance Bank [56], Madison-Qingdao Metabolomics Consortium Database [57] and the Birmingham Metabolite Library [58], contain libraries of metabolite spectra that can be downloaded and processed in a similar manner to experimental samples to aid in the metabolite identification. Commercially available software packages containing proprietary spectral libraries are also available, including Chenomx NMR Suite [59] and KnowItAll Metabolomics Edition, which facilitate compound identification without peak picking and querying the databases. For additional NMR-based metabolomics databases and software please see this recent review [60]. To identify unassigned compounds, standard proton NMR can be complemented with two-dimensional NMR approaches such as J-resolved spectroscopy, correlation spectroscopy, total correlation spectroscopy and heteronuclear correlation spectroscopy to resolve the connectivity between signals and increase signal dispersion [49]. Ultimately compound assignments are confirmed by spiking samples with solutions containing the putative compounds.

Challenges and Limitations

Among the most difficult challenges in metabolomics is the annotation of unknown metabolic signals. The Metabolomics Standards Initiative (MSI) has issued a variety of suggestions for reporting minimal experimental parameters to ensure that metabolomic data can be used and reproduced by other laboratories [61]. Importantly, the identification of metabolites must always be based on at least two orthogonal physicochemical characteristics, such as retention index and mass spectrum. Identifications that are based on authentic chemical standards are generally more trustworthy than annotations based on calculated characteristics. Nevertheless, the metabolome itself is an unrestricted entity that clearly comprises more than the suite of known compounds that can be found in classical textbooks or purchased from chemical manufacturers. The metabolome cannot be simply computed from reconstructed biochemical pathways due to enzymatic diversity, substrate ambiguity, and variation in regulatory mechanisms. Hence, the finding of many unknown signals in metabolomic surveys comes as no surprise to biochemists. The sheer complexity of natural products, including isomeric compounds, renders the use of accurate masses and database queries insufficient for annotation of metabolites. Instead, novel algorithms are needed to score metabolic signals based on all available information, from calculated physicochemical characteristics to presence in biochemical databases. Such algorithms might ultimately boost the quality of metabolomic data in a similar way as SEQUEST did for proteomic analysis. Currently, no software is available to perform this much-needed task.

Important Notes and Common Problems to Avoid

Although metabolomic workflows are similar to proteomic workflows, several problems remain to be solved to facilitate metabolomics experiments. These problems include the need for fast and efficient quenching of metabolic reactions, the lack of clean-up steps or fractionation before analysis, and the identification and validation of metabolites. No method is currently available to remove non-metabolites (clean-up) without affecting the amount or types of metabolites in a sample. Similarly, the best way to fractionate metabolites is not currently known. Fractionation of metabolites may result in metabolite loss and degradation, and may introduce bias in metabolites detected as well as their concentrations. Currently, the best method for comprehensive metabolite determination of a complex sample is to subject the sample to several different analytical methods and combine results. To accurately determine differences between experimental and control samples, an adequate number of samples and replicates needs to be analyzed so that the normal biological variation between the sample sources (such as cells/tissues/organs) can be determined and taken into account. As with proteomics, metabolomics analysis of unfractionated samples is limited by the wide concentration range of metabolites (or proteins in the case of proteomics) present in cells and tissues. One often overlooked problem is that many metabolites may be altered if enzymatic activity is not completely stopped when samples are collected. The sample preparation method can also affect metabolites that are sensitive to oxidative changes such as NADH and NADPH.

The type of mass spectrometer utilized for the metabolomics experiment is also important, because some mass spectrometers are better suited to detect certain types of metabolites. Many secondary metabolites are easier to analyze by LC-MS methods, whereas most primary metabolites can readily be quantified by GC-MS procedures. It is also important that researchers realize that procedures that worked for one organism or tissue may not be adequate for other sample types.

When using MS-based metabolomics, it is important to realize that ion intensities and chromatographic retention times often suffer from temporal drift. To minimize internal variation in an experimental study set, sample analysis should be randomized and samples all investigated on the same mass spectrometer in the same series of runs. The presence of significantly more unknown metabolites than known metabolites is also a major problem for metabolomics. Unlike proteomics in which the number of theoretical proteins is determined from the genome, the predicted total number of metabolites in humans is unknown. The human metabolome database (HMDB, http://www.hmdb.ca/statistics, 08/20/2015) lists the total number of detected and quantified metabolites as 3,001, while the total number of theoretical metabolites having associated proteins (enzymes and transporters) as 22,138, and the total number of expected metabolites at 38,220. It is best to use regularly updated databases as annotations change frequently. Once the reconstructed network is completed, it is essential to manually check the network for incorrect annotations. Methods are available to test for inconsistencies in the network [62]. Problems are also associated with missing or incorrect annotation of protein subunits, protein complexes, and isozymes resulting in some proteins which are enzymes not being designated as enzymes.

While many metabolomics experimental workflows exist, few online resources exist to share or obtain metabolomics data sets. Recently, a metabolite Atlas framework and interface was set up to allow online access to raw mass spectrometry data together with information about the molecules detected [63]. Importantly, this interface allows integration with systems biology tools that permit metabolomics data to be linked to biological models [63].

Integrating Results

Analysis and Visualization Software

Functional data from proteomic and metabolomic experiments can be visualized in reconstructed metabolic networks using several programs including PathwayTools Omics Viewer [63]. PathwayTools permits the visualization and analysis of genes, enzymes and metabolites and allows the user to add experimental values of enzymes and metabolites to any available full pathway map. Many full pathway maps are currently available [64]. Other popular visualization programs include Cytoscape [65] and VANTED (Visualization and Analysis of Networks containing Experimental Data) [66]. Cytoscape is an open source software platform that allows visualization of molecular interaction networks and biological pathways and the integration of these networks. Besides visualization, VANTED can be used to analyze biological networks. VANTED can combine medium- to large-scale experimental data sets from different time points/conditions on networks. Software utilized to visualize networks is shown in Table 16.1.

Metabolic Modeling

Genome-scale metabolic models allow detailed and quantitative predictions of organism behavior. Although reconstructing genome-scale metabolic models is complex, resources such as the Model SEED (http://www.theseed.org/models/) [74], and MetaMerge [75] are available to facilitate the creation of these models. Model SEED is an online resource that significantly speeds up the time and effort required to create new metabolic models. Users can utilize the RAST annotation system to annotate genome sequences, which are then automatically sent to the Model SEED to create the metabolic model. The metabolic network created by Model SEED includes networks of metabolic reactions, the gene-protein-reaction associations for each reaction, and a model of metabolism that can be simulated using Flux Balance Analysis (FBA). Note that these models take a minimum of 2 days to reconstruct. The usefulness of this approach is that models can be predicted for genetic deletion animals by using genomes with the gene removed [74]. MetaMerge is a resource that integrates two existing metabolic network models into a single metabolic network model [75]. It is likely that within the next decade metabolic reconstructions for most genomes will be widely available. These

Table 16.1 Table showing metabolomic resources

Program	Website	Features of software/program	Type of resource
MeltDB 2.0	https://meltdb.cebitec.uni-bielefeld.de/	Platform for the analysis and integration of data from metabolomics experiments with transcriptomics or proteomics data	Data analysis platform [67]
Metabolomic Analysis and Visualization Engine (MAVEN)	http://genomics-pubs.princeton.edu/mzroll/	Open source cross platform LC-MS metabolomics data analyser	Data analysis platform [68]
MarVis-Pathway tool	http://marvis.gobics.de/	Platform for functional annotation of filtered/combined (cross-omics) data sets or selected clusters	Data analysis platform [69]
MetaMapp	http://metamapp.fiehnlab.ucdavis.edu/	Visualizes mass spectrometry based metabolomics datasets as network graphs in Cytoscape	Data analysis platform [70]
MetaboAnalyst	http://www.metaboanalyst.ca/	Online pipeline for high-throughput metabolomics studies. Commonly used procedures for metabolomic data processing, normalization, multivariate statistical analysis, and data annotation	Data analysis platform [71]
BiGG	http://bigg.ucsd.edu/	A knowledge base of biochemically, genetically, and genomically organized genome-scale metabolic network reconstructions	Database
Kyoto Encyclopedia of Genes and Genomes (KEGG)	http://www.kegg.jp/	Bioinformatics database containing information on genes, proteins, reactions, and pathways	Database
MetaCyc	http://www.metacyc.org/	Curated database of experimentally determined metabolic pathways. Contains 2260 pathways from 2600 different organisms (as of 09/04/15)	Database od known pathways
Human Metabolome Database (HMDB)	http://www.hmdb.ca/	Online database containing chemical, clinical, and molecular biology/biochemistry data about small molecule metabolites in the human body. The database contains 41,993 metabolite entries (as of 09/04/15)	Database

Metlin	https://metlin.scripps.edu/	Repository of metabolite information and tandem mass spectrometry data designed to facilitate metabolite identification in metabolomics	Database
metaTIGER	http://www.bioinformatics.leeds.ac.uk/metatiger/	Contains a collection of metabolic profiles and phylogenomic information on a taxonomically diverse range of eukaryotes which provides novel facilities for viewing and comparing the metabolic profiles between organisms	Database
ERGO	https://ergo.integratedgenomics.com/	A subscription-based service developed by Integrated Genomics. It integrates data from every level including genomic, biochemical data, literature, and high-throughput analysis into a comprehensive user friendly network of metabolic and nonmetabolic pathways	Tool for Metabolic Modeling
IOMA (Integrative omics-metabolic analysis)	http://www.cs.technion.ac.il/~tomersh/methods.html	Matlab program for integrating quantitative proteomic and metabolomic data with genome-scale metabolic network models to predict metabolic flux	Tool for Metabolic Modeling [72]
Pathway Tools	http://bioinformatics.ai.sri.com/ptools/	Has several components. Allows construction of pathway/genome databases. MetaFlux component can generate a quantitative metabolic model from that pathway/genome database using flux-balance analysis	Tool for Metabolic Modeling [63]
ProMeTra	https://prometra.cebitec.uni-bielefeld.de/cgi-bin/login.cgi	Allows results from metabolomics to be combined with transcriptomics or proteomics data	Tool for Metabolic Modeling [73]
KEGGtranslator	http://www.cogsys.cs.uni-tuebingen.de/software/KEGGtranslator/	Converts KEGG files into multiple different formats suitable for modeling	Tool for Metabolic Modeling
Model SEED	http://www.theseed.org/	Automated online resource to create a draft metabolic model	Tool for Metabolic Modeling

(continued)

Table 16.1 (continued)

Program	Website	Features of software/program	Type of resource
MetaMerge	The full code of the MetaMerge algorithm available at http://www.genomebiology.com/2012/13/1/r6/additional	Semi-automatic combining of a pair of existing metabolic network reconstructions into a single metabolic network model	Tool for Metabolic Modeling
Cytoscape	http://www.cytoscape.org/	Open source platform for visualizing complex networks and integrating networks	Network visualization tool
VANTED (Visualization and Analysis of Networks containing Experimental Data)	https://immersive-analytics.infotech.monash.edu/vanted/	Used for visualization and analysis. Users can create and edit networks, as well as map experimental data onto networks	Network visualization tool
Pajek	http://mrvar.fdv.uni-lj.si/pajek/	Program for analysis and visualization of very large networks	Network visualization tool
CellNetAnalyzer (CNA)	http://www2.mpi-magdeburg.mpg.de/projects/cna/cna.html	MATLAB based program with a graphical user interface that allows the exploration of structural and functional properties of metabolic, signaling, and regulatory networks	FBA computation
openCOBRA Project	http://opencobra.sourceforge.net/openCOBRA/Welcome.html	Open source flux analysis program	FBA computation
Flux Balance Analysis (FBA)	http://systemsbiology.ucsd.edu/Downloads/FluxBalanceAnalysis	Flux analysis program	FBA computation
Matlab	http://www.mathworks.com	Statistical programming environment	Integrated multivariate data analysis
R	https://www.r-project.org/	Free software for statistical computing and graphics	Integrated multivariate data analysis
Statistica	http://www.statsoft.com	Commercial software which allows data analysis, data management, data visualization, and data mining solutions	Integrated multivariate data analysis

genome-scale metabolic models allow a comprehensive understanding of intracellular molecular mechanisms and are helpful in predicting phenotypes in the investigated organism. The quantitative and qualitative behavior of such networks can be carried out by FBA [76], kinetic modeling using differential equations [77], or by Elementary Mode Analysis [78]. These methods all allow the identification of subpathways that can operate at a steady state. For instance, FBA examines the flow of metabolites through a metabolic network using a mathematical approach, and is widely used for genome-scale metabolic network reconstructions [79]. FBA allows the rate of production of metabolites to be predicted, as network reconstructions contain all of the known metabolic reactions in an organism. The fluxes determined by the models may be compared with experimental data, and may yield predictive models of biochemical networks exposed to different conditions. Advantages of FBA include the requirement for less intensive input data than needed for traditional model construction, and the speed of performing simulations using FBA, which can be carried out for thousands of reactions in a few seconds on typical laboratory computers. Programs that are utilized to determine FBA are shown in Table 16.1.

Integration of Proteomics and Metabolomics Data

To integrate proteomic and metabolomic data, it is best to perform integrated analysis to determine biological processes or pathway associations. The two main approaches to integrate proteomic and metabolomics data are: (1) a data-driven approach that determines pathway associations directly from the available data, and (2) a knowledge-based approach, which relies on existing information about metabolic and proteomic pathways.

A few programs are now available to integrate –omic data, termed integrative omics-metabolic analysis (IOMA) [72], which quantitatively integrate proteomic and metabolomic data with genome-scale metabolic models to more accurately predict metabolic flux distributions. Using metabolomic and proteomic data IOMA (which utilizes a knowledge-based approach) was able to correctly predict the effect of different gene knockouts on metabolic fluxes in *Escherichia coli* [72]. IOMA was also shown to successfully predict the metabolic state of human erythrocytes [72]. Alternatively, the ProMeTra software allows results from metabolomics to be combined with transcriptomics or proteomics data [73]. After combining the draft network with a reference network such as found in the KEGG database, the greedy algorithm [80] is used to resolve discrepancies with experimental data (Fig. 16.1).

Manual integration of -omic results is more common (a type of a data-driven approach) and very time consuming but has been shown to be a powerful approach to uncover important biological pathways. Commonly a pathway analysis tool such as the Ingenuity Pathway Analysis program (IPA®, www.ingenuity.com/), Pathway commons (http://www.pathwaycommons.org/), BioCyc (http://biocyc.org/ov-expr.shtml), Database for Annotation, Visualization and Integrated Discovery (DAVID, https://david.ncifcrf.gov/home.jsp), Reactome (http://www.reactome.org/), METACORE

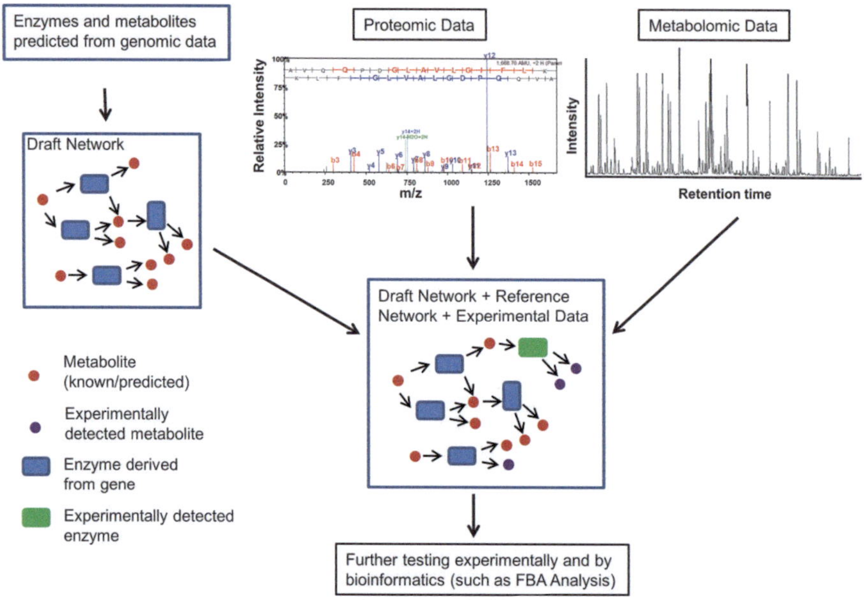

Fig. 16.1 Illustrative example of using proteomics and metabolomics data to create an integrative model. Genome data is used to create a draft network. However, the proteomic and metabolomic data may contain enzymes and/or metabolites that are not present in the draft network. Combining the draft network with a reference network such as that found in the KEGG database results in an improved network. The use of the common greedy algorithm which calculates the minimal sets of reactions that are needed to make the network compliant with experimental data results in a final integrated network. Metabolites levels associated with highly versus lowly detected enzymes can be used to predict which reactions are active and which are only partially active or inactive. Any hypotheses resulting from the integrated network can be tested experimentally or by bioinformatics

(http://lsresearch.thomsonreuters.com/), or MetaboAnalyst (http://www.metaboanalyst.ca/) is used to analyze the metabolomic and proteomic data individually and then the results are manually combined. For example, the manual integration of transcriptomic, proteomic, and metabolomics data was utilized to show that Cyclosporine A induced stress in human renal epithelial cells is not directly linked to its primary pharmacology [17].

Although several statistical analysis programs are available to determine differences between proteins and metabolites, few programs are currently available to do statistical analysis of the integrated dataset. Enzymes and metabolites that are part of the same pathway are expected to show some correlation (such as a similar pattern of up or down regulation). The correlation between enzymes and metabolites in a network can be determined quantitatively using their Pearson correlation [80]. For time series data, Granger causality testing can be used to detect significant cause and effect associations between proteins and metabolites [80]. Statistical analysis of integrated data can also be carried out using several statistical analysis programs including R (http://www.r-project.org/), MatLab, and SPSS (SPSS Inc.) [80].

Examples of Using Proteomics and Metabolomics to Address Biologically Relevant Questions

Proteomics and metabolomics in combination have been used in several studies to better understand cardiac pathophysiology (Table 16.2). Proteomic and metabolomic analysis of atherosclerotic vessels from apolipoprotein E-deficient mice showed potential associations of immune-inflammatory responses, oxidative stress, and energy metabolism [81]. De Souza, et al. investigated the molecular mechanisms involved in congestive heart failure (CHF) using a combined proteomic and metabolomic approach [12]. Protein extracts from left atrial tissue of CHF and control dogs were analyzed by tandem MS, and high-resolution NMR spectroscopy was used to measure metabolite levels. MS identified several changes in structural proteins and decreases in antioxidants and heat shock proteins. Upregulation of enzymes such as malate dehydrogenase (DH), α-/β-enolase and pyruvate dehydrogenase suggested metabolic stress, which was confirmed by metabolomic analysis. Metabolomics suggested that energy was used less efficiently in CHF hearts, and that oxidative stress potentially due to metabolic disturbances may be contributing to the depletion of antioxidant enzymes. Table 16.2 shows a summary of integrated metabolomics and proteomic studies used in cardiovascular research and the major study findings.

A combined proteomic and metabolomic approach was used to investigate the metabolic processes altered in human atrial fibrillation (AF) [9]. Atrial tissue from AF patients and controls was analyzed by NMR spectroscopy and mass spectrometry. Metabolomics revealed increases in beta-hydroxybutyrate, a substrate in ketone body metabolism, as well as ketogenic amino acids. Proteomic findings showed that 3-oxoacid transferase was differentially expressed in AF. Together proteomic and metabolomic results strongly suggested a role of ketone bodies in AF.

By integrating proteomic and metabolomic results using a data-driven approach, Mayr et al. gained greater insight into the role of PKCδ vascular smooth muscle cells (SMCs) than either method would have revealed individually [8]. Using 2-dimensional gel electrophoresis followed by mass spectrometry analysis, approximately 30 proteins were identified that were differentially expressed between PKC$\delta^{+/+}$ and PKC$\delta^{-/-}$ cells. Many of the changes were in proteins involved in energy metabolism, including glycolytic enzymes, triose phosphate isomerase and phosphoglycerate kinase, which were increased in PKCδ deficient SMCs. PKC$\delta^{-/-}$ cells also showed elevated expression of isocitrate dehydrogenase and glucose-6-phosphate dehydrogenase, enzymes that have been linked to glutathione (GSH) metabolism. Metabolomics revealed that compared to PKC$\delta^{+/+}$ SMCs, PKC$\delta^{-/-}$ SMCs have higher levels of GSH, lower levels of alanine, a marker for glycolytic activity, higher lactate, indicative of impaired glucose metabolism, and lower creatine levels. Taking the proteomic and metabolomic results together, the authors suggested that PKCδ depletion disrupts glucose metabolism, which affects energy reserves, leading to upregulation of other energy production pathways. Disrupted glucose metabolism appeared to contribute to increased GSH levels, which are likely responsible in part

Table 16.2 Summary of integrated proteomic and metabolomic studies in cardiovascular research

Focus of study	Types of data integrated	Major findings	References
Protein and metabolite changes in apolipoprotein $E^{-/-}$ mouse aorta during atherogenesis progression	Proteomics and metabolomics	Increased oxidative stress, metabolic disturbances, (e.g., increased fatty acid oxidation), and immune inflammatory responses were observed in apolipoprotein deficient mice	[81]
Molecular mechanisms of atrial profibrillatory remodeling in CHF	Proteomics and metabolomics	Changes in structural proteins (e.g., fragmentation of desmin and filament), oxidative stress, upregulation of heat shock proteins, and metabolic disturbances (e.g., shift toward alpha-ketoacid metabolism) were observed in CHF	[12]
Metabolic processes during persistent atrial fibrillation	Proteomics and metabolomics	Increased ketolytic energy metabolism appears to serve the increased energy demand in atrial fibrillation	[9]
Cardioprotective adaptation process during long-term myocardial hybernation	Proteomics metabolomics, and transcriptomics	Alterations in anaerobic glycolysis, hypoxia-inducible factor (HIF)-1 activation, and reduced phosphorylation of cardiac troponin I were observed in hibernating hearts. Changes in transcript and proteins levels occurred to maintain steady metabolite levels	[11]
The role of PKCδ in vascular smooth muscle cells	Proteomics and metabolomics	PKCδ is a key kinase involved in regulating glucose metabolism, redox state, and cytoskeletal rearrangements	[8]
Mechanisms of PKCε-mediated cardioprotection using transgenic mice with constitutively active or dominant negative PKCε	Proteomics and metabolomics	The expression of enzymes involved in energy metabolism was affected by changes in PKCε activity. Depletion of energy metabolites in response to I/R injury was slower with active PKCε. PKCε modulates glucose metabolism in the heart	[82]
Characterization of stem cell antigen 1-positive (Sca-1(+)) progenitors in vascular adventitia of apolipoprotein $E^{-/-}$ mice	Proteomics and metabolomics	Apolipoprotein $E^{-/-}$ deficiency affects progenitor cell behavior. Expression of enzymes involved in glucose metabolism was altered, leading to faster glucose consumption and reduced interleukin-6 secretion	[83]

Proteomics was carried out by 2DE and MS, metabolomics was carried out using NMR spectroscopy, and transcriptomic data was acquired by qPCR for the observed resistance of PKCδ$^{-/-}$ SMCs to oxidative stress-induced cell death. By combining proteomic and metabolomic findings, the authors were able to deduce much more about the effects of PKCδ depletion in SMCs and provide a mechanistic explanation for the increase in neointima formation observed in PKCδ deficient mice [8]. Mayr, et al. also investigated the mechanisms of PKCε-mediated cardioprotection using transgenic mice with constitutively active or dominant negative PKCε using 2DE and MS proteomics coupled with NMR metabolomics [82].

Transcriptomic, metabolomic, and proteomics datasets were analyzed to investigate the cardioprotective adaptive process in myocardial hibernation [11]. This combined approached showed that anaerobic glycolysis was affected and hypoxia-inducible factor (HIF)-1 activation played a role in metabolic alterations. Overall, the results showed that changes occurred at the transcript and protein levels to maintain relatively stable metabolite levels. These examples demonstrate the benefits of integrating -omics approaches to better understand cardiac biology.

Beyond the cardiovascular field, the integration of multiple –omic techniques have been adopted in other fields. The investigation of the systemic response in wild-type *Arabidopsis thaliana* and a starch-deficient mutant (phosphoglucomutase-deficient) to abiotic temperature stress, analysis of combined proteomic and metabolomic data revealed specific metabolite-protein co-regulation of this process [84]. Integrated proteomic and metabolomic investigations of kidney fibrosis in a rat model revealed changes in complement and coagulation cascades, regulation of actin cytoskeleton, and the MAPK signaling pathway [85]. Another study utilizing proteomic and metabolomics data to determine the effect of hyperosmotic stress on human conjunctival epithelial cells discovered that activated glycerophosphocholine synthesis and O-linked β-N-acetylglucosamine glycosylation are key pathways in ocular surface cells under hyperosmotic stress [86].

Conclusion

The synergy through integration of data from different -omic platforms, such as proteomics and metabolomics, is beginning to provide a more comprehensive view of complex biological systems. This holistic view of an organism, organ, or disease state will be essential to understand complex regulatory processes in normal physiology and pathophysiology. Although to date a limited number of published studies have integrated multiple -omic data sources, these studies indicate that such integration reveals greater insight into systemic responses of the biological system being investigated, beyond simply the sum of its parts. As more powerful and user friendly software platforms are developed that allow better and faster integration of proteomic and metabolic data, integrative -omics will become a common method to enhance our understanding of biological processes.

References

1. Fiehn O. Metabolomics–the link between genotypes and phenotypes. Plant Mol Biol. 2002;48(1–2):155–71. Epub 2002/02/28.
2. Nicholson JK, Lindon JC, Holmes E. 'Metabonomics': understanding the metabolic responses of living systems to pathophysiological stimuli via multivariate statistical analysis of biological NMR spectroscopic data. Xenobiotica. 1999;29(11):1181–9. Epub 1999/12/22.
3. Zajac J, Shrestha A, Patel P, Poretsky L. The main events in the history of diabetes mellitus. In: Poretsky L, editor. Principles of diabetes mellitus. 2nd ed. New York: Springer; 2010. p. 3–16.
4. Ideker T, Galitski T, Hood L. A new approach to decoding life: systems biology. Annu Rev Genomics Hum Genet. 2001;2:343–72. Epub 2001/11/10.
5. Vinayavekhin N, Homan EA, Saghatelian A. Exploring disease through metabolomics. ACS Chem Biol. 2010;5(1):91–103. Epub 2009/12/22.
6. Shah SH, Kraus WE, Newgard CB. Metabolomic profiling for the identification of novel biomarkers and mechanisms related to common cardiovascular diseases: form and function. Circulation. 2012;126(9):1110–20. Epub 2012/08/29.
7. Lee J, Jung Y, Park JY, Lee SH, Ryu DH, Hwang GS. LC/MS-based polar metabolite profiling reveals gender differences in serum from patients with myocardial infarction. J Pharm Biomed Anal. 2015;115:475–86. Epub 2015/08/25.
8. Mayr M, Siow R, Chung YL, Mayr U, Griffiths JR, Xu Q. Proteomic and metabolomic analysis of vascular smooth muscle cells: role of PKCdelta. Circ Res. 2004;94(10):e87–96. Epub 2004/05/08.
9. Mayr M, Yusuf S, Weir G, Chung YL, Mayr U, Yin X, et al. Combined metabolomic and proteomic analysis of human atrial fibrillation. J Am Coll Cardiol. 2008;51(5):585–94. Epub 2008/02/02.
10. Blanchet L, Smolinska A, Attali A, Stoop MP, Ampt KA, van Aken H, et al. Fusion of metabolomics and proteomics data for biomarkers discovery: case study on the experimental autoimmune encephalomyelitis. BMC Bioinform. 2011;12:254. Epub 2011/06/24.
11. Mayr M, May D, Gordon O, Madhu B, Gilon D, Yin X, et al. Metabolic homeostasis is maintained in myocardial hibernation by adaptive changes in the transcriptome and proteome. J Mol Cell Cardiol. 2011;50(6):982–90. Epub 2011/03/01.
12. De Souza AI, Cardin S, Wait R, Chung YL, Vijayakumar M, Maguy A, et al. Proteomic and metabolomic analysis of atrial profibrillatory remodelling in congestive heart failure. J Mol Cell Cardiol. 2010;49(5):851–63. Epub 2010/07/27.
13. Alonso A, Marsal S, Julia A. Analytical methods in untargeted metabolomics: state of the art in 2015. Front Bioeng Biotechnol. 2015;3:23. Epub 2015/03/24.
14. Vu TN, Laukens K. Getting your peaks in line: a review of alignment methods for NMR spectral data. Metabolites. 2013;3(2):259–76. Epub 2013/01/01.
15. Vu TN, Valkenborg D, Smets K, Verwaest KA, Dommisse R, Lemiere F, et al. An integrated workflow for robust alignment and simplified quantitative analysis of NMR spectrometry data. BMC Bioinform. 2011;12:405. Epub 2011/10/22.
16. De Meyer T, Sinnaeve D, Van Gasse B, Tsiporkova E, Rietzschel ER, De Buyzere ML, et al. NMR-based characterization of metabolic alterations in hypertension using an adaptive, intelligent binning algorithm. Anal Chem. 2008;80(10):3783–90. Epub 2008/04/19.
17. Wilmes A, Limonciel A, Aschauer L, Moenks K, Bielow C, Leonard MO, et al. Application of integrated transcriptomic, proteomic and metabolomic profiling for the delineation of mechanisms of drug induced cell stress. J Proteomics. 2013;79:180–94. Epub 2012/12/15.
18. Fuhrer T, Zamboni N. High-throughput discovery metabolomics. Curr Opin Biotechnol. 2015;31:73–8. Epub 2014/09/10.
19. Vinayavekhin N, Saghatelian A. Chapter 30: Untargeted metabolomics. In: Ausubel FM, et al., editor. Current protocols in molecular biology. New Jersey: John Wiley & Sons, Inc; 2010. Unit 30.1 1–24. Epub 2010/04/08.
20. Dunn WB, Wilson ID, Nicholls AW, Broadhurst D. The importance of experimental design and QC samples in large-scale and MS-driven untargeted metabolomic studies of humans. Bioanalysis. 2012;4(18):2249–64. Epub 2012/10/11.

21. Lindsey ML, Gomes AV, Smith S, de Castro Brás LE. How to design a cardiovascular proteomics experiment. 2015.
22. Lindsey ML, Mayr M, Gomes AV, Delles C, Arrell DK, Murphy AM, et al. Transformative impact of proteomics on cardiovascular health and disease: a scientific statement from the American Heart Association. Circulation. 2015;132(9):852–72. Epub 2015/07/22.
23. Naz S, Vallejo M, Garcia A, Barbas C. Method validation strategies involved in non-targeted metabolomics. J Chromatogr A. 2014;1353:99–105. Epub 2014/05/09.
24. Roberts LD, Souza AL, Gerszten RE, Clish CB. Chapter 30: Targeted metabolomics. In: Ausubel FM, et al., editor. Current protocols in molecular biology. New Jersey: John Wiley & Sons, Inc; 2012. Unit 30.2 1–24. Epub 2012/04/04.
25. Mast C, Lyan B, Joly C, Centeno D, Giacomoni F, Martin JF, et al. Assessment of protein modifications in liver of rats under chronic treatment with paracetamol (acetaminophen) using two complementary mass spectrometry-based metabolomic approaches. J Proteomics. 2015;120:194–203. Epub 2015/03/31.
26. Teul J, Garcia A, Tunon J, Martin-Ventura JL, Tarin N, Bescos LL, et al. Targeted and non-targeted metabolic time trajectory in plasma of patients after acute coronary syndrome. J Pharm Biomed Anal. 2011;56(2):343–51. Epub 2011/06/24.
27. Dunn WB, Broadhurst D, Begley P, Zelena E, Francis-McIntyre S, Anderson N, et al. Procedures for large-scale metabolic profiling of serum and plasma using gas chromatography and liquid chromatography coupled to mass spectrometry. Nat Protoc. 2011;6(7):1060–83. Epub 2011/07/02.
28. Beckonert O, Keun HC, Ebbels TM, Bundy J, Holmes E, Lindon JC, et al. Metabolic profiling, metabolomic and metabonomic procedures for NMR spectroscopy of urine, plasma, serum and tissue extracts. Nat Protoc. 2007;2(11):2692–703. Epub 2007/11/17.
29. Sellick CA, Hansen R, Stephens GM, Goodacre R, Dickson AJ. Metabolite extraction from suspension-cultured mammalian cells for global metabolite profiling. Nat Protoc. 2011;6(8):1241–9. Epub 2011/07/30.
30. Want EJ, Masson P, Michopoulos F, Wilson ID, Theodoridis G, Plumb RS, et al. Global metabolic profiling of animal and human tissues via UPLC-MS. Nat Protoc. 2013;8(1):17–32. Epub 2012/12/12.
31. Zhu ZJ, Schultz AW, Wang J, Johnson CH, Yannone SM, Patti GJ, et al. Liquid chromatography quadrupole time-of-flight mass spectrometry characterization of metabolites guided by the METLIN database. Nat Protoc. 2013;8(3):451–60. Epub 2013/02/09.
32. Blatherwick EQ, Svensson CI, Frenguelli BG, Scrivens JH. Localisation of adenine nucleotides in heat-stabilised mouse brains using ion mobility enabled MALDI imaging. Int J Mass Spectrometry. 2013;345–347:19–27.
33. Folmes CD, Nelson TJ, Martinez-Fernandez A, Arrell DK, Lindor JZ, Dzeja PP, et al. Somatic oxidative bioenergetics transitions into pluripotency-dependent glycolysis to facilitate nuclear reprogramming. Cell Metab. 2011;14(2):264–71. Epub 2011/08/02.
34. Leon Z, Garcia-Canaveras JC, Donato MT, Lahoz A. Mammalian cell metabolomics: experimental design and sample preparation. Electrophoresis. 2013;34(19):2762–75. Epub 2013/02/26.
35. Dettmer K, Nurnberger N, Kaspar H, Gruber MA, Almstetter MF, Oefner PJ. Metabolite extraction from adherently growing mammalian cells for metabolomics studies: optimization of harvesting and extraction protocols. Anal Bioanal Chem. 2011;399(3):1127–39. Epub 2010/12/03.
36. Sellick CA, Hansen R, Maqsood AR, Dunn WB, Stephens GM, Goodacre R, et al. Effective quenching processes for physiologically valid metabolite profiling of suspension cultured Mammalian cells. Anal Chem. 2009;81(1):174–83. Epub 2008/12/09.
37. Sellick C, Knight D, Croxford A, Maqsood A, Stephens G, Goodacre R, et al. Evaluation of extraction processes for intracellular metabolite profiling of mammalian cells: matching extraction approaches to cell type and metabolite targets. Metabolomics. 2010;6(3):427–38.
38. Lei Z, Huhman DV, Sumner LW. Mass spectrometry strategies in metabolomics. J Biol Chem. 2011;286(29):25435–42. Epub 2011/06/03.

39. Kind T, Wohlgemuth G, Leedo Y, Lu Y, Palazoglu M, Shahbaz S, et al. FiehnLib: mass spectral and retention index libraries for metabolomics based on quadrupole and time-of-flight gas chromatography/mass spectrometry. Anal Chem. 2009;81(24):10038–48. Epub 2009/11/26.
40. Babushok VI, Linstrom PJ, Reed JJ, Zenkevich IG, Brown RL, Mallard WG, et al. Development of a database of gas chromatographic retention properties of organic compounds. J Chromatogr A. 2007;1157(1–2):414–21. Epub 2007/06/05.
41. Chen CT, Shih YR, Kuo TK, Lee OK, Wei YH. Coordinated changes of mitochondrial biogenesis and antioxidant enzymes during osteogenic differentiation of human mesenchymal stem cells. Stem Cells. 2008;26(4):960–8. Epub 2008/01/26.
42. Wilson ID. Chapter 5: Chromatographic and electrophoretic separations combined with mass spectrometry for metabonomics. In: Lindon JC, Nicholson JK, Holmes E, editors. The handbook of metabonomics and metabolomics. Amsterdam: Elsevier Science B.V; 2007. p. 149–69.
43. Xu F, Zou L, Ong CN. Multiorigination of chromatographic peaks in derivatized GC/MS metabolomics: a confounder that influences metabolic pathway interpretation. J Proteome Res. 2009;8(12):5657–65. Epub 2009/10/20.
44. Patti GJ, Yanes O, Siuzdak G. Innovation: metabolomics: the apogee of the omics trilogy. Nat Rev Mol Cell Biol. 2012;13(4):263–9. Epub 2012/03/23.
45. Smith CA, O'Maille G, Want EJ, Qin C, Trauger SA, Brandon TR, et al. METLIN: a metabolite mass spectral database. Ther Drug Monit. 2005;27(6):747–51. Epub 2006/01/13.
46. Dunn W, Erban A, Weber RM, Creek D, Brown M, Breitling R, et al. Mass appeal: metabolite identification in mass spectrometry-focused untargeted metabolomics. Metabolomics. 2013;9(1):44–66.
47. Bajad SU, Lu W, Kimball EH, Yuan J, Peterson C, Rabinowitz JD. Separation and quantitation of water soluble cellular metabolites by hydrophilic interaction chromatography-tandem mass spectrometry. J Chromatogr A. 2006;1125(1):76–88. Epub 2006/06/09.
48. Buescher JM, Moco S, Sauer U, Zamboni N. Ultrahigh performance liquid chromatography-tandem mass spectrometry method for fast and robust quantification of anionic and aromatic metabolites. Anal Chem. 2010;82(11):4403–12. Epub 2010/05/04.
49. Ross A, Schlotterbeck G, Dieterle F, Senn H. Chapter 3: NMR spectroscopy techniques for application to metabonomics. In: Lindon JC, Nicholson JK, Holmes E, editors. The handbook of metabonomics and metabolomics. Amsterdam: Elsevier Science B.V; 2007. p. 55–112.
50. Brown FF, Campbell ID, Kuchel PW, Rabenstein DC. Human erythrocyte metabolism studies by 1H spin echo NMR. FEBS Lett. 1977;82(1):12–6. Epub 1977/10/01.
51. Bales JR, Higham DP, Howe I, Nicholson JK, Sadler PJ. Use of high-resolution proton nuclear magnetic resonance spectroscopy for rapid multi-component analysis of urine. Clin Chem. 1984;30(3):426–32. Epub 1984/03/01.
52. Lindon JC, Nicholson JK. Spectroscopic and statistical techniques for information recovery in metabonomics and metabolomics. Annu Rev Anal Chem. 2008;1:45–69. Epub 2008/07/19.
53. Beckonert O, Coen M Fau KHC, Keun Hc Fau WY, Wang Y Fau ETMD, Ebbels Tm Fau HE, Holmes E Fau LJC, et al. High-resolution magic-angle-spinning NMR spectroscopy for metabolic profiling of intact tissues. Nat Protoc. 2010;5(6):1019–32.
54. Wishart DS. Advances in metabolite identification. Bioanalysis. 2011;3(15):1769–82.
55. Wishart DS, Knox C, Guo AC, Eisner R, Young N, Gautam B, et al. HMDB: a knowledgebase for the human metabolome. Nucleic Acids Res. 2009;37(Database issue):D603–10. Epub 2008/10/28.
56. Ulrich EL, Akutsu H, Doreleijers JF, Harano Y, Ioannidis YE, Lin J, et al. BioMagResBank. Nucleic Acids Res. 2008;36(Database issue):D402–8. Epub 2007/11/07.
57. Cui Q, Lewis IA, Hegeman AD, Anderson ME, Li J, Schulte CF, et al. Metabolite identification via the Madison metabolomics consortium database. Nat Biotechnol. 2008;26(2):162–4. Epub 2008/02/09.
58. Ludwig C, Easton J, Lodi A, Tiziani S, Manzoor S, Southam A, et al. Birmingham metabolite library: a publicly accessible database of 1-D 1H and 2-D 1H J-resolved NMR spectra of authentic metabolite standards (BML-NMR). Metabolomics. 2012;8(1):8–18.

59. Weljie AM, Newton J, Mercier P, Carlson E, Slupsky CM. Targeted profiling: quantitative analysis of 1H NMR metabolomics data. Anal Chem. 2006;78(13):4430–42. Epub 2006/07/01.
60. Ellinger JJ, Chylla RA, Ulrich EL, Markley JL. Databases and software for NMR-based metabolomics. Current Metabolomics. 2013;1(1):15–27.
61. Sansone Sa Fau FT, Fan T Fau GR, Goodacre R Fau GJL, Griffin J Fau HNW, Hardy Nw Fau K-DR, Kaddurah-Daouk R Fau KBS, et al. The metabolomics standards initiative. Nature Biotech. 2007;25:846–8.
62. Sauro HM, Ingalls B. Conservation analysis in biochemical networks: computational issues for software writers. Biophys Chem. 2004;109(1):1–15. Epub 2004/04/03.
63. Karp PD, Paley SM, Krummenacker M, Latendresse M, Dale JM, Lee TJ, et al. Pathway Tools version 13.0: integrated software for pathway/genome informatics and systems biology. Brief Bioinform. 2010;11(1):40–79. Epub 2009/12/04.
64. Caspi R, Altman T, Billington R, Dreher K, Foerster H, Fulcher CA, et al. The MetaCyc database of metabolic pathways and enzymes and the BioCyc collection of Pathway/Genome Databases. Nucleic Acids Res. 2014;42(Database issue):D459–71. Epub 2013/11/15.
65. Shannon P, Markiel A, Ozier O, Baliga NS, Wang JT, Ramage D, et al. Cytoscape: a software environment for integrated models of biomolecular interaction networks. Genome Res. 2003;13(11):2498–504. Epub 2003/11/05.
66. Junker BH, Klukas C, Schreiber F. VANTED: a system for advanced data analysis and visualization in the context of biological networks. BMC Bioinform. 2006;7:109. Epub 2006/03/08.
67. Kessler N, Neuweger H, Bonte A, Langenkamper G, Niehaus K, Nattkemper TW, et al. MeltDB 2.0-advances of the metabolomics software system. Bioinformatics. 2013;29(19):2452–9. Epub 2013/08/07.
68. Clasquin MF, Melamud E, Rabinowitz JD. Chapter 14: LC-MS data processing with MAVEN: a metabolomic analysis and visualization engine. In: Editoral board, Baxevanis AD, et al., editors. Current protocols in bioinformatics. 2012. Unit 14.11. Epub 2012/03/06.
69. Kaever A, Landesfeind M, Feussner K, Mosblech A, Heilmann I, Morgenstern B, et al. MarVis-Pathway: integrative and exploratory pathway analysis of non-targeted metabolomics data. Metabolomics. 2015;11(3):764–77. Epub 2015/05/15.
70. Barupal DK, Haldiya PK, Wohlgemuth G, Kind T, Kothari SL, Pinkerton KE, et al. MetaMapp: mapping and visualizing metabolomic data by integrating information from biochemical pathways and chemical and mass spectral similarity. BMC Bioinform. 2012;13:99. Epub 2012/05/18.
71. Xia J, Sinelnikov IV, Han B, Wishart DS. MetaboAnalyst 3.0-making metabolomics more meaningful. Nucleic Acids Res. 2015;43(W1):W251–7. Epub 2015/04/22.
72. Yizhak K, Benyamini T, Liebermeister W, Ruppin E, Shlomi T. Integrating quantitative proteomics and metabolomics with a genome-scale metabolic network model. Bioinformatics. 2010;26(12):i255–60. Epub 2010/06/10.
73. Neuweger H, Persicke M, Albaum SP, Bekel T, Dondrup M, Huser AT, et al. Visualizing post genomics data-sets on customized pathway maps by ProMeTra-aeration-dependent gene expression and metabolism of Corynebacterium glutamicum as an example. BMC Syst Biol. 2009;3:82. Epub 2009/08/25.
74. Jorgenrud B, Jalanko M, Helio T, Jaaskelainen P, Laine M, Hilvo M, et al. The metabolome in finnish carriers of the MYBPC3-Q1061X mutation for hypertrophic cardiomyopathy. PLoS One. 2015;10(8):e0134184. Epub 2015/08/13.
75. Altmaier E, Fobo G, Heier M, Thorand B, Meisinger C, Romisch-Margl W, et al. Metabolomics approach reveals effects of antihypertensives and lipid-lowering drugs on the human metabolism. Eur J Epidemiol. 2014;29(5):325–36. Epub 2014/05/13.
76. Varma A, Palsson BO. Stoichiometric flux balance models quantitatively predict growth and metabolic by-product secretion in wild-type Escherichia coli W3110. Appl Environ Microbiol. 1994;60(10):3724–31. Epub 1994/10/01.
77. Fell DA. Metabolic control analysis: a survey of its theoretical and experimental development. Biochem J. 1992;286(Pt 2):313–30. Epub 1992/09/01.

78. Schuster S, Fell DA, Dandekar T. A general definition of metabolic pathways useful for systematic organization and analysis of complex metabolic networks. Nat Biotechnol. 2000;18(3):326–32. Epub 2000/03/04.
79. Orth JD, Thiele I, Palsson BO. What is flux balance analysis? Nat Biotechnol. 2010;28(3):245–8. Epub 2010/03/10.
80. May P, Christian N, Ebenhoh O, Weckwerth W, Walther D. Integration of proteomic and metabolomic profiling as well as metabolic modeling for the functional analysis of metabolic networks. Methods Mol Biol. 2011;694:341–63. Epub 2010/11/18.
81. Mayr M, Chung YL, Mayr U, Yin X, Ly L, Troy H, et al. Proteomic and metabolomic analyses of atherosclerotic vessels from apolipoprotein E-deficient mice reveal alterations in inflammation, oxidative stress, and energy metabolism. Arterioscler Thromb Vasc Biol. 2005;25(10):2135–42. Epub 2005/08/27.
82. Mayr M, Liem D, Zhang J, Li X, Avliyakulov NK, Yang JI, et al. Proteomic and metabolomic analysis of cardioprotection: interplay between protein kinase C epsilon and delta in regulating glucose metabolism of murine hearts. J Mol Cell Cardiol. 2009;46(2):268–77. Epub 2008/11/26.
83. Mayr M, Zampetaki A, Sidibe A, Mayr U, Yin X, De Souza AI, et al. Proteomic and metabolomic analysis of smooth muscle cells derived from the arterial media and adventitial progenitors of apolipoprotein E-deficient mice. Circ Res. 2008;102(9):1046–56. Epub 2008/04/05.
84. Wienkoop S, Morgenthal K, Wolschin F, Scholz M, Selbig J, Weckwerth W. Integration of metabolomic and proteomic phenotypes: analysis of data covariance dissects starch and RFO metabolism from low and high temperature compensation response in Arabidopsis thaliana. Mol Cell Proteomics MCP. 2008;7(9):1725–36. Epub 2008/05/01.
85. Cao H, Zhang A, Sun H, Zhou X, Guan Y, Liu Q, et al. Metabolomics-proteomics profiles delineate metabolic changes in kidney fibrosis disease. Proteomics. 2015;15(21):3699–710. Epub 2015/08/11..
86. Chen L, Li J, Guo T, Ghosh S, Koh SK, Tian D, et al. Global metabonomic and proteomic analysis of human conjunctival epithelial cells (IOBA-NHC) in response to hyperosmotic stress. J Proteome Res. 2015;14(9):3982–95. Epub 2015/08/12.

Chapter 17
Clinical Cardiovascular Proteomics

Gemma Currie, Peter Matt, and Christian Delles

Abstract Proteomics has the potential to be translated from a research environment to clinical practice. In the first instance the discovery of novel disease pathways and definition of druggable targets are promises of clinical proteomics. In parallel, clinical proteomics will define new protein-based biomarkers for molecular definition of disease, diagnosis of disease and prediction of events. In cardiovascular medicine the potential applications are manifold and examples are already available for conditions throughout the cardiovascular continuum from early risk factors to intermediate traits and advanced disease, all of which have been subject to proteomic studies. Despite the recent progress most of the available data do not fulfil criteria for novel biomarkers for cardiovascular diseases that are clinically applicable. Studies have been small, findings have not been reproduced in independent cohorts, plausible links to pathophysiology are not always present and sophisticated technical and bioinformatic requirements in proteomics pose challenges to translation of research findings to clinical cardiovascular medicine. Better standardisation of experiments, coordinated research efforts and close collaboration between clinicians and basic sciences will help to ask the right questions and provide the right answers and solutions in the near future.

Keywords Proteomics • Cardiovascular disease • Atherosclerosis • Coronary artery disease • Heart failure • Hypertension • Chronic kidney disease • Biomarkers • Diagnosis • Prediction

G. Currie, MB, ChB • C. Delles, MD, FRCP, FAHA, FBHS (✉)
Institute of Cardiovascular and Medical Sciences, BHF Glasgow Cardiovascular Research Centre, University of Glasgow, 126 University Place, Glasgow G12 8TA, UK
e-mail: Christian.Delles@glasgow.ac.uk

P. Matt, MD, FETCS
Aortic Research Group, Division of Cardiac Surgery, University Hospital Basel, University of Basel, Basel, Switzerland

G. Agnetti et al. (eds.), *Manual of Cardiovascular Proteomics*,
DOI 10.1007/978-3-319-31828-8_17

Introduction

Cardiovascular diseases (CVD) are the major cause of morbidity and mortality in Europe, North America and other Western societies. In Europe a north-east to south-west gradient in mortality from CVD remains, with age-standardised mortality rates ranging from 731 to 1763 per 100,000 [1]. In the US, CVD accounted for 31.3 % of all deaths based on 2011 mortality data [2]. Recently, incidence and prevalence of CVD have also increased in developing countries, making CVD a truly global epidemic [3]. Consequently CVD are associated with significant costs to health care systems that are expected to rise further in face of the increasing prevalence of cardiovascular risk factors, namely obesity and the metabolic syndrome [3].

Against the background of a global CVD epidemic clinical research currently focusses on early and accurate diagnosis of CVD and prediction of adverse outcome in patients with CVD. A number of cardiovascular risk scores based upon a range of traditional risk factors have been developed and validated. It has, however, been recognised that modifications to these risk scores are required to better reflect risk profiles of specific populations. The Framingham model has been found to provide reasonable estimates of cardiovascular risk in Northern American populations but over-estimates cardiovascular risk in some European populations [4]. The SCORE model with its global European and national versions is a prime example of such adaptations to cardiovascular risk prediction across European regions [5]. Despite increasingly precise predictive power on a population basis, risk scores do not necessarily perform well in individual subjects or in groups of subjects who share certain characteristics and may have risk factors but also protective factors that are not represented in risk scores. Some of these factors are poorly understood or unknown, but may include genetic factors, diet, physical activity and other lifestyle factors.

Against this background there is an unmet clinical need for stratified cardiovascular medicine. This will include accurate diagnosis of CVD and accurate risk prediction in individual subjects with subsequent targeted preventative therapy. This approach is considered highly cost effective not only due to prevention or delay of onset of CVD related organ damage but also due to the potential savings on diagnostic procedures and drug therapy in subjects at low cardiovascular risk [6]. In this chapter we will therefore focus on potential clinical applications of proteomics and will review recent developments in the wider context of CVD biomarker research.

Biomarkers of Cardiovascular Disease

A biomarker is defined as "a characteristic that is objectively measured and evaluated as an indicator of normal biological processes, pathogenic processes, or pharmacologic responses to a therapeutic intervention" [7]. The common denominator of recent studies is that the most promising biomarkers for cardiovascular risk reflect early subclinical organ damage. For example, urinary proteomic studies have demonstrated that changes in extracellular matrix composition are a hallmark of CVD and that such changes are represented by fragments of matrix proteins in urine [8, 9]. These changes occur many years before the onset of overt disease and years before signals from other biomarkers can be detected. Another recent example is the finding that raised serum C-terminal propeptide of type I procollagen levels indicate increased myocardial collagen synthesis and precede and parallel the development of cardiac fibrosis [10].

Multiple Roles of Biomarkers

Biomarkers can play multiple roles and provide useful clinical information at different levels (Table 17.1). In general terms, Frank and Hargreaves [18] have classified biomarkers into biomarkers of the natural history of disease (type 0 biomarkers); biomarkers of the biological effect of a therapeutic intervention (type 1 biomarkers); and surrogate markers for certain aspects of disease that can be used as outcomes of clinical trials and for regulatory purposes (type 2 biomarkers).

We will explore the role of proteomics as biomarker for CVD in more detail in this chapter but will first outline the development of clinically useful biomarkers in general.

Phases of Biomarker Development

The development of novel biomarkers for clinical use typically follows a number of defined stages that have been described in detail elsewhere [19]. In brief the development process can be described in four phases:

1. **Proof of concept.** This phase is characterised by mechanistic *in vitro* and *ex vivo* studies, studies in experimental models and small-scale studies in humans (typically patients with a well-defined condition and healthy controls) to demonstrate the pathophysiological relevance and ability of a novel biomarker to differentiate health from disease.
2. **Validation in independent cohorts.** Initial data will then be replicated and validated in larger cohorts with adjustment for confounding factors and often using a prospective study design.
3. **Demonstration of incremental value over established markers.** This is one of the most critical steps in the development of a biomarker. Novel biomarkers would normally only be regarded as clinically useful if they provide information over and above already existing diagnostic tools. However, biomarkers can also be clinically useful if they provide similar information to existing diagnostic

Table 17.1 Overview of biomarkers in clinical medicine

Type of biomarker	Explanation	Examples
Disease defining biomarkers	Used in some diseases as a condition to establishing a diagnosis	Antineutrophil cytoplasmic antibody (ANCA)-associated vasculitis is defined by the presence of ANCA and a diagnosis of these diseases can only be made if ANCA are present
Diagnostic biomarkers	Used in clinical practice to reach a diagnosis so that appropriate therapy can be initiated. Whilst often linked to disease processes diagnostic biomarkers do not necessarily specifically reflect individual pathophysiological pathways	Raised serum levels of cardiac troponins indicate in the first instance myocardial cell death but do not provide disease specific information [11]. Raised cardiac troponin levels can therefore be sensitive for many conditions including myocardial infarction, myocarditis or pulmonary embolism with right ventricular strain but specificity is limited and depends on additional clinical and biochemical information
Predictive biomarkers	Used to identify patients at risk of developing disease and/or adverse disease outcomes	Most predictive biomarkers are also diagnostic biomarkers with lower thresholds being used in clinically asymptomatic stages for prediction of future overt disease. High sensitivity assays have facilitated the use of established diagnostic biomarkers such as cardiac troponins or C-reactive protein to predict future disease and events [12, 13]
Biomarkers for therapy monitoring	Used to monitor the effect of treatment and to adjust therapeutic regimens depending on biomarker response	This concept is fully established in oncology, infectious disease and inflammatory diseases where tumour markers, inflammatory markers and levels of autoantibodies help monitoring treatment effects and are often the immediate target of drug or other therapies. In cardiovascular medicine serial measurements of cardiac troponins can probably help to monitor the effect of reperfusion strategies in acute myocardial infarction [14, 15] and Brain Natriuretic Peptide (BNP) may have a potential to monitor treatment in patients with heart failure [16] although such strategies remain controversial [17]

tests at lower cost, can be analysed more easily or provide information beyond the specific disease of interest.

4. **Biomarker-guided therapy.** The ultimate usefulness of a novel biomarker can only be proven in prospective trials stratifying patients into a biomarker positive and a biomarker negative group to explore if outcomes associated with specific

therapies are different between the groups. This is a step that not many circulating biomarkers have reached, although biomarkers in a wider sense have of course been used to define entry criteria for participation in clinical trials, e.g. reduced left ventricular ejection fraction as a prerequisite for a patient to be included in heart failure trials.

Pre-test Probability and the Right Test for the Right Patient

Most diagnostic and predictive biomarkers will not be 100 % sensitive and 100 % specific for the condition of interest. Appropriate cut-off values will have to be defined that lead to an appropriate balance between sensitivity and specificity depending on the intended use of a biomarker. Whilst this can be achieved relatively easily for diagnostic biomarkers the establishment of thresholds for predictive biomarkers is far more challenging.

Predictive tests therefore tend to categorise individuals into risk groups (*e.g.* low, intermediate and high). Guidelines then focus on the low (no treatment required) and the high risk groups (treatment required) whereas guidance for those at intermediate risk remains vague. It is often this intermediate risk group that will be subject to additional biomarker analysis in the hope that subjects can later be reclassified into either of the two extreme groups [20]. Irrespective of the primary risk assessment method it is clear that pre-test probability must be considered before biomarkers are assessed in research studies or further diagnostic tests are performed in clinical practice. For any novel biomarker a precise definition of its indication in the diagnostic work-up is of paramount importance.

Proteomics

Proteomics is the analysis of a large number of proteins or polypeptides in tissue and body fluids [21]. Other chapters of this book have defined the field in more detail. In addition we would like to refer to relevant reviews on clinical proteomics in the literature [22–24]. By quantifying a range of peptides and peptide fragments, proteomics has the potential of simultaneously detecting changes in the many pathways involved in the pathogenesis of CVD.

Like any other biomarker, proteomics-based biomarkers can be diagnostic, predictive and/or disease defining. Whilst we will discuss examples in the following sections in detail we would like to emphasise that genomics in the first instance describes a potential and may play a particularly important role in risk prediction, whereas proteomics describes the actual state of the organism that results from the complex interplay between genetic and environmental factors, and may therefore play a key role in the description of disease processes and the definition of disease at a molecular level.

Platforms for Clinical Proteomics

Proteomic platforms have been described in more detail elsewhere in this book (see Chaps. 6, 7, 8 and 9 for details). Clinical proteomics makes use of the same platforms and techniques but has its own specific requirements:

- **Targeted vs non-targeted proteomic techniques.** In biomarker discovery non-targeted techniques are generally preferred as they will allow an unbiased approach. Non-targeted methods will also have advantages in the molecular definition of disease as the resolution will be much higher than with the limited number of proteins that can be detected with targeted techniques. Once an array of relevant biomarkers has been discovered and validated the development of targeted assays will be reasonable. These can be single-marker platforms such as ELISAs or multiplexing arrays.
- **Throughput.** Especially for clinical routine but also for larger-scale clinical studies high throughput of samples is required. Sample runs of several hours with additional preparation and separation steps would not be acceptable in clinical practice.
- **Reproducibility and Good Laboratory Practice.** Especially in clinical diagnostics measurements have to be reproducible and standardised and follow the principles of Good Laboratory Practice. Research-oriented laboratories often underestimate the logistics required to offer a reliable clinical service.
- **Simplicity and compatibility.** Proteomics is generally considered a "high tech" approach requiring specialist skills. Widespread application in clinical practice can only happen if procedures are compatible with routine analysers in clinical biochemistry laboratories.
- **Costs.** Proteomic assays have to be cost efficient in order to be implemented in clinical practice. Full economic costings taking account of infrastructure, service, consumables and salaries as well as detailed health economic considerations regarding costs and benefits of biomarker analysis including the effect of false positive and false negative results on health care systems have to be conducted.

Currently only targeted proteomic approaches tend to fulfil these criteria and it is up to the proteomic community to pave the way for translation of more comprehensive techniques into clinical practice.

Sample Types for Clinical Proteomics

As discussed in more detail elsewhere in this book (Chaps. 2 and 3) a large number of tissues and biofluids can be subject to proteomic analysis. Clearly, invasive tests (needle biopsies, endoscopic sampling, surgical sampling) are more acceptable in case of severe and life threatening conditions where precise diagnosis is important, *e.g.* malignant diseases. For risk prediction and in fact for most diagnostic purposes in clinical medicine, non or minimally invasive sampling of body fluids (e.g. blood, saliva or urine) is preferable. In this chapter we will focus on plasma and urinary proteomics and will discuss a number of examples in more detail. We will mention tissue proteomics where less invasive studies are not available or where they are not sufficiently robust.

Plasma or serum are generally the biofluids of choice as the blood is in direct contact with relevant organs and will contain proteins originating from them. Disadvantages of plasma include the high protease activity resulting in altered protein content and composition after sampling; the complexity of the plasma proteome with concentrations of plasma proteins from picomolar to millimolar range; and protein binding, particularly to the highly abundant albumin [25]. For clinical studies in humans there are advantages of urine over blood for the purpose of proteomics. Urine contains polypeptides originating from a large number of biochemical pathways within the body but final concentrations in the urine also depend on the glomerular filter and tubular secretion and absorption processes. In contrast to blood, protease activity in urine is low so that urine is generally considered to be much more stable. Analyses are technically less demanding and the data can thereby be more robust in urine compared to plasma [26].

The Potential of Proteomics in Cardiovascular Disease

The fact that the primary manifestation of CVD is often an acute ischaemic event such as myocardial infarction or stroke coupled with the limited predictive value of conventional cardiovascular risk factors means that identification of individuals "at risk" of an event remains a major challenge and novel tools for accurate risk assessment on an individual basis are required.

Atherosclerosis is common to many cardiovascular conditions, and it is now widely understood that narrowing of the vessel lumen and subsequent ischaemia are preceded by subclinical alterations in several key pathways including: inflammation; oxidative stress; thrombosis; and vascular remodeling which remain only partially unravelled. In view of this complex pathophysiology a single biomarker is unlikely to perform adequately for early diagnosis, prognosis and therapeutic monitoring. Clinical application of proteomic strategies to CVD management could potentially offer a number of key advantages:

- Proteomics offers the unique opportunity to disentangle the cellular processes that precede transition to overt CVD in a non-targeted and therefore non-biased manner;
- Such enhanced understanding of the perturbations at cellular level which precede overt CVD brings potential to identify novel therapeutic targets and allow an integrated view of disease pathogenesis;
- A "multi-marker" approach including potential biomarkers involved in each of the key pathways listed above may improve the accuracy of cardiovascular risk prediction;

- Evaluating changes which are present early in the disease process may aid early identification and treatment of "at risk" individuals with more subtle manifestations of CVD.

Despite the potential advantages that could accompany translation of such approaches into clinical practice, proteomic research itself remains largely within the "preclinical" stage. Making the leap from the specialised laboratory into the clinic requires that a proposed biomarker should outperform the current state-of-the-art; provide robust and reproducible results; be tested in an adequately-powered prospective study; and offer cost-effectiveness, targets which many proteomic biomarkers have yet to achieve.

Examples of Proteomic Studies in Cardiovascular Diseases

Atherosclerosis

Despite the fact that it is the common precursor to many CVDs, there is relatively little published data on application of proteomics in clinical studies of human atherosclerotic disease.

At a cellular level a number of groups have reported outcomes of proteomic studies following stimulation with pro-atherosclerotic factors, giving insight into mechanisms underlying disease. The LC-MS platform has been used to compare the effects of oxidised versus native low density lipoprotein (LDL) on monocytic cells. Proteins upregulated following stimulation with oxidised LDL included cathepsins, proteoglycans and urokinase-type plasminogen activator receptor [27]. Endothelial cells are a key component of the atherosclerotic plaque and work on human umbilical vein endothelial cells (HUVEC) in the presence and absence of pro-inflammatory stimulation using 2D gel electrophoresis combined with MALDI-TOF MS identified 35 altered proteins as a result of pro-inflammatory conditions [28]. Vascular smooth muscle cells (VSMCs) are another integral player in the development of atherogenesis and 2D gel studies have revealed that activated cells demonstrate altered chaperone phosphorylation, thereby revealing their potential role in VSMC activation [29].

One key limitation of using cultured cells for such studies is that the culture process itself can result in a phenotypically altered cell: although we can glean information on how these cells respond to stimuli, we cannot probe their precise proteome as expressed within the arterial lumen. A more useful approach may be to focus on the proteomics of atherosclerotic plaques themselves, a technique employed by a number of groups using tissue obtained from carotid artery surgery. Early work demonstrated a significant increase in secreted proteins from complex plaques in comparison to healthy vessel using 2D-gel electrophoresis [30]. Later studies have identified typical plaque proteins including superoxide dismutase (SOD) and osteoglycin [31]. A number of studies have now reported reduced secretion of heat shock protein 27 (HSP27) from complex ruptured plaques and correlation with lower plasma HSP27 levels in patient with known atherosclerosis [32–34]. Despite being among the more extensively validated potential biomarkers to emerge from plaque proteomics, HSP27 has yet to make the leap into the clinical sphere and its utility as predictor of later cardiovascular events remains to be confirmed [35].

Taking this work onto a larger scale and moving towards clinical application, the Athero-Express study has established a bio-bank cohort including plaques and other vascular specimens from almost 3000 patients with longitudinal follow up for manifestations of cardiovascular disease. Early studies within this cohort highlighted osteopontin (OPN) as one plaque biomarker potentially predictive of cardiovascular events [36]. This finding was further validated where patients with plaque OPN level within the highest quartile were at increased risk of cardiovascular events. Fatty acid binding protein 4 (FABP4) is another plaque biomarker to emerge from this cohort [37] which has been proposed as a potential treatment target [38].

Hypertension

Hypertension remains one of the most important risk factors for CVD worldwide and its incidence and prevalence continue to increase. In contrast to rarer causes of the disease where pathophysiological mechanisms are relatively well-understood, the mechanisms underpinning "essential hypertension" which accounts for the majority of cases remain to be disentangled. It is generally accepted that essential hypertension is the result of interaction between multiple aberrant physiological pathways as well as being influenced by genetic and environmental factors [39–41]. Use of proteomic strategies to identify a broad range of altered proteins and peptides seems ideally suited to investigate some of these as yet unknown mechanisms, yet large scale clinical studies are lacking. To date most work has centred on animal models of hypertension, with a focus on associated conditions such as left ventricular hypertrophy (LVH) rather than hypertension as a unique entity.

In advance of dissection particular disease pathways it is useful to establish strain-specific proteomic profiles of these animal models. For example, a number of markers have been shown to be differentially expressed in the left ventricular proteome of Dahl salt-sensitive and salt-resistant rats [42]. Early work using ventricular homogenates and 2D gel electrophoresis coupled to MALDI-TOF MS investigated the myocardial proteome in early versus late stage hypertension in spontaneously hypertensive rats (SHR) in comparison to control animals. Thirteen differentially expressed proteins were identified in SHR before the onset of sustained hypertension, including proteins associated with glycolysis, oxidative stress and cell metabolism. Seven proteins were differentially expressed in late stage hypertension, presumably due to established disease [43]. Other groups have used transverse aortic constriction as a model of pressure overload in animal work. Proteins identified from 2D- gel electrophoresis coupled to MALDI-TOF MS using LV homogenates from these animals include fatty acid binding proteins, actin and myosin as well as collagen fragments which appear to be a central component of many proteomic studies [44]. Perhaps more exciting in these discovery experiments is the identification of proteins with as yet unknown functions, providing novel directions for future research [44].

In addition to mechanistic insights and increased diagnostic accuracy, many proteomic studies demonstrate altered peptide patterns in response to various drug therapies, highlighting the potential this tool has to offer as a therapeutic monitoring strategy. Using 2D-gel electrophoresis coupled to MALDI-TOF MS with LV homogenates from SHR and their normotensive counterparts it has been shown that antihypertensive therapy led to partial reduction in oxidative stress through altered NF-κB activation [45]. A later project using the same proteomic platform to study LV homogenates from rats following induction of LVH by abdominal aortic banding showed that along with reduction in blood pressure and prevention of LVH, treatment with the angiotensin receptor blocker Telmisartan was accompanied by proteome changes. Peptides specifically up-regulated in LVH included actin and myosin fragments; changes which were attenuated following antihypertensive therapy [46].

Coronary Artery Disease

Clinically silent disease processes such as atherosclerosis and hypertension can often be precursors to overt cardiovascular disease, a number of which have also been the focus of proteomic research in recent years.

Despite advances in both pharmacological and interventional management in recent years, coronary artery disease (CAD) remains one of the world's leading causes of mortality. Traditional clinical factors routinely assessed by healthcare practitioners can be of limited value in determining risk on an individual patient basis, and certain patient populations can develop extensive and life-threatening CAD in relative clinical silence [47]. Identification of biomarkers highlighting individuals with early stage or asymptomatic disease would allow clinicians to employ more aggressive and timely intervention in those deemed to be at higher risk, and a number of human studies focussing on clinical utility of proteomics for diagnosis of CAD with sample types that can be collected noninvasively have been published.

Plasma Studies

While discovery of novel biomarkers has historically been based on a "candidate" approach in which suspect proteins are assessed on an individual basis, proteomic-based biomarker discovery has gained traction; particularly identification of low-abundance proteins that may highlight as yet unknown disease pathways is an exciting concept. LC-MS analysis of plasma samples from 53 patients with angiography-proven CAD and 53 healthy control subjects identified 95 peptides which were differentially expressed in those with CAD [48]. Proteins that were upregulated could be grouped broadly into classes including: proteolytic processes; immune mechanisms; growth factors; and haemostasis. A number of complement components were identified in cases, which seems plausible as complement activation has been demonstrated in both stable and ruptured atherosclerotic plaques [49]. Proteolytic enzymes such as cathepsin S, which has been associated with atherosclerosis [50], were also identified in plasma of CAD patients; and fibrinogen γ chain was among the identified proteins associated with coagulation and thrombosis. It has also been shown that some alterations in the plasma proteome persist for many months after an acute event. In samples obtained over a 180 day period following myocardial infarction a number of proteins belonging to functional groups associated with cardiovascular pathology including coagulation; lipid metabolism and inflammation were shown to be persistently altered [51]. In fact the plasma proteome continued to change in the months following the acute event. This highlights the ongoing metabolic disturbance following an ischaemic event but also raises the question of the influence of drug treatment on the plasma proteome and illustrates the potential of proteomic techniques in monitoring treatment response.

Urine Studies

Although the majority of proteomic studies have been cross sectional in nature and remain within the exploratory phase, CAD is an example of a condition where proteomic biomarkers have been evaluated for prognostic as well as diagnostic information. Analysis of urine samples from human subjects with angiography proven CAD compared to controls using capillary electrophoresis coupled to ESI-TOF MS resulted in development of a 15 peptide disease specific "signature" with high sensitivity [8]. Taking this work into a more "real-life" clinical situation, a further project used both urine and plasma samples from subjects presenting acutely with chest pain of who underwent angiography at a later date during the course of their work-up. This time a panel of 17 urinary peptides separated patients who were subsequently proven to have CAD at angiography from those without significant disease. The majority of these peptides were identified as collagen type I and III fragments, findings which were confirmed by immunohistochemistry of human aortic tissue samples. No discriminatory peptides were identified using plasma samples, perhaps as a result of latent protease activity [52]. Data obtained from these previous studies was combined to further increase the diagnostic accuracy of this approach. Using CE-MS to analyse 586 spot urine samples from individuals with CAD and healthy controls a 238-peptide CAD-specific pattern was developed which proved to offer a greater degree of diagnostic accuracy than the smaller panels [9]. Component peptides included fragments of alpha-1-antitrypsin; collagens type I and III; and fibrinogen-alpha chain; all of which relate to key molecular mechanisms underlying CAD. Peptide abundance was unchanged in the shorter term with angiotensin receptor blockade but altered over long-term treatment towards a "healthier" pattern, suggesting that the panel is insensitive to drug treatment itself, but may be reflective of the longer term beneficial effects of interventions [9].

As discussed earlier in this chapter, guidelines for biomarker implementation suggest testing in longitudinal prospective studies measuring hard clinical endpoints [19, 24], a target that can be challenging to achieve due to the length of time taken to carry out such projects. An alternative option for prospectively testing a biomarker's performance is application to existing clinical cohorts or bio-banks. Using this strategy the CAD238 panel has been applied to baseline urine samples collected during a large multicentre randomised controlled blood pressure trial [53]. Samples were obtained from individuals who later suffered a coronary event and matched to individuals who remained CAD-free over 5 years of follow up. There was a trend towards a healthier peptide pattern at baseline in the healthy control samples, but more interestingly on survival analysis the CAD panel was predictive of later cardiovascular events in patients who were asymptomatic at baseline [54]. The potential of multi-marker proteomic panels for detection of subclinical disease is evident, it remains to be seen whether the CAD238 panel can outperform other cardiovascular biomarkers and inform early therapeutic interventions.

Heart Failure

Tissue Studies

Although no large scale clinical studies using proteomic biomarkers in heart failure have been published to date, a number of projects have focussed on human samples rather than animal or cell models taking the techniques a step closer to the clinical realm.

A decade ago the first study mapping the human heart proteome was published [55]. Here researchers used a variety of separation and quantification technologies including 2D-gel electrophoresis, SELDI and various other MS methodologies in an effort to catalogue and identify human heart proteins – a platform from which to begin exploration of the healthy heart as well as different disease states. This work resulted in identification of 388 peptides from 110 unique proteins. The vast majority of these were mitochondria-associated, whilst a much smaller number than expected were involved in muscle contraction and regulation of heart rate [55]. It is clear that the heart has high energy demands, and so the large number of mitochondrial proteins may not be entirely unexpected. In addition the processing of tissue prior to peptide separation and identification may also have an effect of the proteome-a factor that should be borne in mind when considering the translation of any such results into clinical practice.

Studies using samples of diseased human heart have also been performed. Using 2D-gel electrophoresis coupled to MALDI-TOF MS a comparison of the proteome of failing myocardium in patients with arrhythmogenic right ventricular cardiomyopathy (ARVC) revealed 35 proteins commonly altered in failing hearts, 5 of which were altered by more than 1.5-fold. These proteins belong to a range of functional classes including: cytoskeletal and myofibrillar proteins; stress proteins; energy metabolism; and antioxidants. The most altered protein was identified as heat shock protein 70 (HSP70), belonging to the stress protein family and shown to be increased 1.64-fold in failing compared to non-failing hearts, and across different underlying aetiologies including dilated and ischaemic cardiomyopathy [56]. This discovery based approach highlights the utility of these studies in identification of potential novel biomarkers, however as discussed above one must consider the influence of pre-analytical processing on the human heart proteome, as well as the invasive nature of tissue sampling. Utilisation of samples which require less pre-analytical processing and can be collected in a minimally invasive manner is more appealing when considering clinical application of these techniques.

Plasma Studies

Human plasma samples have been central to a number of proteomic-based heart failure studies. A number of these projects focus on detection of left ventricular remodeling following myocardial infarction [57, 58], a strong but clinically silent predictor of subsequent heart failure [59]. As with many other cardiovascular diseases there are a number of mechanisms implicated in this process including fibrosis, proteolytic activity and apoptosis and its pathogenesis remains incompletely understood making exploratory proteomic work an attractive prospect for discovery of novel disease pathways [59]. In one such study human plasma samples from patients displaying varying degrees of remodeling, and with sequential echocardiographic follow up data available, were processed using SELDI-TOF technology coupled with LC-MS. Individuals with the greatest degree of remodeling were characterised by elevated post-translational variants of α1-chain of haptoglobin (Hpα1) which were not readily detectable by Western blot or ELISA, whilst traditional clinical markers such as creatine kinase (CK) levels did not discriminate between those with high versus low remodeling [57]. Although certainly a useful strategy for discovery of novel disease biomarkers, these finding must be reproduced and validated in an independent cohort to prove truly clinically relevant.

The majority of the plasma proteome is composed of around 20 highly abundant proteins. Analysis of the "deep" plasma proteome, which refers to the many thousands of proteins masked by these more highly abundant proteins, remains challenging as there is no readily available amplification technique for low-abundance proteins and no analytical platform is able to separate and quantify all plasma proteins [60]. In addition, we now know that immunodepletion of highly abundant proteins such as albumin can simultaneously deplete multitude of additional albumin-bound proteins [61]. Combinatorial peptide ligand library (CPLL) is an alternative approach through which investigators can simultaneously dilute high-abundance proteins and concentrate low-abundance proteins [62], and has been used to investigate the deep plasma proteome in patients with remodelling post myocardial infarction. Using human plasma samples CPLL has been shown to be reproducible and offer improved intensity of low abundance proteins; in fact in one study the majority of differentially expressed peptides between patients with and without remodelling were not detectable in plasma samples prior to CPLL processing. In particular, the N-terminal human albumin fragment was shown to be significantly down-regulated in high remodelling patients over a 1-year follow up period [58] and has been proposed as a potential biomarker of left ventricular remodelling.

Urine Studies

Recent years have seen escalating interest in diastolic dysfunction; a condition with a prevalence of up to 27 % of asymptomatic patients undergoing echocardiography [63] for which no widely applicable screening test or effective treatment is currently available. Clinical proteomic work has been performed with the aim of identifying novel biomarkers for early diagnosis of the disease and novel disease mechanisms which could inform treatment strategies. Urine represents perhaps the most easily collectable and stable sample type for proteomic studies, and requires little in the way of pre-analytical processing. Through analysis of the urinary proteome using CE-MS, a panel of 85 urinary peptides was compiled which distinguished hypertensive patients with asymptomatic diastolic dysfunction from control subjects with 69 % sensitivity and 94 % specificity [64]. The majority of these peptides were identified as varying fragments of collagens which certainly seems physiologically plausible given that heart failure is characterised by increased interstitial deposition and cross-linking of collagen [65]. Although combining a larger number of biomarkers results in improved accuracy and robustness, these findings will again require further validation and prospective testing in a larger scale independent cohort.

Cardiovascular Surgery

The potential of clinical proteomics has also been evaluated in the context of cardiovascular surgery. Pioneering work has been done on investigating heart failure, in particular dilated cardiomyopathy [66, 67]. Most forms of heart failure are preceded by a clinically silent period of myocardial remodelling which can be reversible if detected early and appropriate therapeutic measures are introduced. Beyond this stage progressive exhaustion of myocardial energy resources; alterations in the cardiomyocytes; and changes in the extracellular matrix can lead to more severe heart failure. The optimal time for an invasive cardiovascular intervention such as myocardial revascularization, heart valve reconstruction or replacement (mitral and/or aortic valve) can be defined as the time point in the course of the disease at which all myocardial changes are still completely reversible [68]. Clinical parameters such as symptoms, echocardiography or haemodynamic factors do often not precisely determine the point at which an individual transitions to irreversible disease.

Proteomic techniques might provide an important contribution in determining the precise molecular pathways but also the timeline that leads to this specific point of no return in the development of heart failure

Heart transplantation has been the gold standard for patients with severe chronic heart failure. Long-term outcome after heart transplantation mainly depends on the development of cardiac allograft vasculopathy and rejection. While right ventricular endomyocardial biopsy serves as an invasive source for diagnostic analysis, proteomic techniques might serve as a powerful non-invasive tool monitoring the cardiac allograft and detecting patterns associated with acute and/or chronic rejection [69]. Shortage of donor organs has led to increased use of mechanical left ventricular assist devices. These devices have become an important tool to stabilize the condition of patients with refractory end-stage heart failure awaiting transplantation, or even as definitive therapy without further transplantation. It has been suggested that in some instances mechanical unloading of the heart provides enough improvement in heart function that transplantation is no longer necessary [66, 68]. Again, proteomic techniques could provide a suitable diagnostic, monitoring or even prognostic tool.

A number of studies exploring proteomic biomarkers as tools to determine the optimal time-point for invasive therapeutic interventions, e.g. aortic valve replacement/reconstruction, mitral valve replacement/reconstruction and thoracic aortic replacement have already been published [68, 70]. Clinicians mainly focus on serological biomarker discovery; identifying peripheral markers which could then serve as an indicator for the disease, the prognosis and the likely response to the therapeutic intervention.

Biomarkers of upcoming events such as myocardial ischemia or infarction, heart failure, aortic aneurysm rupture or dissection are rare but much needed. Such biomarkers would allow risk stratification for patients who stand to benefit most from further diagnostics and monitoring. Again, proteomics could play a key role in this process. A preliminary study on ascending aortic aneurysms revealed that high-throughput proteomic analysis identified biomarkers which could potentially indicate imminent aortic dissection or rupture [71]. Another study found that proteomic patterns in preoperatively collected serum samples discriminated between patients who developed multisystem organ failure after thoraco-abdominal aortic repair and those who did not [72]. Clinical translation of such biomarkers, if they can be validated in large cohorts, could translate into a meaningful increase in survival through facilitating earlier intervention in "at-risk" individuals.

From a surgical point of view, it is fascinating that proteomic methods have the potential to improve invasive procedures through characterising molecular changes which result from specific interventions. For example, a recent study on the use of deep hypothermia (<25 °C) during aortic surgery revealed, using iTRAQ labeling tandem mass spectrometry, that a significant upregulation of complement activation occurs in those with normothermic cardiopulmonary bypass compared to those with deep hypothermia [73]. In contrast, it also showed that rewarming the patient potentially exerts a significant effect on the plasma proteome as well. This could at least in part explain some of the difficulties seen in those with deep hypothermic surgery, e.g. hemorrhagic complications, neurological complications, systemic inflammatory response syndrome.

Stroke

Prompt and accurate diagnosis is essential for the timely management of stroke, particularly in the context of ischaemic events where there is a narrow window of opportunity for administration of thrombolytic therapy. The initial diagnosis is made based on clinical presentation, however even following specialist assessment by a stroke physician up to 20 % of patients are found to have an alternative diagnosis [74]. The ability to accurately confirm the diagnosis at an early stage in situations where there is a degree of diagnostic doubt would facilitate timely and appropriate intervention; use of proteomics is one possible approach to this unmet clinical need however its potential remains largely unexplored in human subjects.

The spontaneously hypertensive stroke-prone rat (SHRSP) is a useful animal model for exploration of complex cerebrovascular pathology [75]. Early proteomic work using this model highlighted increased urinary excretion of several acute phase proteins prior to the onset of cerebrovascular disease, suggesting that a picture of increased systemic inflammation may precede clinical events [76]. 2D-gel electrophoresis identified increased abundance of transferrin; α2-HS-glycoprotein; α1-antitrypsin and fragments of albumin in these animals [76]. Over a decade later, examination of brain homogenates from Wistar rats exposed to middle cerebral artery occlusion using gel electrophoresis with MALDI-TOF MS offered some mechanistic insights into the pleiotropic effects of statin therapy in stroke patients. Simvastatin was associated with reduced heat shock protein 75 (HSP75) in these animals, a finding which was later confirmed in plasma from patients following acute stroke [77], identifying another potential biomarker worthy of more detailed exploration in clinical studies.

A recent study used CE-MS to analyse urine samples from patients within 24 h of ischaemic stroke compared to healthy controls with the aim of developing a biomarker model to aid in the diagnosis of acute stroke. The resultant panel of 35 urinary peptides, including uromodulin and a number of collagen fragments, was able to separate stroke cases from controls with 93 % specificity [78]. Not only do these findings indicate that urinary proteomics could become a useful tool in early diagnosis of acute stroke, but they demonstrate again the concept that classifiers based on a larger number of biomarkers can offer a high degree of diagnostic accuracy.

Chronic Kidney Disease

The prevalence of chronic kidney disease (CKD) continues to rise fuelled by increasing prevalence of conditions such as diabetes and hypertension, however only a small proportion of patients progress to end stage renal disease (ESRD) with many dying from associated conditions before reaching renal replacement therapy

(RRT) [79]. The presence of CKD is a strong independent predictor of adverse clinical outcomes and cardiovascular disease in particular [80, 81]. Even the earliest, preclinical stages of disease confer an increase in cardiovascular morbidity and mortality [82]. The current clinical definition of CKD relies upon detecting reduction in glomerular filtration rate (GFR) based on serum creatinine or an increase in protein or albumin leak. However by the time such changes are detectable according to current clinical thresholds, substantial and irreversible structural damage has already occurred in relative clinical silence [83, 84]. Intervention with drug therapies which block the renin-angiotensin system (RAAS) is currently recommended as first line therapy for slowing progression of CKD [85, 86] however studies of early or more aggressive intervention have generally been disappointing due to a large burden of side effects and adverse events [87]. Development of tools for early diagnosis of CKD and prediction of outcomes to allow targeted intervention for those who stand to benefit most in terms of minimising risk of both ESRD and cardiovascular disease remains an unmet clinical need. Proteomic research appears to be a promising strategy to address this issue.

Plasma Studies

The molecular mechanisms involved in CKD are complex and likely include coagulation defects; vascular calcification; oxidative stress; inflammation and endothelial dysfunction [88, 89]. Many previous studies have reported changes in individual plasma proteins in CKD, however a larger scale approach has the potential to bring together disparate molecular evidence providing an integrated view of CKD pathogenesis. Application of untargeted plasma proteome analysis using LC-MS in samples from patients with CKD recently confirmed modifications to a number of these pathways known to be integral to CKD pathogenesis, but also identified increased lysosome C and leucine-rich-alpha-2 glycoprotein which are related to vascular disease and heart failure [90] therefore contributing to new knowledge of the link between the kidney and cardiovascular disease.

Urine Studies

Urine has proved a particularly useful medium in proteomic studies of CKD. It is more stable than blood in terms of proteolytic degradation, requires less pre-analytical processing and is largely representative of renal pathophysiology [91]. An array of urinary biomarkers for accurate diagnosis of CKD was first developed in 2010 from urine samples from 230 patients with CKD and 379 healthy control subjects analysed by using the CE-MS platform. A panel of 273 peptides associated with CKD was identified and combined using support vector machine-based modelling to create a specific classifier (CKD273). Validation in a blinded test set resulted in accurate separation of cases and controls with high sensitivity and specificity [92]. To date, this is the proteomic classifier whose diagnostic utility has been most rigorously tested in independent studies, both case–control and longitudinally. Its component peptides include many derived from collagen types I and III, uromodulin, alpha-1-antitrypsin and osteopontin and there is significant overlap with other cardiovascular proteomic panels, likely reflecting the multiple common mechanistic pathways underpinning these conditions and providing a framework for moving towards mechanistic hypotheses.

The prognostic potential of CKD273 has been evaluated in a number of studies. In a cohort of patients attending renal outpatient clinics with diagnosis of CKD, baseline CKD273 classifier score was significantly higher in those who reached the primary endpoint of ESRD or death during the subsequent 3.6 year follow up [93]. Of course guidance also requires that proposed biomarkers outperform current clinical practice. A recent analysis of CKD273 in a large cross sectional multicentre cohort study showed that CKD273 performed better that urinary albumin excretion (UAE) as a predictor of CKD progression. Adding CKD273 to a combination of eGFR and UAE significantly improved prediction of CKD risk [94], suggesting that CKD273 has the potential to outperform the current state-of-the-art, although large scale prospective studies to formally assess this are lacking.

Diabetic Nephropathy

Diabetic nephropathy is the most common precipitant of ERSD in the western world and has therefore attracted specific interest in CKD research. Microalbuminuria (MA) is the earliest clinical index of DN and is also associated with significantly increased cardiovascular disease burden [95–97] even at levels below the currently accepted diagnostic threshold [98]. The pathogenesis of DN is complex and incompletely understood, sharing multiple common underlying mechanistic pathways with cardiovascular disease. Multimarker proteomic approaches are therefore well-suited to investigate the disease mechanisms behind DN and to identify potential biomarkers for early diagnosis, and are in fact advancing beyond the discovery and validation stage into prospective clinical studies in this field. The foresight of many research groups has been pivotal to the discovery of disease-specific biomarkers and the availability of sequential samples from large diabetes cohorts with long follow-up has allowed researchers to compare the proteome of patients many years before clinically detectable renal insufficiency develops.

CE-MS analysis of the urinary proteome has been the prominent proteomic strategy for DN biomarker discovery and a number of disease-specific peptide panels have been described [99–101]. Although developed in samples from patients with CKD of diverse aetiology the CKD273 classifier has proved particularly promising in this arena. Not only has it been shown to distinguish CKD from healthy controls [92], but the panel appears to accurately identify DN with high consistency across multiple centres, independent of confounding factors such as age and gender [102] and to remain stable even after 3 years of storage [101]. As with a number of other disease-specific panels, transition towards a "healthier" proteome has been demonstrated following 2 years of treatment according to evidence based guidelines [103].

This classifier has also been demonstrated to have some prognostic promise, identifying normoalbuminuric subjects likely to progress to overt DN in advance of clinical detection of MA in separate clinical cohorts with between 3 and 15 years follow up data available [101, 104]. In terms of outperforming the current state-of-the-art, one study has shown that CKD273 had greater sensitivity for detection of DN in type 2 diabetes compared to MA [101]. Its capability for early diagnosis of DN and utility in informing therapeutic decision-making now being prospectively tested for the first time in a large scale randomised clinical trial [105].

Summary and Perspectives

Like other novel approaches to unravel the pathophysiology of CVD and to translate such findings to clinical practice clinical cardiovascular proteomics is still characterised by a number of limitations and weaknesses. The major challenges at this time are small sample sizes of clinical proteomic studies, failure to verify results independently, and lack of expected clinical utility; which all have the potential to lead to over-interpretation of data and unjustified enthusiasm. On the clinical side, pathophysiological concepts and novel biomarkers that lack of appropriate interventions, do not take patient factors and patient wishes into account and have an unfavourable cost effectiveness will not be considered useful in the management of patients with CVD. Nevertheless proteomic strategies have matured in recent years and some clinical conditions may soon benefit from the potential that proteomics can offer whereas other conditions will need considerably more time to see clinically relevant proteomic strategies (Table 17.2).

Table 17.2 Current state of proteomic biomarker development in cardiovascular medicine

Condition	Proof of concept studies in humans	Biomarkers validated in independent cohort	Incremental value above current best practice	Demonstrated clinical utility	Impact on clinical outcomes tested prospectively in RCT
Atherosclerosis	✓	(✓)	×	×	×
Hypertension	×	×	×	×	×
Heart failure	✓	×	×	×	×
Stroke	✓	×	×	×	×
Coronary artery disease	✓	✓	×	×	×
Chronic kidney disease	✓	✓	(✓)	(✓)	(✓)

RCT randomised clinical trial
✓indicates the presence of at least two independent high quality studies
(✓) indicates the availability of one high quality or a number of lower-quality studies
× indicates the absence of substantial research output in this area

The proteomic community has made significant progress in defining minimal standards for reporting of data (minimum information about a proteomics experiment; MIAPE [106] and criteria for implementation of proteomic biomarkers into clinical practice have been defined [22]. It is now important that individual groups adhere to such criteria and that high-quality studies will be designed in collaboration between clinicians and proteomic experts to address clinically relevant questions. Recent examples have demonstrated the potential of proteomics to revolutionise clinical medicine and it is now the responsibility of the clinical proteomic community to make it work.

References

1. Muller-Nordhorn J, Binting S, Roll S, Willich SN. An update on regional variation in cardiovascular mortality within Europe. Eur Heart J. 2008;29(10):1316–26.
2. Mozaffarian D, Benjamin EJ, Go AS, Arnett DK, Blaha MJ, Cushman M, de Ferranti S, Despres JP, Fullerton HJ, Howard VJ, Huffman MD, Judd SE, Kissela BM, Lackland DT, Lichtman JH, Lisabeth LD, Liu S, Mackey RH, Matchar DB, McGuire DK, Mohler III ER, Moy CS, Muntner P, Mussolino ME, Nasir K, Neumar RW, Nichol G, Palaniappan L, Pandey DK, Reeves MJ, Rodriguez CJ, Sorlie PD, Stein J, Towfighi A, Turan TN, Virani SS, Willey JZ, Woo D, Yeh RW, Turner MB. Heart disease and stroke statistics--2015 update: a report from the American Heart Association. Circulation. 2015;131(4):e29–322.
3. Abegunde DO, Mathers CD, Adam T, Ortegon M, Strong K. The burden and costs of chronic diseases in low-income and middle-income countries. Lancet. 2007;370(9603):1929–38.
4. Hense HW, Schulte H, Lowel H, Assmann G, Keil U. Framingham risk function overestimates risk of coronary heart disease in men and women from Germany–results from the MONICA Augsburg and the PROCAM cohorts. Eur Heart J. 2003;24(10):937–45.
5. Conroy RM, Pyorala K, Fitzgerald AP, Sans S, Menotti A, De BG, De BD, Ducimetiere P, Jousilahti P, Keil U, Njolstad I, Oganov RG, Thomsen T, Tunstall-Pedoe H, Tverdal A, Wedel H, Whincup P, Wilhelmsen L, Graham IM. Estimation of ten-year risk of fatal cardiovascular disease in Europe: the SCORE project. Eur Heart J. 2003;24(11):987–1003.
6. Lloyd-Jones DM. Cardiovascular risk prediction: basic concepts, current status, and future directions. Circulation. 2010;121(15):1768–77.
7. Biomarkers Definitions Working Group. Biomarkers and surrogate endpoints: preferred definitions and conceptual framework. Clin Pharmacol Ther. 2001;69(3):89–95.
8. Zimmerli LU, Schiffer E, Zurbig P, Good DM, Kellmann M, Mouls L, Pitt AR, Coon JJ, Schmieder RE, Peter KH, Mischak H, Kolch W, Delles C, Dominiczak AF. Urinary proteomic biomarkers in coronary artery disease. Mol Cell Proteomics. 2008;7(2):290–8.
9. Delles C, Schiffer E, von zur Muhlen C, Peter K, Rossing P, Parving H, Dymott J, Neisius U, Zimmerli L, Snell-Bergeon J, Maahs D, Schmeider R, Mischak H, Dominiczak A. Urinary proteomic diagnosis of coronary artery disease: identification and clinical validation in 623 individuals. J Hypertens. 2010;28(11):2316–22.
10. Ho CY, Lopez B, Coelho-Filho OR, Lakdawala NK, Cirino AL, Jarolim P, Kwong R, Gonzalez A, Colan SD, Seidman JG, Diez J, Seidman CE. Myocardial fibrosis as an early manifestation of hypertrophic cardiomyopathy. N Engl J Med. 2010;363(6):552–63.
11. Skeik N, Patel DC. A review of troponins in ischemic heart disease and other conditions. Int J Angiol. 2007;16(2):53–8.
12. Omland T, de Lemos JA, Sabatine MS, Christophi CA, Rice MM, Jablonski KA, Tjora S, Domanski MJ, Gersh BJ, Rouleau JL, Pfeffer MA, Braunwald E. A sensitive cardiac troponin T assay in stable coronary artery disease. N Engl J Med. 2009;361(26):2538–47.

13. Kaptoge S, Di AE, Pennells L, Wood AM, White IR, Gao P, Walker M, Thompson A, Sarwar N, Caslake M, Butterworth AS, Amouyel P, Assmann G, Bakker SJ, Barr EL, Barrett-Connor E, Benjamin EJ, Bjorkelund C, Brenner H, Brunner E, Clarke R, Cooper JA, Cremer P, Cushman M, Dagenais GR, D'Agostino Sr RB, Dankner R, Davey-Smith G, Deeg D, Dekker JM, Engstrom G, Folsom AR, Fowkes FG, Gallacher J, Gaziano JM, Giampaoli S, Gillum RF, Hofman A, Howard BV, Ingelsson E, Iso H, Jorgensen T, Kiechl S, Kitamura A, Kiyohara Y, Koenig W, Kromhout D, Kuller LH, Lawlor DA, Meade TW, Nissinen A, Nordestgaard BG, Onat A, Panagiotakos DB, Psaty BM, Rodriguez B, Rosengren A, Salomaa V, Kauhanen J, Salonen JT, Shaffer JA, Shea S, Ford I, Stehouwer CD, Strandberg TE, Tipping RW, Tosetto A, Wassertheil-Smoller S, Wennberg P, Westendorp RG, Whincup PH, Wilhelmsen L, Woodward M, Lowe GD, Wareham NJ, Khaw KT, Sattar N, Packard CJ, Gudnason V, Ridker PM, Pepys MB, Thompson SG, Danesh J. C-reactive protein, fibrinogen, and cardiovascular disease prediction. N Engl J Med. 2012;367(14):1310–20.
14. Apple FS, Henry TD, Berger CR, Landt YA. Early monitoring of serum cardiac troponin I for assessment of coronary reperfusion following thrombolytic therapy. Am J Clin Pathol. 1996;105(1):6–10.
15. Sherwood MW, Kristin NL. High-sensitivity troponin assays: evidence, indications, and reasonable use. J Am Heart Assoc. 2014;3(1):e000403.
16. Januzzi JL, Troughton R. Are serial BNP measurements useful in heart failure management? Serial natriuretic peptide measurements are useful in heart failure management. Circulation. 2013;127(4):500–7.
17. Desai AS. Are serial BNP measurements useful in heart failure management? Serial natriuretic peptide measurements are not useful in heart failure management: the art of medicine remains long. Circulation. 2013;127(4):509–16.
18. Frank R, Hargreaves R. Clinical biomarkers in drug discovery and development. Nat Rev Drug Discov. 2003;2(7):566–80.
19. Hlatky MA, Greenland P, Arnett DK, Ballantyne CM, Criqui MH, Elkind MS, Go AS, Harrell Jr FE, Hong Y, Howard BV, Howard VJ, Hsue PY, Kramer CM, McConnell JP, Normand SL, O'Donnell CJ, Smith Jr SC, Wilson PW. Criteria for evaluation of novel markers of cardiovascular risk: a scientific statement from the American Heart Association. Circulation. 2009;119(17):2408–16.
20. Wang TJ. Assessing the role of circulating, genetic, and imaging biomarkers in cardiovascular risk prediction. Circulation. 2011;123(5):551–65.
21. Wilkins MR, Pasquali C, Appel RD, Ou K, Golaz O, Sanchez JC, Yan JX, Gooley AA, Hughes G, Humphery-Smith I, Williams KL, Hochstrasser DF. From proteins to proteomes: large scale protein identification by two-dimensional electrophoresis and amino acid analysis. Biotechnology (N Y). 1996;14(1):61–5.
22. Mischak H, Ioannidis JP, Argiles A, Attwood TK, Bongcam-Rudloff E, Broenstrup M, Charonis A, Chrousos GP, Delles C, Dominiczak A, Dylag T, Ehrich J, Egido J, Findeisen P, Jankowski J, Johnson RW, Julien BA, Lankisch T, Leung HY, Maahs D, Magni F, Manns MP, Manolis E, Mayer G, Navis G, Novak J, Ortiz A, Persson F, Peter K, Riese HH, Rossing P, Sattar N, Spasovski G, Thongboonkerd V, Vanholder R, Schanstra JP, Vlahou A. Implementation of proteomic biomarkers: making it work. Eur J Clin Invest. 2012;42(9):1027–36.
23. Yassine H, Borges CR, Schaab MR, Billheimer D, Stump C, Reaven P, Lau SS, Nelson R. Mass spectrometric immunoassay and MRM as targeted MS-based quantitative approaches in biomarker development: potential applications to cardiovascular disease and diabetes. Proteomics Clin Appl. 2013;7(7–8):528–40.
24. Lindsey ML, Mayr M, Gomes AV, Delles C, Arrell DK, Murphy AM, Lange RA, Costello CE, Jin YF, Laskowitz DT, Sam F, Terzic A, Van EJ, Srinivas PR. Transformative impact of proteomics on cardiovascular health and disease: a scientific statement from the American Heart Association. Circulation. 2015;132(9):852–72.
25. Jacobs JM, Adkins JN, Qian WJ, Liu T, Shen Y, Camp DG, Smith RD. Utilizing human blood plasma for proteomic biomarker discovery. J Proteome Res. 2005;4(4):1073–85.

26. Good DM, Thongboonkerd V, Novak J, Bascands JL, Schanstra JP, Coon JJ, Dominiczak A, Mischak H. Body fluid proteomics for biomarker discovery: lessons from the past hold the key to success in the future. J Proteome Res. 2007;6(12):4549–55.
27. Fach EM, Garulacan LA, Gao J, Xiao Q, Storm SM, Dubaquie YP, Hefta SA, Opiteck GJ. In vitro biomarker discovery for atherosclerosis by proteomics. Mol Cell Proteomics. 2004;3(12):1200–10.
28. Gonzalez-Cabrero J, Pozo M, Duran MC, de Nicolás R, Egido J, Vivanco F. The proteome of endothelial cells. Methods Mol Biol. 2007;357:181–98.
29. Boccardi C, Cecchettini A, Caselli A, Camici G, Evangelista M, Mercatanti A, Rainaldi G, Citti L. A proteomic approach to the investigation of early events involved in vascular smooth muscle cell activation. Cell Tissue Res. 2007;328(1):185–95.
30. Duran MC, Mas S, Martin-Ventura JL, Meilhac O, Michel JB, Gallego-Delgado J, Lazaro A, Tunon J, Egido J, Vivanco F. Proteomic analysis of human vessels: application to atherosclerotic plaques. Proteomics. 2003;3(6):973–8.
31. de la Cuesta F, Barderas MG, Calvo E, Zubiri I, Maroto AS, Darde VM, Martin-Rojas T, Gil-Dones F, Posada-Ayala M, Tejerina T, Lopez JA, Vivanco F, Alvarez-Llamas G. Secretome analysis of atherosclerotic and non-atherosclerotic arteries reveals dynamic extracellular remodeling during pathogenesis. J Proteomics. 2012;75(10):2960–71.
32. Martin-Ventura JL, Duran MC, Blanco-Colio LM, Meilhac O, Leclercq A, Michel JB, Jensen ON, Hernandez-Merida S, Tunon J, Vivanco F, Egido J. Identification by a differential proteomic approach of heat shock protein 27 as a potential marker of atherosclerosis. Circulation. 2004;110(15):2216–9.
33. Lepedda AJ, Cigliano A, Cherchi GM, Spirito R, Maggioni M, Carta F, Turrini F, Edelstein C, Scanu AM, Formato M. A proteomic approach to differentiate histologically classified stable and unstable plaques from human carotid arteries. Atherosclerosis. 2009;203(1):112–8.
34. Donners MM, Verluyten MJ, Bouwman FG, Mariman EC, Devreese B, Vanrobaeys F, van Beeumen J, van den Akker LH, Daemen MJ, Heeneman S. Proteomic analysis of differential protein expression in human atherosclerotic plaque progression. J Pathol. 2005;206(1):39–45.
35. Kardys I, Rifai N, Meilhac O, Michel JB, Martin-Ventura JL, Buring JE, Libby P, Ridker PM. Plasma concentration of heat shock protein 27 and risk of cardiovascular disease: a prospective, nested case-control study. Clin Chem. 2008;54(1):139–46.
36. de Kleijn DP, Moll FL, Hellings WE, Ozsarlak-Sozer G, de Bruin P, Doevendans PA, Vink A, Catanzariti LM, Schoneveld AH, Algra A, Daemen MJ, Biessen EA, de Jager W, Zhang H, de Vries JP, Falk E, Lim SK, van der Spek PJ, Sze SK, Pasterkamp G. Local atherosclerotic plaques are a source of prognostic biomarkers for adverse cardiovascular events. Arterioscler Thromb Vasc Biol. 2010;30(3):612–9.
37. Peeters W, de Kleijn DP, Vink A, van de Weg S, Schoneveld AH, Sze SK, van der Spek PJ, de Vries JP, Moll FL, Pasterkamp G. Adipocyte fatty acid binding protein in atherosclerotic plaques is associated with local vulnerability and is predictive for the occurrence of adverse cardiovascular events. Eur Heart J. 2011;32(14):1758–68.
38. Furuhashi M, Tuncman G, Gorgun CZ, Makowski L, Atsumi G, Vaillancourt E, Kono K, Babaev VR, Fazio S, Linton MF, Sulsky R, Robl JA, Parker RA, Hotamisligil GS. Treatment of diabetes and atherosclerosis by inhibiting fatty-acid-binding protein aP2. Nature. 2007;447(7147):959–65.
39. Delles C, Neisius U, Carty D. Proteomics in hypertension and other cardiovascular diseases. Ann Med. 2012;44 Suppl 1:S55–64.
40. Padmanabhan S, Delles C, Dominiczak A. Genetic factors in hypertension. Arch Med Sci. 2009;5(2A):S212–9.
41. Staessen J, Wang J, Bianchi G, Birkenhager W. Essential hypertension. Lancet. 2003;361:1629–41.

42. Grussenmeyer T, Meili-Butz S, Roth V, Dieterle T, Brink M, Winkler B, Matt P, Carrel TP, Eckstein FS, Lefkovits I, Grapow MT. Proteome analysis in cardiovascular pathophysiology using Dahl rat model. J Proteomics. 2011;74(5):672–82.
43. Jin X, Xia L, Wang L, Shi J, Zheng Y, Chen W, Zhang L, Liu Z, Chen G, Fang N. Differential protein expression in hypertrophic heart with and without hypertension in spontaneously hypertensive rats. Proteomics. 2006;6:1948–56.
44. Lindsey M, Goshorn D, Comte-Walters S, Hendrick J, Hapke E, Zile M, Schey K. A multidimensional proteomic approach to identify hypertrophy associated proteins. Proteomics. 2006;6:2225–35.
45. Gallego-Delgado J, Lazaro A, Osende J, Esteban V, Barderas M, Gomez-Guerrero C, Vega R, Vivanco F, Egido J. Proteomic analysis of early left ventricular hypertrophy secondary to hypertension: Modulation by antihypertensive therapies. J Am Soc Nephrol. 2006;17:S159–64.
46. Liu L, Wang W, Meng X, Gao J, Wu H, Wang P, Wu W, Wang L, Ma L, Zhang W. Left ventricular hypertrophy induced by abdominal aortic banding and its prevention by angiotensin receptor blocker Telmisartan – a proteomic analysis. J Physiol Biochem. 2010;66:329–38.
47. Scognamiglio R, Negut C, Ramondo A, Tiengo A, Avogaro A. Detection of coronary artery disease in asymptomatic patients with type 2 diabetes mellitus. J Am Coll Cardiol. 2006;47(1):65–71.
48. Donahue MP, Rose K, Hochstrasser D, Vonderscher J, Grass P, Chibout SD, Nelson CL, Sinnaeve P, Goldschmidt-Clermont PJ, Granger CB. Discovery of proteins related to coronary artery disease using industrial-scale proteomics analysis of pooled plasma. Am Heart J. 2006;152(3):478–85.
49. Laine P, Pentikainen MO, Wurzner R, Penttila A, Paavonen T, Meri S, Kovanen PT. Evidence for complement activation in ruptured coronary plaques in acute myocardial infarction. Am J Cardiol. 2002;90(4):404–8.
50. Sukhova GK, Shi GP, Simon DI, Chapman HA, Libby P. Expression of the elastolytic cathepsins S and K in human atheroma and regulation of their production in smooth muscle cells. J Clin Invest. 1998;102(3):576–83.
51. Darde VM, de la Cuesta F, Dones F, Alvarez-Llamas G, Barderas MG, Vivanco F. Analysis of the plasma proteome associated with acute coronary syndrome: does a permanent protein signature exist in the plasma of ACS patients? J Proteome Res. 2010;9:4420–32.
52. von zur Muhlen C, Schiffer E, Zurbig P, Kellman M, Brasse M, Meert N, Vanholder R, Dominiczak A, Chen Y, Mischak H, Bode C, Peter K. Evaluation of urine proteome pattern analysis for its potential to reflect coronary artery atherosclerosis in symptomatic patients. J Proteome Res. 2009;8:335–45.
53. Oparil S. The ASCOT blood pressure lowering trial. Curr Hypertens Rep. 2006;8(3):229–31.
54. Brown C, McCarthy N, Hughes A, Sever P, Stalmach A, Mullen W, Dominiczak A, Sattar N, Mischak H, Thom S, Mayet J, Stanton A, Delles C. Urinary proteomic biomarkers to predict cardiovascular events. Proteomics Clin Appl. 2015;9:610–7.
55. Westbrook JA, Wheeler JX, Wait R, Welson SY, Dunn MJ. The human heart proteome: Two-dimensional maps using narrow-range immobilised pH gradients. Electrophoresis. 2006;27(8):1547–55.
56. Wei YJ, Huang YX, Shen Y, Cui CJ, Zhang XL, Zhang H, Hu SS. Proteomic analysis reveals significant elevation of heat shock protein 70 in patients with chronic heart failure due to arrhythmogenic right ventricular cardiomyopathy. Mol Cell Biochem. 2009;332(1–2):103–11.
57. Pinet F, Beseme O, Cieniewski-Bernard C, Drobecq H, Jourdain S, Lamblin N, Amouyel P, Bauters C. Predicting left ventricular remodeling after a first myocardial infarction by plasma proteome analysis. Proteomics. 2008;8:1798–808.

58. Fertin M, Beseme O, Duban S, Amouyel P, Bauters C, Pinet F. Deep plasma proteomic analysis of patients with left ventricular remodeling after a first myocardial infarction. Proteomics Clin Appl. 2010;4:654–73.
59. St John Sutton M, Sharpe N. Left ventricular remodeling after myocardial infarction. Pathophysiology and therapy. Circulation. 2000;101:2981–8.
60. Righetti P, Boschetti E, Lomas L, Citterio A. Protein equalizer technology: the quest for a "democratic proteome". Proteomics. 2006;6:3980–92.
61. Zhou M, Lucas DA, Chan KC, Issaq HJ, Petricoin III EF, Liotta LA, Veenstra TD, Conrads TP. An investigation into the human serum "interactome". Electrophoresis. 2004;25(9):1289–98.
62. Guerrier L, Thulasiraman V, Castagna A, Fortis F, Lin S, Lomas L, Righetti PG, Boschetti E. Reducing protein concentration range of biological samples using solid-phase ligand libraries. J Chromatogr B Analyt Technol Biomed Life Sci. 2006;833(1):33–40.
63. Kuznetsova T, Herbots L, Lopez B, Jin Y, Richart T, Thijs L, Gonzalez A, Herregods M, Fagard R, Diez J, Staessen J. Prevalence of left ventricular diastolic dysfunction in a general population. Circ Heart Fail. 2009;2:105–12.
64. Kuznetsova T, Mischak H, Mullen W, Staessen J. Urinary proteome analysis in hypertensive patients with left ventricular diastolic dysfunction. Eur Heart J. 2012;33:2342–50.
65. Burlew B, Weber K. Cardiac fibrosis as a cause of diastolic dysfunction. Herz. 2002;27:92–8.
66. McGregor E, Dunn MJ. Proteomics of the heart: unraveling disease. Circ Res. 2006;98(3):309–21.
67. Birks EJ, Owen VJ, Burton PB, Bishop AE, Banner NR, Khaghani A, Polak JM, Yacoub MH. Tumor necrosis factor-alpha is expressed in donor heart and predicts right ventricular failure after human heart transplantation. Circulation. 2000;102(3):326–31.
68. Matt P, Carrel T, White M, Lefkovits I, Van EJ. Proteomics in cardiovascular surgery. J Thorac Cardiovasc Surg. 2007;133(1):210–4.
69. Zerkowski HR, Grussenmeyer T, Matt P, Grapow M, Engelhardt S, Lefkovits I. Proteomics strategies in cardiovascular research. J Proteome Res. 2004;3(2):200–8.
70. Oda T, Matsumoto KI. Proteomic analysis in cardiovascular research. Surg Today. 2016;46(3):285–96.
71. Black KM, Masuzawa A, Hagberg RC, Khabbaz KR, Trovato ME, Rettagliati VM, Bhasin MK, Dillon ST, Libermann TA, Toumpoulis IK, Levitsky S, McCully JD. Preliminary biomarkers for identification of human ascending thoracic aortic aneurysm. J Am Heart Assoc. 2013;2(6):e000138.
72. Feezor RJ, Baker HV, Xiao W, Lee WA, Huber TS, Mindrinos M, Kim RA, Ruiz-Taylor L, Moldawer LL, Davis RW, Seeger JM. Genomic and proteomic determinants of outcome in patients undergoing thoracoabdominal aortic aneurysm repair. J Immunol. 2004;172(11):7103–9.
73. Oda T, Yamaguchi A, Yokoyama M, Shimizu K, Toyota K, Nikai T, Matsumoto K. Plasma proteomic changes during hypothermic and normothermic cardiopulmonary bypass in aortic surgeries. Int J Mol Med. 2014;34(4):947–56.
74. Libman RB, Wirkowski E, Alvir J, Rao TH. Conditions that mimic stroke in the emergency department. Implications for acute stroke trials. Arch Neurol. 1995;52(11):1119–22.
75. Okamoto K, Yamori Y, Nagaoka A. Establishment of the stroke-prone spontaneously hypertensive rats (SHR). Circ Res 1974;33/34:143–53.
76. Sironi L, Tremoli E, Miller I, Guerrini U, Calvio AM, Eberini I, Gemeiner M, Asdente M, Paoletti R, Gianazza E. Acute-phase proteins before cerebral ischemia in stroke-prone rats: identification by proteomics. Stroke. 2001;32(3):753–60.
77. Campos-Martorell M, Salvador N, Monge M, Canals F, Garcia-Bonilla L, Hernandez-Guillamon M, Ayuso MI, Chacon P, Rosell A, Alcazar A, Montaner J. Brain proteomics identifies potential simvastatin targets in acute phase of stroke in a rat embolic model. J Neurochem. 2014;130(2):301–12.

78. Dawson J, Walters M, Delles C, Mischak H, Mullen W. Urinary proteomics to support diagnosis of stroke. PLoS One. 2012;7(5):e35879.
79. Keith D, Nichols G, Gullion C, Betz-Brown J, Smith D. Longitudinal follow up and outcomes among a population with chronic kidney disease in a large managed care organisation. Arch Intern Med. 2004;164:659–63.
80. Go A, Chertow G, Fan D, McCulloch C, Hsu C. Chronic kidney disease and the risks of death, cardiovascular events and hospitalization. N Engl J Med. 2004;351:1296–305.
81. Hemmelgarn B, Manns B, Lloyd A, James M, Klarenbach S, Quinn R, Wiebe N, Tonelli M, for The Alberta Kidney Disease Network. Relation between kidney function, proteinuria and adverse outcomes. JAMA. 2010;303:423–9.
82. Di Angelantonio E, Chowdhury R, Sarwar N, Aspelund T, Danesh J, Gudnason V. Chronic kidney disease and risk of major cardiovascular disease and non-vascular mortality: prospective population based cohort study. Br Med J. 2010;341:c4986.
83. Perrone R, Madias N, Levey A. Serum creatinine as an index of renal function: new insights into old concepts. Clin Chem. 1992;38:1933–53.
84. Fioretto P, Steffes M, Mauer S. Glomerular structure in nonproteinuric insulin-dependent diabetic patients with various levels of microalbuminuria. Diabetes. 1994;43:1358–64.
85. KDIGO. KDIGO 2012 clinical practice guideline for evaluation and management of chronic kidney disease. Kidney Int Suppl. 2013;3:73–90.
86. National Institute for Health and Care Excellence (NICE). Type 2 diabetes in adults: management. https://www.nice.org.uk/guidance/ng28. Accessed 2 May 2016.
87. Navaneethan SD, Nigwekar SU, Sehgal AR, Strippoli GF. Aldosterone antagonists for preventing the progression of chronic kidney disease: a systematic review and meta-analysis. Clin J Am Soc Nephrol. 2009;4(3):542–51.
88. Afsar B, Turkmen K, Covic A, Kanbay M. An update on coronary artery disease and chronic kidney disease. Int J Nephrol. 2014;2014:767424.
89. Nguyen-Khoa T, Massy ZA, De Bandt JP, Kebede M, Salama L, Lambrey G, Witko-Sarsat V, Drueke TB, Lacour B, Thevenin M. Oxidative stress and haemodialysis: role of inflammation and duration of dialysis treatment. Nephrol Dial Transplant. 2001;16(2):335–40.
90. Glorieux G, Mullen W, Duranton F, Filip S, Gayrard N, Husi H, Schepers E, Neirynck N, Schanstra J, Jankowski J, Mischak H, Argiles A, Vanholder R, Vlahou A, Klein J. New insights in molecular mechanisms involved in chronic kidney disease using high-resolution plasma proteome analysis. Nephrol Dial Transplant. 2015;30(11):1842–52.
91. Mullen W, Delles C, Mischak H, members of the EuroKUP COST Action. Urinary proteomics in the assessment of chronic kidney disease. Curr Opin Nephrol Hypertens. 2011;20:654–61.
92. Good DM, Zurbig P, Argiles A, Bauer HW, Behrens G, Coon JJ, Dakna M, Decramer S, Delles C, Dominiczak AF, Ehrich JH, Eitner F, Fliser D, Frommberger M, Ganser A, Girolami MA, Golovko I, Gwinner W, Haubitz M, Herget-Rosenthal S, Jankowski J, Jahn H, Jerums G, Julian BA, Kellmann M, Kliem V, Kolch W, Krolewski AS, Luppi M, Massy Z, Melter M, Neususs C, Novak J, Peter K, Rossing K, Rupprecht H, Schanstra JP, Schiffer E, Stolzenburg JU, Tarnow L, Theodorescu D, Thongboonkerd V, Vanholder R, Weissinger EM, Mischak H, Schmitt-Kopplin P. Naturally occurring human urinary peptides for use in diagnosis of chronic kidney disease. Mol Cell Proteomics. 2010;9(11):2424–37.
93. Argiles A, Siwy J, Duranton F, Gayrard N, Dakna M, Lundin U, Osaba L, Delles C, Mourad G, Weinberger K, Mischak H. CKD 273, a new proteomics classifier assessing CKD and its prognosis. PLoS One. 2013;8(5):e62837.
94. Schanstra JP, Zurbig P, Alkhalaf A, Argiles A, Bakker SJ, Beige J, Bilo HJ, Chatzikyrkou C, Dakna M, Dawson J, Delles C, Haller H, Haubitz M, Husi H, Jankowski J, Jerums G, Kleefstra N, Kuznetsova T, Maahs DM, Menne J, Mullen W, Ortiz A, Persson F, Rossing P, Ruggenenti P, Rychlik I, Serra AL, Siwy J, Snell-Bergeon J, Spasovski G, Staessen JA, Vlahou A, Mischak H, Vanholder R. Diagnosis and prediction of CKD progression by assessment of urinary peptides. J Am Soc Nephrol. 2015;26(8):1999–2010.

95. Mogensen C. Microalbuminuria predicts clinical proteinuria and early mortality in maturity onset diabetes. N Engl J Med. 1984;310:356–60.
96. Deckert T, Feldt-Rasmussen B, Borch-Johnsen K, Jensen T, Kofoed-Enevoldsen A. Albuminuria reflects widespread vascular damage. The Steno hypothesis. Diabetologia. 1989;32(4):219–26.
97. Bruno G, Merletti F, Bargero G, Novelli G, Melis D, Soddu A, Perotto M, Pagano G, Cavallo-Perin P. Estimated glomerular filtration rate, albuminuria and mortality in type 2 diabetes: the Casale Monferrato study. Diabetologia. 2007;50:941–8.
98. Ninomiya T, Perkovic V, de Galan BE, Zoungas S, Pillai A, Jardine M, Patel A, Cass A, Neal B, Poulter N, Mogensen CE, Cooper M, Marre M, Williams B, Hamet P, Mancia G, Woodward M, MacMahon S, Chalmers J. Albuminuria and kidney function independently predict cardiovascular and renal outcomes in diabetes. J Am Soc Nephrol. 2009;20(8):1813–21.
99. Rossing K, Mischak H, Dakna M, Zurbig P, Novak J, Julian B, Good D, Coon J, Tarnow L, Rossing P, on behalf of the PREDICTIONS Network. Urinary proteomics in diabetes and CKD. J Am Soc Nephrol. 2008;19:1283–90.
100. Alkhalaf A, Zurbig P, Bakker SJ, Bilo HJ, Cerna M, Fischer C, Fuchs S, Janssen B, Medek K, Mischak H, Roob JM, Rossing K, Rossing P, Rychlik I, Sourij H, Tiran B, Winklhofer-Roob BM, Navis GJ. Multicentric validation of proteomic biomarkers in urine specific for diabetic nephropathy. PLoS One. 2010;5(10):e13421.
101. Zurbig P, Jerums G, Hovind P, Macisaac RJ, Mischak H, Nielsen SE, Panagiotopoulos S, Persson F, Rossing P. Urinary proteomics for early diagnosis in diabetic nephropathy. Diabetes. 2012;61(12):3304–13.
102. Siwy J, Schanstra JP, Argiles A, Bakker SJ, Beige J, Boucek P, Brand K, Delles C, Duranton F, Fernandez-Fernandez B, Jankowski ML, Al KM, Kunt T, Lajer M, Lichtinghagen R, Lindhardt M, Maahs DM, Mischak H, Mullen W, Navis G, Noutsou M, Ortiz A, Persson F, Petrie JR, Roob JM, Rossing P, Ruggenenti P, Rychlik I, Serra AL, Snell-Bergeon J, Spasovski G, Stojceva-Taneva O, Trillini M, von der Leyen H, Winklhofer-Roob BM, Zurbig P, Jankowski J. Multicentre prospective validation of a urinary peptidome-based classifier for the diagnosis of type 2 diabetic nephropathy. Nephrol Dial Transplant. 2014;29(8):1563–70.
103. Andersen S, Mischak H, Zurbig P, Parving H, Rossing P. Urinary proteome analysis enables assessment of renoprotective treatment in type 2 diabetic patients with microalbuminuria. BMC Nephrol. 2010;11:29.
104. Roscioni SS, de Zeeuw D, Hellemons ME, Mischak H, Zurbig P, Bakker SJ, Gansevoort RT, Reinhard H, Persson F, Lajer M, Rossing P, Lambers Heerspink HJ. A urinary peptide biomarker set predicts worsening of albuminuria in type 2 diabetes mellitus. Diabetologia. 2013;56(2):259–67.
105. Lindhardt M, Persson F, Currie G, Pontillo C, Beige J, Delles C, von der Leyen H, Mischak H, Navis G, Noutsou M, Ortiz A, Ruggenenti PL, Rychlik I, Spasovski G, Rossing P. Proteomic prediction and Renin angiotensin aldosterone system Inhibition prevention Of early diabetic nephRopathy in TYpe 2 diabetic patients with normoalbuminuria (PRIORITY): essential study design and rationale of a randomised clinical multicentre trial. BMJ Open. 2016;6(3):e010310.
106. Taylor CF, Paton NW, Lilley KS, Binz PA, Julian Jr RK, Jones AR, Zhu W, Apweiler R, Aebersold R, Deutsch EW, Dunn MJ, Heck AJ, Leitner A, Macht M, Mann M, Martens L, Neubert TA, Patterson SD, Ping P, Seymour SL, Souda P, Tsugita A, Vandekerckhove J, Vondriska TM, Whitelegge JP, Wilkins MR, Xenarios I, Yates III JR, Hermjakob H. The minimum information about a proteomics experiment (MIAPE). Nat Biotechnol. 2007;25(8):887–93.

Chapter 18
Concluding Remarks: Proteomics AD 2025

Ian Wright, Giulio Agnetti, and Jennifer E. Van Eyk

Abstract Mass Spectrometry has potentially been one of the most powerful tool available to scientists. The future for mass spectrometry lies not only in deep technological advances in its inherent performance but also in increasing its robustness and ease of use, and by reductions in size and cost, that will make it increasingly accessible to more users.

The power that Mass Spectrometry can bring to science will increasingly be maximized through its combination with other data rich "omics" tools.

The authors describe how this landscape is changing today and will change in a future that is becoming increasingly described in terms of translational, precision and personalized medicine which will all require smarter, more "dynamic" diagnostic tools.

Keywords Mass Spectrometry (future of) • Proteomics • Cardiology • Data Independent Acquisition (DIA) • Data Dependent Acquisition (DDA) • "Static/Dynamic diagnostics" • Cardiology

I. Wright, BSc, CBiol. MRSB
Advanced Clinical Biosystems Research Institute,
Cedars-Sinai Medical Center, Los Angeles, CA, USA

Strategic Innovations LLC, Earleville, MD, USA
e-mail: ian.wright@cshs.org

G. Agnetti, PhD, FAHA
Laboratory of Cardiovascular Biochemistry, Division of Cardiology, Department of Medicine, The Johns Hopkins University School of Medicine, Baltimore, MD, USA

Department of Biomedical and Neuromotor Sciences (DIBINEM), University of Bologna, Bologna, Italy

J.E. Van Eyk, PhD, FAHA, FISHR
The Advanced Clinical Biosystems Research Institute, Heart Institute, Cedars-Sinai Medical Center, Los Angles, CA, USA

Department of Medicine, Johns Hopkins University School of Medicine, Baltimore, MD, USA

G. Agnetti et al. (eds.), *Manual of Cardiovascular Proteomics*,
DOI 10.1007/978-3-319-31828-8_18

A New Picture for Health in 2025

Science is increasingly challenged to interpret the complex biology of the human body and process it in such a way that we as scientists, clinicians and patients can make decisions on our health and well being. We are at the beginning of a new era in which we can utilize the opportunities offered by that complexity rather than be overwhelmed by it. This newfound optimism comes from the coincident emergence of new sets of analytical tools that allow us to dive deeper into an individual's "omes" that composedly make up each of our cell, organs and ultimately, self. The challenge, and previously unattainable reward, will come from our ability to make sense of that deep quantitative molecular information and combine the results multi modally, or as Eric Topol [1] would say, across the "Panome" to provide a new understanding of how our bodies work, or occasionally do not work.

Mass spectrometry (MS) and data computation is at the center of this technical revolution and it has developed into a key tool set in providing crucial information on at least three of the major components of that panome: the metabolome, the microbiome and the third and arguably most important, the proteome. This new capability allows us to phenotypically describe any biological (physiological and pathological) state on any given day and under any specific set of conditions. MS driven techniques when applied to good clinical questions will give us new insight and assistance in predicting, diagnosing, monitoring and treating disease.

Cardiovascular disease (CVD) remains our largest health issue however the application of MS and the development of complimentary computational tools is starting to help us understand the, additive impact and interrelatedness of nutrition and lifestyle, diabetes, autoimmunity, inflammation and oncology to heart disease and will hopefully open up new solutions and promising routes to effective therapies.

The development of a capability to better understand the underlying molecular and cellular mechanisms involved and consequently the process underlying cardiovascular disease open up the potential to plan new studies involving much larger populations, adding specific attempts to readdress our knowledge of gender and ethnicity differences that have often been poorly served so far.

Combining MS based techniques that are used for broad discovery experiments (Data Independent Acquisition (DIA, also referred to as SWATH©) and Data Dependent Acquisition (DDA, also referred to as shotgun)) with those that target specific proteins (e.g. SISCAPA©, Selected Reaction Monitoring or SRM, also termed Multiple Reaction Monitoring or MRM, and its reiteration called Parallel Reaction Monitoring or PRM – see Chaps. 9 and 10 for detailed description) also allow for translational programs. This provides an opportunity to speed up and make safer our attempts to bring this new found knowledge to the wider public. These new approaches and technologies could also help understand the impact of changes related to cultural aspects of diet and exercise, and therefore our ability to motivate populations to alter lifestyles, comply with treatment or even develop the treatments best suited to a particular individual. These MS methods also have the advantage of being able to be used on a larger variety of sample types, e.g. dried blood spots, urine, cells and tissue, in a way that traditional immunoassay's can't. This will allow individuals to relate results their own personalized normal reference ranges and utilize the greater diagnostic and predictive value that will come from longitudinal measurements.

These MS-based technologies and an increasing number of new applications, will also improve the financial effectiveness of treatment by allowing us to better understand the risk vs. benefit associated with any individual's healthcare. By better understanding the role of proteins and their post-translational modifications (PTMs) we will gain insights into disease and "wellness" that have been hidden to us previously. An analysis of disease versus normal physiological PTMs can be used, as could different protein isoforms, to add specificity and dynamism at the biological and personalized level. Understanding molecular regulation via PTMs and being able to target and quantify them at the site specific level will open up numerous applications from drug selection to diagnostics. The advent of individualized and personalized understanding of what is 'normal' may better help us persuade the population generally, and the health economic funders, specifically, to take wellness seriously. This knowledge will greatly increase the effectiveness and development of 'smart' solutions for example: fitness watches, mail-in sample types, remote sampling and testing in commercial environments or remote diagnostic testing and sampling testing at pharmacies or at home. For those readers who are also plane travellers, it may come as a surprise to discover that the machines commonly used at every airport for swab tests are indeed MS based.

The preceding chapters in this book have described in detail the MS based techniques and approaches that will be the foundation for this "information" revolution. We are in an era of transition in the area of health research and clinical application where the emphasis, which has been placed so far largely on genes and transcripts, is now moving to their products, protein and their regulation. We propose that this cultural revolution, largely made possible by technological advances, will allow us to correlate much better our intimate molecular make up with our phenotype, both in health and disease. The contribution of genetic medicine in discovering new pathogenetic pathways and in allowing the early detection of genetic abnormalities, particularly in pediatric medicine, is indisputable. However, with the major health concerns worldwide, including CVD, being largely sporadic and episodic in nature and only partly dependent on genetic mutations, we believe that this revolution is necessary. Moreover, while gene therapy has high promise because of the phenotypical effect of life-style, diet and aging, it may not be widely applicable. This is to say, that these environmental aspects, rather than genes, likely play an important role in many modern diseases that affect westernized societies, including CVD. In order to decypher the intimate mechanisms underlying these fundamental process we have to move on from genes and transcripts, to proteins, their modifications and folding. We believe that the key elements to draw a new picture of diseases that are relevant for our modern societies reside at this level and that these new proteomics technologies will greatly enhance our power to interpret and manipulate new pathophysiological mechanisms.

The capability of using diagnostic and data processing computational power to capture, analyze and process huge amounts of genomic, epigenetic, proteomic (including isoform and PTM assessment), metabolic and lipidomic, data, amongst others, will revolutionize our approach to discovery science and applied medicine. The traditional approach has been to look for simple solutions, simple pathways, perfect, single, in vitro diagnostic tests. The benefit of being able to look deeper into the proteome through the development of MS techniques such as DIA to better understand the impact of PTMs of proteins combined with the ability to look broader across "-omics" by developing and integrating new bioinfomatic tools will allow us to move from a traditional "static "to a much more useful "dynamic" diagnostics and provide essential insight into how we actually function (or don't).

The reality is that traditional diagnostics may have gone about as far as it can. Biology is hugely complex, the solutions to biology problems are hugely complex; only by embracing and making sense of that complexity can we deliver what is needed. If personalized diagnostics and therapeutics are to occur, we will need the additional layers of -omic information to tease out the variants within any population. Details of an individual's redox status for example, can explain why one version of a drug would be effective and another wouldn't. This is crucial today where late failure of a pharmaceutical development is expensively unacceptable and companies look more towards diagnostics to select those patient groups that precisely derive the most benefit.

MS has the great potential to make a difference through its ability to offer a gold standard methodology to actually identify and measure, in a very selective way, the molecules being studied. MS technology now has the ability to multiplex assays in a very efficient way and the capability to relatively and absolutely quantify analytes. On the contrary, classical immunoassays have limitations in their identification capabilities either because of the undesired interference from auto antibodies or their, sometimes problematic reliance on antibody interaction with complex sites on proteins.

A New Picture of Proteomics

The move from population diagnostics based on single protein diagnostic markers of limited utility to individualized diagnostics, most likely using panels of markers, which can better describe the dynamic phenotype will shape the next 20 years. Pharma companies describe how up to 80 % of their current drug development pipeline being dedicated to a "Personalized Medicine" approach. Until now this has been mostly a genomic driven activity but there seems to be a general acceptance that the next big drive will come from incorporating proteomics into that model. The combination of precision medicine specifically impacting the correct analyte or pathway targets and personalized medicine where we specifically target those individuals who derive the best risk/benefit form a treatment creates something new that has been referred to as individualized medicine.

Understanding the relationship within and between an individual's proteomes (and its individual sub-proteomes such as nucleome, and including PTMs), microbiome, metabolome, and lipidome will help us moving from the limited predictive power of genetic medicine and allow us to get early indications of new issues, imminent relapse of existing issues and appropriate response to treatment. This also starts to get to what a picture of what an individual's wellness looks like especially if we can incorporate any cyclical changes that are perfectly normal for a given individual or combinations of small but longitudinal changes to an individual's –panome, using the sum knowledge of the significant multi-omic parameters that reflect an individual's status at any given time, In addition these changes may be highly significant for that individual but may be lost when measured against what is "normal" for a population.

Add to this the potential to combine other information from other sources, for example imaging e,g, CPT/PET or staining techniques into the algorithm and we move forward significantly. Importantly it should be feasible to have imaging reagents that are selective to specific disease "status" based on defined "panomic" phenotypes. However, all of this will be of little value if we do not also see an increased effort to tackle the computational and bioinfomatic hurdles of making sense of this huge amount of data.

As we better understand the underlying processes between diseases it becomes clear that inflammatory and metabolic, changes impact, reflect and drive many diseases. The links between, for example, cardiovascular, neurological and oncological disease become better understood as we acquire better tools, better ability to process large data and better designed population studies. Even for perhaps surprising and previously unexplained findings such as the increased risk of later life heart disease in women due to something as relatively common, such as premature childbirth [2]. Significant institutional and commercial support of this "new" approach to science will be required to open the doors to true individualized medicine.

These advances, in turn, pose additional challenges for regulatory agencies who will also need to move "fast and nimble" so as not to stifle the opportunity offered by these new tools and knowledge. Handling complex quantitative data analysis may have to happen "behind the curtains" to enable this new paradigm to translate into normal clinical practice. Fortunately, the trend towards seamlessly managing large data might provide the input tools that also allow the increased safety and security which regulatory groups are challenged to uphold in order to safeguard the population as a whole. The potential for using an individual as their own "normal" range may prevent the rare cases of abnormal results that a traditional normal "population" reference range based test might miss and the regulatory authorities and traditional diagnostic companies spend huge effort and time trying to assure against.

Many of the new MS advances and technologies that are currently available are addressed in this book. For those at the horizon there is something to be learned from the collective experience of proteomic scientists. A good example could be

SELDI MALDI TOF ™, which we briefly address in the introductory chapter as an example of a technology that was overpromised and under delivered. Many new technologies are promoted before their limitations are widely enough tested on real samples/studies. Therefore, this current chapter is also meant to guide informed decision on future proteomic and MS investments. We believe that by merging traditional approaches with cutting-edge, MS technologies we will be able to harness the power of both new and established technologies, and that this marriage of innovation and tradition has great promise. A few technologies have followed this trend, beginning with DDA/shotgun and DIA which demonstrated the ability to consistently measure large number of proteins or SRM assays which can do so quickly for a limited number of proteins when applied to highly focused experiments. When coupled with well-established approaches such as affinity capture [3] this will provide an ability to target protein complexes or increase the number of PTMs that can currently be studied into a higher throughput method. All of which give more dimensional information about an individual and the "panomic" phenotypic state.

On the other hand, this technology also addresses a pressing limitation in translational research, where the limited availability of control tissue is often an obstacle. As it is now possible to create, and share, the electronic version of an organ's proteome, it is conceivable that this new knowledge, when shared, will enable many researchers to overcome this limitation by exploiting the limited number of existing samples which are collected to proteomic standards [4]. These are just two examples that make us confident that MS using data dependent acquisition (DDA also known as shotgun) will be workable.

DIA and DDA/shotgun methods have had a profound effect on proteomics and in fact, driven it, along with computational software for handling the complex MS data. Yet, the future will be transformed with both DIA and targeted method becoming major players. With various iterations and commercial editions to come, it will continue a profound impact on our understanding of complex diseases (e.g. heart failure) in the next decade as it is the only method that allows novel changes at an individual sample level to be fully understood.

Another set of technologies which resulted from the combination of MS with existing, established technologies is MALDI TOF imaging [5]. This approach results from the combination of a MALDI source to image proteins, peptides or small molecules in tissue sections. Classical imaging techniques, which employ antibodies tagged with fluorophores, are limited by the cross-talk between these reporters. In fact 3–4 antigens can be monitored routinely in tissue samples using conventional immunostaining. In theory, by creating targeted approaches such as inclusion lists that "filter" whatever diagnostic peptide we may be interested in, the number of antigens that can be detected by a MALDI source is virtually unlimited. In this case the limitation would be sensitivity. This burden is currently being addressed by Scott Tanner (DVS Sciences Inc.), Gary Nolan (Stanford) and colleagues who pioneered the research on metal-labeled antibodies applied to flow-cytometry. In their seminal study they were able to use antibodies labeled with transition metals to monitor tens of different antigens at once in single cells [6, 7] pushing past the number of simultaneous measures per cell compared to traditional flow cytometry approaches. This was made possible by yet another exciting technology, named Mass Cytometry. As the name cleverly suggests this results by the combination of conventional flow cytometry with MS and allows the detection of multiple antigens with little interference due to the resolution of MS detectors and the distinct and numerous transition metals that can be used to tag antibodies. These are only few examples of the wonderful advances that we are expecting to become widely available to the scientific and clinical community of the next decade.

Technical Advances in Mass Spectrometry

Many of the major MS manufacturers are also changing their focus as they realize that for MS to really impact clinical practice they must not only improve on their ability to provide drive to deeper data through increasing important factors such as sensitivity to measure lower abundant, but extremely important proteins, but they must also seriously tackle the features that will enable the uptake in the clinical laboratory. Topics such as improved ease of use, size, robustness, increased automation, and the use of robotics were what drove the last revolution in clinical laboratories with the advent of large automatable random access clinical chemistry and immunoassay systems. The manufacturers of MS instrumentation, as the ultimate detection methodology, will also need to solve these issues if they are going secure a winning position in the clinical lab community.

It is interesting to notice how the business market has changed over the last decades in that, at the beginning of proteomics the emphasis was placed by manufacturer's on improving performance. That is, the market advantage thus far was a measure of instrument performance; hence new MS instruments would generally provide better measurements in terms of speed, sensitivity, resolution, etc. with some key investment into key computational analysis (both at handling MS data but also in the "data to knowledge" aspects). However, very much like the computer industry, we now reached a stage in the clinical and medical community, where the level of some elements of performance outran the need of many users. Products that were designed to target pioneering discovery labs are still required, but if there is a movement towards using MS as a "common" laboratory tool new tools will need to be developed.

From a user's perspective, at least, it appears that the manufacturing industry is now expanding their targeting to intermediate segments of the market, for which middle-line instruments would serve just as good. In fact, during the initial boom of MS applied to biomedical sciences, a few laboratories worldwide could afford to invest on highly performing instrumentation. As normal, scientists who pioneered the early benefits of MS, benefited from having access to cutting-edge technology and build their scientific productivity on the use of instruments and technologies that very few other labs could access.

The pioneering work is not finished and each new technology development, be it MS instrument platform, computational or new tools for enrichment and sample preparation opens up new door for gaining biological and clinical insight. There are absolutely still technological advances required e.g. in increasing sensitivity to allow us to get to the lower abundant proteins and variants required to complete the phentotype picture. Therefore, it is likely that new exciting technologies will still be tested at cutting-edge labs as it has been so far, while in the meantime the more established technologies, which were cutting-edge just a few years back, become the norm for the scientific community at large.

It is also interesting to see how older technologies are re-purposed and return to make a remarkable impact. Many proteomic scientists who made the history of proteomics had their first encounter with MALDI-TOF mass spectrometers (see Chaps. 1 and 2 for details) via use of 2D gel electrophoresis – even when others were pushing for non-gel based approaches using ESI instrumentation. Early in the days of manual isotopic spectral de-convolution (i.e. the most abundant isotopic peak in a spectrum was chosen manually), proteins were identified by peptide mass fingerprinting (PMF), which does not require tandem MS that nowadays represent the norm. This early approach provides less confidence in the identification of complex protein mixtures as peptides are not "sequenced" but identified on the basis of their mass alone. Although this may be a limitation for complex samples, MALDI-TOFs are robust and incredibly fast. A MS spectrum can be acquired in less than a second for PMF (plus sample preparation which is now streamlined). It is therefore noticeable that these "old" instruments came back to notoriety in recent years for the fast identification of pathogens in clinical laboratories [8]. This is a lesson to be learned that may apply to many other techniques: as much as new is generally better, "old" isn't always bad. However, the quality control and reproducibility of the sample preparation, data analysis and other process controls are now in place, hopefully reducing possibility of error or mistaken identification.

Bioinformatics

Similar to many fast developing technologies, proteomics has evolved so fast that it is difficult, but important, to take stock and look at the bigger pictures. There is an additional challenge with proteomics in that MS is only one attribute that is needed for successful mapping and quantification of proteins, let alone dealing with additional complexity. Bioinformatics, specifically, needed to be developed simultaneously to handle the MS data itself, multi-level biostatics, storage, and then the interpretation of the data especially when considering the new amount and variety of PTMs that are now routinely discovered during an average shotgun proteomic investigation. In part, this has also enabled the success of targeted proteomics by focusing the depth of MS approaches on a smaller, and highly enriched protein sample [9]. Targeted approaches arguably allow the deepest level of detail and it is why these were deemed by Nature Methods to be "method of the Year" in 2013.

Indeed, as much as many proteomic scientists are devoted to their science their strength may not always be statistical analysis. It requires a thorough, extensive training to learn how to deal with all the technical aspects of a proteomic investigation, let alone MS. It has been an understandable issue of all –omics that bioinformatics has not always been allowed to take a share of the driving. Over the years, proteomic scientists, like genomic scientists before them, are increasingly understanding that partnering with bionformaticians is key to ensuring that maximum benefit is gained form these new technologies. For instance, by shortening the list of proteins that it is meaningful to discuss in an article, or to be considered in a clinical study. All of these aspects have fostered an ongoing discussion and as much as there is a desperate need for bioinformaticians that are able to understand the unique opportunities and limitations of protein biochemistry (as opposed to that of nucleic acids), there are also encouraging signs, such as the recent release of the R coding specifically for Proteomics [10]. "R" is an open-source biostatistical tool, recently re-equipped with a Windows GUI, but also a worldwide project for the implementation and diffusion of biostatistics. Therefore, scientist can tap into the freely shared and collective knowledge of many programmers and biostatistician and freely download a modular suite which allows almost infinite data manipulations, plots generation, and statistical tools.

For the future it will be essential that we see increased development and the accessibility to biostatistical tools in three areas:

(i) Bioinformatics that enables MS technology, e.g. by allowing the data generated by new approaches (such as DIA) to produce maximum benefit and is easily utilized
(ii) Bioinformatics that enables the combining of data from multiple disciplines or -omics (panomics)
(iii) Bioinformatics that allows us of link all of this information and processes it in a way that truly provides useful tools and education for clinicians that can drive changes to clinical practice

Our ability to combine metadata from various -omic modalities and also link that information to other clinical data will revolutionize how we approach solving complex biological and medical conditions. Add the ability to do this across many clinical and research centers through consortia and partnership and we begin to really harness the power of data in a way that we couldn't previously do. Our knowledge of how and why our bodies act and react has gaps that we can start to fill in by improvements in these three areas.

Proteomics 2025

As for the approaches described in this book, we are sure that they will continue to change and develop and open up new, unpredictable possibilities. The classical bottom-up approaches, defined by the fact that proteins are digested prior to MS (see Chap. 7 for details), although valuable in some circumstances, may lose some the information at the intact protein level. A 40-year old technique: two dimensional gels still has a role in the context of small labs and targeted proteomics especially when involving PTMs and isolated protein complexes [11]. Although importantly, top-down approaches, which imply the analysis of intact protein by MS are rapidly evolving and will have a key role to play in both understanding cross talk between PTM and binding partners or protein complexes.

Targeted proteomics, both by mean of traditional biochemical methods combined to proteome-wide analyses or by targeted MS approach (such as SRM/ MRM/ pRM) will become increasingly useful and accessible at the point that it may replace western blot and potentially some ELISA allowing analysis of partial or complete pathways rather than a smaller subset. The role of DIA as a hybrid approach will continue to be developed (see Chaps. 10 and 11).

The use of induced pluripotent stem cells (IPSCs) combined with MS-based analysis and characterization is rapidly becoming a major tool in drug discovery and development. Again the complex biology of controlling and monitoring IPSC derived applications can benefit from the type of information that MS can provide. IPSCs are creating so much interest as they are expandable in culture, renewable and able to made into a number of different cell and organoids. It is clear that this relatively young science will revolutionize many basic and applied scientific approaches, especially in the field of therapeutic and pharmaceutical development. A joint approach combining IPSC and proteomics can only help move this faster, deeper and with more certainty.

The Future for MS in the Clinical Laboratory

While these and other emerging approaches will be helpful for the translation of MS derived discoveries into actionable clinical decisions there are no short-cuts to understand and cure disease. However, a multi-omic approach, including proteomics, will help to define biology and move it into higher throughput methods required for translation medicine to be successful. The clinical laboratory offers the biggest rewards but posses the biggest challenges for MS. Sensitivity still needs to be improved upon if we are going to be able to look at all the proteins we want to look at. We still need improvement by a few orders of magnitude of sensitivity if we truly want to look at targeted low abundance proteins or peptides. Some of the newer immunoassay platform companies will continue to push the boundaries of sensitivity for difficult or very low abundance biomarkers such as cardiac troponin I or interleukins and the MS focused companies will need to catch up in that department.

Furthermore, MS instruments will need to be made much easier to use and become much more robust if they are going to be widely adopted in a modern clinical laboratory world challenged by issues in adequate staffing and funding. Most clinical labs will gladly trade off a highly flexible research–mode instrument for a detection technology that is never "down" and able to be used by all of their rotating staff. Many of the manual process that are currently used with MS allow for unacceptable error to be introduced and will need to be automated to improve both performance and costs. Improvements to sample processing procedures and in the ability to link MS, as a detection technology, to automated robotic systems will be required. It will most likely require improvement or removal of protein/peptide

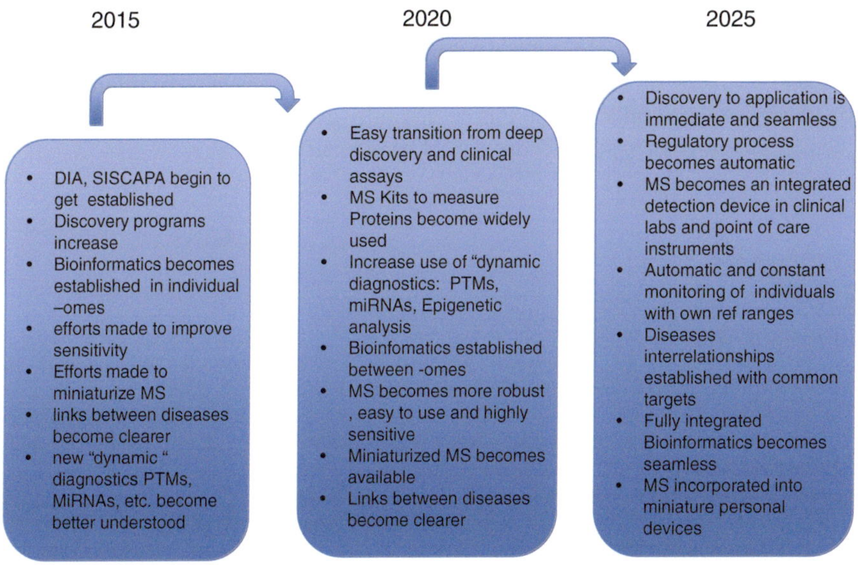

Fig. 18.1 MS Roadmap to 2025

separation methods, such as liquid chromatography steps online with MS instrument to further simplify and reduce components of time and errors.

A new business model that creates more opportunity for funding the acquisition of equipment through reagent kits will be required if MS going to be championed by the companies that currently have the major share of business in clinical laboratories. It is expected that the traditional MS companies and, or, the major IVD companies will find a model that will change that paradigm.

It is interesting to lay out what will have to happen if this roadmap to 2025 is to actually happen as we have described here- Fig. 18.1. Of course, this is only a potential route and one that illustrates the development and maturation of the omics field as it moves towards clinical application and use. There will be surprises along the way including, hopefully new disruptive technologies which are not yet foreseen.

Concluding Remarks

Finally, while looking forward, it sometimes pays to take a quick look back and no one has predicted better what was required as has John Yates at The Scripps Research Institute who eruditely and correctly predicted 10 years ago: "What we need is to ensure that the cutting-edge technological developments in proteomics labs disseminate to all levels of the research community. What we need, in short, is the democratization of mass spectrometry" [12]. MS is a remarkable tool which is unique in being able to reach out to the broader scientific community focused on discovery as well as the clinical and translational science. However this will require MS and computation to move in two different directions, each with different criteria. It can and hopefully will do this successfully as the result will be an improvement in the efficiency and effectiveness of health care delivery.

References

1. Topol E. Chapter 5. In: The patient will see you now. New York: Basic Books; 2015.
2. Heida KY, Velthuis BK, Oudijk MA, Reitsma JB, Bots ML, Franx A, van Dunne FM, on behalf of the Dutch Guideline Development Group on Cardiovascular Risk Management after Reproductive Disorders. Cardiovascular disease risk in women with a history of spontaneous preterm delivery: a systematic review and meta-analysis. Eur J Prev Cardiol. 2015;3(3):253–63.
3. Collins BC, Gillet LC, Rosenberger G, Rost HL, Vichalkovski A, Gstaiger M, Aebersold R. Quantifying protein interaction dynamics by SWATH mass spectrometry: application to the 14-3-3 system. Nat Methods. 2013;10:1246–53.
4. Guo T, Kouvonen P, Koh CC, Gillet LC, Wolski WE, Rost HL, Rosenberger G, Collins BC, Blum LC, Gillessen S, Joerger M, Jochum W, Aebersold R. Rapid mass spectrometric conversion of tissue biopsy samples into permanent quantitative digital proteome maps. Nat Med. 2015;21(4):407–13.
5. Gessel MM, Norris JL, Caprioli RM. MALDI imaging mass spectrometry: spatial molecular analysis to enable a new age of discovery. J Proteomics. 2014;107:71–82.
6. Bendall SC, Simonds EF, Qiu P, Amir ED, Krutzik PO, Finck R, Bruggner RV, Melamed R, Trejo A, Ornatsky OI, Balderas RS, Plevritis SK, Sachs K, Pe'er D, Tanner SD, Nolan GP. Single-cell mass cytometry of differential immune and drug responses across a human hematopoietic continuum. Science. 2011;332:687–96.
7. Agnetti G. Mass spectrometry goes with the flow: mass cytometry and its potentials in regenerative medicine. Circ Cardiovasc Genet. 2012;5:379–80.
8. Randell P. It's a MALDI but it's a goodie: MALDI-TOF mass spectrometry for microbial identification. Thorax. 2014;69:776–8.
9. Marx V. Targeted proteomics. Nat Methods. 2013;10:19–22.
10. Gatto L, Christoforou A. Using R and bioconductor for proteomics data analysis. Biochim Biophys Acta. 1844;2014:42–51.
11. Oliveira BM, Coorssen JR, Martins-de-Souza D. 2DE: the phoenix of proteomics. J Proteomics. 2014;104:140–50.
12. Yates III J. Proteomics: power to the people! Scientist. 2005;19:30–1.

Index

G. Agnetti et al. (eds.), *Manual of Cardiovascular Proteomics*,
DOI 10.1007/978-3-319-31828-8

MIX
Papier aus verantwortungsvollen Quellen
Paper from responsible sources
FSC® C105338

If you have any concerns about our products,
you can contact us on
ProductSafety@springernature.com

In case Publisher is established outside the EU,
the EU authorized representative is:
Springer Nature Customer Service Center GmbH
Europaplatz 3, 69115 Heidelberg, Germany

Printed by Libri Plureos GmbH
in Hamburg, Germany